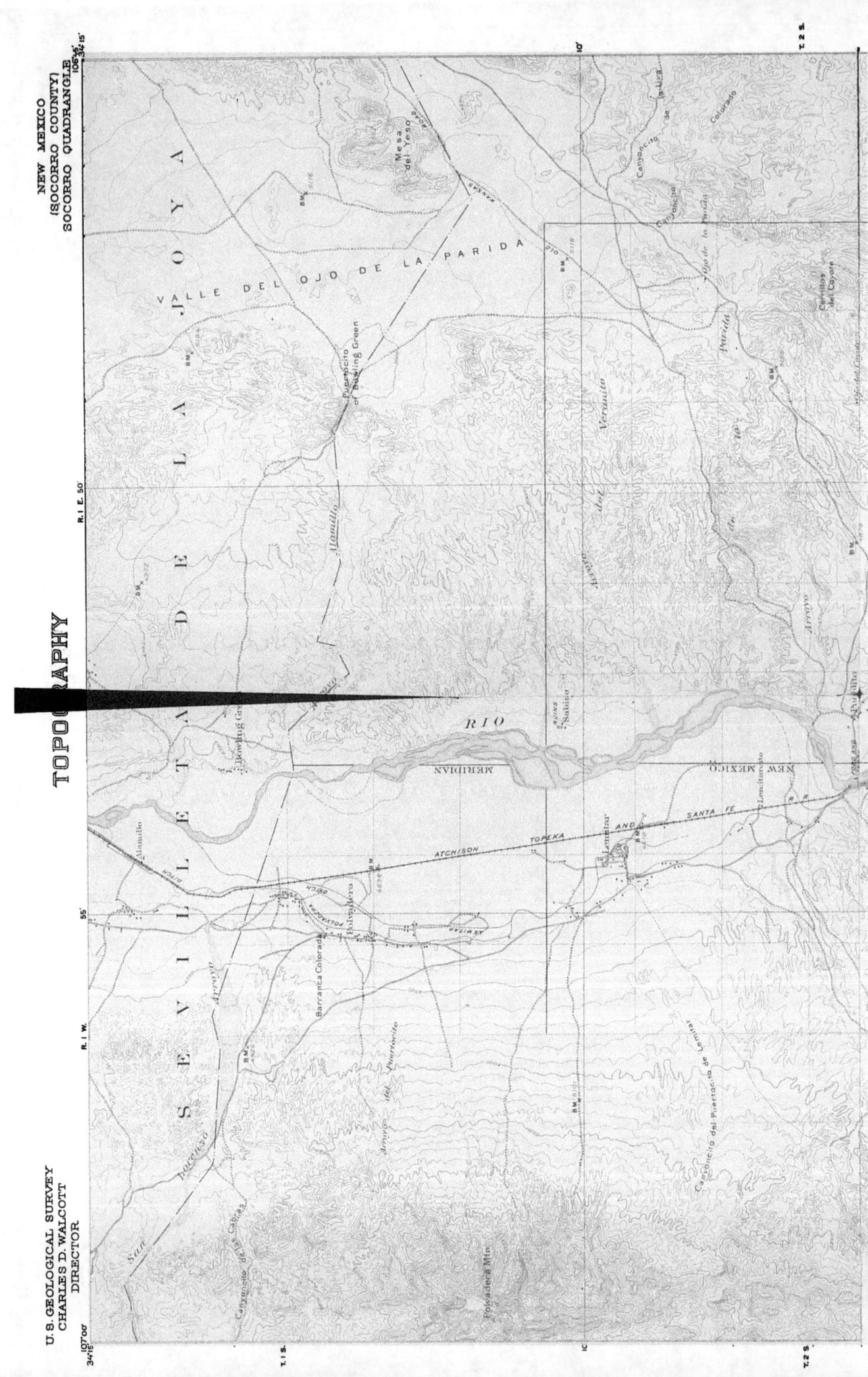

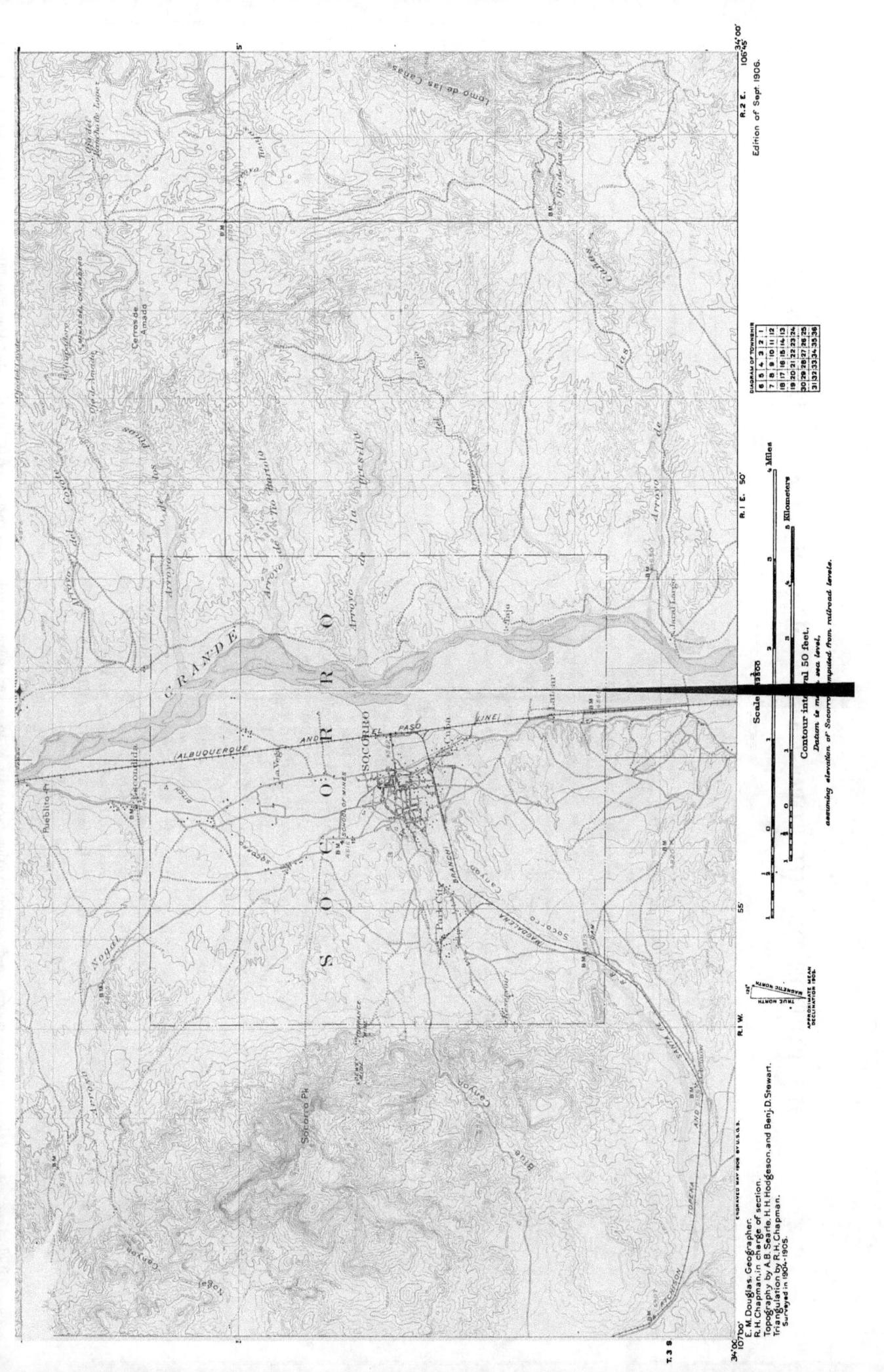

DESCRIPTION OF THE TOPOGRAPHIC MAP OF THE UNITED STATES

The United States Geological Survey is making a topographic map of the United States. This work has been in progress since 1882, and about three-tenths of the area of the country, excluding outlying possessions, has been mapped. The mapped areas are widely scattered, nearly every State being represented, as shown on the progress map accompanying each annual report of the Director.

This great map is being published in atlas sheets of convenient size, which are bounded by parallels and meridians. The four-cornered division of land corresponding to an atlas sheet is called a *quadrangle*. The sheets are of approximately the same size: the paper dimensions are 20 by 16½ inches; the map occupies about 17½ inches of height and 11¼ to 16 inches of width, the latter varying with latitude. Three scales, however, are employed. The largest scale is 1:62500, or very nearly one mile to one inch; i. e., one linear mile on the ground is represented by one linear inch on the map. This scale is used for the thickly settled or industrially important parts of the country. For the greater part of the country an intermediate scale of 1:125000, or about two miles to one inch, is employed. A third and still smaller scale of 1:250000, or about four miles to one inch, has been used in the desert regions of the far West. A few special maps on larger scales are made of limited areas in mining districts. The sheets on the largest scale cover 15' of latitude by 15' of longitude; those on the intermediate scale, 30' of latitude by 30' of longitude; and those on the smallest scale, 1° of latitude by 1° of longitude.

The features shown on this map may, for convenience, be classed in three groups: (1) *water*, including seas, lakes, ponds, rivers and other times, are shown, not by full lines, but by lines of dots and dashes. Ponds which are dry during a part of the year are shown by oblique parallel lines. Salt-water marshes are shown by horizontal ruling interspersed with tufts of blue, and fresh water marshes and swamps by blue tufts with broken horizontal lines.

Relief is shown by contour lines in *brown*. Each contour passes through points which have the same altitude. One who follows a contour on the ground will go neither uphill nor downhill, but on a level. By the use of contours not only are the shapes of the plains, hills, and mountains shown, but also the elevations. The line of the seacoast itself is a contour line, the datum or zero of elevation being mean sea level. The contour line at, say, 20 feet above sea level is the line that would be the seacoast if the sea were to rise or the land to sink 20 feet. Each a line runs back up the valleys and forward around the points of hills and spurs. On a gentle slope this contour line is far from the present coast line, while on a steep slope it is near it. Thus a succession of these contour lines far apart on the map indicates a gentle slope; if close together, a steep slope; and if the contours run together in one line, as if each were vertically under the one above it, they indicate a cliff. In many parts of the country are depressions or hollows with no outlets. The contours of course surround these, just as they surround hills. Those small hollows known as sinks are usually indicated by hachures, or short dashes, on the inside of the curve. The contour interval, or the vertical distance in feet between one contour and the next, is stated at the bottom of each map. This interval varies according to the character of the area their descriptions, as well as the descriptions and geodetic coordinates of triangulation stations, are published in the annual reports and bulletins of the Survey. The publications pertaining to specified localities may be had on application.

The works of man are shown in *black*, in which color all lettering also is printed. Boundaries, such as State, county, city, land-grant, reservation, etc., are shown by broken lines of different kinds and weights. Cities are indicated by black blocks, representing the built-up portions, and country houses by small black squares. Roads are shown by fine double lines (full for the better roads, dotted for the inferior ones), trails by single dotted lines, and railroads by full black lines with cross lines. Other cultural features are represented by conventions which are easily understood.

The sheets composing the topographic atlas are designated by the name of a principal town or of some prominent natural feature within the district, and the names of adjoining published sheets are printed on the margins. The sheets are sold at five cents each when fewer than 100 copies are purchased, but when they are ordered in lots of 100 or more copies, whether of the same sheet or of different sheets, the price is three cents each.

The topographic map is the base on which the facts of geology and the mineral resources of a quadrangle are represented. The topographic and geologic maps of a quadrangle are finally bound together, accompanied by a description of the district, to form a folio of the Geologic Atlas of the United States. The folios are sold at twenty-five cents each, except such as are unusually comprehensive, which are priced accordingly.

Applications for the separate topographic maps

mountains, hills, valleys, cliffs, etc.; (3) *culture*, i. e., works of man, such as towns, cities, roads, railroads, boundaries, etc. The conventional signs used for these features are grouped below. Variations appear in some maps of earlier dates.

All water features are shown in *blue*, the smaller streams and canals in full blue lines, and the larger streams, lakes, and the sea by blue water-lining. Certain streams, however, which flow during only a part of the year, their beds being dry at other feet; in a mountainous region it may be 200 feet. Certain contours, usually every fifth one, are accompanied by figures stating elevation above sea level. The heights of many definite points, such as road corners, railroad crossings, railroad stations, summits, water surfaces, triangulation stations, and bench marks, are also given. The figures in each case are placed close to the point to which they apply, and express the elevation to the nearest foot only. The *exact* elevations of bench marks and accompanied by the cash or by post-office money order (not postage stamps), and should be addressed to—

THE DIRECTOR,

United States Geological Survey,

Washington, D. C.

July, 1905.

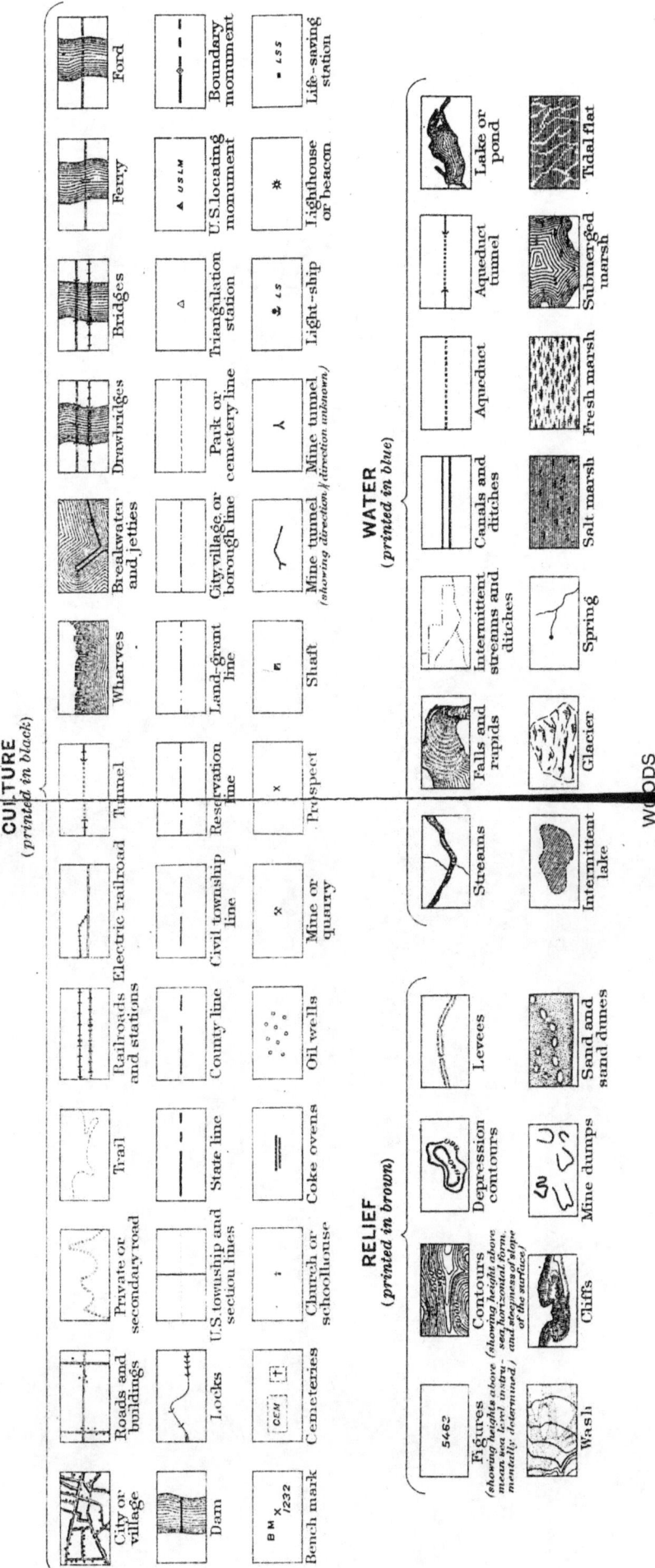

DEPARTMENT OF THE INTERIOR
UNITED STATES GEOLOGICAL SURVEY
GEORGE OTIS SMITH, DIRECTOR

PROFESSIONAL PAPER 68

THE ORE DEPOSITS OF NEW MEXICO

BY

WALDEMAR LINDGREN, LOUIS C. GRATON
AND CHARLES H. GORDON

WASHINGTON
GOVERNMENT PRINTING OFFICE
1910

CONTENTS.

	Page.
PART I. GENERAL FEATURES	15
Introduction	15
Historical sketch of mining and smelting in New Mexico	17
Metal production	19
Geology	25
Epitome of geologic history in its relation to topography	25
Pre-Cambrian rocks	26
Distribution	26
Petrographic character	27
Paleozoic and Mesozoic sediments	29
Tertiary and Quaternary events	32
Total thickness of pre-Tertiary sedimentary formations	34
Intrusive rocks of early Tertiary age	35
Distribution	35
Composition and structure	36
Dike rocks	40
Relation to mineral deposits	40
Mode of intrusion	40
Contact metamorphism	41
Tertiary and Quaternary lavas	42
General features	42
Rhyolite	42
Andesite, latite, and trachyte	43
Basalt	44
Connection with mineral deposits	46
Review of igneous activity	46
Metal deposits	46
Geographic distribution	46
Metals contained	47
Geologic distribution	47
Pre-Cambrian deposits	48
General character and distribution	48
The ores	49
Metasomatic processes	50
Genesis	51
Contact-metamorphic deposits	51
General features and distribution	51
The ores	51
Development and production	53
Structure	53
Oxidation	55
Genesis	56
Veins connected with intrusive rocks of early Tertiary age, exclusive of replacement veins in limestone	56
General relation and distribution	56
The ores	57
Structure and geologic relations	58
Metasomatic processes in the wall rocks	59
Oxidation	59
Age of the veins	59
Review of districts	60
Copper deposits due to oxidizing surface waters	61

CONTENTS.

PART I. GENERAL FEATURES—Continued.
    Geology—Continued.
        Metal deposits—Continued.

	Page.
Vein and replacement deposits in limestone, exclusive of contact-metamorphic deposits	62
General features and distribution	62
The ores	63
Structural features	64
Metasomatic processes	66
Genesis	67
Veins connected with volcanic rock of Tertiary age	67
General features and distribution	67
Country rock	68
Structure	69
The ores	69
Oxidation	70
Ore shoots	70
Metasomatic processes	71
Genesis	71
The hot springs at Ojo Caliente and their deposits	72
Lead and copper veins of doubtful affiliation	74
Placers	74
Copper deposits in sandstone	76
General features	76
The ores	77
Genesis	78
Minerals occurring in New Mexico	80
PART II. DETAILED DESCRIPTIONS	82
Taos County, by L. C. Graton and W. Lindgren	82
General features	82
Rio Hondo (Twining) district	83
Geology	83
Ore deposits	84
Red River district	84
Location	84
General geology	85
Mineral deposits	85
General features	85
Black Copper mine	86
Anaconda group	86
Other properties near Red River	86
Highland Chief	87
Jay Hawk mine	87
Golden Treasure mine	87
Other prospects on Black Mountain	87
Bitter Creek and Independence mine	88
Placers	88
Anchor district	88
La Belle district	89
Copper Hill or Picuris district	89
Glenwoody district	91
Colfax County, by L. C. Graton	91
General features	91
Elizabethtown district	92
Location and history	92
General geology	93
Mineral deposits	95
General features	95
Aztec mine	96
Black Horse mine	97
Montezuma mine	98
Rebel Chief mine	98
French Henry Mountain	99
Moreno Centennial mine	100

CONTENTS.

PART II. DETAILED DESCRIPTIONS—Continued.
  Colfax County, by L. C. Graton—Continued.
    Elizabethtown district—Continued.

	Page.
Mineral deposits—Continued.	
Red Bandana mine	100
Denver mine	100
Legal Tender mine	100
Alabama mine	100
Other vein deposits	101
Ajax mine	101
Iron Mountain	102
Baldy tunnel	103
Copper park	104
Placers	104
Cimarroncito district	105
Location and general geology	105
Mineral deposits	106
General features	106
Thunder mine	107
Garst mine	108
Anaconda mine	108
Contention mine	108
Mora and San Miguel counties, by L. C. Graton and W. Lindgren	108
General features	108
Copper deposits of the Santa Fe Range and upper Pecos River	110
Geology	110
Iron deposits	112
Copper deposits in Paleozoic sandstones	112
Copper and gold deposits in pre-Cambrian rocks	112
General features	112
Hamilton mine	113
Rociada district	114
General features	114
Mineral deposits	115
Prospects near Rociada	115
Developments at Hadley	116
Tecolote district	116
General features	116
Mineral deposits	117
Blake mine	117
Bonanza mine	120
Other deposits	121
Conclusions	122
San Juan and McKinley counties, by W. Lindgren	123
Rio Arriba County, by L. C. Graton	124
General features	124
Hopewell and Bromide districts	124
Location and history	124
Geology	125
Mineral deposits	126
General features	126
Hopewell district	126
General geology	126
Jaw Bone mine	128
Iron Mountain	128
Dixie Queen mine	128
Red Jacket mine	129
Mineral Point mine	129
Croesus mine	129
Other prospects	129
Placers	130
Bromide district	130
General geology	130
Sixteen to One mine	131

PART II. DETAILED DESCRIPTIONS—Continued.
    Rio Arriba County, by L. C. Graton—Continued.
        Hopewell and Bromide districts—Continued.
            Mineral deposits—Continued.

	Page.
Bromide district—Continued.	
Whale mine	131
Pay Roll mine	131
Admiral group	131
Tampa mine	132
Bromide mine	132
Dillon Tunnel	133
Strawberry mine	133
Other prospects	133
Valencia County, by F. C. Schrader	134
General features	134
Copper deposits of the Zuni Mountains, N. Mex	134
Description of the region	135
The rocks	136
Pre-Cambrian rocks	136
"Red Beds"	137
Lavas	137
Copper deposits	138
General occurrence	138
Deposits in the pre-Cambrian rocks	138
Deposits in the "Red Beds"	139
Developments	140
Other deposits	140
Sandoval County, by F. C. Schrader and L. C. Graton	140
General features	140
Copper deposits of the Sierra Nacimiento, N. Mex	141
Introduction	141
Description of the region	141
The rocks	142
Copper deposits in the "Red Beds"	143
General statement	143
Nacimiento district	143
General features	143
Copper Glance mine	144
Bluebird claim	145
Eureka mine	145
San Miguel mine	146
Facilities	147
Gallina district	147
Abiquiu "mines"	149
Associated minerals	149
Origin of the deposits	149
Copper deposits in the pre-Cambrian rocks	149
Cochiti or Bland district	150
Location and history	150
General geology	151
Mineral deposits	153
General features	153
Albemarle mine	158
Washington mine	159
Lone Star mine	159
Iron King mine	160
Crown Point mine	161
Other properties	161
Copper prospect	162
Bernalillo and Torrance counties, by W. Lindgren	162
Santa Fe County, by W. Lindgren	163
General features	163

PART II. DETAILED DESCRIPTIONS—Continued.
    Santa Fe County, by W. Lindgren—Continued.

	Page.
Cerrillos district	164
General features and geology	164
Mineral deposits	166
Dolores or Old Placer mining district	167
General features and geology	167
Mineral deposits	168
Placers	168
Quartz veins	169
Contact-metamorphic deposits	169
San Pedro or New Placer district	170
General features and geology	170
Mineral deposits	171
General features	171
Gold-bearing veins	172
Lead deposits	172
San Pedro copper mine	172
Placers	174
Lincoln County, by L. C. Graton	175
General features	175
Nogal district	176
Location	176
Geology	176
Ore deposits	177
American, Helen Rae, and Iowa and New Mexico mines	177
Hopeful and Vera Cruz mines	178
Other prospects	179
White Oaks district	179
Jicarilla district	183
Otero County, by L. C. Graton	184
General features	184
Jarilla district	184
Tularosa district	187
Socorro County (eastern and western parts), by L. C. Graton	190
General features	190
Mogollon district	191
Location and history	191
General geology	192
Ore deposits	194
General statement	194
Structure and rock alteration	195
Mineralogy	197
Genesis	198
Value	198
Mining developments	199
Estey district	201
Iron deposits in the northern Sierra Oscura	203
Dona Ana County, by W. Lindgren	205
General features	205
Organ Mountains	205
General features	206
Geology	206
Mineral deposits	209
General features	209
Gold and silver veins in granitic rocks	210
Lead-silver veins in limestone	210
Contact-metamorphic deposits	211
Sierra and central Socorro counties, by C. H. Gordon	213
General relations	213
Location and extent	213
Previous investigations	214
Climate and vegetation	217

CONTENTS.

PART II. DETAILED DESCRIPTIONS—Continued.
  Sierra and central Socorro counties, by C. H. Gordon—Continued.

	Page.
General relations—Continued.	
Culture	218
Mining districts	218
Physiographic features	218
Drainage	218
Mountains	220
Bolson plains	221
Geology	222
General relations	222
Pre-Cambrian rocks	225
Cambrian system	225
Shandon quartzite	225
Mimbres limestone	226
Devonian system	228
Percha shale	228
Mississippian series ("Lower Carboniferous")	228
General features	228
Lake Valley limestone	229
Kelly limestone	231
Pennsylvanian series ("Upper Carboniferous")	231
General features	231
Magdalena group	232
Manzano group	233
Nomenclature	234
Triassic (?) system	236
Cretaceous system	236
Tertiary system	237
Quaternary deposits	237
Palomas gravel (Pleistocene)	237
Recent deposits	238
Igneous rocks	238
Andesite and latite	238
Rhyolite	238
Basalt	238
Intrusive rocks	238
Mineral deposits	239
Introduction	239
Socorro Peak district	239
Historical note	239
Geology	239
Ore deposits	240
Canyoncito district	240
Geology	240
Ore deposits	241
San Lorenzo district	241
Magdalena district	241
Geographic relations	241
Historical sketch	241
Topographic relations	242
General geologic relations	243
Stratigraphy	243
Pre-Cambrian rocks	243
Carboniferous rocks	245
Post-Carboniferous sediments	246
Igneous rocks	247
Structural features	248
Ore deposits	250
Silver Mountain (Water Canyon) district	258
Geographic relations	258
Mineral deposits	258

CONTENTS.

PART II. DETAILED DESCRIPTIONS—Continued.
   Sierra and central Socorro counties, by C. H. Gordon—Continued.
      Mineral deposits—Continued.

	Page.
Hop and Mill canyons	258
Geographic relations	258
Geology	258
Mineral deposits	259
Rosedale district	259
Geographic relations	259
Historical note	259
Geology	259
Ore deposits	260
Black Range and Apache districts	260
Geographic relations	260
Historical note	260
Geology	262
Mineral deposits	262
Hermosa (Palomas) district	266
Geographic relations	266
Geology	266
The ores	267
Kingston and adjoining districts	268
Geographic relations	268
Historical note	268
Geology	268
Ore deposits	269
Bromide No. 1 (Tierra Blanca) district	271
Geographic relations	271
Geology	271
Ore deposits	271
Carpenter district	272
Geographic relations	272
Geology	272
Ore deposits	272
Las Animas (Hillsboro) district	272
Geographic relations	272
Historical statement	273
Geology	273
Ore deposits	275
Lake Valley district	276
Situation and history	276
Geology	277
Ore deposits	277
Pittsburg district	282
Geographic relations	282
Historical statement	282
Geology	282
Placers	283
Caballos district	284
Geography	284
Geology	284
Ore deposits	284
Luna County, by C. H. Gordon and W. Lindgren	285
General features	285
Cooks Peak district	287
Geographic relations	287
Discovery and production	287
Geology	288
Ore deposits	288
Florida Mountain district	289
Victorio district	290
General features and geology	290
Mineral deposits	290

PART II. DETAILED DESCRIPTIONS—Continued.
    Luna County, by C. H. Gordon and W. Lindgren—Continued.

	Page
Tres Hermanas mining district	292
General features and geology	292
Mineral deposits	293
Grant County, by L. C. Graton, W. Lindgren, and J. M. Hill	295
General features	295
Pinos Altos district	297
Location and history	297
Geology	297
Mineral deposits	298
General features	298
Pacific mine	298
Mountain Key mine	299
Cleveland group	300
Other properties	301
Placers	301
Chloride Flat district	301
Location and history	301
Geology and mineral deposits	302
Fleming Camp	304
Santa Rita and Hanover districts	305
Location and history	305
General geology	306
The rocks	308
Structure and sequence	309
Ore deposits	311
General features	311
Veins	311
Contact-metamorphic deposits	312
Deposits of secondary enrichment	315
Central district	317
Georgetown district	318
Lone Mountain district	320
Burro Mountain district	321
Location and history	321
Geology and mineral deposits	321
General features	321
Copper deposits	322
Turquoise deposits	324
Black Hawk district	324
Telegraph district	325
Gold Hill district	326
Steeple Rock district	327
Kimball district	328
San Simon district	329
General features and geology	329
Granite Gap mine	330
Deposits between Granite Gap and Steins Pass	331
Lordsburg district	332
General features	332
Mine descriptions	334
Eureka (Hachita) district	335
General features	335
Geology	336
Mines at old Hachita	336
Copper Dick mine	337
Summary	337
Sylvanite district	338
Location	338
History	338
Geography	339
Geology	339

PART II. DETAILED DESCRIPTIONS—Continued.
    Grant County, by L. C. Graton, W. Lindgren, and J. M. Hill—Continued.
        Sylvanite district—Continued.

	Page.
Ore deposits	340
Detailed descriptions	341
Apache No. 2 district	343
Topography	343
Geology	343
Mineral deposits	344
Fremont district (Luna and Grant counties)	345
General features	345
Geology	345
Ore deposits	346
Mine descriptions	346
Other counties	348
Index	349

ILLUSTRATIONS.

	Page.
PLATE I. Topographic map of New Mexico	In pocket.
II. Map of New Mexico showing locations and names of metal-mining districts	46
III. Section of fissure vein, Pinos Altos district; natural size	58
IV. A, Section of oxidized copper ores in sandstone, Zuni Mountains; B, Section of oxidized ore, Anderson mine, Apache No. 2 district; C, Section of chalcocite replacing coal, Nacimiento district	78
V. A, Taos Range; B, Valley of Rio Grande at Glenwoody	82
VI. A, View from San Miguel mine; B, South edge of Mesa Blanca Capulin; C, Portion of Las Tusas Mesa	148
VII. A, White Oaks, Lincoln County; B, Stope in South Homestake mine, Nogal district	178
VIII. A, Nannie Baird and I mines, Jarilla district; B, Open cut, Republic mine, Hanover district	184
IX. A, View from point near summit of divide between Cooney Canyon and Mogollon; B, Last Chance silver-gold mine, Mogollon	190
X. A, View across Cooney Canyon, Mogollon district; B, Granite spires of Los Organos, west slope of Organ Mountains	204
XI. A, Modoc mine and mill, Organ Mountains; B, Stephenson-Bennett mine, Organ Mountains	210
XII. Geologic sketch map of parts of San Mateo Mountains and the Black Range	224
XIII. Correlation table; central Socorro, Sierra, and Luna counties	226
XIV. A, Coyote Buttes; B, Socorro Peak	234
XV. A, South side of Blue Canyon, Socorro Mountains; B, Blue Canyon, looking east across the Rio Grande valley	240
XVI. A, Magdalena Mountains; B, View on Arroyo de la Parida	242
XVII. Sketch map of Magdalena mining district	252
XVIII. Thin sections of ore from Graphic mine, Kelly	252
XIX. A, Kelly mining camp; B, North entrance to Graphic mine	254
XX. A, Cliffs of rhyolite tuff in Chloride Canyon, Black Range; B, North end of San Mateo Mountains	258
XXI. Sketch map of Kingston, Carpenter, and Tierra Blanca districts	268
XXII. A, Santa Rita mine; B, Sylvanite camp	314
FIGURE 1. Map of New Mexico showing location of contact-metamorphic deposits and related veins and replacement deposits of early Tertiary age	52
2. Map of New Mexico showing location of late Tertiary veins in or near flows of rhyolite, andesite, and latite	68
3. Map of New Mexico showing location of copper deposits in sedimentary rocks, mainly "Red Beds"	76
4. Diagrammatic section through Cimarron Range, at Elizabethtown	93
5. Section through Rebel Chief mine, Elizabethtown district	99
6. Section at mouth of main tunnel, Iron Mountain, Elizabethtown district	103
7. Sketch of thin section of granitic phase of porphyry at Cimarroncito	107
8. Section through Blake mine, Tecolote district	118
9. Section from Las Tusas Mesa to Mount Sedgwick, Zuni Mountains	135
10. Reconnaissance map of Nacimiento district	144
11. Map of Cerrillos, Ortiz, and San Pedro districts	164
12. Sketch of vertical section, northwest wall of Cooney Canyon	194
13. Sketch showing approximate location of the principal veins of Mogollon district	196
14. Sketch map of the northern part of the Organ Mountains, near San Augustin Pass	206
15. General section of geologic formations in Sierra County and central Socorro County	223
16. Profile section from Kingston to Jornada del Muerto	224
17. Geologic column at Kelly, Magdalena district	244
18. Generalized profile section across the Magdalena Range at Kelly	249
19. Sketch showing part of workings of Kelly mine, Magdalena district	256
20. Cross section of Kelly mine, Magdalena district	257
21. Sketch map of the Black Range and Apache districts	261
22. Sketch map showing geologic features of Hillsboro district	273
23. Topographic and geologic map of Lake Valley district	278
24. Section across Lake Valley district	279

ILLUSTRATIONS.

	Page.
FIGURE 25. Map of Lake Valley mines	280
26. Cross section of deposits at Lake Valley	281
27. Sketch map of Pittsburg mining district	283
28. Map of portions of Grant and Luna counties	286
29. Sketch map showing relative position of principal claims in Victorio mining district	291
30. Geologic section west of Silver City	303
31. Diagrammatic section across Hanover Creek	312
32. Sketch map of Sylvanite district	338
33. Map of Fremont district	345

THE ORE DEPOSITS OF NEW MEXICO.

By WALDEMAR LINDGREN, LOUIS C. GRATON, and CHARLES H. GORDON.

PART I. GENERAL FEATURES.

INTRODUCTION.

A general reconnaissance of the ore deposits of New Mexico was undertaken by the United States Geological Survey in 1905. The plan of the work provided for no detailed examinations, but it was intended to cover the whole Territory, partly in order to obtain information regarding a great number of little-known districts concerning which inquiries were often received in the Geological Survey, partly in order to study systematically a large field of the Rocky Mountains, and to obtain from such a comprehensive study generalizations for the science of ore deposits. It is believed that both these results have been attained. The miner will find in this volume the local information which he needs; the student of mineral deposits will find in the general part conclusions concerning a broad metallogenetic field, in which epochs of mineralization have recurred since the earliest times. It is believed that the most important result of the work consists in the recognition of the various epochs and the determination of their relations and the mineralogical and structural characteristics which distinguish them.

The field work in the northern and southern parts of the Territory was done by L. C. Graton and Waldemar Lindgren, Mr. Graton undertaking the larger part of the detailed work. To C. H. Gordon was assigned the study of the south-central part, including the whole of Sierra County and the central part of Socorro County. In addition special reports were made by F. C. Schrader on the deposits in the Zuni Mountains, in the Nacimiento Range, in the northern part of the Sierra Oscura, and in the northern part of the Sierra de los Caballos. In 1909 J. M. Hill contributed reports on the Fremont and Sylvanite districts. B. S. Butler has extended valuable aid in the examination of thin sections. The chemists of the Survey laboratory have also assisted by furnishing many analyses of rocks.

Those who peruse the descriptions of the mining districts should bear in mind that the amount of space devoted to any district is not to be regarded as a measure of its relative importance. The exigencies of travel over wide areas served by uncertain or intermittent transportation, coupled with the necessity of covering the whole field in a limited time, resulted in a disproportionate division of the time spent in the investigation of various camps. Furthermore, the geology and other features of some districts are comparatively simple and can be described briefly, whereas in other districts, some of them of less commercial importance, the geologic relations are such as to demand more extended description.

It should also be remembered that many statements in these descriptions are based on information supplied by residents of the region. The very large proportion of idle and consequently inaccessible mines and the small amount of time that could be devoted to each district made it necessary to rely on such information. Though it is undoubtedly reliable on the whole, the writers have endeavored to differentiate throughout the volume the statements thus dependent on information received from others and those resulting from their own investigations.

As the report covers so large an area it has not been possible, in all respects, to bring it up to date so far as mining operations are concerned. The report is, however, not intended as a detailed account of present mining developments, but as a permanent guide to the geologic structure of the districts. From year to year the results of the mining operations are recorded in the annual publication of the Survey entitled "Mineral resources of the United States."

The Territory of New Mexico has received much attention from geologists since 1846 and 1847, when Wislizenus [a] passed through it; at that time it was still a part of the Republic of Mexico.

In 1853 and 1854 it was visited by W. P. Blake and Thomas Antisell.[b] J. S. Newberry[c] and Jules Marcou made further explorations in 1859 in connection with the Macomb expedition. The most detailed work was undertaken between 1875 and 1880 by the Wheeler Survey, of which the geologists were J. J. Stevenson,[d] E. E. Howell, G. K. Gilbert,[e] and O. Loew. Stevenson examined the northern part of the Territory, while the investigations of Gilbert were confined to the southwestern part.

Somewhat later C. E. Dutton[f] published a detailed report on the northwestern part. Much work has been done since then by private persons. C. L. Herrick published a number of papers on the geology of the central part during the last years of the nineteenth century. Many papers have also been contributed by C. R. Keyes, by D. W. Johnson, and by W. T. Lee. The literature is very extensive, and a complete review will not be attempted in this volume, especially as most of it refers to purely geologic questions and has little to do with the mineral deposits of the Territory, concerning which the published data are in fact meager. In the portion contributed by C. H. Gordon, however, will be found a summary comprising the most important publications. (See pp. 214–217.)

A preliminary report on the geology and mineral deposits of New Mexico was published by the present authors in 1906.[g]

This volume does not attempt to deal comprehensively with the general geology of the Territory. Little will be found in it concerning the plateau province of the northwestern part or the flat-lying formations occurring along the eastern and southeastern boundaries.

Attention is confined mainly to the belt containing the mineral deposits which extends from the northern boundary diagonally across the Territory in a southwesterly direction to the southwest corner. This belt is about 100 miles wide in the northern part, but spreads in the southwest corner so as to occupy a width of about 200 miles.

The literature concerning these ore deposits consists chiefly of papers written by individuals and published in the Transactions of the American Institute of Mining Engineers or elsewhere. Among the contributors are W. P. Blake, B. Silliman, jr., Ellis Clark, C. L. Herrick, D. W. Johnson, C. R. Keyes, H. W. Turner, M. B. Yung, and R. S. McCaffery. Little detailed work has been done. In 1904 F. A. Jones published a volume of 349 pages, entitled "New Mexico mines and minerals," in which for the first time a résumé was given of the historical, geologic, and mining features of the various camps of the Territory. This book of reference is quoted frequently in the following pages; it contains a great amount of valuable material, including a list of minerals which forms the base of a similar table in this report. The geologic features are not always adequately treated and are in places erroneously stated, but the work is not claimed to be a geologic treatise. Much information is contained in R. W. Raymond's Statistics of mines in the States and Territories of the Rocky Mountains, issued from 1869 to 1876.

A map showing the distribution of the mining districts of New Mexico was given in the report of the governor of the Territory for 1903. In a report on the production of gold and silver in the Western States for 1907[h] the location of the camps was shown on the general

---

[a] Wislizenus, A., Memoir of a tour to northern Mexico: S. Misc. Docs. No. 26, 30th Cong., 1st sess., 1848, 141 pp., with map.

[b] Report of explorations and surveys to ascertain route for a railroad from the Mississippi River to the Pacific Ocean, vol. 3, pp. 119, with geologic map.

[c] Geological report by J. S. Newberry. Report of the exploring expedition from Santa Fe, N. Mex., to the junction of the Grand and Green rivers of the Great Colorado of the West in 1859 under command of Capt. J. N. Macomb, 1876, pp. 152.

[d] Report upon geological examinations in southern Colorado and northern New Mexico during the years 1878 and 1879, by John J. Stevenson. Rept. U. S. Geog. Surveys W. 100th Mer., vol. 3, Supplement, Geology, 1881, pp. 3–406, with atlas sheets.

[e] Report on the geology of portions of New Mexico and Arizona examined in 1873 by G. K. Gilbert: Rept. U. S. Geog. Surveys W. 100th Mer., vol. 3, Geology, 1875, pp. 503–566, with atlas sheets.

[f] Dutton, C. E., Mount Taylor and the Zuñi plateau: Sixth Ann. Rept. U. S. Geol. Survey, 1885, pp. 105–198.

[g] Bull. U. S. Geol. Survey No. 285, 1906, pp. 74–86.

[h] Mineral Resources U. S. for 1907, U. S. Geol. Survey, 1908.

map of the mining districts of the western United States. The New Mexico part of this map is reproduced in Plate II. Plate I (in pocket) is a general topographic map of the Territory, compiled with great care by Gilbert Thompson from the best available sources.

HISTORICAL SKETCH OF MINING AND SMELTING IN NEW MEXICO.

The history of mining and metallurgy in New Mexico is of considerable interest and reaches much farther back in time than that of the other States of the Cordilleran province. A chapter on this subject is contained in Jones's "New Mexico mines and minerals," mentioned above. Valuable material for such a history is also found in Bancroft's "History of the Pacific States" and in a book by Ex-Governor L. Bradford Prince, entitled "Historical sketches of New Mexico."

Unlike the more northerly States, New Mexico was at the time of the discovery of America settled by tribes of semicivilized Pueblo Indians, who lived in well-constructed cities and who practiced agriculture by the aid of irrigation. Mining was not extensively developed among them, but they knew the value of turquoise, which occurs in various places in the Territory, and which was used by them for ornaments and probably also in some way as money. There is also some evidence that they used gold for ornaments and that they probably obtained some of it from placers within the region. The first white man who set foot in the Territory is believed to have been Álvar Núñez Cabeza de Vaca, who reached it from the East in 1534. An expedition headed by Coronado explored the northern part of the Territory in 1540 and 1541. Between 1580 and 1680 the Indians were converted to Christianity by Spanish monks, and a number of missions were established.

The early Spanish explorers speak of turquoise as being obtained in the country, and it is now a generally accepted belief that the quarries containing turquoise in the Los Cerrillos Hills were worked prior to the advent of the Spaniards. There is also some evidence that the turquoise mines of the Burro Mountains and the Hachita Range were similarly worked at a very early date. The early missions were established principally by the Jesuits, and it is believed that some mining was done by the Indians under their direction. It is thought that gold was obtained from the gravels near Taos and there is some evidence of early attempts to open silver mines. Traces of such ancient mining for silver are found in the Los Cerrillos district, at a prospect near Ojo Caliente, and possibly also in the Hachita Range. The evidence is, however, not conclusive.

In 1680 the Pueblo Indians revolted against the oppressive rule of the Jesuits, and the Spaniards were obliged to abandon the country, returning, however, about twenty years later. It is stated that on their return it was expressly stipulated by the Indians that the Spaniards should not again engage in mining, but only in agricultural pursuits. At any rate little mining was done during the eighteenth century. F. A. Jones [a] mentions a document found in Santa Fe under date of 1713, which refers to an old covered-up mine in the "Sierra de San Lazora." About the end of the eighteenth century the copper mines at Santa Rita, in the southwestern part of the Territory, were discovered, and they are said to have had a large production for a time. It is worthy of note that these mines have been almost continuously worked up to the present day.

The era of placer mining in New Mexico began in 1828, when the Old Placers in the Ortiz Mountains south of Santa Fe were discovered. Though they were worked with the most primitive methods, a great deal of gold was taken out during the next ten years. In 1839 the still richer New Placers at the foot of the San Pedro Mountains were discovered. The production soon decreased, but it is safe to say that more or less placer gold has each year since 1828 been obtained from one or both of these districts. The total production of these placers probably amounts to several million dollars. The earliest metal mining from lodes, except the work at Santa Rita, was undertaken near the Old Placers when, in 1833, the Ortiz gold quartz vein was discovered.

Aside from these undertakings there was little activity in utilizing the mineral wealth of the region. New Mexico was incorporated as a Territory of the United States at the close of the

---

[a] Op. cit., p. 13.

Mexican war in 1846. There was little change, however, in the condition of the mining industry until 1860, the year which may be said to mark the beginning of the modern epoch of mining in the Territory. This is not surprising in view of the isolation of the region and the lack of knowledge of and facilities for the reduction of metals. The country, although rich in mineral resources, contained few large gold and silver veins similar to those which were worked at a very early date in Mexico. The lead mines of the Organ Mountains were discovered in 1849 and the production of this metal was maintained on a small scale for the ten years following the discovery. Here, however, as well as in the whole southwestern portion of the Territory, the hostility of the Apache Indians was a serious drawback to mining operations. In 1859 or 1860 placer gold is said to have been discovered at Pinos Altos, Grant County, by California pioneers, but the Apache Indians soon forced the prospectors to leave.

Bancroft[a] states that the census reports of 1860 mention only one silver and three copper mines, all in Dona Ana County, employing 390 workmen and producing $212,000. But the governor, in his message of 1861–62, alludes to 30 gold lodes at Pinos Altos, employing 300 men and paying $40 to $250 per ton, to rich gold placers near Fort Stanton, and to work at Placer Mountain near Santa Fe, besides the copper mines at Santa Rita and Hanover. All work was suspended during the Confederate invasion of 1861–62 and subsequent unchecked Indian raids until 1863, when the industry was revived in a small way. Lode mining at Pinos Altos was resumed in 1867.

In 1866 the silver-lead mines of the Magdalena Mountains were discovered by J. S. Hutchason. In the same year prospectors from Colorado found placer gold in the northern part of the Territory at Elizabethtown in Colfax County, and in that district operations on a larger or smaller scale have continued until the present day.

The placers in the White Mountain district of Lincoln County, formerly known as the Sierra Blanca region, were found by Americans in 1865, and quartz lodes were located soon afterwards, but little systematic prospecting was done until about fifteen years later.

The surveying and construction of the Southern Pacific Railroad and the Atchison, Topeka and Santa Fe Railway through the central and southern parts of the Territory, from 1879 to 1882, brought a large number of prospectors and miners. This was an epoch of great activity. Practically all the mining districts now worked were then discovered and developed. It was essentially an epoch of silver mining; copper and the other base metals were little sought for, and even gold was less considered than silver. A great number of mills and small smelters were erected; many of the latter were built at inconvenient places and operated on unsuitable ores without much metallurgical knowledge. Rich silver ores were discovered, milled, or smelted at Silver City and Georgetown; and it is reported that over $3,000,000 in silver was obtained in a short time at Chloride Flat, near Silver City. Stamp mills were built at Pinos Altos. Rich silver-lead ores were extracted from the Victorio district near Deming. Many new discoveries were made in the Mimbres Mountains, or Black Range, for many years haunted by hostile Indians. In 1877 placers and gold-quartz veins were found near Hillsboro, and during the few following years the Hermosa, Kingston, Apache, and Cuchillo Negro districts were prospected and a considerable amount of silver ores extracted.

In 1878 were found the phenomenally rich silver mines of Lake Valley, which in a few years yielded a total of 5,000,000 ounces of silver, but which were soon worked out. The Mogollon Mountains, in western Socorro County, close to the Arizona line, contain one of the richest districts of the Territory, though one of the most difficult of access. This vicinity, now known as the Cooney district, was first prospected in 1875, but not until about twenty years later was much progress made in the mining and reduction of the rich gold and silver ores. Small lead smelters were built at Socorro, Los Cerrillos, Hachita, and a number of other places. The Socorro plant was built about 1885, and in 1887 another lead smelter was erected at El Paso, Tex.

As indicated above, success did not always accompany these enterprises; a great number of them failed; at other places the rich silver ores of the croppings were soon worked out. All this,

---

[a] Op. cit., p. 649.

aided by the depreciation in silver, produced a natural reaction, so that the mining industry of the Territory was, during the last decade of the nineteenth century, in a more or less pronounced state of inactivity.

New discoveries were, however, made from time to time. Among them may be noted the Abe Lincoln mine, in the White Oaks district, where a valuable gold-quartz vein was exploited from 1890 to recent years. During the same period the gold-silver veins of Cochiti, in Sandoval County, were worked, and produced heavily for some time. Placers were discovered near the Rio Grande at Pittsburg, in Sierra County. Grant County has always been the most prominent and permanent producing region, chiefly from the Pinos Altos and Santa Rita mines. Between 1875 and 1885 a number of mills for amalgamating gold and silver ores with or without roasting were in operation in Grant County. Later a smelter was operated for many years by Senator Hearst, and his associates, of San Francisco. Several attempts were made to utilize the low-grade copper ores in sandstone, in the Sierra Nacimiento, the Sierra Oscura, and other places.

The exhaustion of the rich silver ores and the failure of many of the small reduction plants was followed by a period of more or less apathy, during which, however, the modern phases of the metal industry were gradually unfolding. With the dawn of the twentieth century a strong demand for the baser metals, especially copper, lead, and zinc sprang up. The smelters at El Paso and Socorro about 1901 fell under the control of the American Smelting and Refining Company and the Socorro plant was dismantled, the El Paso plant being enlarged to permit the treating of copper ores. The copper plant at Douglas, Ariz., built in 1902, provided a market for low-grade copper ores in the southwestern part of the Territory. The old smelter at Silver City was discontinued about the same time, and though recently a copper plant was reestablished at that place, the small smelters seem to find it difficult here as elsewhere to compete with the large plants.

Extensive low-grade copper deposits were opened in the Burro Mountain district about 1900 and concentrating plants were erected to treat the ores. The lead ores of the Magdalena Range in Socorro County were found to be associated with ores rich in zinc, and an important mining industry sprang up at Kelly; the zinc ores were shipped to eastern points and large concentrating works are now being built. A small lead furnace is in intermittent operation at Deming, and a local copper-matting plant has been built at Jarilla, in the Oro Grande district. Another small copper plant has been in intermittent operation in the San Pedro Mountains, Santa Fe County.

In the meanwhile the gold and silver mining industry shows comparatively little advance, the production remaining at a few hundred thousand dollars in gold and a similar number of ounces of silver. The larger part of this output is at present derived from the Cooney district, in the Mogollon Mountains.

It would not be fair to close this brief review without some reference to the production of coal, although this report does not include a treatment of the coal resources of New Mexico. The northern part of the Territory contains large fields of bituminous coal of Cretaceous age, the most important being those at Raton and Dawson, in the northeastern part, and those near Gallup, in the northwestern part. Near the Colorado boundary line and connecting with the Durango field in that State are other important and lately developed coal fields. The first recorded production of coal in New Mexico was in 1882, when it was 157,092 short tons. A constant increase in the production was noted from that date until 1907, when the output attained a maximum of 2,628,959 tons. In 1908 there was a slight decrease.

METAL PRODUCTION.

The earliest metal production in the Territory of which records remain came from the Santa Rita copper mines, from which heavy shipments were made to Chihuahua early in the nineteenth century. No exact figures are available. In 1828 the gold production from the Old Placers began and in 1839 the New Placers were discovered. For several years an output of some importance was maintained and it is not improbable that $3,000,000 or more was extracted up to 1880. No authentic or exact records remain; the available information is given under "Mining history" or in the detailed descriptions of the localities. After 1850 lead production began from the adobe furnaces of the Organ Mountains and about 1870 lead bullion was shipped from the Magdalena mines. Gold production from the placers of Pinos Altos began about 1860 and from the Elizabethtown diggings in 1866; but here too the records are mainly from hearsay. The earliest records in government publications date from 1869, but are doubtless only approximate.

Bancroft[a] states that the yield of gold and silver from 1864 to 1868 was between $125,000 and $250,000 annually. In the period from 1869 to 1874 the annual production was probably about $500,000. From 1875 to 1880 the yearly output is given by the same authority as $400,000.

The metal statistics in the following tables have been compiled by J. M. Hill of the United States Geological Survey. The first table gives the data contained in R. W. Raymond's reports, and in the earlier reports of Ross Browne. The silver has been converted into ounces; the figures for this metal are obviously incomplete.

Production of precious metals in New Mexico from 1869 to 1875.

[Compiled from the reports of Ross Browne and R. W. Raymond.]

Year.	County.	Gold.	Silver (ounces).	Year.	County.	Gold.	Silver (ounces).
1869	Colfax	$131,000		1872	Colfax	$100,000	
	Dona Ana		15,152		Grant	350,000	
	Grant	170,000			Other counties	50,000	
	Santa Fe	10,000					
	Other counties	166,102		1873		500,000	
						500,000	
		477,102	15,152	1874		500,000	
1870	Colfax	370,000		1875	Colfax	55,000	
	Grant	112,400			Grant	25,000	181,450
	Santa Fe	18,000			Other counties	20,000	
	Taos	8,000					
						100,000	181,450
		508,400			Grand total	2,985,502	
1871	Colfax	210,000					
	Dona Ana		7,576				
	Grant		68,182				
	Other counties	190,000					
		400,000	75,758				

It is unfortunate that no official records exist for the period 1876 to 1880, which may be said to represent the beginning of lode mining in the Territory. The records of the United States Mint for gold and silver begin in 1881 and continue up to the present time. The records of the production of copper and lead collected by the United States Geological Survey begin in 1882 and 1883.

The second table gives the production of gold, silver, copper, lead, and zinc. The figures for gold and silver are taken from the final figures in the reports of the Director of the Mint. Since 1905 these figures have been obtained in cooperation with the United States Geological Survey. The copper and lead data, with exceptions noted, are given from the records as they appear in the "Mineral resources of the United States," published by the Geological Survey.

The production of zinc ores began in 1903. The larger part of the zinc ores of New Mexico are used for the manufacture of paint. The figures give the equivalent of metallic zinc, calculated from the zinc content of the ore, less probable loss in smelting or manufacturing.

The third table presents the production by counties so far as recorded. The data are taken from the reports of the Director of the Mint up to 1903. Those from 1904 to 1908 are copied from the mine reports in "Mineral Resources of the United States."

---

[a] History of New Mexico, p. 749.

METAL PRODUCTION.

Production of gold, silver, copper, lead, and zinc in New Mexico from 1880 to 1907, inclusive.

Year.	Gold.	Silver (fine ounces).	Copper (pounds).	Lead (pounds).	Zinc (pounds).	Year.	Gold.	Silver (fine ounces).	Copper (pounds).	Lead (pounds).	Zinc (pounds).
1880	$125,000					1896	$475,800	687,800	2,701,664	6,922,000	
1881	185,000					1897	356,500	539,500	701,892	18,246,000	
1882	150,000	1,395,348	869,498			1898	539,000	425,300	1,592,371	11,594,000	
1883	280,000	2,205,427	823,511	4,800,000		1899	584,100	503,300	3,935,441	9,712,000	
1884	300,000	2,325,581	59,450	12,000,000		1900	832,900	434,300	4,169,400	b 7,210,420	
1885	800,000	2,325,581	79,839	10,000,000		1901	688,400	563,400	9,629,884	2,248,000	
1886	400,000	1,782,946	558,385	a10,000,000		1902	531,100	457,200	6,614,916	1,482,000	
1887	500,000	1,778,947	283,664	14,650,000		1903	244,600	180,700	7,300,832	1,226,000	c 1,856,297
1888	602,000	928,125	1,631,271	a10,000,000		1904	381,900	214,600	5,368,666	2,726,000	c 17,991,780
1889	1,000,000	1,130,000	3,686,137	9,528,000		1905	265,800	354,900	5,334,192	2,464,000	c 15,142,254
1890	850,000	1,300,000	850,034	b 1,404,376		1906	266,200	453,400	7,099,842	1,280,000	c 17,292,655
1891	905,000	1,325,000	1,233,197	b 1,351,030		1907	330,000	599,500	10,140,140	3,914,000	c 750,085
1892	950,000	1,075,000	1,188,796	b 8,937,500		1908	306,300	400,900	4,991,351	1,172,000	c 3,576,516
1893	913,100	458,400	280,742	b 2,680,000							
1894	567,751	632,183	31,884	5,946,000			14,822,651	25,172,138	81,300,718	167,573,326	56,609,587
1895	492,200	694,800	143,719	6,080,000							

a Estimate (W. L.).
b From reports of the Director of the Mint.
c From mine reports published in Mineral Resources of the United States.

Production of metals in New Mexico by counties.

Year.	County.	Gold.	Silver (ounces).	Copper (pounds).	Lead (pounds).	Zinc (pounds).
1881	Colfax	$150,000				
	Grant	25,000	213,179			
	Santa Fe	10,000				
		185,000	213,179			
1882	Bernalillo		3,876			
	Colfax	20,000				
	Dona Ana	20,000	697,674			
	Grant	35,000	329,457			
	Lincoln	40,000	19,380			
	Santa Fe	25,000	11,627			
	Socorro	10,000	333,334			
		150,000	1,395,348			
1883	Colfax	25,000				
	Dona Ana	85,000	949,613			
	Grant	110,000	930,232			
	Lincoln	24,000	7,752			
	Santa Fe	15,000	7,752			
	Socorro	6,000	310,078			
	Taos	15,000				
		280,000	2,205,427			
1884	Colfax	50,000				
	Dona Ana	5,000	34,884			
	Grant	105,000	825,581			
	Lincoln	40,000				
	Santa Fe	14,000				
	Sierra	19,000	1,000,000			
	Socorro	67,000	465,116			
		300,000	2,325,581			
1885	Bernalillo	9,500	393			
	Colfax	195,000	34,882			
	Dona Ana	8,000	9,303			
	Grant	259,000	872,091			
	Lincoln	55,000	2,326			
	Santa Fe	12,000	10,075			
	Sierra	118,000	910,854			
	Socorro	150,000	503,876			
		806,500	2,343,800			
1888	Grant	3,762	297,909			
	Lincoln	470,697				
	Other counties	128,178	796,066			
		602,637	1,093,975			
1889	Bernalillo	2,300	960	41,597		
	Colfax	100,000	10,853	29,412		
	Dona Ana	4,956	64,031		1,863,158	
	Grant	517,795	595,945	682,891	1,554,580	
	Lincoln	209,709	6,783			
	Mora	3,000	12,790	10,084	65,790	
	Santa Fe	37,190	70,717	4,515,731	1,628,080	
	Sierra	113,875	634,639	112,050	810,474	
	Socorro	147,500	69,245		3,415,790	
		1,136,325	1,465,963	5,391,765	9,337,872	

Production of metals in New Mexico by counties—Continued.

Year.	County.	Gold.	Silver (ounces).	Copper (pounds).	Lead (pounds).	Zinc (pounds).
1890	Colfax	$16,500	2,404	14,285		
	Dona Ana	500	22,092		107,232	
	Grant	292,218	441,917	323,811	255,997	
	Lincoln	227,500	1,937			
	Mora	850	2,945	2,857		
	Santa Fe	3,200	1,665	200,000	69,826	
	Sierra	190,400	671,938	97,142	36,159	
	Socorro	4,800	168,313		935,162	
		735,968	1,313,211	638,095	1,404,376	
1891	Dona Ana	700	20,294		201,637	
	Grant	314,365	466,465	476,924	528,046	
	Lincoln	340,025	6,131			
	Santa Fe	15,000	12,785	923,076	87,723	
	Sierra	217,213	681,836	184,615	14,620	
	Socorro	45,000	116,692		511,694	
	Other counties	10,000	8,115	11,539	7,310	
		942,303	1,312,318	1,596,154	1,351,030	
1892	Dona Ana	650	91,118		1,100,000	
	Grant	326,600	296,949	86,207	4,770,000	
	Lincoln	256,126	8,000			
	Santa Fe	45,000	91,428		750,000	
	Sierra	285,000	549,544	262,517	142,500	
	Socorro	65,800	66,770		2,125,000	
	Other counties	10,000	6,857	12,931	50,000	
		989,176	1,110,666	361,655	8,937,500	
1893	Dona Ana	10,000	24,806		1,250,000	
	Grant	232,000	181,938			
	Lincoln	205,000	3,876			
	Santa Fe	15,000	4,651			
	Sierra	329,000	162,791	272,728	105,000	
	Socorro	115,000	19,380		1,250,000	
	Taos	23,000	3,876			
	Other counties	10,000	4,651		75,000	
		939,000	405,969	272,728	2,680,000	
1894	Bernalillo	36,000	24,000			
	Colfax	78,000	27,000		66,667	
	Dona Ana	21,000	3,000		166,667	
	Grant	185,000	26,000	26,000	660,000	
	Lincoln	148,500	12,000			
	Santa Fe	28,600	4,560		103,334	
	Sierra	192,000	77,500	110,000	833,334	
	Socorro	93,450	30,000			
	Other counties	20,000	10,000		33,334	
		802,550	214,060	136,000	1,863,336	
1895	Bernalillo	38,250	28,932			
	Colfax	175,000	27,517			
	Dona Ana	15,000	91,738		853,664	
	Grant	172,000	39,748	51,258	230,184	
	Lincoln	195,000	7,182			
	Sierra	150,000	84,094		152,438	
	Santa Fe	17,800	3,055		76,214	
	Socorro	180,000	71,868			
		943,050	354,134	51,258	1,312,500	
1896	Bernalillo	35,000	32,837			
	Colfax	160,000	37,314			
	Dona Ana	18,000	104,479			
	Grant	186,920	129,023			
	Lincoln	60,000	17,913			
	Sierra	180,000	129,853			
	Santa Fe	25,000	7,463			
	Socorro	150,000	74,618			
		814,920	533,500			
1897	Bernalillo	21,850	21,314			
	Colfax	95,670	14,758			
	Dona Ana	5,340	16,717		234,325	
	Grant	96,057	161,657	1,154,162	9,984,056	
	Lincoln	39,580				
	Sierra	65,750	75,201			
	Santa Fe	20,466	2,055			
	Socorro	125,700	141,732		4,483,294	
	Taos	2,376	276	1,120		
		472,789	433,710	1,155,282	14,701,675	
1898	Bernalillo	53,760	30,757			
	Colfax	97,900	9,662			
	Dona Ana	4,350	11,135		123,368	
	Grant	86,240	165,081	3,267,452	9,369,818	
	Lincoln	49,750	524			
	Sierra	357,220	87,651	233,030		
	Santa Fe	58,800	4,238			
	Socorro	103,820	179,011		2,501,828	
		811,840	488,059	3,500,482	11,995,014	

METAL PRODUCTION.

Production of metals in New Mexico by counties—Continued.

Year.	County.	Gold.	Silver (ounces).	Copper (pounds).	Lead (pounds).	Zinc (pounds).
1899	Bernalillo	$53,760	30,040			
	Colfax	97,900	9,422			
	Dona Ana	4,350	10,859		108,614	
	Grant	96,740	194,046	2,960,411	9,169,302	
	Lincoln	49,970	510	158,108		
	Santa Fe	58,800	4,132			
	Sierra	377,000	85,478			
	Socorro	116,680	187,788		2,216,337	
		855,200	522,275	3,118,519	11,494,253	
1900	Bernalillo	218,259	109,206	4,889		
	Colfax	32,277	1,640	2,991		
	Dona Ana	2,873	6,532	712,385	179,976	
	Grant	413,704	231,726	6,506,517	3,742,964	
	Eddy	289	542	82,048	6,447	
	Santa Fe	2,593	23			
	Socorro	45,549	49,582	811,135	664,753	
	Sierra	37,991	111,142	64,757	2,615,953	
	Lincoln	72,183	467	8,132	327	
	Unknown	58,370	19,962			
		884,088	530,821	8,192,854	7,210,420	
1901	Bernalillo	193,919	123,254			
	Colfax	56,190	51,459			
	Dona Ana	3,655	15,275	515,152	165,116	
	Grant	277,055	181,564	5,682,169	1,860,465	
	Lincoln	74,468	405	12,122		
	Santa Fe	1,720	47			
	Socorro	28,057	67,945	727,273	534,884	
	Sierra	29,615	227,887	60,607	2,580,186	
	Other	51,675	100,000	249,474	1,300,256	
		716,354	767,836	7,246,797	6,440,907	
1902	Bernalillo	44,182	35,427	1,000		
	Colfax	3,552	73			
	Dona Ana	3,400	192		9,950	
	Grant	78,710	48,513	7,251,757	464,840	
	Lincoln	50,607	11,520			
	Luna		10,382		711,825	
	Otero	3,200	146	117,520		
	Rio Arriba	452	5,807	5,740		
	San Miguel			3,500		
	Santa Fe	6,341	1,808	3,250	32,131	
	Sierra	21,705	25,114	56,400	82,965	
	Socorro	42,056	146,503	540,000	1,189,004	
		254,205	285,485	7,979,167	2,490,915	
1903	Bernalillo	217	269		16,190	
	Colfax	98,427	2			
	Dona Ana		9,671	611,796	147,610	
	Grant	89,555	89,763	6,130,970	4,040,581	
	Lincoln	46,449	301			
	Luna		6,168		1,355,965	
	Otero	2,914	285	3,204		
	Rio Arriba	145				
	Sandoval	2,109	930			
	San Miguel	827	799	74,303		
	Santa Fe	3,514	183	9,596	4,012	
	Sierra	6,584	7,500	7,556	2,596	
	Socorro	14,841	84,805	700,000	1,050,000	1,856,297
	Taos	4,031	1,280	100,000		
		269,613	201,956	7,637,425	6,616,954	1,856,297
1904	Colfax	94,277	67			
	Dona Ana		24,101	40,000	1,581,488	
	Grant	68,428	74,793	4,428,508	179,142	
	Lincoln	28,596	2,576		2,261	
	Luna	1,695	8,549	16,000	671,772	
	Otero	2,619	164	14,400		
	Rio Arriba	484	52	846		
	Sandoval	1,899	1,035			
	San Miguel	450	450	24,900		
	Santa Fe	5,427		2,308		
	Sierra	74,596	17,055	16,700		
	Socorro	101,672	85,471	425,508	688,209	17,991,780
	Taos	1,787	240	3,000		
		381,930	214,553	4,972,170	3,122,872	17,991,780
1905	Colfax, Otero, Rio Arriba, Taos	34,598	276	53,602		
	Dona Ana	6,710	21,776	96,058	327,707	30,000
	Grant	46,007	86,629	5,291,222	321,035	257,203
	Lincoln	7,475	757	14,404	7,511	
	Luna		5,199		463,956	225,000
	Santa Fe, San Miguel, Valencia	6,243	42	8,900		
	Sierra	99,042	8,760	46,664		
	Socorro	117,435	245,753	615,175	390,000	14,630,051
		317,510	369,192	6,126,025	1,510,209	15,142,254

Production of metals in New Mexico by counties—Continued.

Year.	County.	Gold.	Silver (ounces).	Copper (pounds).	Lead (pounds).	Zinc (pounds).
1906	Colfax	$14,492	93			
	Dona Ana		34,051	434,000	1,207,193	
	Grant	99,809	163,987	6,388,830	710,895	144,656
	Lincoln	16,463	1,946			
	Luna	413	11,265		831,193	103,836
	Otero	423	533	133,166		
	San Miguel, Santa Fe	15,464	68	2,119		
	Sierra	18,048	8,102	5,995		
	Socorro	127,907	271,082	64,560	238,088	17,044,163
		293,019	491,127	7,028,670	2,987,369	17,292,655
1907	Colfax	11,914	83			
	Dona Ana	6	25,612	776,125	675,189	
	Grant	113,042	224,279	8,046,315	1,394,390	73,729
	Lincoln	41,417	721			
	Luna	13,741	8,633		1,022,773	
	Mora, Rio Arriba, San Miguel, Taos	1,704	1,404	62,765		
	Otero	14,811	2,135	679,480		
	Santa Fe	17,979	17,635	1,231,260		
	Sierra	8,224	2,250	5,425	9,548	
	Socorro	105,426	422,744	188,645	707,981	676,356
	Others	1,718	48			
		329,982	705,544	10,990,015	3,809,881	750,085
1908	Colfax	11,274	66			
	Dona Ana		9,283	21,150	379,697	
	Grant	45,682	95,477	5,242,767	244,589	251,070
	Lincoln	29,712	5,472	1,195		
	Luna	3	1,077	5,934	127,535	
	Otero	35,690	4,351	723,907		
	Rio Arriba and Taos	18,445	604	85,341	99,794	
	Santa Fe	2,031	6			
	Sierra	18,018	8,589	12,874	6,075	
	Socorro	136,890	280,091	19,462	16,073	3,325,446
	Unapportioned	1,012	28			
		298,757	405,044	6,112,630	873,763	3,576,516

Source of gold and silver in New Mexico 1904–1908.

[Fine ounces.]

Year.	Placers.		Dry or siliceous ores.		Copper ores.		Lead ores.	
	Gold.	Silver.	Gold.	Silver.	Gold.	Silver.	Gold.	Silver.
1904	7,228		6,861	75,684	4,137	79,369	250	59,500
1905	4,805	662	10,372	295,484	44	19,962	138	53,084
1906	1,297	195	12,486	304,917	365	118,548	27	67,467
1907	936	173	12,035	542,252	2,327	115,323	666	47,796
1908	1,122	153	9,545	364,335	3,757	24,768	28	15,788

Tonnage of ore in New Mexico, 1904–1908.

[Short tons.]

Year.	Dry or siliceous ores.	Copper ores.	Lead ores.	Zinc and zinc-lead ores.	Total.
1904					108,106
1905					145,629
1906	26,724	126,330	19,001	34,636	207,691
1907	46,486	164,849	5,873	2,784	219,992
1908	31,642	93,359	3,879	6,554	135,434

Number of producing mines in New Mexico, 1904–1908.

Year.	Placer.	Lode.	Total.
1904	24	80	104
1905	21	52	73
1906	19	52	71
1907	23	104	127
1908	26	108	134

The production of gold from 1869 to 1875 was in round numbers $3,000,000. From 1876 to 1879 it may be estimated at $1,500,000. From 1880 to 1908 the records show $14,800,000, making a total of $19,300,000 since 1869. The earlier placer production from the Old Placers, New Placers, Colfax County, Pinos Altos, and other placers may conservatively be estimated at $11,000,000, making an admittedly approximate total production for the Territory of $30,000,000.

The production of silver from 1882 to 1908, inclusive, was 25,200,000 ounces. The earlier production, from 1870 to 1882, came mainly from Socorro, Grant, and Sierra counties, and in some years it was certainly large. A conservative estimate would probably be about 10,000,000 ounces, which would make the approximate total production of the Territory 35,000,000 fine ounces.

GEOLOGY.

EPITOME OF GEOLOGIC HISTORY IN ITS RELATION TO TOPOGRAPHY.

The pre-Cambrian rocks of New Mexico tell a story, dimmed by antiquity, of periods of sedimentation on an unknown basement; of granitic intrusions tremendous in scale; of igneous and dynamic metamorphism. The "historical" records may be said to begin with the Cambrian period, when the Rocky Mountain core of northern New Mexico, and in fact the whole north-central part, was a land area subject to degradation, the sediments produced by which accumulated in the Cambrian strata of southern New Mexico. The northern land area was submerged at the end of the Mississippian (lower Carboniferous) and sedimentation went on, with some interruptions, until the close of Cretaceous time, when the Territory was covered by a mantle of sediments, perhaps 10,000 feet in thickness.

At the beginning of the Tertiary period igneous intrusive activity began; laccoliths, stocks, sills, and dikes of monzonite and quartz monzonite, with corresponding porphyries, were forced into the sediments, evidently bulging them in places, but rarely breaking the tough crust and rarely reaching the surface.

These intrusions were accompanied by a general continental uplift, raising the whole Territory 3,000 to 10,000 feet above sea level. Dislocations outlining the principal ranges accompanied this crustal movement. In the prolongation of the Rocky Mountains of Colorado the sediments were domed and then cut by vertical faults, along which subsidence took place. After erosion these conditions would produce the impression of a vertical upthrust of the pre-Cambrian rocks. This north-central part now forms the highest mountain region of the Territory, rising to elevations of 13,000 feet. South of Glorieta, where the Rocky Mountains proper dip below the Cretaceous sediments, the beds were subjected to stresses which produced monoclinal blocks with more or less pronounced fault scarps. The principal disturbances probably outlined the present valley of the Rio Grande and are marked by a series of sharply accentuated north-south ranges of apparently tilted blocks, such as the Sandia, Manzano, Oscura, San Andreas, and Organ ranges on the east side and the Nacimiento, Limitar, Magdalena, Cristobal, Caballos, and Cuchillo Negro ranges on the west. Some of the scarps face east, others west. Here also the apparent tilting may be the result of doming, faulting, and subsidence. At the same time was outlined the easternmost chain of the New Mexico ranges, which is separated from the Organ, San Andreas, and Oscura chain by the structural depressions of the Sacramento Valley. Like a graceful festoon this chain extends into New Mexico from trans-Pecos Texas and contains three units, the Guadalupe and Sacramento Mountains and the Sierra Blanca, all of them with gentle easterly slopes and steep western scarps. At the north the Sierra Blanca merges into a Cretaceous plateau; on the east side of the chain lie the almost level Tertiary strata of the Llano Estacado, affected only by a slight continental uplift. This plain is separated from the ranges by erosional scarps.

The northwestern part of New Mexico participated in the general uplift, but suffered only slight deformation. This is the plateau province proper, characterized by gently dipping Cretaceous strata that have been sculptured by erosion into terraces and scarps. It contains the broad uplift of the Zuni Plateau, in which erosion has exposed the older strata down to the pre-Cambrian. It is surmounted by some flows of Tertiary and Quaternary lavas. On the east it is limited by the ancient land mass of pre-Cambrian rocks, the Hopewell Mountains, and farther south by the westward-facing scarp of the first of the monoclines of central New Mexico, the Sierra Nacimiento.

In central New Mexico the plateau province extends far to the east; it is generally considered to cease at the Rio Grande, but in reality the high plateaus continue for some distance east of the local interruption by the Manzano and Oscura ranges until, near the eastern boundary of the Territory, they finally merge into the Great Plains.

Northeastern New Mexico is commonly referred to as a part of the Great Plains, but it is in reality, as pointed out by Hill,[a] an eroded plateau of Cretaceous rocks, surmounted by basaltic flows.

The Mimbres or Black Range, which lies west of the Rio Grande, is supposed to mark the southeastern limit of the plateau province. In structure it appears to show some relationship with the Rocky Mountain ranges of northern New Mexico. Its appearance is that of an upthrust of pre-Cambrian rocks, flanked on both sides by Paleozoic rocks dipping away from the core.

The extreme southwest corner of New Mexico embraces a part of a province foreign to the Territory as a whole—that of the Arizona desert ranges, numerous and small, trending northward and separated by desert basins. That these ranges are post-Cretaceous admits of little doubt. Probably they were outlined during the same early Tertiary deformation that produced the ranges of the Rio Grande valley. They differ from the latter by a far less marked monoclinal structure. They were probably outlined by faults, but few of the dislocations are conspicuous in their present topography.

The orogenic movements which outlined the present topography tended to create lake basins. Thus a large Eocene lake existed in northwestern New Mexico and another of later Tertiary age in the upper valley of the Rio Grande.

Ever since the uplifts and deformations erosion has been actively endeavoring to modify the scarps of the ranges and trench the plateaus. From the bulging domes of sediments over laccolithic intrusions it has carved the mountain groups of the Cerrillos, Ortiz, and San Pedro.

About the middle of the Tertiary, after erosion had been at work for a long time, masses of lava began to pour out over the southern part of the plateau province. They flooded the Black Range and the country westward to the Arizona line. At centers of eruption there rose above this plateau great piles of volcanic rocks, such as the Mogollon Range in the west, the San Mateo Mountains north of the Black Range, and the Valles Mountains northwest of Albuquerque. The andesitic and rhyolitic eruptions ceased, but shortly afterwards, in late Tertiary and early Quaternary time, basalt began to issue from generally inconspicuous vents and covered large areas in the upper Rio Grande and on the Cretaceous plateau in the northeast.

In diminishing volume these flows continued to a recent time, but the deformational and igneous history of the Territory really ends at the beginning of the Quaternary. Since that time the only important changes in the topography have been those effected by erosional agencies, in reducing the bulk of the mountains, in building enormous débris fans, in draining lakes, and in deepening canyons. A brief glacial epoch left its imprints on the highest range of the Territory between Santa Fe and the Colorado boundary line.

PRE-CAMBRIAN ROCKS.

DISTRIBUTION.

Pre-Cambrian formations do not occupy large areas in New Mexico. The largest masses of these rocks are found in the north and constitute on the whole the southward extension of the Sangre de Cristo Range of Colorado. The terminal point of this belt of pre-Cambrian rocks is some 20 miles south of Santa Fe, and its total length is about 120 miles. The areas have been outlined with fair completeness by J. J. Stevenson,[b] who refers to the pre-Cambrian ranges as

---

[a] Hill, R. T., Notes on the Texas-New Mexico region: Bull. Geol. Soc. America, vol. 3, 1892, pp. 85–100.
[b] U. S. Geog. and Geol. Surveys W. 100th Mer., vol. 3, supplement, 1881, maps.

axes of uplift and distinguishes the Culebra axis, which forms the main continuation of the Sangre de Cristo group; the Mora axis, which lies farther west; and the Sante Fe axis, which begins near Taos and ends some distance south of Santa Fe. South of Santa Fe the pre-Cambrian exposures are scattered, and in general, small. An isolated area, surrounded by Carboniferous rocks, is found in the Zuni Mountains, west of Albuquerque. Other pre-Cambrian areas lie along the west and south sides of the Sandia Range. Narrow strips of old granitic rocks underlie the limestones and quartzites of the Oscura, San Cristobal, and Caballos ranges; they occur also in the Magdalena Range west of Socorro, in the Florida Range southeast of Deming, in the Burro Range southwest of Silver City, and at several places along the Gila River and in the Peloncillo Range along the boundary line of Arizona.

PETROGRAPHIC CHARACTER.

Most of the available information concerning the pre-Cambrian of northern New Mexico is due to Stevenson, whose explorations were undertaken in 1878 and 1879. The rocks were referred to the Archean, and are described as granites, gneisses, and schists, together with some quartzites and other probably sedimentary rocks. Some doubt has been cast on these observations, which were not substantiated by petrographic examinations. C. R. Keyes [a] mentions the supposedly sedimentary rocks and believes that none of them are of clastic origin. His paper, however, contains little new and authentic evidence.

In general, reddish orthoclase and microcline granites predominate in the pre-Cambrian of New Mexico. In many places they are schistose, but over large areas the schistosity may be developed only in traces. Dioritic rocks, locally transformed into amphibolitic schists, appear in smaller masses in nearly all districts. Rocks of tuffaceous origin, partly stratified, were noted in the Magdalena Mountains. In a number of places, especially in northern New Mexico, incontrovertible evidence was collected during the present investigations showing the existence of an extensive sedimentary series consisting of quartzites, mica schist of clastic origin, and some limestones. These rocks form irregular masses, intruded and greatly metamorphosed by the granitic rocks. No basement of this sedimentary series has thus far been recognized.

According to present knowledge, then, there are great similarities between the pre-Cambrian of Colorado and that of New Mexico. The oldest rocks observed are quartzites, mica schist, and limestones. They have been intruded by enormous masses of a normal granite, and this granite has in turn been intruded by dikes and masses of dioritic rocks. In some places these dioritic rocks are cut by pegmatite dikes and a later granite. Schistosity in various degrees has been produced in both sediments and igneous rocks. The observations of the present investigations are summarized in the following paragraphs:

In the Rociada district, 20 miles northwest of Las Vegas, the pre-Cambrian rocks consist of gneisses and schists, the latter interstratified with crystalline limestone and quartzite, striking N. 45° W. and dipping 60° SW.

At Picuris, southwest of Taos, the prevailing rock is a granite or granitic gneiss, with more or less of basic intrusives. There is also a strongly metamorphosed sedimentary series, consisting of conglomeratic quartzites, knotty schists, and slates, containing garnet, andalusite, staurolite, sillimanite, and corundum. Similar altered rocks are found at Glenwoody, on the Rio Grande.

The isolated pre-Cambrian area of the Hopewell and Bromide districts, west of the Rio Grande and Tres Piedras, consists chiefly of gneissic granite, which in places cuts a dark dioritic gneiss. A number of dikes, probably also of pre-Cambrian age, cut these granular rocks, and have themselves been more or less metamorphosed. The dikes range from diorite porphyry to quartz monzonite porphyry; others are granite porphyries. A series of prominent hills, north and west of Hopewell, consists of very dense and massive gray quartzite, which is undoubtedly of sedimentary origin. An isolated and prominent outcrop of the same rock, surrounded by basalt, occurs about 30 miles south of the Hopewell district, on the road between Barrancas and

---

[a] The fundamental complex beyond the southern end of the Rocky Mountains: Am. Geologist, vol. 36, 1905, pp. 112-122.

Ojo Caliente. In the eastern part of the same district streaks and lenses of impure crystalline limestone occur, interfoliated with biotite-chlorite schist.

In the range rising just east of Santa Fe, coarse-grained reddish granitic rocks prevail, traversed by a rough sheeting trending northward. The lowest foothills consist, however, of micaceous schist dipping 30° W. and thoroughly injected by dikes of granite and pegmatite. Several miles above the city occur masses of schistose amphibolite with biotite, which probably are altered diorite or diabases. These contain a few pegmatite dikes. Near the divide biotite schists, striking N. 50° E. and standing almost vertical, were observed. Associated with them are fine-grained reddish granites of gneissoid structure.

Along the upper Pecos River smaller outcrops of pre-Cambrian rocks appear, extending a few miles south from the Hamilton mine. At that mine the rock is a dark fine-grained amphibolite, which a short distance farther south merges into normal red granite. South of Macho Creek dioritic rocks prevail. The last outcrops of the Santa Fe Range are found in Apache Canyon, along the Santa Fe Railway, where the rock exposed is a red granite.

Extensive exposures of pre-Cambrian rock were observed at Arroyo Hondo, about 20 miles north of Taos. At the mouth of the canyon there are streaks of amphibolite with flat westerly dip, alternating with white gneissoid granite. An extensive area of amphibolite schist extends from a point 5 miles above the mouth of the canyon up to the Twining copper mines. This amphibolite is intruded by a coarse light-gray granite. The strike of the schist is N. 45° E.

The pre-Cambrian rocks in the vicinity of the Sandia Mountains were not examined in detail, as this district contains no mineral deposits of importance. Yung and McCaffery[a] described the pre-Cambrian rock that underlies the Paleozoic strata of the Sandia Mountains as a coarse pink granite. They also mention a line of hills extending from the San Pedro Mountains in a southwesterly direction to Tijeras Canyon, south of the Sandia Mountains. These hills are said to consist of a remarkably pure bluish-white quartzite, mica schist, and some pink granite. It is likely that this is the largest area of pre-Cambrian sediments in the Territory, and the rocks are probably of the same age as the somewhat similar rocks in the Hopewell district described above, but are apparently less disturbed by subsequent intrusions. G. B. Richardson observed outcrops of a similar quartzite in Abo Canyon, east of Belen.

The isolated area of pre-Cambrian rocks near Copperton, in the Zuni Mountains, consists essentially of granites and gneisses, the schistosity trending on the whole a little north of west and the dip being at steep angles to the south. F. C. Schrader distinguishes a reddish massive granite, locally containing tourmaline, in the eastern part of the area. Granite porphyry appears in the western part, and is believed to be the most recent of the series. It consists essentially of orthoclase or perthite and quartz, with small foils of biotite. The phenocrysts of quartz are prominent, and a fluidal structure is observed in the surrounding groundmass. The rock appears to be of an effusive type. In the northwestern part of this pre-Cambrian area, in the Copper Hill district, a micaceous granitoid gneiss prevails, and there are smaller masses of diorite and amphibolite.

In the Sierra Nacimiento, situated in Sandoval County about midway between the Zuni Mountains and the Hopewell district, an axial core of pre-Cambrian rocks is exposed. The prevailing granite is of two types. A red granite predominates and closely corresponds to the tourmaline granite of the Zuni Mountains; the other type is a dark-gray rock. Both are in part schistose, the gray granite more so than the red. Both are cut by dikes of younger fine-grained granitic rock, whose intrusion took place at a date later than that of the schistosity of the granite.

Pre-Cambrian areas extending as narrow north-south belts are reported from the western foot of the Sierra Oscura, probably also along the Sierra San Andreas. Similar exposures are to be seen along the western foot of the Caballos and San Cristobal ranges. So far as has been ascertained, reddish granite, in part schistose, forms the prevailing rock and is in places accompanied by smaller masses of diorite or amphibolite schist. The exposures of pre-Cambrian rocks

---

[a] Yung, M. B., and McCaffery, R. S., Ore deposits of the San Pedro district, New Mexico: Trans. Am. Inst. Min. Eng., vol. 33, 1903, pp. 350–362.

in the Magdalena Mountains are somewhat difficult to differentiate from possibly later intrusions. The pre-Cambrian, however, contains remains of a bedded pyroclastic series, probably a diabase tuff, and dioritic rocks occur with this. The tuffs are intruded by dikes of red granite and pegmatite.

Outcrops of coarse-grained granite, underlying the Paleozoic rocks, are found in the Florida Mountains southeast of Deming.

A core of pre-Cambrian rocks is exposed in the Burro Mountains southwest of Silver City. The prevailing rock is a granite with associated pegmatite; a dioritic rock, in places somewhat gneissic, is present in smaller amounts. This region is near the previously studied Clifton quadrangle, Arizona, in which the pre-Cambrian rocks consist of a reddish massive granite, inclosing remnants of an earlier formation of sedimentary sericite schist, to which the name Pinal schist was given.

The Franklin Mountains lie just south of New Mexico in Texas, but are the direct continuation of the Organ Mountains. The succession of rocks in the Franklin Mountains has been studied by G. B. Richardson,[a] who finds that Upper Cambrian beds overlie, with erosional unconformity, two distinct pre-Cambrian formations. The lower one (Lanoria quartzite) consists of about 1,800 feet of light and dark quartzites and subordinate slates. The upper pre-Cambrian formation is a mass of rhyolite porphyry, over 1,500 feet thick, with a basal conglomeratic member ranging in thickness from a few inches to about 400 feet. The Lanoria quartzite may be of the same age as that exposed in the Hopewell district and in Tijeras Canyon.

The quartz monzonite of the Organ Mountains, which by most geologists has been regarded as pre-Cambrian, is shown by the observations recorded in this volume to be intrusive into the Paleozoic limestone.

It is obviously impossible to connect all these observations so as to form a consistent theory in regard to the pre-Cambrian history of New Mexico. They indicate, however, that the pre-Cambrian contains a sedimentary formation, chiefly quartzitic, invaded and broken, in some places almost to obliteration, by enormous masses of normal, usually reddish, microcline granite. In most but not all places a schistosity, varying greatly in intensity, has been produced in this granite. At a few places the granite also breaks through or contains remnants of older greenstone tuffs, amphibolites, and rhyolites. The granite is in turn intruded by aplite, abundant pegmatite, and some masses or dikes of diorite.

It is believed that these pre-Cambrian sediments correspond in age with those which are embedded in red granite in various parts of Colorado and that they perhaps also should be correlated with the quartzitic schists of the Pinal formation of southeastern Arizona.[b]

It is not believed that these sediments of New Mexico should be correlated with the Grand Canyon series of Walcott,[c] which is thought to have been deposited after the great invasion of granite.

PALEOZOIC AND MESOZOIC SEDIMENTS.

The pre-Cambrian history of the Territory, though imperfectly known, indicates a period of sedimentation followed by mountain building and igneous intrusion, which was in turn succeeded by long-continued erosion that exposed the intrusive cores. At the base of the Paleozoic section is a strongly pronounced unconformity. In general it has a fundamental appearance, as, for instance, where sandstones and limestones rest horizontally on vertical gneisses.

In the whole of northern New Mexico the Pennsylvanian or upper Carboniferous rests directly on the pre-Cambrian, and these conditions appear to continue as far south as Socorro. Cambrian rocks were first definitely recognized in New Mexico in 1905 by C. H. Gordon,[d] who found fossils of this age in the Caballos Range, along the thirty-third parallel. The thickness

---

[a] Tin in the Franklin Mountains, Texas: Bull. U. S. Geol. Survey No. 285, 1906, pp. 146-149. El Paso folio (No. 166) Geol. Atlas U. S., U. S. Geol. Survey, 1909.

[b] Van Hise, C. R., and Leith, C. K., Pre-Cambrian geology of North America: Bull. U. S. Geol. Survey No. 360, 1909, pp. 768-779.

[c] Walcott, C. D., Algonkian rocks of the Grand Canyon of the Colorado: Jour. Geology, vol. 3, 1895, pp. 312-330; Fourteenth Ann. Rept. U. S. Geol. Survey, pt. 2, 1894, pp. 497-524.

[d] Gordon, C. H., and Graton, L. C., Lower Paleozoic formations in New Mexico: Am. Jour. Sci., 4th ser., vol. 21, 1906, pp. 390-395.

of the Cambrian quartzite and shales, which here underlie the Ordovician limestone, is only 55 feet. In the Mimbres Mountains, near Kingston, the thickness is 75 feet, and in the Florida Range, southeast of Deming, 135 feet of quartzites were measured in the same stratigraphic position. Toward the west, near Silver City, L. C. Graton measured a thickness of nearly 1,100 feet of quartzitic sandstones, including some strata of limestone. Across the Arizona line in the same latitude Waldemar Lindgren[a] found 200 feet of quartzites, with some shales, resting on granite underneath the Ordovician and obtained near the top of these rocks some fossils suggestive of the Cambrian. In the Franklin Mountains G. B. Richardson[b] found 150 feet of sandstone containing Upper Cambrian fossils. At Bisbee, in southern Arizona F. L. Ransome measured 400 feet of nonfossiliferous, probably Cambrian sandstone, covered by 750 feet of fossiliferous Middle Cambrian limestone. This is the only known occurrence of heavy Cambrian limestone in this region, although it is possible that some of the limestone which in southwestern New Mexico covers the quartzite may also be of Cambrian age.

All this indicates beyond doubt that during the period preceding the Cambrian the whole of western New Mexico, so far as available exposures permit a judgment, was a land area which gradually became planed down from a very high to a more moderate relief. During the Cambrian period the sea appears to have advanced northward, but only to about latitude 33° 30', where the Cambrian deposits seem to thin out. The greatest thickness of sediments is found at Silver City. In the Caballos Range there appears to exist a gradual transition between quartzitic sandstone and underlying granite, suggestive of secular decay of the granite, preceding deposition.

It is interesting to note that the Ordovician, Silurian, Devonian, and lower Carboniferous (Mississippian) are likewise lacking in northern New Mexico, and that they begin to appear prominently at about latitude 33° 30'; some exposures of the Mississippian have been recognized at about the latitude of Socorro, or 34°.

It is possible, of course, that deposits of these periods may once have covered northern New Mexico and that they were removed by an epoch of erosion between the Mississippian and the Pennsylvanian, but it does not seem likely, especially as so extensive a removal would hardly have been effected without some evidence of structural unconformity between the two principal divisions of the Carboniferous. It is therefore thought probable that the northern part of western New Mexico, now constituting the southern extension of the Rocky Mountains and a part of the plateau province, was a land area during the larger part of the Paleozoic era, until covered by the Pennsylvanian sea, and that even then certain parts of the Rocky Mountain system in New Mexico remained above water.

The Ordovician rests conformably on the Cambrian and consists of 450 to 1,200 feet of usually cherty limestone, very poor in fossils. It has been recognized in the Caballos Range, in the Mimbres Range, and at Silver City and other places.

The Silurian is recognized at Silver City, at Lake Valley, and probably also at Hillsboro as thin beds of limestone and quartzite, but does not seem to be present everywhere between these places. It lies conformably above the Ordovician and at Silver City can not be differentiated from it. In the Franklin Range of Texas Richardson measured 1,000 feet of Silurian limestone.

The Devonian is present in an area extending from the Mimbres Mountains westward to the Arizona line, and everywhere appears as a thin formation of dark clay shales, usually calcareous in the upper portion. The maximum thickness (465 feet) was noted at Silver City. In the Mimbres Range, where Gordon has given the name Percha shale to the formation, it is only 200 feet thick, and its upper part contains a rich and characteristic Devonian fauna. The Devonian is not present in the Franklin Mountains of Texas, but appears with a thickness of 200 feet at Clifton, Ariz. The Devonian shale is believed by Gordon to overlie the Ordovician limestone with an unconformity of erosion, but the shale and underlying limestone are conformable at Georgetown, Lone Mountain, and Silver City, and also at Clifton, Ariz. At

---

[a] Clifton folio (No. 129), Geol. Atlas U. S., U. S. Geol. Survey, 1905.
[b] Tin in the Franklin Mountains, Texas: Bull. U. S. Geol. Survey No. 285, 1906, pp. 146-149.

any rate the Devonian period in southern New Mexico was characterized by shallow, muddy deposits, uniform over a large area.

The Mississippian, or lower Carboniferous, has been recognized at several places south of latitude 34°. W. T. Lee found limestone of this age in the Ladrones Range and Gordon believes, on the basis of evidence collected by C. L. Herrick, that the lower part of the section in the Magdalena Mountains belongs to this series. Characteristic Mississippian faunas were found by Gordon at Kingston and Hillsboro, and the horizon has for some time been known to be represented at Lake Valley, where a thickness of over 200 feet of limestone has been measured. Rocks of the same age are also present in the Silver City district. Gordon states that at Hillsboro these limestones rest upon the eroded surface of the Devonian calcareous shales, but farther west there is no evidence of unconformity.

The Pennsylvanian, or upper Carboniferous, is deposited with a considerable thickness over the whole Territory and reaches its maximum in the country between Santa Fe and Las Vegas. As far south as the latitude of Socorro the Pennsylvanian consists in large part of sandstones and shales in repeated alternation with some limestone beds. But south of this line the pure limestones prevail and at the same time the total thickness appears to diminish. Everything indicates near-shore conditions in the northern part of the Territory, where some land areas probably existed even at that time.

In northern and central New Mexico there appear to be two divisions of the Pennsylvanian, the upper in certain parts possibly including the Permian. According to W. T. Lee and C. H. Gordon [a] the Pennsylvanian of central New Mexico is divisible into two groups, the lower called the Magdalena group and the upper called the Manzano group. The Manzano consists in fact of the "Red Beds" of Carboniferous age.

The average thickness of the Magdalena group in the vicinity of Socorro is 1,500 feet. This group is subdivided into (1) a lower formation called the Sandia, consisting chiefly of shales, limestones, and sandstones, the first two predominating, with a thickness ranging from 500 to 700 feet; and (2) an upper formation called the Madera limestone, 300 to 500 feet thick, consisting of blue limestone with some shale. Both formations are well exposed in Socorro and Bernalillo counties.

The Manzano group, best exposed on the east side of the Rio Grande near Socorro, consists of 2,000 feet of red and variegated sandstones, shales, limestones, and gypsiferous beds. Its upper part is a bed of blue limestone 300 to 500 feet thick, with Pennsylvanian fossils, which Lee has named the San Andreas limestone. Beneath this is the Yeso formation, consisting of 500 to 1,000 feet of yellow, pink, and white sandstones and shales, with gypsum and some limestone. The basal formation of the Manzano group, to which the name Abo sandstone has been given, is composed of dark-red sandstones interstratified with sandy shales. There are distinct unconformities produced by erosion at the top and the bottom of the Manzano group.

South of Socorro County, Lee has traced the red beds of the Manzano group down to Rincon. In the Franklin Mountains, according to G. B. Richardson, the Carboniferous "Red Beds," and in fact all "Red Beds," are absent and the Pennsylvanian is represented by the Hueco limestone, 3,000 feet thick. Still farther east on the boundary line between Texas and New Mexico the Hueco is covered by 2,500 feet of Guadalupian [b] sandstone and limestone. Near Silver City and in the southwest corner of the Territory the Pennsylvanian consists almost exclusively of limestones of moderate thickness. The "Red Beds" are absent and Cretaceous rocks carrying Benton fossils rest with distinct unconformity on the eroded Carboniferous formations.

In the Zuni Mountains, according to Dutton, there are 1,650 feet of Pennsylvanian and Permian "Red Beds" resting directly on the pre-Cambrian rocks. In the Mora uplifts there are, according to J. J. Stevenson, 3,276 feet of upper Carboniferous, consisting of sandstones, shales, and limestones in rapid alternation. In the upper Pecos Valley, above Pecos, there is, according to Lindgren, about 4,000 feet of the same series which apparently can not be further divided.

---

[a] Gordon, C. H., Note on the Pennsylvanian formations in the Rio Grande valley: Jour. Geology, vol. 15, 1907, pp. 805–816. Lee, W. T., and Girty, G. H., The Manzano group of the Rio Grande valley: Bull. U. S. Geol. Survey No. 389, 1909.

[b] Richardson, G. B., Paleozoic formations in trans-Pecos Texas: Am. Jour. Sci., 4th ser., vol. 25, June, 1908.

At neither place does this series comprise any "Red Beds." Several coal seams, at least one of which is workable, are present. Between Pecos and Glorieta, the relations being especially well exposed in La Cueva Creek, this lower series is covered by a thick mass of "Red Beds" and coarse grits, containing some interbedded limestone, evidently of Paleozoic age. These coarse grits are beyond doubt derived from the pre-Cambrian mass of Thompson and Penacho peaks, just to the northwest, and their extensive development demonstrates the existence of a land area along the Santa Fe range and probably also an epoch of erosion within Pennsylvanian time. Above these grits and "Red Beds" are similar strata of uncertain age which underlie the Glorieta Mesa and which are capped by a yellow sandstone, outcropping at the edge of the escarpment; the sandstone has been regarded as of Dakota age. The total thickness of the "Red Beds" in the upper Pecos Valley, as measured by Newberry, is 1,350 feet.

Triassic "Red Beds" have not been demonstrated to exist near Glorieta or along the Mora uplift, nor do they occur in the southern part of the Territory. They are present, however, as shown by Newberry, in the Sierra Nacimiento and at Abiquiu.

The Cretaceous system is strongly developed over almost the entire Territory, but detailed consideration of it does not fall within the scope of this report. In the Mora uplift and its eastern foothills Stevenson measured 5,850 feet of Cretaceous. In the Zuni uplift Dutton estimated 3,700 feet as the aggregate thickness of the rocks of that period. Near Gallup F. C. Schrader[a] measured from 4,300 to 5,700 feet of Cretaceous. Along Galisteo Creek D. W. Johnson and W. T. Lee each measured 4,315 feet of Cretaceous beds.

Over the central part of the Territory the Cretaceous rests on "Red Beds," probably with unconformity by erosion. In the extreme north (north of Taos Pass, according to Stevenson) the Cretaceous rests on the Carboniferous. Near Elizabethtown the Cretaceous, according to Graton, rests directly on the pre-Cambrian, but the character of the contact is not established beyond doubt.

In the southwestern part of New Mexico, near Silver City, the Cretaceous rocks rest on the upper Carboniferous with a pronounced unconformity. The total thickness is not known, but probably is far less than in the northern part of the Territory. At Clifton, Ariz., a few hundred feet of similar Cretaceous rocks (Colorado) are exposed; still farther south the Cretaceous appears in great thickness and in the Hachita Range it is probably rather extensively developed. At Bisbee, in southeastern Arizona, Ransome has measured 4,000 feet of Cretaceous rocks, which, however, belong to the Comanche or Lower Cretaceous series, and which continue, developing on a large scale, into Mexico.

During the Cretaceous period the sea covered almost the whole of the Territory, probably more continuously so than in any other period save the early Pennsylvanian. A mantle of pliable shales with sandstone beds, probably averaging about 3,000 feet in thickness, extended then practically over the whole Territory.

TERTIARY AND QUATERNARY EVENTS.

At the close of Cretaceous time the long-maintained condition of quiescence and scarcely broken periods of deposition ceased. Here, as elsewhere in the Rocky Mountain region, the deposition of the Cretaceous beds was followed by an epoch of igneous activity and of mountain building, during which the most important of the mineral deposits of the Territory were formed. Intrusions of large masses of monzonitic magmas seem to have constituted the first step; they were forced in underneath the pliable and tough mantle of Cretaceous sediments, bulging it in laccolithic fashion. Their distribution corresponds with the zone of orogenic disturbance and with the belt of mineral deposits extending across the Territory in a north-northeast and south-southwest direction. The marine conditions ceased, but during the Eocene lake basins were probably established in the northwestern part of New Mexico, possibly also elsewhere.

Mountain building accompanied and succeeded intrusion. While conditions still remained quiescent in eastern and northwestern New Mexico, tremendous forces were at work in the

---

[a] Schrader, F. C., The Durango-Gallup coal field: Bull. U. S. Geol. Survey No. 285, 1906, pp. 243-257.

southward extension of the Rocky Mountain region. The old pre-Cambrian core in the north seems to have been forced upward by faulting, or by warping followed by faulting. Farther south the sediments were broken along north-south lines and the characteristic New Mexican monoclinal ranges were created, the prototype of the "Great Basin structure," which in fact is far more characteristically developed in this region than in the area from which it receives its name. Knowledge concerning the mechanics of this mountain building is somewhat uncertain, and the purposes of this report do not require their detailed discussion. It is sufficient to say that the Sandia, Oscura, San Andreas, Caballos, Organ, and other ranges, which lie in the southern continuation of the Rocky Mountain uplift, clearly show this combination of a fault scarp, facing east or west, with a monocline sloping gently in the opposite direction. Folding is absent or only slightly developed.

During the same epoch the desert ranges of the southwestern part of New Mexico were outlined by intrusions and by faulting, though here the monoclines are far less conspicuous and erosion seems to have been the most active factor in producing the present mountain forms.

A general uplift of the plateau province, and to a less degree of the whole Territory, accompanied this orogenic disturbance, the major features of the present drainage system were outlined, and of course erosion at once began its work of exposing the deeper strata and the intrusive rocks.

A second epoch of igneous activity, distinctly separate from the first epoch of intrusion, began, probably in the middle Tertiary, and vast masses of andesitic and rhyolitic flows were extruded. The flows, which in places are 2,000 feet or more thick, rest on the old sedimentary rocks, beveled by erosion, on the earlier intrusive rocks, or on the pre-Cambrian granites and gneisses. Naturally these piles of lava spreading as vast plateaus or gentle slopes from their foci of eruption were at once attacked by erosion, the work of which is well shown in the canyons of Gila and San Francisco rivers and in the consequent drainage of the Valles Mountains, north-northwest of Albuquerque. These mountains contain an important center of rhyolitic eruptions of middle Tertiary age. The San Mateo and probably other ranges in Socorro County have a similar history. There are several minor areas in the ranges of the extreme southwest, but the largest field of rhyolites, latites, and andesites is that extending from the vicinity of Socorro to the south end of the Mimbres Range and thence westward to Silver City, the Mogollon Mountains, and over into Arizona.

In the later Tertiary (Miocene epoch) a lake of large extent appears to have existed in the upper Rio Grande Valley and its deposits are known as the Santa Fe marl. Some of the deposits assigned to this age are probably subaerial, as held by D. W. Johnson, who examined them near Cerrillos, but there can be little doubt that lacustrine conditions prevailed at other places farther north.

Toward the close of the Tertiary there was a third outbreak of igneous activity. Basalts were erupted in large quantities and spread over vast areas, covering the Santa Fe marl, the eroded older sediments, or the intrusive rocks. Such basalt flows cover considerable areas in various parts of the Territory; for instance, at Raton, in the foothills east of the Mora uplift, along the northern course of the Rio Grande, and at many points farther south along the same river. There are also comparatively recent basalt flows along the Mexican boundary line between El Paso and Tres Hermanas and at a number of other places.

Basaltic eruptions continued during the Quaternary period and such eruptions of minor extent have taken place very recently. During the early part of the Quaternary there was a heavy accumulation of land deposits, forming masses of coarse gravels (known as the Palomas gravel), which filled many of the structural troughs to a depth of about 1,000 feet. Basalt flows were poured out again after the deposition of these gravels, and smaller flows of Recent age, possibly erupted only a few hundred years ago, are found in eastern Socorro and Otero counties, and western Dona Ana and Valencia counties.

Since the main part of the basaltic outburst deep erosion has taken place. The whole upper canyon of the Rio Grande has been excavated, and in the central part of the Territory the streams have cut into the Palomas gravel to a depth of 800 feet. Their cutting was accompanied by the formation of gravel terraces, as well shown along the Rio Grande near Albuquerque.

TOTAL THICKNESS OF THE PRE-TERTIARY SEDIMENTARY FORMATIONS.

In the northwestern part of the Territory Dutton [a] measured 3,700 feet of Cretaceous sediments, including Laramie. Underneath these were found 3,150 feet of supposedly Jurassic and Triassic rocks, mainly sandstones (parts of these are probably really Carboniferous) and at the base of the section 1,650 feet of Pennsylvanian, including 450 feet of Permian. Older formations are absent. The total thickness of all these beds amounts to 8,500 feet.

In the same region Shaler [b] noted from 2,500 to 6,200 feet of Cretaceous rocks, which he subdivided from top to bottom into the Laramie formation, Lewis shale, Mesaverde formation, Mancos shale, and Dakota sandstone.

In the Mora uplift, northwest of Las Vegas, J. J. Stevenson [c] measured 5,850 feet of Cretaceous, underlain by 3,276 feet of Pennsylvanian, a total thickness of 9,126 feet.

In the central part of the Territory, from Cerrillos to Socorro, according to the writings of C. H. Gordon W. T. Lee, and D. W. Johnson, the approximate thicknesses would be as follows:

Section of pre-Tertiary sedimentary rocks in central New Mexico.

	Feet.
Cretaceous:	
Galisteo sandstone	1,500
Madrid formation	1,500
Montana and Colorado	650
Dakota (?) sandstone	120
Jurassic (?):	
Morrison (?) formation	210
Pennsylvanian (upper Carboniferous):	
Manzano group	2,000
Magdalena group	1,300
Mississippian (lower Carboniferous), only near Socorro	125

This indicates a total thickness for the Mesozoic sediments of about 4,000 feet and for the Pennsylvanian of 3,300 feet.

At Kingston, in the Mimbres Range, Gordon measured the following section.

Section at Kingston.

	Feet.
Pennsylvanian:	
Manzano group	600–800
Magdalena group	815
Mississippian: Lake Valley limestone	105
Devonian: Percha shale	200
Ordovician (?): Mimbres limestone	445
Cambrian: Shandon quartzite	75

At Silver City Graton measured the following beds:

Section at Silver City.

	Feet.
Cretaceous shale and sandstone	880?
Pennsylvanian limestone	90
Mississippian limestone	370
Devonian shales	465
Silurian limestone	250
Ordovician limestone	620?
Cambrian quartzite	1,072
	3,047?

---

[a] Sixth Ann. Rept. U. S. Geol. Survey, 1885, pp. 111-183.
[b] Shaler, M. K., A reconnaissance survey of the western part of the Durango-Gallup coal field of Colorado and New Mexico: Bull. U. S. Geol. Survey No. 316, 1907.
[c] U. S. Geog. Surveys W. 100th Mer., vol. 3, Supplement, 1881, p. 77.

In the Franklin Mountains, Texas, north of El Paso, G. B. Richardson found these rocks:

Section in Franklin Mountains, Texas.

	Feet.
Pennsylvanian: Hueco limestone	3,000
Silurian: Fusselman limestone	1,000
Ordovician: El Paso and Montoya limestones	1,250
Cambrian: Bliss sandstone	150
	5,400

These data show that the Cretaceous in the northwestern and central parts of the Territory probably averages 4,000 feet in thickness; that the Paleozoic formations in the north-central part are about 5,000 feet in thickness, diminishing in the central part to about 3,500 feet. In the south-central and southwestern parts the total thickness of the Paleozoic column is from 3,000 to 5,400 feet. The Cretaceous is largely eroded in this region, but the probabilities are that it decreases somewhat in thickness toward the southwest.

It is a fair inference that over a large part of the mineral belt of New Mexico, at the close of the Cretaceous, there was deposited a blanket of sediments from 6,000 to 9,000 feet in thickness. This important conclusion will be used in the following discussions of the igneous activity.

INTRUSIVE ROCKS OF EARLY TERTIARY AGE.

DISTRIBUTION.

The intrusive rocks of post-Paleozoic age appear as isolated bodies in a belt stretching across the Territory from north to south. There is a considerable number of these bodies, and they take the form of stocks, laccoliths, sheets, and dikes. The rhyolites, andesites, and basalts rest in places on their eroded surfaces, but the flow rocks do not themselves contain any intrusive masses. The intrusive bodies are found in pre-Cambrian granites and schists and all parts of the great conformable sedimentary series from Cambrian up to and including the latest Cretaceous. The intrusion, therefore, falls in the very latest Cretaceous or, more probably, in the earliest Tertiary; for an uplift and an epoch of considerable erosion must have intervened between the intrusion of these bodies far below the surface and the outflowing of the lavas, which is considered to have begun in middle Tertiary time. In some places decisive evidence of the age of the intrusion is lacking; for instance, the quartz monzonite of the Organ Mountains can only be designated post-Carboniferous. It is probable, however, that this mass also was intruded during early Tertiary time.

The belt in which the intrusive rocks are exposed extends from northern Taos and Colfax counties in a south-southwest direction, approximately following the Rio Grande. In the southwest the belt widens and reaches from the Organ Mountains, in Dona Ana County, to the Arizona boundary line, adjoining Grant County. The area occupied by these intrusive rocks is small in the aggregate, compared to the vast stretches covered by the flows.

Small intrusive bodies occur in the Red River district, in Taos County. In the Cimarron Range, in western Colfax County, they occupy a belt 35 miles long and several miles wide, in Cretaceous rocks.[a] Farther south intrusive rocks appear in the Cerrillos Hills, and in the Ortiz, San Pedro, and South mountains, Santa Fe County; at Cochiti, Sandoval County; in the Magdalena Range, and probably at many places in the Oscura and San Andreas ranges, Socorro County. Several smaller areas are exposed by erosion in the Black Range, Sierra County, from Hermosa to Cooks Peak. One of the largest intrusions, at least 15 miles long and 8 miles wide, was observed in the Organ Range; this mass has been erroneously described by several geologists as pre-Cambrian. East of this range lies the small intrusive body of the Jarilla Hills. In the desert ranges of Luna and southern Grant counties are many small intrusions in Carboniferous and Cretaceous strata. In northern Grant County, near Silver City, Hanover, and Pinos Altos, several rather large areas have been laid bare by erosion along the southern margin of the plateau province.

---

[a] Marked "trachyte and rhyolite" on the geologic map of the Wheeler Survey.

Somewhat outside of the belt described above are the intrusive masses which break through Cretaceous strata in the White Mountains, in Lincoln County.

COMPOSITION AND STRUCTURE.

A remarkable uniformity of composition characterizes these intrusive rocks. With few exceptions they are intermediate between granite and diorite. A few are granodiorites or granodiorite porphyries, and occasionally a true diorite may be encountered, but most of them correspond closely to monzonite or quartz monzonite or their porphyries. From north to south the following general characteristics may be noted. In the Red River district, Taos County, the intrusive rock is a greatly altered monzonite porphyry. At Elizabethtown, Colfax County, there are dikes and stocks of light-gray monzonite porphyry with prominent feldspar crystals in a darker fine-grained groundmass. Hornblende and biotite occur sparingly. A similar rock is found in the Cimarroncito district. At Cochiti, Sandoval County, the monzonite is exposed below rhyolite flows; it is a grayish-green granular rock with orthoclase, albite, and oligoclase in large individuals, a little interstitial quartz, and some biotite; apatite, zircon, and much magnetite are the accessories. The structure is hyphidiomorphic granular.

In the Cerrillos Hills, on the Santa Fe Railway, monzonites of a rather basic type intrude Upper Cretaceous shales as dikes and sheets and laccolithic masses. Analyses 1 and 2 of the table on page 39 show the principal types. The rock of No. 2 is a dark-gray monzonite porphyry with phenocrysts of andesine or labradorite and augite in a microcrystalline, rather coarse groundmass of orthoclase; it contains much magnetite. Details are given in the description of the Cerrillos mining district (p. 164).

A short distance to the south rise the small Ortiz and San Pedro mountain groups. Both are distinctly laccolithic, showing a ring of uptilted sedimentary rock, and in the San Pedro group the sediments markedly arch over the intrusive rock. The Ortiz mass is intruded in Upper Cretaceous shales and sandstones; at San Pedro the intrusion has taken place mainly in Carboniferous rocks. The igneous rocks of the Ortiz Mountains are hornblende monzonite and monzonite porphyries with rather less orthoclase than usual. A sheet projecting to the Madrid coal mine from the main mass is a dark-gray, almost granular rock with closely crowded small feldspar crystals and needles of black hornblende not over 4 millimeters long. The microscope shows fine phenocrysts of hornblende, augite, orthoclase, and andesine in a microcrystalline groundmass of small stout prisms of orthoclase, embedded in larger grains of another feldspar, probably oligoclase; apatite and magnetite are accessories. A number of careful analyses of rocks from the Ortiz Mountains have lately been published by J. H. Ogilvie, but unfortunately with incomplete descriptions. Five of these analyses are quoted at the end of this chapter.

The prevailing rock at San Pedro is more acidic. (See analysis 9, p. 39.) It contains white closely crowded phenocrysts of andesine, 2 to 3 millimeters in length, and well-defined needles of dark-brown hornblende in a dark-gray microcrystalline groundmass of quartz and orthoclase. Chlorite and epidote are among the secondary minerals. The rock may be calculated to contain approximately 19 per cent quartz, 17 per cent orthoclase, 40 per cent albite, 8 per cent anorthite, the last two combining to Ab_5An_1; it is classed as a granodiorite porphyry.

In the Socorro Mountains, half a mile north of the Torrance mine, a small mass of granular hornblende monzonite is intruded in limestone. In the Magdalena Range there are many intrusives of this kind, mainly in the north end of the range near Kelly; they form irregular masses in limestone or in pre-Cambrian granites and greenstones. An intrusive mass northeast of the Graphic mine forms an elliptical area half a mile in its largest diameter. This is a gray rock of granular structure containing orthoclase and andesine, the latter predominating; besides these minerals there are a colorless pyroxene, green hornblende, biotite, and a little interstitial quartz; magnetite and apatite are the accessories. Analysis 6 shows this rock to be a monzonite with about 11 per cent of quartz, 20 per cent of orthoclase, 29 per cent of albite, and 29 per cent of anorthite, the last two combining to andesine.

At the Hardscrabble mine, in the same region, and also three-fourths of a mile west of Water Canyon post-office appear intrusive masses of a more quartzose rock, whose partial analysis is given under No. 12 in the table. These porphyries contain large phenocrysts of orthoclase, andesine, or oligoclase and a few of quartz in a microcrystalline groundmass of orthoclase, plagioclase, and some hornblende, with magnetite, titanite, and apatite as accessories. The analysis would indicate 27 per cent of quartz, 22 per cent of orthoclase, 30 per cent of albite, and 13 per cent of anorthite, the last two combining to andesine (Ab_2An_1). The rock is a granodiorite porphyry.

The intrusive rock which seems to stand in closest relationship to the ore deposits of the Magdalena district differs markedly from the type prevailing in the Territory. This rock cuts off the limestone on the west and its surface exposures are meager. It is a light-gray porphyry with phenocrysts of orthoclase, quartz, albite, hornblende, and ilmenite, in a groundmass made up of minute laths of feldspar in a matrix which probably consists of quartz and orthoclase. Alteration makes exact determination difficult. According to analysis 13 the rock is clearly a granite porphyry and looks very much like the intrusives at Metcalf, near Clifton, Ariz.

There are few exposures of intrusive rocks in the Black Range in Sierra County. Monzonite porphyries were noted in the Kingston and Tierra Blanca districts and granite porphyry in the Carpenter district, all intersecting Paleozoic limestone. At Hillsboro there are several intrusive masses; the largest are to the northeast of the town, where they cut the Ordovician limestones. A partial analysis (No. 8 in the table) shows the rock to be a granodiorite porphyry with roughly 20 per cent of quartz, 14 per cent of orthoclase, 34 per cent of albite, and 23 per cent of anorthite, the last two combining to andesine. A more basic rock of similar type forming the dump of the Ready Pay mine, about 2 miles north-northeast of Hillsboro, contains phenocrysts of orthoclase, soda-lime feldspar (andesine?), and augite in a coarse granular groundmass of orthoclase, with some plagioclase and a little quartz and magnetite. A partial analysis of this rock (No. 2) indicates that it contains no quartz, 29 per cent of orthoclase, 29 per cent of albite, and 32 per cent of anorthite. The rock is a monzonite.

Cooks Peak is a prominent landmark in northern Luna County, forming the south end of the Black or Mimbres Range. It consists of a massive porphyry intruded into the Paleozoic limestones, with dikes of similar porphyry radiating from it into the limestones on the north side. The intrusive rock is a gray porphyry with prominent phenocrysts of andesine, also some of hornblende and biotite; magnetite is relatively abundant; the groundmass is microcrystalline granular and consists of essentially the same minerals together with quartz and orthoclase. A complete chemical analysis is given under No. 10 in the table (p. 39). The norm calculated by B. S. Butler is as follows:

Norm of porphyry of Cooks Peak.

Quartz	16.62	Magnetite	2.55
Orthoclase	17.79	Hematite	1.60
Albite	34.06	Ilmenite	1.37
Anorthite	16.40	Apatite	.34
Diopside	3.89		
Enstatite	3.60		98.22

In the quantitative system the rock is a dacose. By the ordinary classification it is best designated as a granodiorite porphyry, as it contains too much plagioclase to be termed a quartz monzonite.

The Organ Mountains contain an intrusive area of unusual size; it occupies nearly the whole of the range, rising several thousand feet above the débris fans at the foot of the mountains, and appears to be about 15 miles long and from 3 to 8 miles wide. The rock is granular, except near the contacts, where it assumes in places a coarsely porphyritic structure. North of St. Augustine Pass the Paleozoic rocks cover it as a wide, low arch, and laccolithic relations are suggested. The rock is of brownish-gray color in the outcrops and appears like a normal granite. A typical specimen, a complete analysis of which is recorded under No. 7, is reddish-gray, of medium grain, containing some large, ill-developed crystals of orthoclase up to 15 millimeters in length. The mass of the rock has an average grain of 3 or 4 millimeters and consists of anhedrons of orthoclase, rarely with microperthite, laths of oligoclase, some andesine, and small quartz grains. There are small percentages of chestnut-brown biotite and of pale-green augite, in places roughly prismatic and partly converted to greenish hornblende; the accessory minerals are magnetite, apatite, and titanite. The titanite is visible by the naked eye in all specimens.

The rock is a quartz monzonite. The norm may be calculated as follows:

Norm of quartz monzonite of the Organ Mountains.

Quartz	10.50	Ilmenite	2.43
Orthoclase	26.13	Magnetite	3.94
Albite	35.11	Apatite	.93
Anorthite	11.40	Water, calcite, etc	1.65
Diopside	3.25		
Hypersthene	4.66		100.00

In the quantitative system the rock is designated as a monzonose. No pegmatites but many dikes of quartz monzonite porphyry and syenite porphyry accompany this intrusion.

The three peaks of Tres Hermanas, south of Deming, are built up of a mass of coarse granite porphyry, intrusive into Carboniferous limestone. The rock is brownish gray and contains phenocrysts of orthoclase up to 15 millimeters in length, also small foils of biotite and small crystals of dark-green hornblende. Some oligoclase is associated with the orthoclase; the groundmass is micropegmatitic, consisting of quartz and orthoclase. No analysis was made, but the rock is doubtless related to the granite porphyry of the Magdalena district, described above (No. 13 in the table of analyses).

A number of masses of intrusive porphyry are exposed at Silver City, Hanover, Central City, Georgetown, and Lone Mountain; they cut across Cretaceous rocks as well as Paleozoic limestone. At Hanover the relations (fig. 31) strongly suggest that the rock forms laccolithic masses or at least heavy sheets, which dome the Carboniferous limestone. At many places distinct sills were noted and dikes cutting the sedimentary rocks are common.

In the Chloride Flat region, near Silver City, Cretaceous and Devonian strata are intruded by dikes and sills of a gray monzonite porphyry with phenocrysts of black hornblende, with plagioclase, augite, orthoclase, biotite, and a little quartz forming a microcrystalline groundmass.

At Hanover, as well as between Hanover and Silver City, intrusive masses of light-gray quartz monzonite porphyry cut Carboniferous limestone. A partial analysis of the intrusive rock is recorded under No. 11. It is unusually rich in potash and approaches a granite porphyry. Phenocrysts of orthoclase and quartz with a little biotite and hornblende are embedded in a microcrystalline groundmass of orthoclase and quartz. There are also narrow dikes of a somewhat later, dark diorite porphyry, apparently identical with that of Pinos Altos.

At Pinos Altos intrusive masses of similar igneous rock have broken through the Paleozoic sediments and tilted them at various angles. A reddish granular rock, generally called a granite and by many erroneously held to be pre-Cambrian, consists of labradorite, andesite, orthoclase, quartz, green hornblende, and biotite, named in order of abundance. It also contains much magnetite, apatite, zircon, and titanite. Intrusive rocks of other types occur at Pinos Altos, closely related to that just mentioned, but of more basic character. One of these, a porphyry from the Cleveland mine, contains phenocrysts of stout prisms of augite and plagioclase, in a groundmass of laths of plagioclase and a little orthoclase, with grains of augite, apatite, and magnetite. The partial analysis of this rock is shown under No. 4 in the table; the rock appears to be a typical monzonite.

In the Burro Mountain, Telegraph, and Blackhawk districts dikes and stocks of the typical quartz monzonite porphyry of Silver City are abundant, here breaking through pre-Cambrian granites and gneisses. The light-gray porphyry contains phenocrysts of orthoclase and a soda-lime feldspar in a microcrystalline groundmass of quartz and orthoclase. There are also some entirely altered phenocrysts of augite; apatite and rutile were likewise noted.

In the Hachita Range smaller masses of quartz monzonite porphyry and syenite porphyry and lamprophyric dikes break through Carboniferous limestone and Cretacous rocks. In the

Apache Hills, near Hachita, a larger intrusion in the Carboniferous limestone was noted; it contains phenocrysts of quartz, small prisms of soda-lime feldspar, and hornblende replaced by epidote in a microcrystalline groundmass of quartz and feldspar.

At Granite Gap, in the Peloncillo Range, narrow dikes of granite porphyry cut Paleozoic limestone. A few miles farther north and at a point south of Steins Pass the same rocks are cut by larger dikes of monzonite porphyry.

Diorite porphyries of somewhat doubtful age and relationship were noted in the Kimball and Lordsburg districts.

In the White Mountains of Lincoln County, east of the Sierra Oscura and rather outside of the belt of intrusives already described, monzonitic rocks intrude Cretaceous strata. In the Nogal district, near Nogal Peak, the rocks are monzonite porphyries of medium grain and greenish-gray color; they contain phenocrysts of andesine, orthoclase, and augite in a groundmass of feldspar laths and interstitial grains of augite. This rock is cut by darker dikes of diorite porphyry (?).

In the Jicarilla district quartz monzonite porphyry has invaded and domed the sediments. The rock contains phenocrysts of oligoclase, amphibole, and magnetite in a microcrystalline groundmass of quartz in a mesh of orthoclase grains.

In the White Oaks district, in the same region, the Cretaceous shales are broken through by a mass of gray, fine-grained monzonite containing andesine, albite, orthoclase, microperthite, hornblende, biotite, and augite. A partial analysis is recorded under No. 5.

Analyses of intrusive rocks of New Mexico.

	1.	2.	3.	4.	5.	6.	7.	8.	9.	10.	11.	12.	13.
SiO_2	48.21	52.39	52.93	53.99	56.56	57.67	61.12	62.38	62.02	62.95	65.15	67.54	69.32
Al_2O_3	17.96						15.78			15.91			
Fe_2O_3	5.13						2.69			3.30			
FeO	4.47						3.15			1.37			
MgO	4.11						1.90			2.18			
CaO	9.72	6.40	8.50	4.89	4.77	5.89	3.95	4.60	4.62	4.46	1.96	2.65	.53
Na_2O	3.68	3.45	2.72	4.08	5.41	3.47	4.14	4.09	4.76	4.05	2.81	3.63	2.45
K_2O	2.99	4.92	4.71	4.90	4.72	3.54	4.48	2.36	2.84	2.95	5.52	3.82	5.54
H_2O-	1.41						.32			.72			
H_2O+	.21						.56			1.19			
TiO_2	.84						1.30			.67			
ZrO_2							.04						
CO_2							.22						
P_2O_5	.58						.45			.18			
S							.05						
MnO	.31						.09			.08			
BaO	.07						.07			.03			
SrO							.04			.03			
	99.74						100.35			100.7			

1. Mount McKenzie, Cerrillos Hills; George Steiger, analyst.
2. Dump of Ready Pay mine, Hillsboro, No. 444 New Mexico; George Steiger, analyst.
3. Cleveland mine, Pinos Altos, No. 506 New Mexico; W. T. Schaller, analyst.
4. Cash Entry mine, Cerrillos Hills, No. 7 New Mexico; E. C. Sullivan, analyst.
5. North Homestake mine, White Oaks district, No. 967 New Mexico; W. T. Schaller, analyst.
6. Northeast of Graphic mine, Magdalena Range, No. 372 New Mexico; George Steiger, analyst.
7. Near Merrimac mine, 3 miles north of Organ City, Organ Mountains, No. 518 New Mexico; George Steiger, analyst.
8. One and one-half miles northeast of Hillsboro, No. 440 New Mexico; George Steiger, analyst.
9. Montezuma Point, San Pedro copper mine, San Pedro Mountains, No. 28 New Mexico; George Steiger, analyst.
10. Cooks Peak, No. 713 New Mexico; George Steiger, analyst.
11. Santa Rita mine, 300-foot level, No. 572 New Mexico; W. T. Schaller, analyst.
12. Three-fourths of a mile west of Water Canyon post-office, Magdalena Range, No. 410 New Mexico; George Steiger, analyst.
13. South of Waldo tunnel, Graphic mine, Magdalena Range, No. 409 New Mexico; George Steiger, analyst.

The above table, arranged by increasing percentages of silica, shows that only one of the rocks (No. 13) can lay claim to a place among the group of granites. All the rest are distinctly intermediate in composition. Two of them (8 and 10) agree closely in composition with granodiorite; the others are monzonites or quartz monzonites. Naturally the lime decreases with increasing silica, but it is noted that in rocks with 54 to 63 per cent of silica the lime remains generally between 4 and 5 per cent. The sum of the alkalies varies from $6\frac{1}{2}$ to 10 per cent, but there is no constancy in the relation of soda to potash except that neither of them is often below 3 per cent.

By the quantitative system the rocks fall in the rangs of andose, monzonose, and dacose. So far as can be ascertained the composition has no relation to geographic position, rocks of widely differing percentages of silica occurring in the same district. At the same time the rocks undoubtedly present very close family relations, and it seems as if the differentiation from one magma could be easily accounted for by progressive crystallization of the more basic constituents.

Very recently I. H. Ogilvie has published a large number of analyses made of rocks from the single laccolithic mass of the Ortiz Mountains.[a]

Although the analyses are not accompanied by full descriptions of the rocks, five of them are quoted in the following table because they illustrate excellently the various gradations in an intrusive mass of predominatingly monzonitic character. The close relationship of the first four will be easily perceived. Analysis No. 5 shows a type of a normal diorite, which, as a rule, is foreign to the Territory.

Analyses of igneous rocks from Ortiz Mountains.

	1.	2.	3.	4.	5.
SiO_2	63.11	55.04	57.70	51.42	49.09
Al_2O_3	16.75	20.45	19.63	19.40	15.11
Fe_2O_3	2.68	2.09	3.30	3.72	.48
FeO	1.39	2.71	1.60	3.33	7.85
NgO	1.22	1.63	1.21	2.56	7.66
CaO	3.88	5.82	5.14	7.80	11.03
Na_2O	4.76	4.92	5.42	5.28	3.24
K_2O	3.48	4.29	3.74	3.96	.37
H_2O+	1.09	.69	.24	.49	.39
H_2O-	.32	.10	.13	.04	.02
CO_2	None.	None.	None.	None.	None.
TiO_2	.80	1.17	1.10	1.39	3.40
ZrO_2	None.	None.	None.	None.	None.
P_2O_5	.25	.37	.29	.53	.34
S	.03	.04	.06	.03	.06
MnO	.11	.26	.17	.23	.25
BaO	.16	.19	.09	.21	None.
Cr_2O_3					.04
	100.03	99.77	99.77	100.39	99.33

1. Intrusive sheet, west side of mountain. Gray porphyritic rock, phenocrysts of soda orthoclase, green hornblende and quartz; groundmass plagioclase, augite, and magnetite. Laurvikose-lassenose.
2. From spur 7,500 feet in elevation, on southeast side of group. Gray holocrystalline rock, containing orthoclase, two kinds of plagioclase, augite, hornblende, titanite, magnetite, apatite, and a little nepheline. Shoshonose-monzonose-akerose-andose.
3. From mountain 8,000 feet in elevation, on southeast side of group. Mineral composition similar to No. 3. Akerose.
4. High hills on southeast side of group. Porphyritic with phenocrysts of olivine and augite. Groundmass of orthoclase and plagioclase, in places a little nepheline. Essexose.
5. In central part of group. Dark gray, holocrystalline. Mineralogical composition not given. Auvergnose.

In the paper quoted the rocks are termed andesite, dacite, and diorite. According to the nomenclature adopted in this report, 1 is a quartz monzonite; 2, 3, and 4 monzonites, and 5 a diorite. The application of the terms andesite and dacite to coarsely holocrystalline or porphyritic intrusive rocks is much to be deprecated.

DIKE ROCKS.

In all these intrusive masses pegmatitic rocks and dikes are very rare. Porphyritic dikes, containing much quartz and orthoclase, accompany many of the quartz monzonites and in places approach the composition of granite porphyry. Dark dikes are also rather common, but they seem to be like the mother rock only with somewhat greater quantities of dark silicates. Lamprophyric dike rocks were noted at only a few localities; for instance, at the Copper Dick mine and in the Sylvanite district, in the Hachita Range.

RELATION TO MINERAL DEPOSITS.

The intrusive rocks described above have an unmistakably close relationship to ore deposits. Practically every occurrence mentioned is also a mining district of more or less importance.

MODE OF INTRUSION.

Although the intrusives at some places form irregular stocks there is evidence, at almost every one of the localities mentioned above, of a doming of the sedimentary strata in laccolithic

---

[a] Ogilvie, I. H., Some igneous rocks from the Ortiz Mountains, New Mexico: Jour. Geology, vol. 16, 1908, pp. 230-238.

manner, and the rocks also form numerous sheets or sills at various horizons from the Cambrian upward. This doming or arching is noted in the Hanover, Pinos Altos, Nogal, White Oaks, and other districts, and is perhaps best developed in the San Pedro and Organ districts, where the sedimentary roof of the intrusive is in part preserved. All the evidence tends to show that intrusion took place against a heavy load of superincumbent strata and that connection with the surface was not readily established. As most of the intrusives now appear along a zone of great displacement and block mountains, it seems likely that the magmas were forced into the sediments before the great breaks at the border of the plateau province had been formed. As it is positively established that a great epoch of erosion took place between the intrusion of these rocks and the great lava flows of the middle Tertiary, the lavas have no direct relationship with the intrusive rocks. If the intrusive magmas reached the surface as lava flows no definite trace of them has been preserved; it is likely that in most places they did not find their way to the surface. There are, however, in the Pyramid Range and in the northern part of the Peloncillo Range certain obscure and altered rocks, some having the appearance of diorite porphyries and others that of altered and chloritic andesites, which possibly may be found to represent the surface equivalents of the early Tertiary intrusives.

The mode of intrusion of these rocks is better realized when it is considered that at the close of the Cretaceous period practically the whole of New Mexico was covered by a thick mantle of sedimentary rocks, the total thickness of which was 6,000 to 9,000 feet. The Cretaceous part of this section was between 3,000 and 5,000 feet thick and much of it consisted of tough and pliable shales, which formed a heavy mantle which, though elastic and yielding, could not easily be broken by the igneous forces. The figures given also permit some conclusions as to the depth at which the intrusions took place. At Elizabethtown, Cerrillos, Ortiz, and other places the magmas found their horizon of lateral spreading and doming in the Upper Cretaceous strata, perhaps only 2,000 or 3,000 feet at most below the surface of that time. In the Organ Range, in the Tres Hermanas group, and in many other places the magmas were forced in at lower levels, in the Paleozoic strata, but the actual depth below the surface was probably rarely more than 9,000 feet. This result is interesting though by no means new to science. Many years ago Brögger, in his investigations of the Kristiania intrusive district, ascertained that great depth below the surface is not essential for the development of granular rocks and attendant contact metamorphism.

CONTACT METAMORPHISM.

During intrusion the gases liberated from the magma find rapid escape difficult and gradually spread through the surrounding rocks, altering and recrystallizing them, with or without accessions of material from the igneous source. This process, called contact metamorphism, is excellently shown in the rocks surrounding the intrusive bodies of New Mexico, but it is not so intense there as at the contacts of the great batholiths in Idaho or in the Sierra Nevada. Limestone is rapidly and extensively altered in most cases of intrusion, but the best measure of the intensity of the metamorphism is found in its action on shaly argillaceous rocks. In New Mexico these rocks, usually of Cretaceous age, are as a rule little altered, and what change has taken place does not extend far from the contact. Instead of great contact zones extending for a mile or more from the igneous rock, there is only an induration or baking, perhaps with some development of brown mica or epidote. When the shales are calcareous metamorphism takes place far more easily, and diopside, soda-lime feldspars, and various sulphides may develop.

Limestones and dolomites are easily affected by metamorphic agencies, with the abundant development of garnet, epidote, diopside, pyrite, pyrrhotite, chalcopyrite, magnetite, and specularite. Contact action in limestone in this region rarely extends farther than 2,000 feet from the contact. Almost invariably there is a strongly marked selective difference between the various beds, in some places without clearly apparent reason. In beds at the same horizon contact-metamorphic minerals may extend in some strata far beyond a point where no alteration is noticeable in adjacent strata. One bed may, for instance, be converted to coarsely crystalline marble and an adjoining bed, which assuredly also was composed of fairly pure limestone, may be converted into a solid stratum of garnet. In many localities, as in the Organ Mountains and the Tres Hermanas, complete garnetization takes place, the bedding and locally even the fossils being preserved. The Carboniferous limestones usually furnish the best material for contact action, but in the Elizabethtown and Ortiz districts Cretaceous limestones also have suffered far-reaching metamorphism. The process is to some extent capricious and it is possible, as near Silver City, to find places where the limestones show but slight alteration, while not far away, as at Hanover, they have suffered great changes. The deposition of sulphides goes on simultaneously with the rock alteration, and no evidence that it is a distinctly separate and later process can be found.

No observer who has studied these New Mexican occurrences can doubt for a moment that the metamorphism has been accomplished by the accession of great quantities of material, mainly of silica, iron, and sulphur, with smaller amounts of other metals. Further details about contact metamorphism are presented in the discussion of ore deposits (pp. 51-56).

TERTIARY AND QUATERNARY LAVAS.

GENERAL FEATURES.

Igneous rocks which have been poured out as flows over the surface of the ground have a wide distribution in western and northern New Mexico. The southeastern part of the Territory, say from a line drawn from El Paso to the northeast corner, contains few igneous rocks. The effusive rocks may in a general way be designated as andesites, rhyolites, and basalts. The basalt abounds in the northern and central parts. The andesites and rhyolites occur in large areas in the portion of the Territory adjacent to Arizona, which is bounded by a line drawn from Silver City to Lake Valley, thence north to Magdalena, and thence west to the Arizona line.

The flows were poured out over the eroded older formations, which comprise pre-Cambrian granites and gneisses, Paleozoic and Mesozoic sedimentary rocks, and worn-down remnants of intrusive masses of monzonites and monzonite porphyries. Some of the basalt flows rest on Quaternary gravels. The eruptions are believed to have begun in the middle Tertiary and continued well into the Quaternary; smaller eruptions of basalt have probably taken place within the last few hundred years.

The general succession, as in so many places in the Western States, consisted of rhyolite, andesite (or latite), then rhyolite again, and finally basalt. The andesites and rhyolites appear to be exclusively of Tertiary age.

Along the Arizona line, in the Mogollon Range, a rock allied to rhyolite occurs at the base of the series, and the "andesitic" flow forming the larger part of the volcanic mass is unusually basic and clearly related to basalt. The flows here attain a maximum thickness estimated to be several thousand feet. Basaltic rocks alternate with early rhyolite in the sequence of Tertiary eruptions near Clifton, a short distance across the line in Arizona.

In southwestern New Mexico, in the Pyramid and northern Peloncillo ranges, are considerable masses of a greatly altered silicified or chloritized igneous rock, which is described under the Lordsburg and Kimball mining districts. It has been called diorite porphyry or andesite, and much of it seems to have the character of volcanic flows. The age is doubtful, and it may possibly be early Tertiary or Cretaceous.

The full sequence of the middle Tertiary and later eruptions would seem to be as follows: (1) Quartz trachyte or rhyolite (Mogollon district); (2) andesite, basaltic andesite, and latite; (3) rhyolite; (4) basalt.

RHYOLITE.

Rhyolitic eruptions appear on a small scale in the north, in the Red River district, Taos County. Somewhat farther south, west of Santa Fe, in Sandoval County, rises the circular mass of the Valles Mountains, which appear to be a Tertiary rhyolitic volcano surrounded by widespread sheets of massive and tuffaceous rhyolite, also with subordinate andesitic flows. Many of these rocks have been described by J. P. Iddings.[a] This locality is mentioned in more detail in the description of the Cochiti district (p. 150). The San Mateo Mountains, still farther south, 30 miles south-southwest of Magdalena, consist of a similar volcanic pile, chiefly of rhyolite with some underlying andesitic rocks. The Black Range contains much rhyolite, mainly in its highest points as a capping of the andesite and latite flows of that region. This rhyolite extends as far south as Lake Valley, and the rock is very abundant from this locality westward. Thick flows rest on the Paleozoic rocks and porphyries in the vicinity of Pinos Altos and Hanover. In all these occurrences the rhyolite appears to be a normal rock of its kind, usually reddish or brownish and very commonly brecciated or tuffaceous. In the north end of the Black Range the rhyolite overlies flows of latite. Much of it is tuffaceous, and in thin section it is shown to contain large phenocrysts of corroded quartz crystals along with more numerous but smaller fragments of crystals of orthoclase, andesine, and biotite. The groundmass is glassy or spherulitic, locally also microcrystalline. In some of the rock the phenocrysts predominate, giving it the appearance of a nevadite.

In the Mogollon Mountains the rhyolite contains phenocrysts of orthoclase and andesine, with grains of magnetite in a glassy or spherulitic groundmass showing flow structure. Much of it is greatly altered by secondary processes. A partial analysis gave, according to W. T. Schaller, SiO_2, 67.83 per cent; CaO, 2.10 per cent; K_2O, 5.46 per cent; Na_2O, 3.30 per cent.

ANDESITE, LATITE, AND TRACHYTE.

Rocks of andesitic or trachytic type are not common in the northern and eastern parts of the Territory. One of the few occurrences is a small area of andesite, greatly altered by secondary processes, which was noted in the Red River district, Taos County. As stated above, the greatest areas of andesite appear in the southwestern districts from Magdalena southward to Lake Valley and thence westward to the boundary line of Arizona. Rocks classed as andesites also occur in some of the smaller groups of hills rising above the high plains of Luna County; among these are the Victorio Hills, 18 miles west of Deming.

So far as examined, the andesitic rocks have proved to contain a considerable amount of potash, some of them more potash than soda; and there is little doubt that the prevailing type is of intermediate character, placing the rock rather with the latites than with the andesites proper. In connection with the present work the following complete analysis of a representative "andesite" (447b New Mexico) from Hillsboro was made by George Steiger:

Analysis of "andesite" from Bonanza mine, Hillsboro.

SiO_2	54.54	TiO_2	0.86
Al_2O_3	14.66	CO_2	2.19
Fe_2O_3	4.20	P_2O_5	.49
FeO	2.74	S	.01
MgO	3.21	MnO_2	.29
CaO	5.64	BaO	.07
Na_2O	3.47	SrO	.05
K_2O	5.28		
H_2O-	1.10		100.67
H_2O+	1.87		

According to the quantitative system the norm is calculated as follows:

Norm of "andesite" from Bonanza mine, Hillsboro.

Quartz	4.92	Magnetite	6.03
Orthoclase	31.14	Ilmenite	1.67
Albite	29.34	Apatite	1.34
Anorthite	8.90	Calcite	5.00
Enstatite	7.96		
Diopside	1.39		97.69

---

[a] On a group of volcanic rocks from the Tewan Mountains, New Mexico: U. S. Geol. Survey Bull. No. 66, 1890.

This rock is classed as a shoshonose. According to the older nomenclature, it is evidently a normal latite, a rock standing between an andesite and a trachyte. It is to be noted that the phenocrysts are mainly andesine and augite, so that the potash must be contained chiefly in the groundmass. In fact, were it not for the analysis, the rock would unhesitatingly be classed as andesite. The groundmass is trachytic, with some glass. Fluidal structure is commonly noted.

The latites of the Black Range form a thick series of massive flows, tuffs, and breccias and the higher ridges are capped by rhyolite, deposited on the eroded surface of the older rock.

Flow rocks with abundant phenocrysts of orthoclase are not wholly lacking. At Socorro Mountain, near the Torrance mine, the oldest flows consist of an andesite probably related to a latite. This is covered by a rock of trachytic appearance (402 New Mexico). Under the microscope it shows phenocrysts of orthoclase, andesine, and hornblende, in a groundmass composed of glass and microscopic tabular feldspar crystals. A partial analysis by George Steiger gave SiO_2, 65.56; CaO, 3.65; Na_2O, 3.83; K_2O, 3.3. The rock evidently contains an excess of silica, and may be classed as a quartz-bearing latite. This flow is overlain by rhyolite, which is in turn capped by basalt.

It is interesting to note that many of the so-called andesites of the West were originally classed as trachytes. Later, on the basis of determination by phenocrysts, they were termed andesites, and now, on closer examination, they prove to belong to neither class, but to the widespread type of intermediate rocks, characterized by the presence of both orthoclase and soda-lime feldspar.

The thick flows of andesitic rocks of the Mogollon district have not been studied in great detail, as the widespread rock alternation renders it difficult to procure fresh and satisfactory specimens. These lavas generally present a brownish or reddish appearance and are of fine grain. The ferromagnesian silicates are usually decomposed. The rocks appear to be related to the basalts and are somewhat similar to a corresponding series described in the Clifton (Arizona) folio. A partial analysis by W. T. Schaller, subsequently repeated to confirm the relations of the alkalies, gave SiO_2, 48 per cent; CaO, 7.72 per cent; Na_2O, 1.95 per cent; K_2O, 3.28 per cent.

Four andesitic lavas of the Mount Taylor region have been analyzed in the laboratory of the United States Geological Survey.[a] It is worth noting that these analyses show considerable resemblance to the analysis of the Hillsboro rock. All show a high percentage of potash.

BASALT.

Basalt, the most recent of the volcanic flows of New Mexico, is widely distributed in the northern and central parts of the Territory, and in the aggregate covers large areas. The Cretaceous rocks of eastern Colfax and western Union counties, southwest of Raton, are capped by extensive basalt flows and several craters still remain. Other flows lie between the Santa Fe Railway and the foot of the Mora Range, west of Wagon Mound station, on the headwaters of Ocate Creek; these are believed to have issued from the still-preserved Ocate Crater, which reaches 8,902 feet in elevation. Both of these large regions are mapped by J. J. Stevenson on sheets 70A and 70C of the Wheeler maps.

Another great region of basaltic flows follows the Rio Grande from the Colorado line southward almost to Albuquerque, and in many places the thin flows overlie the white, soft Santa Fe marl and form prominent bluffs along the river. Many craters still remain. Scattered basaltic eruptions appear to follow the dislocations along the Rio Grande and may be seen at Albuquerque, Socorro, San Marcial, Elephant Butte, and Rincon, and at several places in the Black Range. At Rincon the basalt eruptions appear to leave the river, which turns south-southeast, but continue southward through the western part of Dona Ana County. Craters and in part very recent flows may be seen northwest and west of El Paso along the Southern Pacific and the El Paso and Southwestern railroads. They are especially well exposed along the latter road about 40 miles west of El Paso.[b] West and east of this median line of the Tertiary area

---

[a] Bull. U. S. Geol. Survey No. 228, 1904, p. 194.
[b] Lee, W. T., Afton craters of southern New Mexico: Bull. Geol. Soc. America, vol. 18, 1907, pp. 211-220.

basalts are much less in evidence. A considerable area of recent flow occurs, however, on the east side of the Oscura Range in northwestern Otero and southwestern Socorro counties; these flows extend northeastward for about 50 miles and are in places 6 miles wide.

In the plateau province of northwestern New Mexico basalt flows are abundant. Among the largest areas are that of Mount Taylor, north of the Santa Fe Railway, in Valencia County, and the great recent lava beds in the central part of the same county, south of the Zuni Plateau. The latter are said to cover about 600 square miles and their exceedingly rough surface gives evidence of the recency of their eruption. Smaller streams of lava followed the water courses to the northeast of this area and may be well observed from the cars on the Santa Fe line.

The age of the basalt flows can be ascertained somewhat more exactly than that of the earlier andesites and rhyolites. Though it is not believed that all the larger basalt flows are of the same age, it is clear that all the basalts are comparatively recent and that there was a considerable interval between the late rhyolite and the basalt. Some evidence is obtained along the canyon of the upper Rio Grande between Espanola and Taos. Here, as indeed is the case as far south as Bernalillo, the basalt flow overlies white, fine-grained sediments, undoubtedly the Santa Fe marl. These loesslike sediments have a decided bedding, the dip of which is in places 30° or more. Between Embudo and Barranca the Denver and Rio Grande Railroad ascends from the river level to the basalt mesa, which is 1,000 feet higher and has an elevation of about 7,000 feet. There are here at least two basalt flows separated by sediments; the thickness of the igneous sheets is from 100 to 200 feet. The upper canyon of the Rio Grande, from a point west of Santa Fe up to Tres Piedras, is cut to a depth of about 1,000 feet below the top of the basalt flows; viewed from a point on the road from Barranca to Taos this precipitous black gash in the level basaltic plateau is most impressive. Furthermore, the exposures along the same road show conclusively that the great débris fans from the Taos Range overlie these basalt flows. From all this evidence it may be concluded that the age of the basalts of this area is early Quaternary or late Tertiary, and that they were erupted at the end of the epoch of the Santa Fe marl. As is well known, the age of that formation—whether late Tertiary or early Quaternary—is a disputed question.

Other evidence is derived from Sierra County, along Rio Palomas, to the west of Rio Grande. Here, according to C. H. Gordon, the Palomas gravel fills the basin of the Rio Grande to a depth of more than 2,000 feet; this wash is derived from the ranges to the west and is believed to date from the early Quaternary. Probably it is of the same age as the great débris from the Taos Range. It has been trenched by Rio Palomas and other streams to a depth of 800 feet or more. Two epochs of eruption are distinguished; during the first epoch vast sheets of basalt were spread over the just-deposited Palomas gravel; succeeding this, as shown by Lee,[a] came an epoch of erosion, followed by another basalt eruption, represented by the San Marcial flow and that farther south, west of Mesilla. Both of these flows have been eroded over considerable areas. Other areas, mentioned above, suggest much more recent eruptions. We have, then, good evidence that the earliest basalts began their eruption in the late Tertiary or earliest Quaternary, and that successive eruptions continued until a very recent date.

So far as examined, these rocks appear to be normal olivine basalts; they are in many places vesicular or scoriaceous and are likely to exhibit a columnar structure. Many of the sheets are thin; few of them exceed 200 feet in thickness, and it is clear that they spread over wide areas from small and inconspicuous craters, many of which are still preserved. A number of basalts from the upper Rio Grande have been described by J. P. Iddings in Survey Bulletin 66. Some of them contain quartz, probably as inclusions from granitic rocks, through which they were erupted.

Four analyses of basalts from the region of the upper Rio Grande have been made in the laboratory of the United States Geological Survey and appear in Bulletin 228.[b] In the same publication (p. 195) are contained analyses of a phonolite from Pleasant Valley, a pyroxene andesite from the Sierra Grande, a plagioclase basalt from San Rafael flow, and a nepheline

---

[a] Lee, W. T., Water resources and their development in the Rio Grande Valley, New Mexico: Water-Supply Paper U. S. Geol. Survey No. 188, 1907, p. 22.
[b] Bull. U. S. Geol. Survey No. 228, 1904, p. 193.

basalt from Ciruella, all in eastern Colfax County. An analcite camptonite near Las Vegas has been described by J. H. Ogilvie.[a]

CONNECTION WITH MINERAL DEPOSITS.

A number of gold, silver, and copper deposits, principally fissure veins, occur in places in the andesitic and rhyolitic series. No metal deposits of any kind have been found in the basalts.

REVIEW OF IGNEOUS ACTIVITY.

The two known periods of igneous activity in the Territory are far apart in time—pre-Cambrian and Tertiary. The results of the former are now mainly apparent in enormous batholiths of intrusive granites, highly acidic and rich in potassium. On the other hand, the Tertiary period is manifested first in smaller intrusions, mainly in the form of laccoliths or sheets, and second in distinctly later but more extensive areas of flow rocks. The intrusive rocks are almost entirely of intermediate composition—monzonites and quartz monzonites. The effusive rocks are latites and latite-andesites, both intermediate rocks, but besides these widespread flows of rhyolite and basalt also appear. Here, as in many other places in the Cordilleran region, the eruption of the basalts persisted almost to Recent time.

The products of the pre-Cambrian igneous period differ so greatly from those of Tertiary time that both could not reasonably have been derived from the same magma basin. On the other hand, there is striking similarity in composition between the generally uniform Tertiary monzonitic intrusive rocks and the latites and latite-andesites of the effusive epoch. The suggestion is justified that they were derived from essentially the same source and that toward the last a differentiation took place in the magma basin, by which the intermediate magma was split into a basic or basaltic and an acidic or rhyolitic portion.

METAL DEPOSITS.

GEOGRAPHIC DISTRIBUTION.

Deposits of metallic ores are rarely found in single occurrences. They cluster characteristically in certain localities and these are designated as "districts." Each district is delimited simply by custom or by regulations of the miners within its confines, and may contain several subdistricts or camps. Several districts constitute a mining region. The metal-mining districts of the Territory are indicated on Plate II, with fair completeness, though a number of localities where simply prospects have been found have not been entered on the map. The mining districts form a belt extending north-northeast and south-southwest from Colfax, Taos, and Rio Arriba counties on the north to Dona Ana, Luna, and Grant counties on the south. The belt attains its greatest width in the southern part of the Territory. The province of the Great Plains on the east and the plateau province on the west and the northwest are poor in valuable ores.

The ore deposits are found in almost all the mountains and ranges in the Territory. (See Pl. I, pocket.) They occur in the Culebra, Taos, Santa Fe, and Las Vegas ranges on the north; in the Zuni uplift; in the Magdalena, San Pedro, Manzano, San Andreas, Oscura, Organ, Mimbres, and Mogollon mountains and in desert ranges in the southwestern part of the Territory. On the other hand the Sandia, Sacramento, and Guadalupe ranges contain few or none. The greatest number of districts are in the southwestern part; in this area the most important deposits extend in a curved line from Socorro through Hermosa, Kingston, Hillsboro, Lake Valley, Cooks Peak, Santa Rita, and Pinos Altos to Silver City and the Mogollon district, near the Arizona line.

The occurrence of metal deposits is by no means in proportion to the size and elevation of the ranges or to the magnitude of structural features. The northern ranges, where uplift and faulting have proceeded on an enormous scale, contain few districts of importance, and, as

---

[a] Jour. Geology, vol. 10, 1902, pp. 500–507.

MAP OF NEW MEXICO
SHOWING LOCATIONS AND NAMES OF METAL MINING DISTRICTS.

noted above, some of the southern ranges are almost wholly barren. These facts are believed to indicate clearly that the general circulation of meteoric waters, always most active in regions of elevation and dislocation, had little or nothing to do with the metallization.

METALS CONTAINED.

There is no regularity in the geographic distribution of the metals. Silver or gold or both are present in nearly all the districts; the copper deposits in sandstone contain, however, only minute quantities of silver, and some contact-metamorphic deposits are also exceedingly poor in the precious metals. As a rule the precious metals are accompanied by copper, lead, and zinc; there are, however, some veins which contain little else but gold. The Territory has yielded perhaps $12,000,000 in placer gold, and also perhaps $18,000,000 from vein deposits, but during recent years the annual yield of gold has rarely exceeded $400,000. Silver was once produced in abundance from a number of rich camps in Sierra and Grant counties, but they soon became exhausted and the present annual silver production is from 400,000 to 600,000 ounces, most of it from the Mogollon district. During earlier years much lead was mined from deposits in the Magdalena Range, at Cooks Peak and elsewhere, but of late these deposits have shown signs of exhaustion; the output from the various mines is now between 2,000,000 and 4,000,000 pounds a year. Copper was formerly unimportant, but the yield has increased steadily and reached 10,000,000 pounds in 1907, chiefly from the districts of Santa Rita, Burro Mountains, and Organ. The production suffered some reduction in 1908. Zinc ores were formerly cast aside, but are now sought after. The Magdalena district, in earlier years a lead producer, yielded in 1904, 1905, and 1906 the equivalent of 12,000,000 to 17,000,000 pounds of metallic zinc; much of the ore is used for the manufacture of zinc-lead paint. In the near future, it is believed, these figures will again be reached. Iron ores, chiefly magnetite, are mined extensively at Fierro, in Grant County, and have been found at other places. More details as to metal production will be found on pages 19–25.

Several metals are absent from the deposits of the Territory. There are no tin ores, though their presence on a small scale is known from the Franklin Mountains at El Paso, Tex., just south of the boundary line. No quicksilver ores occur. Wolframite and scheelite are confined to few localities and small amounts (Apache No. 2, Victorio, White Oaks). Bismuth minerals are only found as rarities. No cobalt and nickel ores have been reported. Arsenic is seldom found. Antimony is more common, but by no means abundant, occurring chiefly in the form of tetrahedrite. Tellurides have been recognized from only two districts, Red River and Sylvanite.

GEOLOGIC DISTRIBUTION.

A great number of types of ore deposits are represented in New Mexico. Among them are copper and iron ores in sedimentary beds; normal, sharply defined fissure veins; mineralized shear zones; lenticular veins in gneiss; replacement veins in limestone; irregular replacement deposits in limestone; contact-metamorphic deposits between limestone and intrusive rocks; and gold placers.

In a general way the pre-Cambrian schists contain disseminated ores in shear zones, and veins of copper and zinc sulphides, with some gold but little silver; the Paleozoic limestones are likely to carry irregular copper, lead, and zinc deposits, many of them greatly enriched by processes of oxidation and some containing much silver; the intrusive porphyries and granitic rocks contain fissure veins with gold, silver, or both, and (in most places) minor amounts of the base metals; finally, the flows of rhyolite and andesite inclose, at some places, fissure veins rich in gold and silver, but ordinarily relatively poor in copper, lead, and zinc.

In attempting to trace the geologic reason for the distribution of the deposits it soon became apparent that the major part of them occur in or near the many, usually small, areas of intrusive rocks—mostly monzonites and quartz monzonites—which extend across the Territory (fig. 1). Almost every one of the intrusions is accompanied by ore deposits. It was also found that certain kinds of ore deposits did not necessarily follow the general belt of these intrusions. Among them are the copper deposits in sandstone (fig. 3); the veins in lava flows (fig. 2), generally in rhyolite and andesite; and certain deposits in pre-Cambrian rocks, mostly in the northern part of the Territory.

In the gold placers, as well as in some of the bedded deposits of iron, the valuable metals were introduced or laid down at the same time as the accompanying rock was formed. These deposits are often called syngenetic. The important ore bodies of most New Mexico mines belong to the class of epigenetic deposits, in which the metals were introduced after the surrounding rocks had been formed.

During the present work particular attention was given to the question of the geologic age of the deposits, in the hope of discerning the main epochs of ore formation. The results have justified this study and show that certain physical characteristics of the deposits in general distinguish the several epochs.

It has been possible to prove that certain deposits in the pre-Cambrian rocks are of pre-Cambrian age, the oldest period of ore formation recognized in New Mexico. As a rule their characteristics differ markedly from those of the later epochs. They are mainly lenticular veins, shear zones, and "fahlbands" in schists, and they carry chiefly gold and copper.

A long time of quiescence followed the pre-Cambrian mineralization. No ore deposits are known to have been formed during the whole of Paleozoic and Mesozoic time, except possibly certain iron deposits in the "Red Beds."

In late Cretaceous or early Tertiary time widespread intrusion took place in the Territory along what now is the mining belt. After these intrusions most of the metal deposits were formed; they consist of contact-metamorphic deposits, veins, shear zones, and silver-lead replacement deposits in limestone. During this epoch gold, silver, copper, lead, and zinc were introduced.

In the latter part of the Tertiary period lavas flooded certain portions of the Territory, and at several places fissure systems in these flows were filled with gold and silver ores. These deposits are much less complex than those of the earlier intrusions and do not carry large amounts of base metals. The deposits of early Tertiary age were soon attacked by erosion and in places gold-bearing gravels resulted from their disintegration. The late Tertiary deposits in lavas have not as a rule yielded rich placer gravels.

The age of the copper deposits in sandstone has not been fully established. Their formation probably began after the general uplift in the early Tertiary, when atmospheric waters began to penetrate the "Red Beds," from which the deposits are believed to have concentrated their copper ores.

The importance of these time distinctions lies in the following facts: The pre-Cambrian comprised periods of vast intrusions and dynamometamorphism, which have not been repeated in New Mexico since that time. Therefore the pre-Cambrian deposits bear the marks of dynamometamorphism and deep-seated processes of ore formation. There was no igneous activity from the beginning of the Paleozoic to the latest part of the Cretaceous, and there was no metamorphism. The intrusions of latest Cretaceous or earliest Tertiary time were not deep seated in the full significance of the word, but were probably forced in at 3,000 to 10,000 feet below the surface. They were followed by ore deposition, the deposits having in general the character of those formed under moderately high pressure and temperature. Lastly came the eruption of the Tertiary lavas, which was followed by a mineralization of fissures relatively close to the earth's surface, and this made the type of the deposits different from those of the others already mentioned.

Thus it happens that each metallogenetic epoch in New Mexico is characterized by distinct types of deposits, corresponding to the physical conditions under which they originated.

PRE-CAMBRIAN DEPOSITS.

GENERAL CHARACTER AND DISTRIBUTION.

Periods of sedimentation, followed by intrusions and metamorphism on a large scale, characterized the long pre-Cambrian era. Present knowledge of its history is very incomplete, but it surely would be surprising if metal deposits had not been formed during these periods. It is well known that metal deposits of considerable variety are found in the pre-Cambrian rocks of the Eastern and Northeastern States; deposits of similar age have been recognized in the Black Hills of South Dakota; and lately the probable pre-Cambrian age of certain deposits in southern Wyoming,[a] in Colorado,[b] and in Arizona [c] has been pointed out.

With these results in mind the deposits in pre-Cambrian rocks of New Mexico were scanned with care. Two were found, at the Hamilton mine and in the Rociada district, both in San Miguel County, which could be satisfactorily proved to antedate the Paleozoic sedimentary series. This accomplished, it was not difficult to establish certain criteria by which the pre-Cambrian deposits could be differentiated even if direct proof of age were not available. In one or two places there may be room for doubt, but in the main the group is well established.

As might be expected, most of these deposits occur in the north, where large areas of pre-Cambrian rocks exist, but some are also found in the west, in the Zuni Mountains, in the plateau province, which otherwise is poor in metal deposits. Needless to say, the pre-Cambrian deposits show no connection with the distribution of later intrusives or flows.

Deposits of this class have been recognized in the upper Pecos and Rociada districts, in San Miguel County; the Hopewell and Bromide districts, in Rio Arriba County; the Arroyo Hondo and Picuris districts, in Taos County; at various places near Santa Fe, in Santa Fe County; in the Zuni district, in Valencia County; and probably in the Gold Hill district, in Grant County.

There is no reason why deposits of pre-Cambrian age should be of one kind; fissure veins, shear zones, contact-metamorphic deposits, and disseminations in schist might well be represented. As a matter of fact, however, only three types have been recognized—quartz-filled fissures, usually of the "lenticular" type; shear zones filled with quartz stringers; and disseminations of sulphides in amphibolitic schist. Transitions exist between the three types.

The pre-Cambrian deposits are likely to differ in two respects from those of later age. First, they bear evidence of deeper-seated origin, as shown by the development of heavy silicates along the walls, especially where the walls consist of amphibolite. This is not universal and some of the veins are characterized by sericitization. As a further consequence of deep-seated origin the quartz filling of fissures is generally massive and glassy. Second, some of them have been subjected to the stresses of dynamometamorphism and show the effect of pressure in following the bending and corrugation of the schists and in consequent lenticular development of the quartz, also in the development of minerals like biotite.

The pre-Cambrian deposits are in gneissoid granites, in gneiss, or in amphibolite. The developments are not extensive; a few shafts have been sunk to a depth of several hundred feet and a great number of small properties have been opened. These deposits are not of great importance in New Mexico and the production is very small. Some ore has been shipped from the upper Pecos, Hopewell, and Bromide districts; in the Bromide mainly from oxidized and enriched ore. The low-grade copper deposits of Frazer Mountain, in the Arroyo Hondo district, yielded some ore during a short time of operation. The greatest production has been won from the placers derived from the eroded and oxidized veins of the Bromide and Hopewell districts. These are reported to have yielded gold valued at several hundred thousand dollars.

THE ORES.

The principal metals contained in these deposits are copper and gold. There is some zinc, but very little lead. Silver is always present in small amounts; in the Bromide district high silver values have been found, probably owing to concentration by oxidation. Silver also occurs in notable amounts in the Picuris district.

---

[a] Spencer, A. C., Copper deposits of the Encampment district, Wyoming: Prof. Paper U. S. Geol. Survey No. 25, 1904.

[b] Lindgren, Waldemar, Notes on copper deposits in Chaffee, Fremont, and Jefferson counties, Colo.: Bull. U. S. Geol. Survey No. 340, 1906, pp. 157-174.

[c] Reid, J. A., Ore deposits of the Cherry Creek district: Econ. Geology, vol. 1, 1905, pp. 417-436. Graton, L. C., Mineral Resources United States for 1906, U. S. Geol. Survey, 1907, p. 389.

In the upper Pecos district chalcopyrite ores accompanied by zinc blende are disseminated in an amphibolite and follow its schistosity. The ore minerals are intergrown with amphibole, biotite, and tourmaline and contain a little gold and silver. At Arroyo Hondo chalcopyrite is disseminated through a broad belt of amphibolite; it is in part intergrown with heavy silicates, but mostly contained in stringers of quartz, calcite, and siderite. At Rociada quartz stringers, many of them with calcite, in gneissoid granite or in micaceous schist contain bornite, chalcocite, pyrite, and galena with some gold. Low-grade copper deposits with some gold and silver form shear zones in the gneiss of the Zuni district. At Picuris, in Taos County, occur veins of glassy quartz in strongly contact-metamorphosed schists; they carry chalcocite with silver and some gold, and somewhat resemble the veins at Virgilina, N. C.

In the Hopewell and Bromide districts two types appear. The first consists of narrow quartz veins, pinching and swelling in "lenticular" manner. In places several of these veins combine in a lode. The minerals are quartz, siderite, calcite, tourmaline, specularite, pyrite, chalcopyrite, molybdenite, and rarely galena; the gold is not free except where the veins are oxidized. The deposits of the second class are "fahlbands" or disseminated deposits in schist. In these the silicification is more intense, but the mineral combination is similar to that of the veins. Both classes conform in direction of extension to the schistosity of the rocks.

METASOMATIC PROCESSES.

As stated above, the pre-Cambrian deposits may be divided into those clearly related to and following more or less complicated fissures, and those containing the ore minerals disseminated in crystalline schist, generally in the direction of the schistosity. The second group may indeed also follow a line of fracture, but if so it is not prominent. The characteristic feature of these deposits is that the ore mineral really forms an integral part of the rock. As best illustrated in the Hamilton mine, on the upper Pecos, the chalcopyrite, zinc blende, etc., are intergrown with the principal mineral of the amphibolite; but with this association appear also a green biotite, some tourmaline, and veinlets of dark quartz. The deposit as a whole has certainly taken part in the general metamorphism which produced the amphibolite from a dioritic or diabasic rock. Whether the metals in the deposit have been introduced from the outside along a healed fissure zone or whether they are simply a concentration of metals contained in the original igneous rock must be left an open question. It is well known, of course, that many basic rocks consolidate with a notable amount of sulphides, particularly of copper, and a concentration of the metallic content could well have taken place during metamorphism. The amphibolite is intruded by a microcline granite, but whether a connection exists between the granite and the deposit must also be left undecided.

The study of the alteration of the wall rock in the veins leads to interesting results. A simple sericitization accompanied by the introduction of small cubes of pyrite is observed at some veins, but in numerous places other forms of alteration appear, suggesting higher temperature and pressure. In many veins biotite develops from amphibolite; tourmaline may appear both in the filling and disseminated in the walls. Quartz is introduced in minute lenses. In the Bromide district garnet, amphibole, and tourmaline have developed along the walls from a country rock of biotite schist. At Rociada quartz, sericite, pyrite, garnet, epidote, and amphibole have been introduced along the walls of gneissoid granite. The partly replaced or altered wall rock rarely contains good values. The following table, for which completeness is not claimed, records the mineral association of the various districts:

Minerals of the pre-Cambrian deposits.

District.	Principal values.	Quartz.	Calcite.	Siderite.	Fluorite.	Tourmaline.	Biotite.	Epidote.	Garnet.	Chlorite.
Hopewell and Bromide	Au, Ag	X	X	X	X	X	X	X	X	X
Picuris	Au, Ag	X								
Arroyo Hondo	Cu, Au, Ag	X	X	X						X
Rociada	Cu	X	X					X	X	X
Upper Pecos	Cu, Au, Ag	X			X	X	X			
Gold Hill	Au	X								
Zuni	Cu	X								

METAL DEPOSITS.

Minerals of the pre-Cambrian deposits—Continued.

District.	Specularite.	Pyrite.	Pyrrhotite.	Chalcopyrite.	Galena.	Zinc blende.	Molybdenite.	Tetrahedrite.	Bornite.	Chalcocite.
Hopewell and Bromide	x	x		x	x	x	x	x		
Picuris		x		x				x	x	x
Arroyo Hondo		x		x					x	
Rociada		x			x	x	x		x	x
Upper Pecos		x	x	x	x				x	
Gold Hill		x								
Zuni		x		x						

NOTE.—Minerals which are products of oxidation omitted.

GENESIS.

The fact that the geologic conditions of remote time are imperfectly known renders it difficult to arrive at definite results as to genesis. Low-grade copper and gold deposits predominate and were in general formed along fissures. In most of the districts (Hopewell, Bromide, Arroyo Hondo, Picuris, and Rociada) they were evidently formed after vast granitic intrusions had shattered an older series of sedimentary rocks, as well as basic igneous rocks. In this respect, then, as well as in many others, they conform with gold-bearing veins of later periods and it would at least be a justifiable hypothesis to seek the origin of their metals in the granitic magma.

CONTACT-METAMORPHIC DEPOSITS.

GENERAL FEATURES AND DISTRIBUTION.

The distribution of the intrusive bodies of early Tertiary age has been described (pp. 35–42). They extend in general in a belt across the Territory, beginning in Colfax County on the north, widening toward the south, and ending in Grant, Luna, and Dona Ana counties. The intrusives take the form of laccoliths, stocks, dikes, and sheets; they were forced into the sedimentary series while it still formed an unbroken thick mantle, and they probably never reached the surface. The rocks are of intermediate composition and comprise quartz monzonite porphyry, monzonite porphyry, granodiorite porphyry, and more rarely quartz monzonite of granitic texture or granite porphyry. Where the intrusive rocks break through the sediment contact metamorphism usually has taken place in varying degree. The metamorphism is not excessive and rarely extends for more than a few hundred feet in horizontal distance. In places only a slight development of epidote in limestone or a hardening or baking of the shale is noted. Stronger action produces garnet rocks from limestone and hornfels from shale. Mineralization usually accompanies the metamorphism and is in fact observed at some points at almost every intrusion of note that adjoins limestone. Few deposits are noted in shale and those only where the shale is calcareous. The general distribution of contact-metamorphic deposits coincides exactly with the area of early Tertiary intrusions.

The districts containing contact-metamorphic deposits are those of Elizabethtown and Cimarroncito, Colfax County; Ortiz and San Pedro, Santa Fe County; Magdalena and Jones Camp, Socorro County; Organ, Dona Ana County; Jarilla, Otero County; Tres Hermanas, Luna County; Hanover, San Simon, Apache No. 2, and Hachita, Grant County. The location of these districts is indicated on Plate II and also in a diagrammatic way on figure 1.

THE ORES.

Copper in the form of chalcopyrite is the most common metal in the contact-metamorphic deposits, but it is usually accompanied by zinc, as zinc blende, and this mineral predominates in the Magdalena district. Iron in the form of magnetite is also of local importance.

Except at one place (the Modoc mine) lead is not prominent, though it now appears so, by reason of secondary enrichment, in the Magdalena and Tres Hermanas districts. As a rule gold is present only in traces and silver to the extent of a few ounces in the primary ore. At Elizabethtown this rule does not hold and the gold values are rather high; the same may be said of some deposits in the Old Reliable mine, Ortiz district, and the Lazarus mine, San Pedro district, but it is doubtful whether these deposits should be regarded as of true contact-metamorphic origin, for the gold is concentrated in little veins or seams in the altered rock.

Calcite is ordinarily present, locally recrystallized in extraordinarily large individuals (San Pedro, Apache No. 2). Quartz commonly forms a small part of the gangue. The most constant gangue mineral is garnet, but it is absent from the Magdalena and Jones Camp districts. The variety most common seems to be andradite; it was positively identified from the San Pedro, Jarilla, and Organ districts. Skeleton crystals and double refraction are often observed, especially where the mineral develops in calcite. Epidote is usually but not invariably present. Wollastonite was noted from the Tres Hermanas, Steins Pass (San Simon), and San Pedro districts. Vesuvianite is reported to be found at San Pedro. Pyroxene is extremely common, especially in the metamorphosed lime shales. It is usually diopside, but hedenbergite—an iron pyroxene—is present at Hanover. Tremolite is less common. Scapolite forms in calcareous shale at Elizabethtown. Ilvaite occurs at Hanover. Magnetite and specularite almost constantly accompany these deposits; close examination will rarely fail to find them.

FIGURE 1.—Map of New Mexico showing location of contact-metamorphic deposits and related veins and replacement deposits of early Tertiary age.

The ore minerals do not exhibit a great variety. Chalcopyrite and zinc blende are almost universally present, the latter in places containing much iron. Galena is subordinate. The primary ores of the Magdalena district contain about 27 per cent of zinc, 4 per cent of copper, and 4 per cent of lead. The Modoc mine, in the Organ Mountains, is an exceptional deposit occurring in limestone at the contact of a mass of andesite of distinctly effusive type. It is characterized by an intimate intergrowth of galena and epidote.

Molybdenite is reported from two districts. Closer search will probably show it to be present in most of the deposits. Scheelite in considerable quantities occurred at the typical contact-metamorphic deposit of the Apache No. 2 district. A mineral related to cosalite or galena-bismuthite was found at the Memphis mine, in the Organ district. Pyrrhotite is not common.

METAL DEPOSITS.

The subjoined table, for which completeness is not claimed, shows the minerals occurring in these deposits, by districts.

Minerals of contact-metamorphic deposits.[a]

District.	Calcite.	Quartz.	Garnet.	Epidote.	Wollastonite.	Diopside.	Tremolite.	Specularite.
Elizabethtown	×		×	×		×	×	×
Cimarroncito	×		×	×				×
Ortiz			×					
San Pedro	×		×		×	×	×	×
Magdalena	×	×		×		×	×	×
Jones Camp								
Jarilla	×	×	×			×	×	×
Organ	×	×	×	×				×
Modoc	×							
Tres Hermanas	×	×	×		×	×		
Hanover	×	×	×	×		×	×	×
San Simon	×		×	×	×	×		
Apache No. 2	×		×					
Hachita (Copper Dick)	×		×			×		

District.	Magnetite.	Pyrite.	Chalcopyrite.	Galena.	Zinc blende.	Molybdenite.	Remarks.
Elizabethtown		×	×				Scapolite and pyrrhotite present.
Cimarroncito	×	×	×				Gold present.
Ortiz							
San Pedro		×	×				Biotite present; vesuvianite reported.
Magdalena	×	×	×	×	×		
Jones Camp	×						
Jarilla		×	×				
Organ	×	×	×		×	×	Bismuth mineral present.
Modoc				×	×		
Tres Hermanas				×	×		Only oxidized zinc ores present.
Hanover	×	×	×		×	×	Hedenbergite and ilvaite present.
San Simon		×	×				
Apache No. 2		×	×				Scheelite present.
Hachita (Copper Dick)			×				

[a] Oxidized copper ores and zinc ores, also chalcocite, generally present.

DEVELOPMENT AND PRODUCTION.

The contact-metamorphic deposits of New Mexico include a number of localities of importance. At San Pedro a large mass of unoxidized chalcopyrite ore has been mined at intervals. There was much high-grade ore, though the larger part yields only a small percentage of copper. The ore is smelted to matte at the same place. When in operation, the output has reached 160 tons of ore a day. The other locality is in the Magdalena district, where for a long time oxidized ores were mined, yielding principally the carbonates of lead (with some silver) and zinc. The primary ore contains chiefly sphalerite, with some chalcopyrite and galena, and the deposits now constitute zinc mines of great importance and heavy production. Both copper and gold are contained in the ores of the Jarilla district, in Otero County.

Copper ores, mostly oxidized, have been mined on a considerable scale in the Organ district and, in the form of chalcopyrite, near Hachita and in the San Simon district. Iron ores with some chalcopyrite, also sphalerite ores, are mined at Hanover. In 1905, about 350 tons of iron ore was mined daily. Galena and oxidized zinc ores have been shipped from Tres Hermanas. Galena ores are also mined at the Modoc mine, in the Organ district.

All this shows that the deposits of this class are of considerable importance and that in several places the primary ores form valuable bodies, without the enriching effect of oxidation.

STRUCTURE.

In the minor deposits the ore may simply form irregular masses at the contact and these rarely extend more than 200 feet into the limestone; but where ores have been formed on a larger scale the mineralization is exceedingly prone to selective action; individual beds, sometimes without clearly apparent reason, prove to be ore carriers, and along them the metamorphism may be carried to considerable distances, even more than 2,000 feet. In such beds no alteration may be shown in the foot and hanging walls. Fissuring is not generally prominent, but a few deposits assume the form of fissure veins, with alteration of the limestone walls to garnet and similar contact-metamorphic minerals. Examples of this form are found in the Hachita district (American vein) and the Jarilla and Pinos Altos districts.

At Elizabethtown calcareous shales are altered, at the contact with thick sills or stocks of monzonite porphyry, to dark crystalline rocks of diopside and scapolite, with some epidote, calcite, amphibole, and magnetite; pyrite, pyrrhotite, and chalcopyrite are also present, intergrown with magnetite. Certain bands of this metamorphosed rock are richer in iron, in places constituting a garnetiferous iron ore, which contains small quantities of gold. Locally, the gold values are much higher, ranging up to $80 a ton. There has been some production from this locality.

Along the margins of the intrusive laccoliths of the Cerrillos Hills and Ortiz Mountains the sedimentary rocks are mainly Cretaceous shales which have suffered some slight metamorphism, but contain no metals. At the Old Reliable mine some probably Cretaceous limestone adjoins the monzonite porphyry. It has been converted to garnet and contains gold, mainly on narrow vertical seams. A somewhat similar occurrence was noted at the Lazarus mine, in the San Pedro Mountains.

At the San Pedro copper mine more than 700 feet of upper Carboniferous rocks have been metamorphosed by the underlying main mass of the laccolith and by dikes extending upward from it. The lower 200 feet of shaly limestone is only partly altered, with local development of garnet and tremolite and a little chalcopyrite and pyrrhotite, but along a certain horizon of purer limestone garnetization has taken place for a distance of half a mile, the thickness of this strongly metamorphosed stratum being about 50 feet. Bunches of chalcopyrite are irregularly distributed in it. Within this zone beds of pure crystalline limestone adjoin wholly garnetized beds. In places the rock consists of a mixture of garnet and coarsely crystalline limestone. On the dip the gently inclined ore beds have been followed for 300 feet. The upper beds of the series consist mainly of somewhat metamorphosed and baked shale and sandstone.

In the Magdalena district the westward-dipping beds of Carboniferous limestone are cut off by an intrusion of granite porphyry, which is best exposed near the Waldo tunnel of the Graphic mine. Metamorphism and mineralization, with pyrite, chalcopyrite, sphalerite, and galena, have occurred at the contact, but the greatest effect has been produced in lower strata (Kelly limestone) at and 30 feet above the so-called "Silver Pipe" limestone stratum. The ore-bearing strata are rather coarsely crystalline, in contrast to the compact, unaltered "Silverpipe" limestone. The ore-bearing beds, which will probably be found to be cut off by the granite porphyry in depth, have been changed to massive zinc blende, chalcopyrite, and galena accompanied by epidote, pyroxene, and tremolite gangue, with some calcite and quartz. Magnetite and specularite are also present. Some of the ore bodies are 40 feet thick. The alteration ceases abruptly above and below these beds and does not everywhere follow the bedding planes. The distance along the dip from the metamorphosing intrusive rock is probably not likely to be less than 2,000 feet. Various explanations have been advanced for these remarkable deposits, but it is held by the senior author of this report that they are without doubt of contact-metamorphic origin.

In the Organ district a narrow zone of contact metamorphism follows the margin of a large body of granular quartz monzonite. This zone is rarely over 200 feet in width. Characteristic are the solid beds of andradite containing ore, chalcopyrite, and its products of oxidation alternating with strata of pure, coarsely recrystallized limestone. The ores are mostly oxidized. The structural relations in the Torpedo mine are not entirely clear; the wholly oxidized ore is here contained in a silicified limestone close to the contact. The beds are of Paleozoic age.

At Tres Hermanas a large mass of granite porphyry intrudes flat Carboniferous strata. At the contact strong garnetization has taken place, but a short distance away there is a succession of little-altered limestone, with hornfels, thin garnet beds, and coarsely crystalline limestone. The primary distribution of zinc blende and galena is masked by oxidation, but the ores appear to reach a distance of 2,000 feet from the intrusive contact.

Near Hachita, in the Apache No. 2 district, a large mass of quartz monzonite porphyry borders flat-lying Paleozoic limestones. The zone of metamorphism is narrow and confined largely to the broken and fractured sediments close to the intrusive rock. The ores are not contained in the garnet rock, of which much is present, but copper minerals are distributed in a coarsely recrystallized limestone.

The relations in the important contact-metamorphic districts of Hanover and Fierro are illustrated in figure 30. Intrusions of quartz monzonite porphyry in the form of sheets and stocks, the latter suggesting a laccolithic development with accompanying doming of the strata, are contained in Carboniferous limestone. The limestone is greatly altered at the contact, but the metamorphism has extended through 200 or 300 feet of the series and probably for a greater distance laterally. The limestone is replaced by garnet, epidote, and hedenbergite, with calcite and quartz; locally magnetite, pyrite, and zinc blende appear in large masses, also a little chalcopyrite. Much of the magnetite is coarse and looks as if it were pseudomorphic after coarse calcite.

OXIDATION.

The oxidation of the contact-metamorphic deposits has progressed somewhat capriciously. Much of the garnetized rock is hard and compact and resists well the attack of oxygenated waters. In most of the districts the sulphides are found a short distance from the surface.

At Tres Hermanas no sulphides were found except a little galena, but here the developments are only superficial. At San Pedro little oxidized ore was discovered. At Hanover and Fierro the sulphides are close to the surface, although other kinds of deposits in the same district are deeply oxidized. The most interesting phenomena relating to oxidation were observed in the Magdalena and Apache No. 2 districts.

When oxidation gets fairly started in a deposit in limestone it effects far-reaching changes, as the solubility of the surrounding rock offers a free field for the metals to arrange themselves in full accord with their affinities for oxygen.

In a certain well-known method of roasting, the copper in pyritic ores is concentrated in the center of a lump of ore, while the iron as oxide forms a shell around it. A similar arrangement takes place in many small deposits subject to oxidation; a core of oxidized copper ores will be found in the interior surrounded by a shell of limonite. A somewhat similar relation seems to govern lead and zinc. In the oxidized ores these metals generally separate, the lead concentrating in distinct bodies, the zinc spreading widely and replacing the surrounding limestone, the appearance and structure of the rock being often perfectly retained.

In the Magdalena district, particularly in the Kelly mine, the opportunities for a detailed study of processes of oxidation were excellent. At the present time the oxidized ore bodies are almost exhausted. The depth of oxidation is about 300 feet, the primary ore consisting of zinc blende with a little pyrite, magnetite, and some galena; in places this sulphide ore, especially at the horizon underneath the "Silver Pipe" limestone, is 15 feet thick (on the 250-foot level). In the oxidized stopes nearer the surface the zinc and lead part company. The oxidized zinc ores form wide stopes in which caves large enough for a man to crawl into are coated with beautiful botryoidal light-green masses of smithsonite. The crust of this material is almost 3 inches thick and underneath is a dark powdery material rich in manganese but also containing much zinc. The lead stopes are much smaller and are composed of almost pure "sand carbonate" with occasional bunches of galena.

A fine instance of the separation of copper and iron is illustrated by a specimen from the Apache No. 2 mining district (Pl. IV, B). The primary ore consists here of a gangue of extremely coarse calcite with a little iron and practically no magnesia. When this is dissolved in acid there remain a skeleton of small films and grains of quartz whose presence would hardly be expected in the apparently homogeneous cleavage pieces. The calcite contains grains of chalcopyrite and small cubes of pyrite. During oxidation this primary low-grade ore becomes surrounded by crusts of secondary calcite, limonite, hematite, malachite, and chrysocolla, as shown in Plate IV, B; the iron and copper separate, the former being deposited in the recrystallized calcite as hydroxide or oxide, while the copper minerals form a thin crust directly adjoining the primary ore and gradually traveling inward as the oxidation progresses. This is exactly what happens under the more intense conditions of artificial oxidation or the roasting of chalcopyrite in metallurgical work. The heat in the presence of oxygen will gradually concentrate the copper sulphide or oxide in the center of the lump, while the ferric oxide forms a shell which can easily be knocked off.

Considered in more detail, the narrow ring of oxidized copper ore consists of alternate narrow bands of malachite and chrysocolla, the latter being due to the quartz distributed microscopically through primary calcite. The malachite always forms the inner zone and projects into the fresh calcite as tufts of slender needles. In places a thin layer of calcite will separate the copper minerals. The wide outer crust consists of recrystallized calcite which only in part has the same orientation as the central cleavage piece of primary ore. This recrystallized calcite contains limonite in flocculent masses, in places distributed concentrically, but does not carry even a trace of copper. At the outer edge of the specimen the limonite changes to dark-brown hematite. Hematite also forms pseudomorphs after pyrite, but probably always passes through the intermediate stage of limonite. There is also some indication that the small quantity of manganese contained in the calcite separates out as pyrolusite along the inner edge of the zone of copper minerals.

GENESIS.

The evidence shows plainly that in contact-metamorphic deposits comparatively pure limestone is partly or wholly replaced by garnet and other silicates and ore minerals. Where the limestones are magnesian, diopside and tremolite will develop. Sometimes two adjacent beds, apparently entirely similar will be transformed, one to granular marble, the other to andradite garnet. Examples of this may be found in the San Pedro, Organ, and Hanover districts. Anyone who wishes can observe these relations; their existence can not be questioned.

The view, then, which would confine contact metamorphism simply to a rearrangement of molecules in a single bed is absolutely contrary to facts. The introduction of silica, iron, and various sulphides into the beds from without seems to be positively proved. There is no evidence that alumina, lime, and magnesia were added.

It is believed that the origin of these substances must be sought in the emanation from the intrusive magmas during the cooling process. Those who seek the source elsewhere must explain why such deposits should always be so closely connected with the intrusive rocks—why similar deposits are not found some distance away in the limestones or in other sediments or in the igneous rock itself. In most instances the intrusive rock, which appears to have been the cause of the ore formation, has remained entirely unaltered and the contacts are perfectly sharp. At a few places the intrusive has been converted into epidote for a short distance from the contact. Disseminated pyrite occasionally occurs near the contact in the intrusive with some accompanying sericitization; but as fissure veins and attendant alteration of the country rock constantly accompany the intrusives as a development later than contact metamorphism it is probable that the observed mineralization of the intrusives belongs to this class of phenomena.

In conclusion attention should be called to the interesting deposits at Jarilla in Otero County and Hachita in Grant County, where garnet and other silicates occur along ore-bearing fissures in limestone; they indicate a transition from the contact-metamorphic deposits to normal veins formed under conditions of less temperature and pressure.

VEINS CONNECTED WITH INTRUSIVE ROCKS OF EARLY TERTIARY AGE, EXCLUSIVE OF REPLACEMENT VEINS IN LIMESTONE.

GENERAL RELATION AND DISTRIBUTION.

The intrusive rocks of early Tertiary age have been reviewed on pages 35–42. It has been shown how stocks, laccoliths, sheets, and dikes of these magmas intruded the whole sedimentary series of the Territory in a belt extending from the north-central to the southwestern part. Practically all the more important intrusive masses contain, in the rock itself or in the surrounding rocks, metal deposits of some kind. Most of them contain both contact-metamorphic deposits and fissure veins. In other words, the distribution of the intrusive rocks corresponds with the distribution of the mining districts.

Named from north to south the intrusive areas which contain both contact-metamorphic deposits and veins are the Elizabethtown district, Colfax County; Ortiz and San Pedro districts, Santa Fe County; Organ district, Dona Ana County; Tres Hermanas district, Luna County; Hanover, Pinos Altos, and Hachita districts, Grant County. Veins without contact-metamorphic deposits occur in the intrusive districts of Cerrillos, White Oaks, Jicarilla, Nogal, Central, Telegraph, Blackhawk, Lordsburg, and Pyramid. In most of the latter group the intrusive rock is not adjoined by Paleozoic limestone.

The developments have not, as a rule, attained great depth. The Lincoln vein, in the White Oaks district, has been opened by a shaft to a depth of 1,300 feet (1905). The shafts of Pinos Altos are only 700 or 800 feet deep. In the other districts less depth has been attained.

THE ORES.

The accompanying table, which is not claimed to be complete, gives in summarized form the most characteristic mineral combinations found in these veins. It will be observed that quartz, pyrite, and gold are the constituents almost always present. In a broad way these veins are therefore auriferous pyritic quartz veins. The veins of Elizabethtown, Ortiz, San Pedro, and White Oaks contain much free gold and have, directly or by secondary concentration as placers, yielded a heavy production. The gold-bearing veins of Pinos Altos have likewise produced heavily, both from placers derived from veins and from the primary ores, which contain little free gold. By far the largest part—perhaps $20,000,000—of the whole gold production of New Mexico has been directly or indirectly derived from these veins. Lode mining has yielded the best results at Elizabethtown, Ortiz, White Oaks, and Pinos Altos; the veins of the San Pedro district have been disappointing. Probably about $10,000,000 in gold has been extracted from the quartz veins. Copper and lead are poorly represented in these veins as a whole, though they are of importance in the Hanover, Central, and Pinos Altos districts.

The silver veins form a smaller group, chiefly represented by the Pinos Altos (Silver Cell mine), Telegraph, Blackhawk, Lordsburg, Pyramid, and Hachita districts, in Grant County, and the Cerrillos district, in Santa Fe County; also certain veins in the Organ district, Dona Ana County. None of these have an output comparable to that of the best gold-bearing veins.

Among the gangue minerals quartz prevails, but calcite is reported from a number of localities. Ankerite and dolomite occur here and there, rhodochrosite at Pinos Altos and Pyramid, siderite at Pyramid. Barite is rather exceptional, but was found in the Pinos Altos, Blackhawk, and Pyramid districts. Fluorite is recorded only from the White Oaks and Telegraph districts.

A number of these districts show mineral combinations which are similar to those of more deep-seated origin; they are summarized in the first part of the table. Tourmaline was found at Cerrillos, White Oaks, and Lordsburg; specularite at Ortiz and Hanover; magnetite at Ortiz and Elizabethtown; garnet at Old Hachita; molybdenite (or wulfenite) at Cerrillos, Hanover, Lordsburg, Pyramid, and Organ. These rarer minerals are nowhere abundant.

Pyrrhotite is of rare occurrence, but was noted in veins in shale at Elizabethtown. Arsenopyrite is very rare and was recognized only at Ortiz. The veins at White Oaks present an unusual association of minerals; the quartz here contains, besides gold and auriferous pyrite, albite, wolframite, tourmaline, fluorite, and gypsum; it is thought that the gypsum is derived from primary anhydrite.

Minerals in veins connected with intrusive rocks of early Tertiary age, exclusive of replacement veins in limestone.

District.	Principal value.	Gold.	Quartz.	Calcite.	Dolomite.	Barite.	Tourmaline.	Fluorite.	Specularite.
Elizabethtown	Au.	×	×	×			×		
Cerrillos	Ag, Pb.		×	×	×		×		
Ortiz	Au.		×	×					×
White Oaks	Au.	×	×				×	×	
Hanover	Cu, Pb, Au, Ag.		×						×
Lordsburg	Ag, Au, Cu.		×	×			×		
Pyramid	Ag, Cu.		×	×		×			
Hachita	Ag, Pb.		×	×					
Sylvanite	Au.	×	×						
Organ	Au, Ag.		×						
San Pedro	Au.	×	×	×					
Nogal	Au.	×	×		×				
Jicarilla	Au.		×						
Tres Hermanas	Pb, Au.		×						
Pinos Altos	Au, Ag, Pb, Cu.		×	×	×	×			
Central	Cu, Pb, Ag.		×		×				
Blackhawk	Ag.				×	×			
Telegraph	Ag.							×	
Fremont	Pb, Au, Ag.	×	×						

District.	Magnetite.	Molybdenite.	Pyrite.	Chalcopyrite.	Galena.	Zinc blende.	Other minerals.
Elizabethtown	×		×	×	×		Pyrrhotite.
Cerrillos			×		×	×	Ankerite, wulfenite.
Ortiz	×	×?	×	×	×		Arsenopyrite.
White Oaks			×				Albite, wolframite, anhydrite(?).
Hanover		×	×	×	×	×	
Lordsburg		×	×		×	×	
Pyramid		×	×		×	×	Rhodochrosite, siderite, wulfenite.
Hachita			×	×	×	×	Garnet, stibnite.
Sylvanite			×	×	×	×	Orthoclase, tetradymite.
Organ	×		×	×	×	×	Tetrahedrite.
San Pedro			×	×			
Nogal			×	×	×	×	
Jicarilla			×				
Tres Hermanas			×		×		
Pinos Altos			×	×	×	×	Rhodochrosite.
Central			×		×	×	Ankerite.
Blackhawk			×				Siderite, horn silver, native silver, and argentite.
Telegraph							
Fremont			×	×	×		

NOTE.—Products of oxidation not included.

Another very peculiar, and in fact unique type, was found in the veins at the new Sylvanite camp, in the Hachita Range. The narrow veins here contain, besides quartz, native gold, pyrite, and chalcopyrite, some tetradymite or telluride of bismuth, and orthoclase in large grains, in part with crystalline outlines and intimately intergrown with quartz. The orthoclase does not appear to be of the adularia type and has not that well-crystallized habit which marks adularia in the veins in late Tertiary volcanic flows. These veins appear to approach more closely to a pegmatite than any others found in the Territory, but they are far from being normal pegmatites.

STRUCTURE AND GEOLOGIC RELATIONS.

While there is no great uniformity in the strike and dip of the veins, the predominance of east-west or northeast-southwest strike can not escape attention; these directions prevail in the Elizabethtown, Cerrillos, Ortiz, San Pedro, Nogal, Organ, Pinos Altos, Hanover, Telegraph, Blackhawk, and Hachita districts. In some other districts, notably Nogal, Jicarilla, and White Oaks, there are also veins of northerly or northwesterly strike. The prevalence of northeasterly and easterly strikes means, of course, that the stresses to which the cooling intruded magmas were subjected were somewhat uniformly distributed. The dips are universally steep, but present no uniformity in direction. On the whole, the veins are fairly regular and none which could with certainty be attributed to contraction during cooling were observed. Compressive stresses appear to account fully for the fissure systems.

There is little of special interest in the structure of the veins. A large proportion are normal, filled fissure veins. Plate III illustrates a fairly typical specimen from one of the Pinos Altos

Natural size

SECTION OF FISSURE VEIN, PINOS ALTOS DISTRICT

veins, which are unusually rich in sulphides. It will be observed that the fissure was clearly an open space, for both walls are lined with a thin coating of quartz crystals; the second stage of the filling proceeded perhaps more rapidly, and the whole inner part of the fissure is filled by an irregularly intergrown mass of quartz, pyrite, chalcopyrite, galena, and zinc blende, leaving some druses in the interior which later were filled with calcite. Other veins are mineralized shear zones of varying width, but on the whole these are not so common as the simpler veins. Examples are found in the Nogal and Jicarilla districts.

The veins are rarely continuous for long distances. A traceable length of 1 mile is unusual.

METASOMATIC PROCESSES IN THE WALL ROCKS.

The wall rocks are almost invariably altered where they adjoin the veins, but the changes are far less conspicuous and widespread than in the Tertiary lavas, in which large areas are subject to propylitization and silicification. In the veins under discussion the alteration, in igneous rocks, occupies only a narrow zone along the filling and consists of sericitization, carbonatization, and silicification accompanied by the introduction of pyrite. A little chlorite is present in many places. No special study has been made of these processes and as during sericitization and carbonatization much silica is necessarily set free, it may be that the cases of silicification reported do not really involve any addition of silica to the rock. The sericitization is not always prominent and in places carbonatization predominates.

OXIDATION.

The water level is in general from 200 to 400 feet below the surface. The Old Abe mine, at White Oaks, where the water level is about 1,300 feet below the surface and the ores are mostly thoroughly oxidized to that depth, is exceptional. In the normal silver-lead replacement deposits in limestone complete oxidation is the rule for considerable depth, but the vein deposits share with the contact-metamorphic deposits a certain capriciousness in extent of oxidation, probably dependent on the impermeability of the ore. The oxidation is furthest advanced in veins with abundant sulphides. At Pinos Altos the water level in 1905 was 450 feet below the shaft collar, but the oxidation ceased 60 feet below the surface; at Elizabethtown the water level was at least 340 feet below the croppings. At Jicarilla the sulphides are near the surface, though the water level is 160 feet deep. In the Hachita Range the water level is very irregular, but sulphides are generally found less than 100 feet below the surface. At Pyramid camp the water level stands at 100 feet; oxidation fails to reach it in many mines, but in some extends far below it. In the silver veins, which usually have a carbonate or barite gangue, oxidation is generally deep. Horn silver is most abundant near the surface; in depth argentite and native silver prevail. In one place native silver was found far below the water level; down to a depth of 740 feet on the incline.

The rich oxidized gold and silver ores generally begin at the surface and there is little evidence of leaching of the croppings. No well-defined zones of secondary sulphides have been observed in these veins.

AGE OF THE VEINS.

Many of the veins of this class intersect Cretaceous shales, generally of Colorado age; and in several places, as at Ortiz and Elizabethtown, the intrusives in which they are contained break through the uppermost Cretaceous strata. From this relation and from the general considerations set forth in the detailed description of the intrusive rocks, it seems safe to attribute a post-Cretaceous age to the deposits. On the other hand, it is evident that they have been subjected to great erosion since their formation; and where the late Tertiary lava flows occur in the same districts the veins are not found in them. Evidence of this kind is reported from the Pinos Altos district, where a vein was followed up to the rhyolite and found to be covered by it. Similar evidence is adduced for the replacement deposits in limestone, which are of the same age as these veins (p. 67). On the whole this class of veins may be regarded as of early Tertiary age. It is not likely that their uppermost parts are preserved at any place.

REVIEW OF DISTRICTS.

At Elizabethtown the veins occur in monzonite porphyry or at the contact of sills of porphyry with Cretaceous shales of Colorado age. They are single fissures or lodes, the latter being more common in porphyry. The greatest traceable length is half a mile. Most prominent is the Aztec vein, which dips at an unusually flat angle, 25° to 30° NE. A production of over $1,000,000 is reported from this vein, largely in coarse free gold. Somewhat similar veins appear in monzonite in the Anchor district.

In the Cerrillos Hills numerous small veins are contained in monzonite or in Cretaceous shales. A northeasterly strike prevails. The veins are narrow and carry principally argentiferous galena and zinc blende. In the Ortiz Mountains, near Dolores, there are veins rich in free gold, probably largely set free by oxidation. Foremost among them is the Ortiz vein, from which much rich gold ore was extracted in early days. The veins have a northeasterly strike and are contained in diorite and monzonite porphyry. The gold veins in the San Pedro Mountains are numerous, but narrow, and contain rich pockets of free gold with pyrite. They are found chiefly in quartz monzonite.

The Jicarilla district contains northwestward-striking veins and shear zones in quartz monzonite porphyry. The Nogal district is characterized by shear zones and brecciated lodes in monzonite porphyry. Some of them strike north-northwest and dip 60° W. Others have a northeasterly strike. Up to 70 per cent of the gold is free. Little silver is present. The greatest depth attained is 400 feet. This district has yielded a moderate production.

The White Oaks district, also in Lincoln County, is currently reported to have produced $2,500,000. The deposits are contained in monzonite, which intrudes Cretaceous shales, and form narrow veins or lodes of greater width, containing mainly quartz and pyrite. A number of unusual minerals are present, as indicated above. One of the principal veins, the Old Abe, trends northward and has been worked down to a depth of 1,380 feet, yielding rich oxidized gold ores with a total production of nearly $1,000,000. The gold ore forms high-grade pockets and shoots. The ore below the seventh level is decidedly poorer than that above. The average grade of the ore is said to be $20 a ton.

In the Organ district the quartz monzonite contains narrow quartz veins, trending eastward and in many places following dikes. Silver veins with galena and tetrahedrite and pyritic gold veins with little free gold are both known. The production is small. The galena veins in limestone are mentioned under the heading of replacement deposits in limestone.

In the Jarilla Hills, in Otero County, some unusual veins are found, which form a transition to contact-metamorphic deposits. They are metamorphosed zones in limestone, close to monzonite porphyry, but follow fractures. Essentially they are quartz-pyrite veins with indefinite boundaries. The walls contain garnet, pyroxene, and specularite. In places porphyry forms one of the walls. The values are in gold and copper; the ores are partly oxidized.

The Hachita Range, in Grant County, near the Mexico boundary, is built of Paleozoic and probably Cretaceous strata, with an enormous number of dikes, sheets, and stocks of monzonite and syenite rocks. The silver veins in limestone or in porphyry dikes carry pyrite, zinc blende, and other sulphides; in some places the limestone walls contain garnet with zinc blende and pyrite. Other deposits form replacements of galena, pyrite, and zinc blende along the bedding planes of limestone. Another class of deposits at Sylvanite camp consists of quartz veins in monzonite, with orthoclase, tetradymite, and native gold. Still others are contact-metamorphic deposits forming garnetized beds of limestone with chalcopyrite in the Cretaceous (?) rocks. The same district also contains peculiar deposits of pyrrhotite and chalcopyrite replacing the cement in a quartzitic sandstone. Although the districts in this range have not proved of great importance, the variety of deposits in the dike-seamed sediments is most interesting.

In the Fremont district the Paleozoic limestone contains veins of argentiferous galena near dikes of porphyry.

The Lordsburg and Pyramid districts are situated in areas of greatly altered rocks of uncertain age and affiliations. In part they appear to be diorite porphyries, in part andesitic surface flows. The deposits in some ways partake of the characteristics of those contained in the Tertiary lavas, though other features, such as the occurrence of tourmaline, suggest deeper-seated conditions. The quartz veins carry chiefly silver and contain chalcopyrite, pyrite, bornite, and a little galena. In places barite, calcite, rhodochrosite, and siderite are present. The shafts have attained only 400 feet in depth. The production is estimated to be several hundred thousand dollars.

Other silver veins are those of the Blackhawk and Telegraph districts; in the former the principal vein strikes N. 67° E. and is opened to a depth of 740 feet on the incline; in the bottom native silver is reported to be found. The veins occur in gneiss, containing porphyry dikes; they are narrow and the values erratic.

The celebrated district of Pinos Altos contains a vein system that strikes northeastward in diorite porphyry and granodiorite porphyry, which metamorphose adjoining limestones. Some of the veins are traceable for 4,000 feet; their average width is 2½ feet, the maximum width about 8 feet. They are filled veins with pyrite, chalcopyrite, zinc blende, and galena in a quartz gangue, with some barite, calcite, etc. The average value of the ore is about $10 in gold and $5 in silver and copper to the ton. The greatest depth attained is 700 feet. The district is currently credited with a production of $5,000,000, a considerable part of which was derived from placers below the oxidized croppings. The production of the veins is estimated at $1,500,000; that of placers derived from the veins at several million dollars. Near by are replacement deposits of zinc blende following fractures in limestone.

The Silver Cell mine, also close by, is a narrow vein in intrusive diorite; it is opened to a depth of 400 feet. Native silver, argentite, and horn silver are said to persist to the deepest level.

The Central district, southwest of Hanover, contains veins of zinc blende, pyrite, and galena in quartz and dolomite gangue; the veins outcrop in porphyry of the Silver City type and in the Cretaceous shale.

The Hanover district is rich in contact-metamorphic deposits and veins. In part the veins occupy northeastward-trending fissures in porphyry of the Silver City type; they contain pyrite, chalcopyrite, zinc blende, and specularite, also molybdenite; silver is present, but there is very little gold. The shafts are 400 feet deep.

The Ivanhoe lode lies along a contact between quartz monzonite and shaly limestone and exhibits both fracture filling and replacement; in places it is 15 feet wide.

COPPER DEPOSITS DUE TO OXIDIZING SURFACE WATERS.

The Santa Rita and the Burro Mountain districts stand in a group by themselves, in which primary deposits of very low grade, undoubtedly belonging to the class of early Tertiary age, were enriched by oxidation to an extraordinary degree so as to constitute copper deposits of great value. At Santa Rita the process had progressed so far as to form rich oxidized copper ores that have been successfully extracted for a century. In the Burro Mountains the concentration yielded a lower grade of chalcocite ore, which could only recently be utilized, when processes of mining and concentration were perfected. The two localities produce the greater part of the copper from New Mexico, each yielding several million pounds annually under favorable conditions.

At Santa Rita a large basin is floored by quartzite (probably of Cambrian age), intruded by quartz monzonite. The surrounding rims are of partly contact-metamorphosed limestone, quartz monzonite, and rhyolite. Most of the copper occurs as native metal in the quartzite, but is found also in the porphyry, especially where kaolinized. Chalcocite also is present, both in the quartzite and the porphyry, with kernels of pyrite. There is no doubt that much of the native copper has been derived from chalcocite and that the latter has resulted from the precipitation of copper sulphate solutions by primary pyrite-forming "stockworks" or irregular seam systems in both rocks. To a considerable extent, however, the native copper was deposited as such from solutions, either in open spaces or as result of replacement of the rock. The principal bodies of native copper were found within 150 feet from the surface. There is said to be little workable copper ore below a depth of 325 feet; the deepest shafts attain 500 feet.

No leached zone of barren rock exists, such as caps many deposits of similar kind. It is believed that the enriching copper solutions were derived from contact-metamorphic deposits in the limestone which once covered the deposits, but which now is eroded. The ores contain no gold or silver. For many years previous to 1908 the work was done mostly by Mexican lessees. Much low-grade ore is said to remain, and development of this was begun on an extensive scale in 1909.

In the Burro Mountain district, southwest of Silver City, a core of pre-Cambrian granitic and gneissoid rock is exposed. This is intruded by masses and dikes of a monzonite porphyry which is kaolinized and in places contains pyrite. In this the copper deposits occur along zones of crushing and fracturing and contain disseminated chalcocite; the ores yield about 4 per cent and are concentrated, the concentrates being shipped. The ore bodies of the principal mine are of great size, some being 50 feet or more in width and several hundred feet in length. The chalcocite replaces primary pyrite, disseminated along the zones of crushing. The pyrite is said to contain about one-third of 1 per cent of copper. A leached zone about 50 feet thick caps the deposits, but contains some malachite and azurite. Below this poor capping are found small masses of rich oxidized ores consisting of cuprite and native copper; underneath these lies the main body of disseminated chalcocite, which extends down for 150 feet or more, and gradually merges into worthless porphyry with disseminated pyrite. The water level stands at 100 feet below the surface. These deposits are similar to many of those in the chalcocitic districts of Arizona, such as Morenci, Globe, and Ray.

VEINS AND REPLACEMENT DEPOSITS IN LIMESTONE, EXCLUSIVE OF CONTACT-METAMORPHIC DEPOSITS.

GENERAL FEATURES AND DISTRIBUTION.

The contact-metamorphic deposits and the vein systems which were formed during and shortly after the intrusions of early Tertiary age have been described (pp. 35–42). Another group of the same affiliation—the deposits in limestone—were, however, reserved for special mention. They easily fall into a separate group, for they furnish a very considerable part of the lead and silver mined in the Territory; also on account of their structural irregularity, a common characteristic of deposits in limestone. Nearly all of them appear in the vicinity of the intrusive masses, but they are entirely distinct in mineral association from the contact-metamorphic deposits; the latter rarely contain any considerable amount of lead and silver.

The districts in which these silver and lead deposits occur are as follows: San Pedro, Hermosa, Kingston, Hillsboro, Sierra Blanca, Lake Valley, Cooks Peak, Organ, Victorio, Florida Mountains, Pinos Altos, Georgetown, Lone Mountain, Chloride Flat, and Granite Gap. All of them are probably primarily connected with fractures which guided the metallic solutions. The deposits at San Pedro, Organ, Granite Gap, Victorio, and Pinos Altos are more or less distinct veins; in the others the extent of the mineralization is determined by certain horizons in the sedimentary series. It will be noted that most of the deposits are in the southern part of the Territory and chiefly in Sierra, Luna, and Grant counties, where the character of the sediments appears to have favored the deposition. (See fig. 1.)

With few exceptions the ores occur in the vicinity of intrusive masses, dikes, and sills of quartz monzonite, quartz monzonite porphyry, or granodiorite porphyry, and at Lake Valley it can be shown that the andesite covers the eroded ore deposits. Only rarely, as at Georgetown, the ore itself extends into the intrusive porphyry; the Tertiary flow rocks where found in the neighborhood are ordinarily not mineralized. The general conclusion reached is that these deposits were formed shortly after the igneous intrusions, at a considerable depth below the surface, and that some of them became exposed at the surface previous to the outflows of the middle Tertiary lavas. In places the beginning of their oxidation can be shown to date back to this period.

Unlike the contact-metamorphic deposits, they do not appear at the immediate contacts of large intrusive masses, but rather some distance away or near minor dikes and sills. This relation is particularly well shown in the Organ district and also in the vicinity of the great porphyry intrusions of Silver City and Hanover. No one can fail, however, to perceive the close genetic connection between these deposits, on one hand, and the contact-metamorphic deposits and the vein systems which cluster around each intrusive mass, on the other. They are simply a special group in which chemical action by the wall rock on the ore-bearing solution resulted in the precipitation of certain mineral combinations.

The production of this group has been very heavy, especially from the oxidized part of the deposits, in which at many localities a remarkable concentration of secondary silver ores has taken place. At present the production is small and comes mainly from Organ, Cooks Peak, and Granite Gap; most of the other districts mentioned are idle, and some appear to be exhausted.

Lake Valley is currently credited with a production of 5,000,000 ounces of silver; Georgetown, 3,000,000 ounces; Chloride Flat, 3,000,000 ounces; Kingston, 5,000,000 ounces; Hermosa, 1,000,000 ounces; Victorio, $1,150,000 in lead, silver, and gold; Cooks Peak, $3,000,000 in lead and silver; Organ, $600,000 in lead; Granite Gap, $600,000 in lead and silver—all these figures being approximate. Probably not less than 20,000,000 ounces of silver have been extracted from these deposits, which made New Mexico famous in the seventies and eighties of the last century.

The developments are extensive in many of the camps, but they are in most places lateral rather than vertical. Few shafts have attained a depth greater than 500 feet.

THE ORES.

The table of mineral association which follows permits at a glance some general conclusions. Gold is generally absent, almost wholly so in some deposits; the Granite Gap and Victorio ores carry some of this metal. Silver is universally present and as a rule is the principal valuable metal. Low silver values are found at Organ and Cooks Peak.

Lead in the form of galena or its products of oxidation is almost universally present, but is by no means always of great economic importance. At Chloride Flat, Lone Mountain, Georgetown, and Lake Valley very little lead is found, and it can not be assumed that the silver ores are due simply to the oxidation of galena. At Granite Gap, Victorio, Cooks Peak, and Organ galena is the most important mineral. Quartz is probably universally present as a replacement of limestone, and jasperoid is also common. Calcite also accompanies the ore. Siderite was noted at Victorio and Granite Gap, ankerite at Lake Valley. Barite is present in one place only, fluorite in two. Vanadinite occurs at three localities and wulfenite indicates primary molybdenite at two. Zinc blende is commonly recorded; chalcopyrite occurs here and there. Pyrite is not found everywhere, though probably always present in small quantities in the primary ore. Argentite, cerargyrite, and native silver are noted in most places, but are not specially abundant in deposits with much lead. Ruby silver is of local occurrence.

There is often some difficulty in determining the original constituents of the primary ore, for the oxidation is everywhere prominent and in many camps the oxidized ores were the only ones which were found profitable to mine. Limonite is universally present, and pyrolusite or other compounds of manganese are very plentiful in some camps, Lake Valley being a conspicuous example. All heavy silicates, magnetite,[a] specularite, and tourmaline, are absent.

---

[a] Except in one place at Chloride Flat, where the mineral doubtless developed as a result of oxidation.

Minerals of veins and replacement deposits in limestone (not contact-metamorphic). [a]

District.	Principal metals.	Quartz.	Calcite.	Ankerite.	Siderite.	Barite.	Fluorite.	Wulfenite.	Vanadinite.	Galena.
San Pedro	Ag, Pb	×	×							×
Organ	Pb, Ag	×	×			×	×	×		×
Granite Gap	Pb, Ag, Au		×							×
Victorio	Pb, Ag	×	×		×					×
Pinos Altos	Pb, Ag, Zn		×							×
Carpenter	Pb, Zn		×							×
Hermosa	Ag		×							×
Kingston	Ag		×							×
Tierra Blanca	Ag		×							×
Hillsboro	Pb, Ag		×							×
Lake Valley	Ag		×	×				×	×	×
Cooks Peak	Pb	×	×				×			×
Georgetown	Ag	×	×							×
Chloride Flat	Ag	×	×							×
Lone Mountain	Ag	×	×					×		

District.	Zinc blende.	Pyrite.	Chalcopyrite.	Argentite.	Horn silver.	Native silver.	Limonite.	Pyrolusite.	Remarks.
San Pedro	×	×	×				×		Veins in limestone. Alabandite reported.
Organ		×	×				×		Veins in limestone. Little silver present.
Granite Gap				×		×	×	×	Veins in limestone, following dikes.
Victorio						×	×		Veins in limestone.
Pinos Altos	×	×	×				×		Veins in limestone, following dikes.
Carpenter		×					×		Shear zones in limestone.
Hermosa		×		×	×	×	×		Below shale, in upper Carboniferous limestone.
Kingston	×		×	×	×	×	×		Below shale in Ordovician limestone.
Tierra Blanca		×		×	×	×	×		Below shale, in lower Carboniferous limestone.
Hillsboro					×	×	×	×	Below shale, in Ordovician limestone.
Lake Valley	×			×	×	×	×	×	Below shale, in lower Carboniferous limestone.
Cooks Peak	×	×					×		Below shale, in Ordovician limestone.
Georgetown									Do.
Chloride Flat				×	×	×	×	×	Below shale, in Silurian limestone.
Lone Mountain		×		×	×	×	×		Below shale, in Ordovician limestone.

[a] Minerals in last four columns are of secondary origin; other products of oxidation are not separately mentioned, but are universally present.

STRUCTURAL FEATURES.

In regard to structural features two divisions may be recognized in this group. In the first the deposit is distinctly related to major fissures in the limestone. To this division belong the Stephenson-Bennett mine in the Organ district and the veins of the Victorio district in Luna County; in part also the Granite Gap mine in Grant County, the Cleveland mine at Pinos Altos, and some others. In the second the ore occurs along certain horizons in the sedimentary series, usually below a prominent shale bed. To this division belong the deposits of Chloride Flat, Cooks Peak, Lake Valley, Kingston, and many others.

In the Organ and Victorio districts the ore follows a well-defined fracture and forms good-sized, in places brecciated, masses; outside of the ore bodies the fracture narrows to a mere seam. The exact age of the Paleozoic limestone or dolomite could not be ascertained. At the Stephenson mine, at Organ, the limestone near the deposit is separated by a fault from the main mass of quartz monzonite of the district. Many minor deposits of galena occur in the district in limestone some distance away from the contact-metamorphic copper deposits. At Victorio no intrusives are seen, but the exposures are very small, Quaternary gravels and late Tertiary andesite covering the larger areas.

At Granite Gap the oxidized ore either follows in a general way a dike of granite porphyry or accompanies east-west fractures. Arsenic and copper are present in oxidized form and the ore also contains gold. The horizon is probably in the lower Paleozoic. At the Lincoln Lucky mine, at San Pedro, Santa Fe County, the ore followed a fracture parallel to the bedding planes of Carboniferous limestone. In the Carpenter district the zinc-lead ore follows shear zones in limestone of Ordovician age, which is cut by masses of granite porphyry.

At the Cleveland mine, near Pinos Altos, the workings are in limestone near diorite porphyry; zinc blende, pyrite, chalcopyrite, and galena with gangue of quartz and calcite have replaced the rock along northeast fractures. The ores form bodies 2 to 10 feet wide.

At Chloride Flat the Paleozoic rocks are cut by dikes of quartz monzonite porphyry, showing only slight contact metamorphism. The ore occurs in Silurian limestone below the Devonian shale either at the contact or extending irregularly down into the limestone. The dip is 35° NE. Thin seams of ore connect the various deposits. Silver chloride is the principal ore. Some native silver and argentite are present. Lead is not abundant.

At Lone Mountain, 6 miles southeast of Silver City, similar geologic relations prevail and the limestones have a dip of 30° NE. Here the main ore horizon is in the Ordovician limestone just below the Devonian shale, the Silurian being probably absent. The deposits do not form beds, but occur in narrow and nonpersistent veins cutting the limestone, which is slightly silicified. Oxidized silver ores prevail. There is no lead and only a trace of copper.

At Georgetown, northeast of Silver City, the Carboniferous and Ordovician limestones again appear and are, as elsewhere, separated by 200 feet of Devonian shale. The dip is 15° S. Dikes and sills of porphyry are present. The ore bodies lie just below the shale and along or near dikes, the occurrence being similar to that at Chloride Flat. Some lead carbonate is present, besides the oxidized silver ores. The limestone is slightly silicified near the deposits. The most regular stope was worked close to the walls of a dike which is 15 feet thick. This stope was 60 feet long, 20 feet high, and 2 to 3 feet wide. The original ore was probably argentiferous galena.

At Cooks Peak a large mass of granodiorite porphyry is intruded into the Paleozoic rocks, which north of the peak dip 25° ESE. The ores occur at the same horizon as in the localities already described, just below the Devonian shale, usually below a silicified zone that follows the contact between limestone and shale. The occurrence of the ore is irregular, in pipes, kidneys, and pockets. One chamber at the El Paso mine was 100 feet long and 35 to 50 feet in cross section, with offshoots extending irregularly. Cores of unaltered galena are commonly surrounded by a shell of carbonate ore and outside of this is limonite. The primary ore consisted of calcite, quartz, fluorite, pyrite, zinc blende, and galena.

At the once celebrated camp of Kingston, now almost idle, the ore horizon is at the same contact between shale and limestone. The flat-lying limestone forms low, gently pitching arches which appear to control the ore deposition. A dike of monzonite porphyry 400 feet wide traverses the district. The ore forms pockets and pipes in the limestone just below a zone of silicification separating it from the shale; it does not extend more than 100 feet below the shale. There are a number of minor fractures in the limestone and ore is found at many of their intersections. One of the ore pipes was followed for 400 feet. The ore is oxidized as usual, contains much native silver, and is contained in a soft talcose gangue. The sulphides of lead, copper, and zinc are, however, present in some of the ore.

Small deposits at Hillsboro containing oxidized ores of lead, manganese iron, molybdenum, and vanadium are found at the same geologic horizon.

At Hermosa, also in Sierra County, the limestones belong in the Sandia formation of the Pennsylvanian (upper Carboniferous); the dip is about 25° NE., and the limestone forms flat arches pitching northward; most of the ore is found under these arches, in the limestone and immediately below a bed of shale which belongs at a higher horizon than the Devonian shale mentioned in the foregoing descriptions.

At the Palomas-Pelican mine pockets and pipes of oxidized silver ore extend irregularly into the limestone from its contact with the shale. The ore is closely connected with fractures, occurring at their intersections. No fractures were observed in the shale. Thin seams of ore connect the various bodies. The ores contain native silver and argentite, also galena, bornite, and pyrite in a talcose clay, an analysis of which is given below. Some of the ore bodies were 20 feet in thickness.

Lake Valley is the most celebrated of all these silver districts. Enormous bodies of almost pure cerargyrite, most famous among which was the "bridal chamber" (fig. 27), were mined here from 1880 to 1885. The total production was 5,000,000 ounces. To-day the camp is practically abandoned. The foothills of the Mimbres Range are here covered by flows of rhyolite and andesite, the erosion of which has in places exposed the Paleozoic sediments. The dip of these beds is 20° SE. The ore constitutes pockets, chambers, and pipes in the lower beds of the Lake Valley limestone (Mississippian, or lower Carboniferous). Some of the ore lies along the contact with an overlying shaly limestone, but it is generally well within the "blue limestone" or along its base at the contact with the "nodular limestone." The shoots follow in the main the bedding of the rocks. No intrusive rocks are here exposed. Ellis Clark,[a] who has given an excellent description of the deposit, believed that a genetic connection existed between the ore and a "porphyrite" which in places caps it. Gordon establishes the identity of this decomposed rock with the andesite which rests on the eroded sediments, and his views are fully shared by the senior writer, who had an opportunity to examine the deposit in 1891, when work was still in progress. The rich ore bodies were found in an old Tertiary erosional depression and everything indicates that the deposits are much older than the andesite. The oxidation and enrichment must have begun far back in the middle Tertiary, and the great richness is undoubtedly due to the exceptional opportunities for concentration. The deposit has been so thoroughly worked over by processes of oxidation that reconstruction of its primary form is difficult.

The Lake Valley deposits, like the other occurrences described above, are probably connected genetically with intrusive porphyries, now hidden underneath the later lava flows. This conclusion is confirmed by an examination of the Tierra Blanca district, 15 miles northwest of Lake Valley. Here we find the silver-bearing horizon identical with that of Lake Valley— that is, in the lower Carboniferous. The ores occur in limestone underneath a bed of shale, and they extend along a narrow band which borders a prominent intrusion of monzonite porphyry.

The Lake Valley ores are in the main siliceous. The so-called flint ores average, according to Clark, 65 per cent of silica, 6 per cent of iron, and 12 per cent of manganese, with 20 ounces of silver to the ton. The high percentage of manganese in the form of pyrolusite (also psilomelane and manganite) is remarkable. In places the manganese increases greatly, and some of these basic ores were very rich in silver. The Bunkhouse ores contained 200 to 500 ounces of silver to the ton, with 8 per cent of silica, 12 per cent of iron, and 24 per cent of manganese, the rest being calcite and gypsum. At the base of the ore series, well down in the "blue limestone," was a deposit containing 30 per cent of silica, 12 per cent of iron, 18 per cent of manganese, and 8 ounces of silver to the ton; this material could not be treated. Some of the ores contained from 1 to 5 per cent of lead. The thoroughly oxidized character of this ore is noteworthy. It was probably derived from a primary ore with much rhodochrosite (perhaps also alabandite) and ankerite and some galena, rich in silver; possibly also it contained primary argentite.

METASOMATIC PROCESSES.

The changes in the limestone close to these deposits are generally few and simple. No silicates are developed, but much silica is introduced, giving rise to cherty and jasperoid rocks, in part also to cellular rocks in which the vugs caused by the solution of the limestone are coated with quartz crystals.

The talcose material accompanying the oxidized ore at Hermosa and other camps is somewhat of a puzzle. It looks like a bleached soft shale and, except for these features, presents considerable similarity to the clay shale overlying the ore-bearing limestone. Both materials were analyzed, as follows:

---

[a] Trans. Am. Inst. Min. Eng., vol. 24, 1894, pp. 138-167. See also Keyes, C. H., Genesis of the Lake Valley silver deposits: Trans. Am. Inst. Min. Eng., vol. 39, 1908, pp. 139-169.

Analyses of shales accompanying the ore at Hermosa Camp.

[Analyst, George Steiger.]

	1.	2.		1.	2.
SiO_2	34.64	49.13	ZrO_2	None.	None.
Al_2O_3	25.58	13.92	CO_2	None.	6.93
Fe_2O_3	.35	.77	P_2O_5	0.06	.03
FeO	3.60	1.87	MnO	.48	.25
MgO	17.47	5.11	BaO	None.	None.
CaO	.75	8.73	SrO	None.	None.
Na_2O	.26	.20	FeS_2	1.70	.86
K_2O		4.25	C	None.	2.00
H_2O-	3.67	1.52			
H_2O+	10.38	a 4.69		99.77	100.92
FeO_2	.83	.66			

a The figure for water at 100°+ is too high by the amount of water equivalent to hydrogen combined with carbon and organic matter. There is no method by which the true amount of water can be determined in this combination.
1. Altered shale, Palomas Chief, near ore.
2. Calcareous normal shale, Palomas camp.

Analysis No. 2 is thoroughly characteristic of a calcareous clay shale. The sample is a fissile, dark-gray, fairly hard shale. The high percentage of potash, the small amount of soda, and the low silica are noteworthy.

Analysis No. 1 is that of a soft fissile material, almost white, and having the appearance of an alteration product derived from the shale of No. 2. At the same time it can not be positively asserted that such was its origin, as no fresh shale was found at the mine. The analysis shows that it contains no calcite and no potash, but consists chiefly of silica, alumina, magnesia, and water. The silica is not sufficient to satisfy the bases; it seems probable that the material consists of about 70 per cent of a hydrous aluminum silicate allied to kaolin and that the remainder is mainly a hydroxide of magnesia, perhaps brucite; possibly it is hydrotalcite. Most likely the substance is the result of the action of magnesian solution from the dolomitic limestone on the overlying clay shale. It is not certain that this is a result of primary metallization; it may rather be caused by oxidizing waters acting on a primary metasomatic product.

GENESIS.

The principal fact of genetic interest is the accumulation of the ore underneath impervious and nonfractured clay shales. This is shown to have taken place in so many districts that it certainly has a genetic significance. It is believed to mean that ascending metallizing solutions, derived from intrusions of quartz monzonite porphyry or allied rocks, were arrested at this level and, becoming stagnant and cooled, were forced to deposit their metallic contents.

VEINS CONNECTED WITH VOLCANIC ROCKS OF TERTIARY AGE.

GENERAL FEATURES AND DISTRIBUTION.

During middle and later Tertiary time eruptions of rhyolite, andesite, and basalt latite took place and extensive areas are covered by these now partly eroded flows. Their distribution has been described in a previous chapter, and it is sufficient to say here that such lava flows appear in northeastern Taos County, in the Valles Mountains of Sandoval County, and in the Mogollon, Socorro, Magdalena, and San Mateo mountains of Socorro County, and that particularly wide areas are covered by them in Sierra and northern Grant counties. (See fig. 2.)

These flow rocks and their associated tuffs contain, in places, gold and silver bearing veins, usually grouped in districts but by no means appearing in such abundance as characterizes the deposits of early Tertiary age associated with the intrusive porphyries. There are vast areas of these volcanic rocks without traces of mineralization. The distribution of the metalliferous districts in them is not, so far as ascertained, connected with the broader structural features.

The evidence of the age of these deposits is, on one hand, that they are contained in fractured volcanic rocks of middle Tertiary age and, on the other hand, that they have suffered much erosion. Some of them can be shown to be older than basalt flows of early Quaternary age. As a matter of fact, however, these veins form a well-defined group, rarely occurring outside of the volcanic areas.

The districts characterized by this group are, named from north to south: Red River, Taos County; Cochiti, Sandoval County; Socorro, Mill Canyon, Rosedale, and Mogollon, Socorro County; Apache, Black Range, and Hillsboro, Sierra County; and Steeple Rock and Kimball, Grant County. Their distribution indicates little regularity; most of them are in the great southwestern area of Tertiary effusives, extending over into Arizona; some districts, like the Mogollon and Steeple Rock, are close to the boundary line. The Rosedale district in the San Mateo Mountains and the Cochiti district in the Valles Mountains are both within Tertiary rhyolitic volcanoes. The Mogollon district also lies in a former center of volcanic activity. All the districts, in truth, appear to be located near eruptive centers.

COUNTRY ROCK.

Many of these deposits are contained in rhyolite or its tuffs and breccias. Some, however, are in andesite, a rock which in this region shows a decided tendency to transitions to latite.

FIGURE 2.—Map of New Mexico showing location of late Tertiary veins in or near flows of rhyolite, andesite, and latite.

At Red River the gold-bearing veins cut rhyolite, andesite, and monzonite porphyry. At Cochiti the veins intersect monzonite upon which rest rhyolite flows; the veins do not appear to extend through the overlying rhyolite. It is believed that they were formed during the early period of eruption of the Valles rhyolitic volcano and that their slightly eroded croppings were covered by later rhyolite flows.

At Socorro the argentiferous veins are in andesite or latite, or in associated tuffs. In the Mogollon district the fissure veins, carrying gold and silver, appear in a horizontal series of volcanic beds aggregating several thousand feet in thickness and comprising older rhyolites, andesites, and basalts; many of the flows are tuffaceous or brecciated.

In the Black Range and Apache districts, in northern Sierra County, a great thickness of andesites and their tuffs (probably including many rocks of the composition of latites) covers the top of the main Mimbres or Black Range. Most of the veins are in andesite, though some are found in rhyolite.

At Hillsboro the auriferous veins are in latite flows. At Carlisle or Steeple Rock and in the Kimball district they are included in a somewhat doubtful flow rock, probably a dacite, looking like the rhyolite of the Mogollon Mountains; associated with this rock is some dark-gray fine-grained andesite, which may be intrusive into the dacite.

STRUCTURE.

It would be strange if strike and dip of the veins were alike in these widely scattered districts, whose fissures were probably determined by purely local conditions of stress. In the Mogollon and the Black Range districts there are two systems of veins, striking respectively north-south and east-west. In the Mogollon district two master veins trend north and south; these are not very productive, but are connected by a richer system of east-west veins.

In the Black Range and Apache districts the north-south veins carry gold and silver with some copper, while the east-west veins contain mainly silver and copper. Some of the gold veins can be traced for several miles, but those trending east and west are less persistent. The deposits at Hillsboro form veins and shear zones striking northeast. At Steeple Rock the strike is northwest or north; at Steins Pass or Kimball the deposit is a brecciated zone along a fault striking north-northwest between diorite porphyry and rhyolite. The fissures at Cochiti are complex fractures and shattered zones with much open space, the strike varying widely from northeast to northwest. No discernible law governs the dip of the veins of this class, though the inclination from the horizontal is nearly always great, being rarely less than 70°. Few veins can be traced for more than a mile; most of them are much shorter; at the same time they are fairly regular and maintain the same dips and strikes.

The internal structures are less regular than in the older veins; the walls are likely to be shattered and broken, the fissures filled with breccias of country rock. In many places these fragments are covered by drusy quartz and ore minerals, giving the veins a decidedly open structure. Most of the sulphide minerals are finely divided instead of massive. Most of the veins are filled cavities, but here and there, as at Cochiti, ore formed by replacement extends far into the shattered walls. The thickness is usually moderate, a few feet at most, though some stopes in the Mogollon district are 20 feet wide and at Cochiti ore bodies were found that were 100 feet or more in width. The only place where the developments have reached 1,000 feet in depth is in the Mogollon district. Few shafts in any other of these districts exceed 700 feet in depth.

THE ORES.

Gold and silver are generally both present; the gold predominates in value at Red River, Carlisle, and Hillsboro. In the Steeple Rock and the Mogollon districts the gold values are as a rule slightly lower than those of silver. Certain veins in the Black Range, Socorro, and Mogollon districts contain almost exclusively silver. Several of the districts, notably the Cochiti and Steeple Rock, have a fairly large production to their credit, but most of them are idle at present, either because of low grade of ore or because of exhaustion of the known ore bodies. One of the districts—the Mogollon—stands, however, preeminent and promises well for the future. From its veins ore to the value of approximately $5,000,000 is reported to have been extracted, and it is at present the most productive gold and silver camp of the Territory. Its production from 1904 to 1908, inclusive, was 83,130 short tons of ore, yielding $506,776 in gold, 1,285,802 fine ounces of silver, and 867,483 pounds of copper, or a total value of $1,434,848.

The base metals are not prominent in the veins of this class, and in general the sulphides make up only a very small percentage of the ore. Lead and zinc are, on the whole, rare; copper, on the other hand, is present in considerable amounts in the Black Range, Hillsboro, Mill Canyon, and Mogollon districts, and in the Mogollon contributes notably to the revenue from the ores. The accompanying table shows at a glance the minerals occurring in these veins, though in some districts the record may not be complete. Quartz is the universal gangue material, but in some places it is accompanied by calcite, fluorite, and barite. Adularia is present in two districts. Pyrite is not a prominent mineral, though probably present everywhere; much of it is finely distributed through quartz and rock instead of occurring in coarser masses as in the older veins. Next to chalcopyrite, bornite is the common copper mineral and is probably primary. Its presence is especially noted in the Mogollon, Black Range, and Apache districts. A telluride, probably petzite, is found at Red River. Finely divided argentite is one of the principal ore minerals at Cochiti, Kimball, and Steeple Rock, probably also in other districts, and it has the appearance of being primary. Free gold is not common in coarse visible form, but, divided in microscopic particles, occurs at many places. Molybdenite was noted as a rarity at Red River. Specularite occurs similarly in one mine in the Mogollon district. Magnetite, tourmaline, garnet, pyroxene, and similar heavy silicates are absent.

The removal by solution of one of the minerals—usually calcite—gives to some ores a peculiar "hackly" or cellular structure. This is especially well shown in the Cochiti, Mogollon, and Kimball districts.

Minerals of late Tertiary veins.[a]

District.	Principal values.	Quartz.	Calcite.	Barite.	Fluorite.	Adularia.	Pyrite.	Chalcopyrite.	Bornite.
Red River [b]	Au, Ag	X			X		X	X	
Cochiti	Au, Ag	X					X	X	
Socorro	Ag	X		X			X		
Mill Canyon	Ag, Cu	X							X
Rosedale	Au, Ag	X					?X		
Black Range and Apache	Au, Ag	X	X	X		X	X	X	X
Hillsboro	Au, Ag, Cu	X	X				X	X	X
Mogollon [c]	Au, Ag, Cu	X	X		X	X	X	X	X
Steeple Rock	Au, Ag	X					X	X	
Kimball	Ag	X	X				X	X	

District.	Chalcocite.	Galena.	Zinc blende.	Telluride.	Tetrahedrite.	Native gold.	Native silver.	Argentite.	Horn silver.
Red River [b]		X	X	X		X			
Cochiti		X	X						
Socorro							X	X	X
Mill Canyon									
Rosedale						X			
Black Range and Apache	X					X	X	X	X
Hillsboro						X			
Mogollon [c]	X			X		X		X	X
Steeple Rock		X	X			X		X	X
Kimball		X	X					X	X

[a] Of secondary products, only argentite and horn silver included. Argentite in part primary.   [b] Molybdenite rare.
[c] Specularite rare. Pyrargyrite in some places.

OXIDATION.

In the high ranges of Red River the zone of oxidation is shallow. At Cochiti the water level stands about 136 feet below the surface, but sulphides remain in the hard and compact ore high above this level. In the Mogollon district the ores are not greatly oxidized, though the general level of operation is high above water level. This shallowness of oxidation is probably due to the rapid erosion in a country of youthful topography. In the Black Range few mines have been carried to water level. At Chloride Creek it was found at a depth of 390 feet, but in some places it is much nearer the surface.

In general, the oxidation has not attained great depth, nor has any notable concentration of secondary sulphides been observed. In the outcrops and upper part of the veins to a depth of about 300 feet an enrichment has taken place, both in gold and silver; as products of oxidation, horn silver and native silver, probably also some argentite, have formed and have in places accumulated owing to their greater solubility. This found expression in the shipments of rich silver ores soon after discovery; later, when sulphide ores of lower grade were encountered, the production slackened or ceased.

ORE SHOOTS.

Few data are available regarding the ore shoots of this group of veins, as in few places are the developments extensive. At Cochiti some of the ore bodies were of great width though of low grade and the values are said to have decreased gradually in depth until the ores became unprofitable. Work ceased at a depth of about 700 feet. Large shoots, 200 feet or more in horizontal extent, are mined in the Mogollon district, one stope attaining 20 feet in width. Some of these shoots appear to continue to a depth of 700 feet or more. No particular rule as to pitch could be established, and the ore bodies do not appear to depend on intersections of fissures.

In the Socorro district the ore was followed for a few hundred feet in depth, but was lost when the vein entered a tuffaceous clay. In the veins of the Steeple Rock district the best ore was found at a depth of 300 feet, below which the values decreased and recovery became difficult.

METASOMATIC PROCESSES.

It is common, in mining districts situated within the Tertiary lava fields, to find a widespread alteration of the rocks, often referred to as a propylitization. Many of the late Tertiary deposits of New Mexico show this phenomenon on a large scale; it is in places accompanied by silicification. With good reason this alteration has been attributed to the widespread permeation of the rocks by thermal waters near the surface, where extensive fracturing permitted such a saturation. This process is well shown in the Mogollon and Black Range districts. In the Mogollon district the rhyolite contains kaolin, calcite, chlorite, and a little epidote; pyrite appears in fine distribution close to the veins. The andesite has suffered similar changes and is bleached; quartz is found in amygdaloid cavities; where the alteration is most intense it is accompanied by the development of pyrite, chalcopyrite, and bornite. Adularia is widespread in small amounts, but there is little sericite. Locally silicification has converted the rock to a jasperoid mass. Though the veins are formed principally by filling, some ore has resulted from the alteration of the country rock.

In the Black Range and Apache districts a widespread propylitic alteration has filled the andesites or latites with calcite, chlorite, epidote, and serpentine. Veinlets of these minerals with pyrite and adularia are common. Similar alteration was observed in the veins of the Hillsboro district. In the Steeple Rock district strong silicification marks the vicinity of the veins which are contained in a rhyolitic rock.

Much alteration has taken place in the rhyolites and andesites of the Red River district. Replacement of the walls of the narrow veins by silica is common, as well as an abundant dissemination of pyrite.

At Cochiti the country rock is a monzonite porphyry; the augite has been decomposed into chlorite, epidote, etc. Along the complicated fracture zones extensive replacement was noted, but the ore rarely extends farther than 10 feet from the fissures. The principal process is a silicification, the original minerals being replaced by fine-grained quartz with some pyrite. The ore resulting from the filling of open spaces is richer than the replacement ore.

GENESIS.

The facts observed in studying this class of veins strongly support the theory of deposition by thermal solutions rising through relatively open fissures and fractured rock masses, near the surface. The mineral association suggests deposition under moderate pressure, and it is believed that in many of the districts the present croppings were but a few hundred feet below the surface of the earth at the time of the ascension of the thermal waters.

Few hot springs are now active near these districts; it is believed they were an ephemeral phase of volcanic activity and that deposition was accomplished within a well-defined epoch. Space does not permit an extended discussion of the hot mineral springs of New Mexico, but it may be pointed out that those of well-defined volcanic affinities, characterized by predominance of sodium carbonate and sodium chloride with silica, boron, and arsenic are few in number. They comprise the Las Vegas, Ojo Caliente, Faywood, and some other springs. The celebrated Jemez Springs are chloride waters of doubtful origin. The Socorro Spring issues on a fault line and it is quite possible that it carries simply somewhat heated surface waters.

A number of springs near the basalt fields of Taos have probably some connection with the early Quaternary eruptions of that region. Probably most of the present hot springs of New Mexico of volcanic affinities are caused by these latest eruptions.

One important piece of evidence bearing on the relations of hot springs and ore deposits has, however, been collected, and this is set forth in the following paragraphs.

THE HOT SPRINGS AT OJO CALIENTE AND THEIR DEPOSITS.

The thermal springs of Ojo Caliente are well known in New Mexico for their curative power and figure in the earliest records of the Territory. They issue in the valley of a tributary to Chama River, Ojo Caliente Creek, which finds its source in the southern slope of the Hopewell Mountains. The easiest access to the springs is from the Barranca station on the Denver and Rio Grande Railroad about 10 miles east-southeast of Ojo Caliente. Barranca is situated on the basaltic plateau 1,000 feet above and 2 miles west of the Rio Grande. Below the basalt flow lie the white, sandy, and even-grained deposits of the Santa Fe marl. The road to the springs soon descends the escarpment of the basaltic mesa, and from this point to the springs extends a wide, sandy valley, merging on the south with the valley of the Chama. It is entirely covered by the Santa Fe marl, which underlies the basalt and which presents the uniform and thinly bedded characteristics of lacustrine beds. At every favorable exposure this thin bedding is observed and it has usually a well-defined dip of 20° to 35°. At the springs, which issue on the west bank of the stream, these lake beds are distinctly tuffaceous, sandy, and gravelly. The tuff appears to be of andesitic character.

Immediately behind the springs rises a bluff of gneissoid reddish rock 100 feet high. After ascending this the trail to the mineral deposit crosses a little flat mostly covered with sand and about 4,000 feet wide. Beyond this rises boldly a range of hills of reddish gneiss, forming a southerly outlier or extension of the Hopewell Mountains, which consist of pre-Cambrian rocks. The highest points are probably at least 1,000 feet above the springs. The trail follows a prominent gulch in these hills and the spring deposit is reached at an elevation of about 500 feet above the springs. A small shaft is located on the brow of the hill, south of the gulch and about 200 feet higher than its bed. A tunnel is driven in the side hill 150 feet lower than the shaft and only a few hundred feet distant horizontally. This tunnel has not yet reached the deposit. Near by are some irregular excavations which are believed to have been made by the aid of fire setting during the early part of Spanish occupancy. These openings, it is stated, were driven on a stringer of the main deposit.

The country rock of the bluff behind the springs and the whole range of hills, so far as examined, is a reddish fine-grained gneiss, striking N. 45° E. and traversed by dikes of coarse pegmatite. The microscope shows it to be a mosaic of quartz and orthoclase with a few larger quartz grains. In places it contains microcline, albite, and microperthite in small grains. The gneissoid structure is shown in the parallel arrangement of small flakes of greenish-brown biotite or small foils of muscovite.

The shaft is sunk on a distinct vein in gneissoid rock. The strike of the vein is N. 70° E. and the dip 70° NNW. Its width is 2 to 3 feet and it is traceable for about 200 feet on each side of the shaft. The walls are fairly defined, but the fissure is full of rock fragments, which are loosely cemented by crusts of colorless fluorite. The mineral is not readily recognized, for besides being colorless it has a coarsely fibrous structure perpendicular to the fragments of rock which it incrusts. It looks somewhat like aragonite. The vein matter is oxidized, containing limonite and oxide of manganese, the latter reported to contain silver. Two stringers filled with fluorite come in from the foot-wall side. From the dump of the tunnel were obtained some little stringers in gneiss, filled with greenish fluorite of normal appearance and occasional crystals of barite (011.110.100 according to W. T. Schaller). The fissure filling is reported to contain gold and silver. The owner, Mr. Antonio Joseph, who also is the proprietor of the springs, states that the best assays return $75 in silver and $30 in gold to the ton, other samples giving, for instance, $4 in gold and $1 in silver. There is no reason to doubt that these figures are authentic. Samples of the crusted fluorite were assayed with great care by Ledoux & Co., of New York, and yielded traces of gold and silver.

About 500 feet southwest of the shaft and directly in the line of the vein, which can be traced in this direction for 200 feet, is a small hill about 75 feet vertically above the shaft. The top of this hill is covered to the extent of about half an acre by a tufaceous hot-spring deposit, probably only a few feet deep. A pit 3 feet deep has been sunk in it. It is a loosely coherent cellular mass, mainly composed of calcite with some limonite. According to Mr. Joseph it contains traces of gold and silver. Three specimens from the surface and from 2 and 2½ feet below the surface were assayed by Ledoux & Co., with special precautions, and yielded as follows per ton: No. 68, gold 0.0008 ounce, silver 0.08 ounce; No. 69, gold 0.0025 ounce, silver 0.1 ounce; No. 70, gold 0.0008 ounce, silver 0.05 ounce. A sample of the tufa was analyzed by George Steiger with the following approximate result: Insoluble, 3 per cent; $Fe_2O_3 + Al_2O_3$, 2.9 per cent; CaO, 50.8 per cent; MgO, none; CO_2 (calculated), 39.6 per cent; P_2O_5, none; fluorine, 0.44 per cent; arsenic, none; barium, none. The determination of fluorine was made with particular care. This composition corresponds to about 89.60 per cent of calcite and 0.9 per cent of fluorite. The bluff of gneiss immediately back of the hot springs was found to contain at several places narrow filled seams of a white mineral which proved to be fluorite.

The conclusions drawn in the field from the facts observed were:

1. That the tufa deposit on top of the hill has been accumulated by the evaporation of hot waters reaching the surface.

2. That the vein formed the conduit through which the hot waters reached the surface and in which calcium fluorite was deposited, while the rest of the lime was held in solution as bicarbonate until reaching the surface. The evidence on this point is strong, but perhaps not conclusive. The very unusual structure of the fluorite itself tends to show that it has been deposited under unusual circumstances and probably close to the surface.

3. That the tufa deposit and the fluorite vein were formed by the Ojo Caliente Springs while issuing at a level several hundred feet above the present springs.

4. That the gulch adjacent to the vein and the whole valley has been eroded since the time when the springs issued at high level.

5. That the fluorite veinlets in the bluff back of the springs and about 100 feet above them were filled during the gradual recession of the thermal waters, keeping step with the excavation of the valley.

It is needless to say that the assays and analysis of the spring deposit strengthened these conclusions.

On returning from the field it was found that a very careful analysis of the waters of the Ojo Caliente Springs had been made a number of years ago, by W. F. Hillebrand. This analysis is subjoined. Several older analyses, evidently of less accuracy and detail, were made by Oscar Loew, analyst of the Wheeler Survey,[a] and they agree well, within limits, with Doctor Hillebrand's work and show that the composition of the springs is approximately uniform. Their temperature is said to vary from 90° to 122° F.

Analysis of water from Ojo Caliente, N. Mex.[b]

[By W. F. Hillebrand.]

Found (parts per million).		Hypothetical combination.		
			Parts per million.	Per cent of total solids.
SiO_2	60.2	LiCl	20.9	0.62
SO_4	151	KCl	59.9	1.76
PO_4	.2	NaCl	305.5	9.01
CO_3	2,153.5	$Na_2B_4O_7$	5.4	.16
B_4O_7	4.2	Na_2SO_4	223.3	6.59
Cl	231.4	Na_2CO_3	1,846.9	54.49
F	5.2	$Ca_3P_2O_8$	.3	.01
Fe_2O_3[c]	1.6	CaF_2	10.7	.32
Al_2O_3	.5	$CaCO_3$	43.0	1.27
Ca	22.8	$SrCO_3$	2.4	.07
Sr	1.4	$MgCO_3$	33.2	.98
Mg	9.5	SiO_2	60.2	1.78
K	31.4	Fe_2O_3	1.6	.05
Na	995.1	Al_2O_3	.5	.01
Li	3.4	CO_2 (bicarbonate)	775.6	22.88
	3,671.4		3,389.4	100.00

[a] Peale, A. C. Lists and analyses of the mineral springs of the United States: Bull. U. S. Geol. Survey No. 32, 1886, p. 195.
[b] Bull. U. S. Geol. Survey No. 113, 1893, p. 114.
[c] State of oxidation unknown; Fe_2O_3 all in sediment.

The water also contains traces of arsenic, nitrates, iodine (?), barium, and ammonium. No organic matter is present. Titanium, bromine, manganese, and sulphides were looked for but not found.

This analysis shows that the water is of the sodium carbonate type with minor amounts of chloride and sulphate of sodium. Particularly noteworthy is the large and unusual content of fluorine, corresponding to nearly 11 parts per million of calcium fluoride. The content in boron is also noteworthy. On the whole, it is a strong mineral water distinctly of the type of volcanic springs—that is, of springs that are common in regions of expiring volcanic activity. No one may question the competency of this water to deposit fluorite and also calcite, on the evaporation of excess carbon dioxide.

The evidence of erosion here adduced tends to show that these springs have been active over a very considerable period of time, possibly since the middle or later part of the Tertiary. The material eroded during the recession of the spring level was probably in large part the tuffs of the Santa Fe formation, which once filled Ojo Caliente Valley. Near Barranca station the top of these lake beds is at an elevation of about 6,800 feet. The elevation at the spring is 6,292 feet (Wheeler); consequently it must be admitted that at one time, probably at the close of the Tertiary, the lacustrine beds reached the level of the tufaceous spring deposit and the fluorite veins.

At the present time the springs issue at the foot of the hill of gneiss in the soft beds of the Santa Fe marl. No deposit of consequence can now form where the springs issue, owing to the manner of their utilization for medicinal purposes.

LEAD AND COPPER VEINS OF DOUBTFUL AFFILIATION.

In the central region of the Territory, in the longitudinal ranges following the Rio Grande, there exist a number of low-grade deposits of copper and lead which do not seem to belong to any of the above-described groups and which, so far as known, have no relationship with igneous rocks. They appear as veins, generally with steep dip and an east-west strike transverse to that of the ranges, and intersect the Paleozoic sediments, but, according to reports, few of them enter the underlying granite; at all events they are rarely productive in it. Little gold or silver is present. None of these districts have achieved great importance. To this group belong the lead veins in the Canyoncito district, north of Socorro. The veins, 3 to 5 feet thick, outcrop in gneiss and contain bunches of galena with quartz, barite, and fluorite.

In the northern part of the Caballos Range the Cambrian quartzite and Lower Ordovician limestone is cut by east-west veins, which contain quartz and a considerable amount of chalcopyrite, with oxidized ores. Most of the ore is found at the quartzite horizon. The upper limestone is traversed by similar veins which contain galena and barite gangue.

Herrick[a] describes similar veins cutting the Paleozoic limestone in the San Andreas Range, 10 miles southeast of Lava station. There are a great number of them and they are very conspicuous; some of them are 5 to 10 feet thick. They do not seem to penetrate into the granite.

Copper has accumulated in the quartzite just above the intersection with the granite. Lead is present at higher horizons. The ores are galena and chalcocite with their products of oxidation; the gangue in some of them is quartz, in others barite and fluorite.

No definite conclusion as to the genesis of these veins seems permissible until they have been studied in more detail. At present they have little economic importance.

PLACERS.

New Mexico has been richer in placer gold than some of the adjoining States, although its output has not been as large as that of Colorado. The gold-bearing gravels seem to be of earlier or later Quaternary age and simply represent the concentration of the débris carried down by erosion from the slopes of the several ranges which contain metallized areas. At

---

[a] Herrick, C. L., Am. Geologist, vol. 22, 1898, pp. 285–291.

certain places, as in the vicinity of the San Pedro Mountains, where erosion has been active without much interruption since Tertiary times, some of the lower gravels may possibly have been accumulated during that age. The Territory contains, so far as known, no well-defined ancient channel systems like those of California, in which the old stream beds have been deeply buried underneath later lava flows.

The principal placer districts are the Old Placers at Dolores and the New Placers at Golden, both in Santa Fe County; the Elizabethtown or Moreno placers, also smaller placers on Cimarroncito and Poñil creeks, Colfax County; the Hopewell placers, Rio Arriba County; the Hillsboro (Las Animas) placers and the more recently discovered Pittsburg placers near Shandon, Sierra County; the Nogal and Jicarilla placers, Lincoln County; and the Pinos Altos placers, Grant County. Less prominent placers are found at many of the gold-mining camps, for instance, at Jarilla, in Otero County; at Sylvanite, in Grant County; on Rio Hondo, Taos County; along Chama River below Abiquiu; along the Rio Grande above Embudo and along Galisteo Creek.

The production of the placers is largely a thing of the past. Two of the richest placer districts in New Mexico, the Old Placers and the New Placers, were discovered by Mexicans in 1828 and 1839, and for many years they yielded a heavy production. Prince[a] says that the former yielded from $60,000 to $80,000 annually in 1832 to 1835, and later not more than $30,000 or $40,000. In 1845 the production of both districts is given as $250,000. In the same year the town of Tuerto sprang up and 2,000 men congregated in these districts. The total production is a matter of conjecture. F. A. Jones estimates that the New Placers have produced $2,000,000 since discovery. The Old Placers probably yielded somewhat less. There is no production at present at the Old Placers, and only a small output at the New Placers near Golden.

The Moreno or Elizabethtown placers yielded heavily in the early sixties and continue to produce somewhat. During a short recent period the production rose to $100,000 a year in consequence of dredge operations, which now are discontinued. The total yield is believed to be $2,500,000.

The total production of placer gold in the Territory to date may be very roughly estimated as between $13,000,000 and $15,000,000. In recent years the production of placer gold has been as follows:

Production of placer gold in New Mexico, 1900–1907.

	Fine ounces.		Fine ounces.
1900	3,628	1904	7,228
1901	2,889	1905	4,805
1902	6,312	1906	1,297
1903	5,544	1907	946

The production at Hillsboro and Pinos Altos was also heavy in early years, but has now dwindled to small proportion. Operations on a fairly large scale were recently in progress at Shandon, on the east side of the Rio Grande at the foot of the Caballos Range, but have now been temporarily suspended. At most of the placer camps, except Elizabethtown and Hopewell and others in close vicinity to rivers, the water supply was short and dry washing has been extensively practiced.

The gravels are generally shallow, 50 feet being an unusual depth. At Elizabethtown some very deep gravels are reported to exist, but they have not been explored. The gravels are most commonly subangular. Much of the gold is of great fineness, above 900.

The placers of Hillsboro are apparently the only ones derived from the late Tertiary deposits. The pre-Cambrian deposits have yielded gold-bearing gravels at Hopewell, in Rio Arriba County, and at Rio Hondo, in Taos County. Almost the whole placer production is derived from the gold-bearing veins connected with the early Tertiary intrusives. The placers

---

[a] Prince, L. B., Historical sketches of New Mexico, 1883, p. 242. See also Bancroft, H. H., History of Arizona and New Mexico, San Francisco, 1889, p. 340.

of the Ortiz and San Pedro mountains (Old Placers and New Placers) have such an origin, the narrow veins being rich in irregularly distributed free gold of high fineness. Similar origin is proved for the gravels at Elizabethtown, Pinos Altos, Jarilla, and Sylvanite.

Considerable amounts of low-grade gravels remain, especially at New Placers, but lack of water seems to prohibit their utilization. In many of these districts costly experiments have been made with dry washers or even amalgamation and cyanide plants, but thus far without success. The only device which thus far holds its own is the simple Mexican "dry blower."

COPPER DEPOSITS IN SANDSTONE.

GENERAL FEATURES.

The sandstones of the "Red Beds" of New Mexico and adjoining States contain at many places disseminated copper minerals. These deposits are found chiefly in northern New

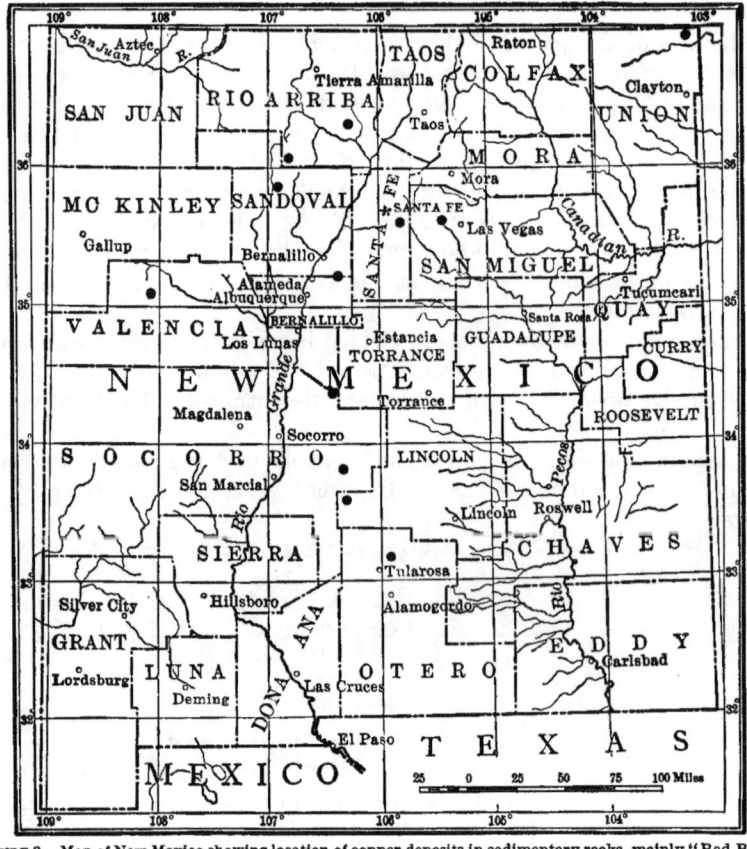

FIGURE 3.—Map of New Mexico showing location of copper deposits in sedimentary rocks, mainly "Red Beds."

Mexico, southern Colorado, northern Arizona, southern Utah, and western Texas. S. F. Emmons has recently well summarized their occurrence.[a] In New Mexico they are especially well developed and have been exploited more than in adjacent States. They attracted notice at an early date, the first occurrence being described by Newberry in connection with explorations for the McComb expedition.

Figure 3 shows the distribution of the more important deposits of this kind in the Territory. A comparison of this diagram with figures 1 and 2 shows plainly that in geographic location they are entirely independent of the precious-metal deposits of New Mexico. The

---

[a] Copper in the "Red Beds" of the Colorado Plateau region; Bull. U. S. Geol. Survey No. 260, 1905, pp. 221-232.

localities are as follows: Zuni Mountains, at Copperton, Valencia County; Sierra Nacimiento, Sandoval County; Abiquiu and Hermosa, Rio Arriba County; Glorieta, Santa Fe County; Tecolote and other districts west of Las Vegas, San Miguel County; Hansonberg and Estey, Oscura Range, Socorro County; Tularosa, Otero County; the east side of the Sandia Range, Bernalillo County; Abo Pass, Torrance County; and several reported occurrences in Union and Guadalupe counties.

The deposits show, in their position, no relationship to igneous rocks of any kind, nor do they follow fault lines or mountain ranges, though it is true that many of them occur in faulted regions. All of them, however, are in sedimentary beds of Paleozoic or Mesozoic age. As is well known, they are particularly common in the "Red Beds," but they occur also in Cretaceous sandstone (Dakota?) above the "Red Beds" (Bonanza mine, Tecolote district), and locally in the Carboniferous limestone below the "Red Beds." According to present views, most of the "Red Beds" form that part of the upper Carboniferous which has been designated the Abo sandstone by Willis T. Lee. The ores do not necessarily occur in the red beds which give rise to the name; more commonly, indeed, they are contained in the light-colored sandstones which are intercalated in it.

The developments are in general small, and the workings mostly surface pits. The deposits have some production to their credit, the larger part of it from richer, picked ore. The greatest production has come from the Sierra Nacimiento and amounts to about 6,300,000 pounds, partly smelted in the district. Lately several smaller plants have been erected near Tecolote and have yielded some copper. The deposit at Tularosa has had a notable production for several years. A little ore has been shipped from the Oscura Range. At several places large amounts of low-grade ore are available, though authorities differ regarding their content. It is believed that there are few large bodies containing over 2 per cent. As the ores generally are mixtures of chalcocite, malachite, and azurite, and as they occur in siliceous gangue, there are difficulties in the way both of concentration and of smelting, aside from the common scarcity of water; and owing to the same conditions many of the leaching processes have proved ineffective.

THE ORES.

The most conspicuous ores are malachite and azurite, but in every locality examined these are merely secondary products of oxidation after chalcocite, which is the principal, often the only primary ore mined. Probably all the chalcocite yields from 1 to 4 ounces of silver; in the low-grade ore the amount of silver is insignificant. The chalcocite is of epigenetic origin; it replaces carbonaceous material in the sediments or the kaolinitic or calcareous cement in the sandstone. The former occurrence is more striking and was observed at an early date. In the Nacimiento and Oscura districts massive nodules and lumps of chalcocite having this origin were abundant and shipments have been made of selected material. A specimen of such ore from the Nacimiento district is shown in Plate IV, C; in its lower part the illustration shows plainly the replacement of the coal by chalcocite, the structure of the wood being retained in some places. The upper part shows less plainly the replacement of the cement in a fine-grained sandstone by chalcocite, the clastic grains remaining in their original form.

The chalcocite is accompanied by small quantities of bornite, pyrite, and chalcopyrite, as described in detail by Mr. Graton under the Tecolote district. Of these only the pyrite shows crystalline form. In the Tecolote district the sandstone contains no coaly material, but the calcareous cement of the rock is replaced by the minerals just enumerated. All four sulphides were apparently introduced simultaneously, and there is no evidence that the chalcocite has replaced primary pyrite. Chalcocite here and there replaces the feldspar of clastic grains and also the disseminated kaolin contained in the feldspar.

The other accompanying minerals are few in number. Barite in flaky crystals was observed on the specimen from the Nacimiento district, and also occurs at Tularosa; gypsum is reported from several places.

In the Oscura Range, at Estey, the ore lies in a 500-foot series of red sandstone with some limestone beds, overlying Carboniferous limestone and dipping 25° E. The primary ore here also is chalcocite, which replaces the calcite cement. Bornite, pyrite, and very little chalcopyrite are sparingly present. The chalcocite also occurs in rounded nodules in a limestone breccia; very commonly the same ore mineral is associated with coaly matter in an arkose traversed by a net of calcite veins. The ores are found at three horizons, the cupriferous beds being 2 to 3 feet thick; ore also occurs in cross fractures and prominent joints. Mr. Graton believes that the faults antedate the metallization and that the solutions rose along the fractures. Other geologists have thought the faults of later origin than the deposits.

An unusual deposit, but undoubtedly belonging to this type, is that of Tularosa in Otero County, described on page 187. The locality is at the base of the Sierra Blanca in a region of red conglomerates and sandstones, probably of upper Carboniferous age. Carboniferous limestone is exposed close by the mine. In this sedimentary series appears a mass of diorite porphyry, probably of intrusive origin. In the sandstones chalcocite replaces the calcareous cement; in the limestone and in the porphyry are veinlets and stringers containing chalcocite, quartz, dolomite, and barite, also a black resinous hydrocarbon, all the minerals being of contemporary origin. Very little pyrite and chalcopyrite are present. There is no evidence of a replacement of pyrite by chalcopyrite.

GENESIS.[a]

It is not the intention to discuss in detail, in this place, the difficult question of the genesis of the copper deposits in the "Red Beds." To do this to best advantage it would be necessary to include the deposits of the adjacent States. The question is even wider than this, for there exists in the Permian and Triassic beds of Europe a series of apparently very similar deposits. An excellent résumé of deposits of this type has lately been given by S. F. Emmons.[b]

It may be pointed out that these accumulations of copper are found in strata which to some extent are land deposits and which contain beds of gypsum and salt, or in general chlorides and sulphates, indicating evaporating closed seas and desert climate. It has been shown that these copper deposits, so far as examined in New Mexico, are beyond all doubt epigenetic. They were assuredly not deposited with the sediments. They contain chalcocite as an original ore mineral, replacing carbonaceous fossil material or calcitic or kaolinitic cement in sandstone, conglomerate, and shale. The chalcocite is accompanied by calcite, locally also by gypsum, and in the Nacimiento and Tularosa districts, by barite. In Stelzner-Bergeat's "Handbook of ore deposits" barite is frequently mentioned in deposits of this kind. W. H. Emmons found that barite accompanied the related deposits of the Cashin mine in Colorado,[c] and in the copper ores of Red Gulch in the same State minute seams of the same mineral were observed cutting chalcocite, which had replaced coal.[d] It is thought to be probable that detailed examinations will result in the discovery of barite as a constant accompanying mineral.

It is further clear that these ores occur chiefly as disseminations in beds, by selective replacement, but also in fractures and fissures. The latter may be in the "Red Beds" proper, but are also, as shown by S. F. Emmons,[e] in the lower part of the Carboniferous, or, as shown by W. H. Emmons in the above-cited paper, in strata which overlie the "Red Beds" proper. Mr. Graton describes (p. 187) the interesting Tularosa occurrence, where the chalcocite occurs both in sandstone of the "Red Beds" and in an underlying, probably intrusive, diorite porphyry.

This group of deposits differs markedly from the ordinary types of veins and replacement deposits of the West. They contain chalcocite as an original ore mineral, deposited when the ores were first formed. They carry no gold and only a very small amount of silver. Finally, they show no relationship to igneous rocks. The single occurrence at Tularosa is counterbalanced by scores of other places where no igneous rocks exist. If heated ascending alkaline

---

[a] The following paragraphs set forth the views of the senior author, which are not wholly accepted by his co-author, Mr. L. C. Graton.
[b] Copper in the "Red Beds" of the Colorado Plateau: Bull. U. S. Geol. Survey No. 260, 1905, pp. 221-232.
[c] Emmons, W. H., Bull. U. S. Geol. Survey No. 285, pp. 125-128.
[d] Lindgren, Waldemar, Notes on copper deposits in Chaffee, Fremont, and Jefferson counties, Colo.: Bull. U. S. Geol. Survey No. 340, 1908, p. 18.
[e] Op. cit., p. 231.

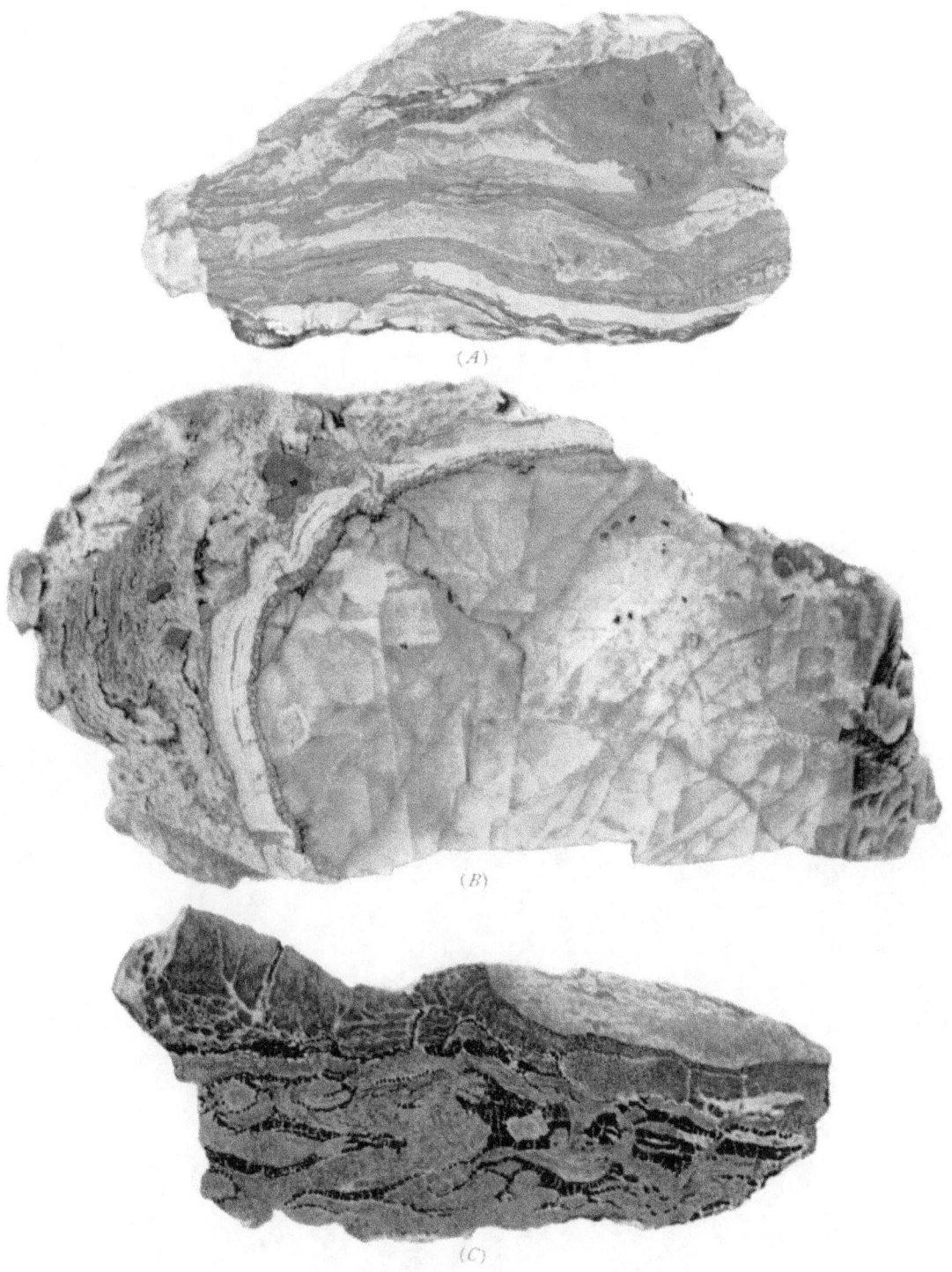

(A) SECTION OF OXIDIZED COPPER ORES IN SANDSTONE, ZUNI MOUNTAINS
(B) SECTION OF OXIDIZED ORE, ANDERSON MINE APACHE NO. 2 DISTRICT
(C) SECTION OF CHALCOCITE REPLACING COAL, NACIMIENTO DISTRICT

Natural size

waters of the sodium carbonate type are admitted as the cause of the ordinary fissure veins and replacement deposits, the same mode of origin can not be consistently attributed to the copper deposits under discussion; it would be necessary to explain why primary chalcocite is absent in the first class and characteristic of the second.

When the preliminary paper on the metal deposits of New Mexico was issued the question of the origin of the copper deposits in sandstone had not been fully considered. It was recognized that many of these deposits occurred along the flanks of uplifts which contained pre-Cambrian copper deposits and a relationship between the two classes was strongly suggested, for instance, in the Zuni Mountains and along the southeastern slope of the Santa Fe and Las Vegas ranges. It was stated[a] that the hypothesis of deposition simultaneously with the sediments by cupriferous sea water seemed improbable and that at least two other possibilities existed—that the copper was deposited by absorption in the clayey sandstone of cupriferous solutions ascending on fault fissures, or that it was leached from older deposits in adjacent ranges which might have been land areas during the accumulation of the "Red Beds," and was carried down by the streams and precipitated by processes of absorption in the sediments near the shores.

A view recently proposed by the senior writer is a modification of this hypothesis.[b] It was suggested by the study of the copper deposits in the "Red Beds" of Fremont County, Colo., where also pre-Cambrian deposits are contained in the adjacent old crystalline rocks. According to this view, copper sulphide minerals contained in the pre-Cambrian land areas were, during the degradation of the inclosing rocks, partly dissolved and thus supplied copper to the surrounding lake or sea, from which the metal was precipitated in the sediments that collected at its bottom, according to the first hypothesis. In part, however, the copper minerals were only disintegrated and were carried into this sea and deposited like other detritus as a part of the sedimentary rocks. When surface waters containing oxygen searched these beds copper would naturally be dissolved as sulphate and its precipitation as chalcocite would follow wherever agents of reduction like carbon were available. The reaction would be $4CuSO_4 + 5C + 2H_2O = 2Cu_2S + 2H_2SO_4 + 5CO_2$.

It is still maintained that the pre-Cambrian deposits supplied the "Red Beds" with minutely divided and widely distributed copper salts, and that the concentration has been effected by atmospheric waters, but it is not thought to be necessary that the depositing waters should have contained oxygen or that the ores should have resulted from solutions of cupric sulphate. In such vast accumulations as the "Red Beds" the descending waters would soon be robbed of their oxygen. The waters of the "Red Beds" are characteristic. They are rich in chlorides and sulphates, chiefly of sodium and calcium. In places where the circulation is deep they may well have a fairly high temperature. From the organic matter in the beds hydrogen sulphide would be added. Many springs and deep wells in the "Red Beds" contain this compound.

Under such circumstances cupric sulphate would probably not remain as such, but would be transformed to sulphide, which again would prove somewhat soluble in water of this type, but might be precipitated by suitable minerals under favorable conditions.

The waters of the "volcanic" type differ greatly from those of the saline beds and are nearly always rich in sodium carbonate and free carbon dioxide. It is believed these two types of water will result in different reactions when deposition takes place, so that, for instance, the waters rich in chlorine will deposit chalcocite, whereas in the "volcanic" type chalcopyrite and bornite will result. There is a possibility of settling this problem by direct experiment.

Briefly, then, the copper deposits in the "Red Beds" are believed to be due to concentration of minute traces of copper in certain strata or on fissures, by circulating atmospheric waters charged with chloride and sulphate.

---

[a] Lindgren, Waldemar, and Graton, L. C., A reconnaissance of the mineral deposits of New Mexico: Bull. U. S. Geol. Survey No. 285, 1906, p. 86.
[b] Lindgren, Waldemar, Bull. U. S. Geol. Survey No. 340, 1908, pp. 170–174. See also Schrader, F. C., Science, new ser., vol. 23, 1906, p. 916.

MINERALS OCCURRING IN NEW MEXICO.

The subjoined table summarizes the minerals of New Mexico, giving their composition and the locality or mode of occurrence:

Minerals of New Mexico.

Mineral.	Chemical composition.	Locality or occurrence.
Actinolite	Metasilicate of calcium-magnesium-iron	In pre-Cambrian schists.
Agate	Hydrated silica	Common in late volcanic rock.
Albite	Sodium-aluminum disilicate (sodium feldspar)	Constituent of granites, etc.
Allophane	Hydrous aluminum silicate	Fierro and Hanover.
Alunogen	Hydrous aluminum sulphate	Upper Gila River.
Amethyst	Silica	Black Range, Great Republic mine.
Andesine	Sodium-calcium aluminum disilicate (sodium-calcium feldspar).	In various igneous rocks.
Andradite	Calcium-iron orthosilicate (a variety of garnet)	Organ, San Pedro, and other districts.
Anglesite	Lead sulphate	Victorio, Cooks Peak district, etc.
Anorthite	Calcium-aluminum disilicate (calcium feldspar)	In various basic igneous rocks.
Anthracite	Carbon	Madrid coal mine.
Apatite	Calcium phosphate	Lake Valley district; common in igneous rocks in microscopic size.
Aragonite	Calcium carbonate	Magdalena, Kingston, and Cooks Peak districts.
Argentite	Silver sulphide	Georgetown, Kingston, etc.
Arsenopyrite	Sulpharsenide of iron	Ortiz and Virginia district; not common.
Asbestos	Hydrated magnesium silicate	Mimbres district.
Asphaltum	Hydrocarbon	Perea grant, Guadalupe County.
Aurichalcite	Basic zinc-copper carbonate	Magdalena district.
Azurite	Basic cupric carbonate	Magdalena, Santa Rita, and other districts.
Biotite	Black mica; orthosilicate of aluminum, iron, magnesium, and potassium.	Common in granites, etc.
Bismuthinite	Bismuth trisulphide	Mogollon (Wilcox) district (?).
Bismuthite	Bismuth carbonate	San Andreas Mountains.
Bloedite	Hydrous magnesium-sodium sulphate	Estancia Salt Lake.
Bornite	Copper-iron sulphide (Cu_3FeS_3)	Black Range, Cooney district, etc.
Bournonite	Copper-lead sulphantimonite	Cerrillos and Central districts.
Brochantite	Basic copper sulphate	Organ district.
Bromyrite	Silver bromide	Tierra Blanca district.
Brookite	Titanium dioxide	Copper Mountain district.
Calamine	Hydrous zinc silicate	Magdalena and Tres Hermanas districts.
Calaverite	Gold-silver telluride	Red River district (?).
Calcite	Calcium carbonate	Common in many deposits.
Carnotite	Hydrous potassium-uranium vanadate	Peralta Canyon, Cochiti district.
Celestite	Strontium sulphate	Sierra Oscura.
Cerargyrite	Silver chloride	Lake Valley, Kingston, and many other districts.
Cerusite	Lead carbonate	Cooks Peak, Magdalena, and other districts.
Cervantite	Antimony trioxide and pentoxide	Central district.
Chalcedony	Hydrous silica	Widely distributed.
Chalcocite	Copper (cuprous) sulphide	Santa Rita, Cooney, Burro Mountain, Tecolote, Sierra Oscura, etc.
Chalcophanite	Hydroxide of manganese and zinc	Magdalena district.
Chalcopyrite	Copper-iron sulphide ($CuFeS_2$)	Common in contact-metamorphic deposits and in veins.
Chlorite	Hydrous silicate of aluminum, iron, and magnesia	Common in decomposed igneous rocks.
Chrysocolla	Hydrous copper silicate	Common in copper deposits.
Chrysolite (olivine)	Magnesium-iron orthosilicate	Zuni Indian Reservation; olivine common in basalts.
Coal (bituminous and lignitic)	Carbon and hydrocarbons	In Carboniferous, Cretaceous, and Tertiary beds.
Copper		Santa Rita, etc.
Corundum (sapphire)	Aluminum sesquioxide	Said to occur in gravel beds near Santa Fe.
Covellite	Copper (cupric) sulphide	Jarilla.
Cuprite	Copper (cuprous) oxide	Santa Rita, etc.
Cyanite	Basic aluminum metasilicate	Said to occur in bromide district.
Descloizite	Basic lead-zinc vanadate	Lake Valley.
Dioptase	Basic copper silicate	Jarilla (?).
Dolomite	Calcium-magnesium carbonate	With limestone and in veins.
Domeykite	Copper arsenide	Pinos Altos (?) and Central (?).
Embolite	Silver chlorobromide	Lake Valley.
Emerald (beryl)	Beryllium-aluminum silicate	Said to occur in gravel beds in the Santa Fe marl.
Enargite	Copper sulpharsenate	Pinos Altos (?) and Central (?).
Endlichite	Lead vanadate-arsenate	Lake Valley and Hillsboro.
Epidote	Basic calcium-iron-aluminum orthosilicate	Many contact deposits and in altered igneous rocks.
Epsomite	Hydrous magnesian sulphate	Estancia Lakes.
Fluorite	Calcium fluoride	Mogollon and many other districts.
Gahnite	Zinc aluminate, a spinel	Cerrillos district.
Galenite	Lead sulphide	Organ, Cooks Peak, Magdalena, San Simon, and many other districts.
Glauberite	Calcium-sodium sulphate	Estancia Lakes.
Gold		Elizabethtown, Hillsboro, Ortiz Mountains, and many other districts.
Graphite	Carbon	Sandia Mountains.
Grossularite	Calcium-aluminum orthosilicate, a variety of garnet.	Present in contact-metamorphic deposits.
Gypsum	Hydrous calcium sulphate	White sands, Abe Lincoln mine, etc.
Halite	Sodium chloride	Estancia and Crater lakes, etc.
Halotrichite	Hydrous aluminum-iron sulphate	Upper Gila River.
Hematite	Iron sesquioxide	Common in contact deposits and in croppings of ore bodies.
Hubnerite	Tungstate of manganese and iron	White Oaks, Victorio, and Parson districts.
Hyalite	Hydrous silica	Central and Cochiti districts.
Hydrozincite	Basic carbonate of zinc	Magdalena and Tres Hermanas districts.
Ilmenite	Iron titanate	Common in eruptive rocks.
Ilvaite	Basic calcium-iron silicate	Fierro and Hanover.
Iodyrite	Silver iodide	Lake Valley.
Kaolin	Hydrous aluminum silicate	Common in many districts.
Labradorite	Calcium-sodium aluminum disilicate (calcium-sodium feldspar).	Common in basalts and other igneous rocks.
Lepidolite	Basic aluminum-potassium-lithium silicate	Copper Hill.
Limonite	Iron hydroxide	Many districts.
Magnetite	Iron oxide (ferric and ferrous)	Fierro, Hanover, and Jones districts; common in contact metamorphic deposits.
Malachite	Basic copper carbonate	Common in copper mines.

Minerals of New Mexico—Continued.

Mineral.	Chemical composition.	Locality or occurrence.
Manganite	Hydrous manganic oxide	Lake Valley.
Marcasite	Iron disulphide	Manzano Mountains (?).
Massicot	Lead-iron oxide	Chloride Flat (?).
Muscovite	White mica	Petaca, Nambe; common in granitic rocks.
Melanotekite	Silicate of lead and manganese	Hillsboro.
Microcline	Potassium-aluminum disilicate (potassium feldspar, triclinic).	Common in granites.
Mallardite	Hydrous manganese sulphate	Lake Valley (?).
Mimetite	Lead arsenate	Socorro (?).
Mirabilite	Hydrous sodium sulphate	Estancia Lakes.
Molybdenite	Molybdenum sulphide	Bromide, Rociada, Red River, Santa Fe, and many other districts.
Octahedrite	Titanium dioxide	Central district (?).
Oligoclase	Sodium-calcium-aluminum disilicate (sodium-calcium feldspar).	In many igneous rocks.
Opal	Hydrous silica	Cochiti and other districts.
Orthoclase	Potassium-aluminum disilicate (potassium feldspar monoclinic).	Common in igneous rocks.
Petzite	Silver gold telluride	Red River (?).
Plumbojarosite	Hydrous sulphate of lead and iron	Cooks Peak.
Polybasite	Silver sulphantimonite	Telegraph district (?).
Proustite	Silver sulpharsenite	Kingston district.
Psilomelane	Manganese dioxide	Caballos Mountains (?), Lake Valley.
Pyrargyrite	Silver sulphantimonite	Kingston and other districts.
Pyrite	Iron disulphide	Common in nearly all deposits.
Pyrolusite	Manganese dioxide	Lake Valley, San Lorenzo district, etc.
Pyrrhotite	Iron sulphide ($Fe_{11}S_{12}$)	Fierro and Hanover districts.
Quartz	Silica	Common in rocks and deposits.
Rhodochrosite	Manganese carbonate	Magdalena and Pinos Altos district.
Rutile	Titanium dioxide	Central district (?); microscopic in many rocks.
Scheelite	Calcium tungstate	Apache No. 2 district, San Andreas Mountains.
Siderite	Iron carbonate	Bromide, Granite Gap, and other districts.
Silver		Kingston, Lake Valley, Virginia, and other districts.
Smithsonite	Zinc carbonate	Cerrillos, Magdalena, Tres Hermanas, and other districts.
Specularite	Iron (ferric) oxide	In many contact deposits.
Sphalerite	Zinc sulphide	In many contact deposits and veins.
Staurolite	Basic iron-aluminum silicate	Picuris district.
Stibnite	Antimony sulphide	Cerrillos and Hachita districts.
Sulphur		Otero Sulphur Springs.
Talc	Hydrous magnesium metasilicate	Common in igneous magnesian rocks.
Tenorite	Copper (cupric) oxide	Santa Rita (?).
Tetradymite	Bismuth telluride	Sylvanite district.
Tetrahedrite	Sulphantimonite of copper	Bromide and Organ districts.
Titanite	Calcium titanosilicate	Microscopic in many igneous rocks.
Tourmaline	Complex silicate of aluminum and boron	Bromide, Cerrillos, Rociada, and other districts.
Tremolite	Calcium-magnesium metasilicate	Common in contact deposits.
Turquoise	Hydrous phosphate of aluminum	Burro, Cerrillos, Hachita, and Jarilla districts.
Uranophane	Hydrous calcium-uranium silicate	Jerome copper mine, Socorro County (?).
Vanadinite	Lead vanadate	Lake Valley, Mimbres mine, Georgetown, Hillsboro (?); Kelly, Socorro County, [a] McGregor mine, Grant County.
Wad	Hydrous manganese oxides	Lake Valley and other deposits.
Willemite	Zinc orthosilicate	Merritt mine, Socorro County, Tres Hermanas, Luna County.
Wollastonite	Calcium metasilicate	Tres Hermanas and other contact deposits.
Wolframite	Iron-manganese tungstate	Victorio district.
Xanthoconite	Silver sulpharsenite	Cerrillos district.

[a] According to Ward's natural science establishment.

PART II. DETAILED DESCRIPTIONS.

TAOS COUNTY.

By L. C. GRATON and W. LINDGREN.

GENERAL FEATURES.

Taos County lies in the most northerly tier of counties in New Mexico, bordering against Colorado on the north. Colfax and Mora counties adjoin it on the east, Rio Arriba County on the west and south. Its area is 2,283 square miles. The western part is mainly a basaltic plateau, trenched to a depth of 1,000 feet by the abrupt black chasm of the Rio Grande. The basaltic plateaus merge on the east into débris fans, in which some fertile alluvial valleys, like that of Taos, have been excavated. East of these gravel slopes the imposing Taos Range (Pl. V, A) rises to snowy heights and is continued south of the Ferdinand Creek Pass by the Mora Range, both of them forming the southern extension of the Sangre de Cristo Range of Colorado. Their summits form the boundary line against Colfax County. The greatest elevations are reached in Taos Peak (13,145 feet) and in Jicarilla Peak, in the southern part of the county (12,944 feet). Southwest of Taos, U. S. Mountain (elevation 10,734 feet) projects nearly to the Rio Grande as a salient from the main ranges. The larger part of the county was mapped by the Wheeler Survey, geologically and topographically (sheets 70 C and 69 D).

Mora Range and the southern part of the Taos Range expose gently dipping Paleozoic strata, bent in low arches and broken by north-south faults. Strips of pre-Cambrian granites and schists are brought up along these faults or exposed underneath the Paleozoic; the principal area extends from the country south of Taos down into the Santa Fe Range. The outlying mass of U. S. Mountain is wholly composed of these old rocks, as is almost the whole northern part of the Taos and Culebra ranges. (See Colfax County, p. 01.)

The Rio Grande traverses the county from north to south. The main water courses draining the western slope of the high ranges empty into this river and are named Red River, Lucero Creek, Pueblo Creek, and Ferdinand Creek.

The mining districts are few in number. The Red River district, containing gold-bearing veins in monzonite porphyry, andesite, and rhyolite, is 30 miles north of Taos, the county seat; the copper deposits in schist of Arroyo Hondo are 15 miles north of the same place; the Picuris or Copper Hill district lies about 12 miles south-southwest of Taos; the Anchor or La Belle district is in the extreme northeastern part of the county. Flaky placer gold is contained in the shallow surface gravels of the débris fans north of Taos, from Lucero Creek and Arroyo Hondo. These placers are said to have been worked by the Spanish padres during the early occupation of the county, and the little gravel piles from these old washings are seen at many places. Even at the present time a small annual output is maintained by Mexican miners.

Some of the bars along the Rio Grande below Cieneguilla are said to contain placer gold, but it is stated that the basalt bowlders in the river interfere with successful working. At Glenwoody, on the same river (Pl. V, B), near the southwestern boundary line of the county, a pre-Cambrian quartzitic rock outcropping on the east side of the stream is said to contain gold in profitable quantities, and it has been stated[a] that 50,000,000 tons are in sight. Even if this rock really does contain the precious metal, it is extremely unlikely that mining it will ever be found profitable.

The production of the county is small, in part because of the lack of transportation facilities.

---

[a] Jones, F. A., Mines and minerals of New Mexico, 1904, p. 159.

A. TAOS RANGE, TAOS COUNTY.
Looking north from Taos.

B. VALLEY OF RIO GRANDE AT GLENWOODY, TAOS COUNTY.
Looking north.

RIO HONDO (TWINING) DISTRICT.[a]

GEOLOGY.

The small but beautiful Taos Valley lies at the western foot of the junction of the Mora and Taos ranges at an elevation of about 7,000 feet. It is formed by the erosion of the great detritus fans near the confluence of Rio Hondo, Lucero Creek, Ferdinand Creek, and Frijole Creek, all draining the western slope of the main range. A few miles southwest of the picturesque old town of Taos the combined streams begin to trench sharply the basalt flows of the Rio Grande Valley, and a short distance farther west they join the Rio Grande in its precipitous gorge a thousand feet deep. On the road from the Barranca Railroad station to Taos, which crosses the river at Glenwoody, the relations of the younger formations are clearly seen. The vesicular and columnar basalt flow is 200 feet thick at the river and is underlain by 700 feet of white, sandy sediments, undoubtedly the Santa Fe marl. They show very sharply marked bedding, usually dipping at angles of 20° or 30°. Between the river and Taos the great débris fans of angular gravel from the Taos Range distinctly overlie the basalts to a depth of several hundred feet. The alluvial valley of Taos is carved in these great aprons of early Quaternary gravel deposits. The Santa Fe marl is held to be Miocene; the basalt to be late Tertiary or early Quaternary.

The Mora Range, from a point far to the south up to Pueblo Creek, presents the appearance of a heavily forested, even-crested block rising to average elevations of 10,000 feet. According to Stevenson, it is built up of gently dipping Carboniferous limestones.

North of Pueblo Creek there is a marked change; an imposing range, with alpine summits and signs of extensive glaciation, rises to elevations of over 13,000 feet (Taos Peak 13,145 feet) and continues, bending somewhat to the north-northeast, to a point beyond the Colorado line. North of Costilla Creek it is known as the Culebra Range. Where it crosses latitude 37° it is about 15 miles wide. The larger part of these mountains was mapped geologically by J. J. Stevenson, under the Wheeler Survey, on a scale of 1:253,440 (sheet 70 A).

The Taos and Culebra ranges are built up almost wholly of pre-Cambrian rocks, mainly granites and schists, except between Taos Peak and Elizabethtown, where the Carboniferous limestones continued northward from the Mora Range lap over the older rocks.

From the town of Taos the road to the Twining district runs northward for about 10 miles over the high débris fans of Lucero Creek and Rio Hondo; near the mouth of Lucero Creek these attain an elevation of 7,700 feet. The mountains present a bold front, suggesting a fault scarp, in which Rio Hondo has incised a deep canyon, very narrow at its mouth, but widening above and at its head opening out into glacial amphitheaters. The length from the mouth of the canyon to Twining, near its head, is about 9 miles. In front of the canyon the creek has cut a trench 200 feet deep into the old débris fans. The pre-Cambrian rocks at the mouth of the canyon consist of flat-lying alternating benches of amphibolite and white gneissoid granite; above this point a coarse light-colored granite prevails. The terminal moraine was located near the mouth of the canyon at an elevation of 7,750 feet. The rock shows an east-west sheeting with steep dip, and along one of these joint planes a narrow quartz vein was found containing pyrite, chalcopyrite, zinc blende, and molybdenite.

At South Fork, 5 miles from the mouth, amphibolite schists reappear, showing some copper stains and injected by white granite. At this place the South Fork mine was operated in 1905. A 10-stamp cyanide mill had been erected, the amphibolite, a quarter of a mile above the mill, on South Fork, being supposed to contain gold, which was to be recovered by the cyanide process. Above South Fork the rock is again a light-colored granite traversed by a number of straight and prominent northwestward-trending dikes of granite porphyry or quartz monzonite porphyry, some of them up to 80 feet wide. Undoubtedly these dikes are of relatively late age and probably belong to the same intrusion as the monzonites on Red River a few miles to the north. Two miles above South Fork these dikes cease, and at the abandoned town of Amizett amphibolite reappears, to continue 2 miles farther upstream to Twining. As elsewhere, the

---

[a] By Waldemar Lindgren.

amphibolite is intruded by coarse light-gray granite. High above Twining on the divide the pre-Cambrian rocks are capped by Carboniferous limestone.

ORE DEPOSITS.

Near Amizett, on the South Fork, and at other places, as noted above, there are small veins trending east and west, many of them situated close to the porphyry dikes which are so prominent in the vicinity. Those at Amizett are said to contain free gold and some of the ore has been worked in arrastres. Claims showing galena and stibnite are mentioned as occurring at the head of the South Fork. All these small deposits probably stand in close genetic connection with the porphyry dike and are believed to be of early Tertiary age.

The principal deposit of the district, at Twining (elevation 8,800 feet), the property of the Frazer Mountain Copper Company, is of a very different type from these small veins and its relationships favor the conclusion that it is of pre-Cambrian age. The property is extensively developed. The works at Twining consist of a small concentrating mill with three rolls and five stamps, as well as a 16-tuyere copper-matte smelter. For flux limestone and scrap iron were used; for fuel a mixture of charcoal and coke. Three charcoal kilns were built near the works. The plant had been in operation and yielded some production in 1904, but has since been idle. Low grade of ore and high costs militated against the success of the enterprise.

The deposit is developed by tunnels. The main tunnel, 300 feet above the smelter, is 1,000 feet long and there are several hundred feet of crosscuts. The croppings of the deposit are 200 or 300 feet higher. The ore body consists of a belt of amphibolite and chloritic schist, with disseminated copper ores. The croppings are reddish brown and are said to be traceable for three-fourths of a mile across the divide. The strike of the schists is N. 55° E. and they dip steeply to the northwest. The greatest width of the ore-bearing belt is 200 feet. One breast in the long tunnel is 60 feet wide, and of this 10 feet is bornite ore of higher grade.

According to an unpublished report by C. L. Herrick an average sample across 150 feet of schists gave 71.9 per cent of silica and 6.3 per cent of copper, also 0.03 ounce of gold and 0.3 ounce of silver to the ton. Another average sample from the "incline shaft" gave 4.70 per cent of copper, as well as 0.36 ounce of gold and 3.20 ounces of silver to the ton.

As best shown on the dump of the long tunnel, the average ore consists of a fine-grained smooth chloritic schist with lenticular masses and veinlets of quartz, calcite, and probably also siderite. The ore minerals occur in these and are chiefly pyrite, chalcopyrite, and bornite. The bornite has the appearance of being formed by secondary sulphide concentration. The deposit as a whole is probably of lower grade than the above analyses would indicate and the conditions for working are unfavorable.

RED RIVER DISTRICT.[a]

LOCATION.

The Red River district lies near the western border of Taos County, about 20 miles south of the Colorado line. It is situated in the heart of the Taos Range, but on the east side of the ridge. The district is drained by Red River, which cuts a deep gorge through the Taos Range and empties into the Rio Grande toward the northwest. The mines and prospects center about the town of Red River, a mining camp situated at an elevation of about 9,000 feet in a beautiful little valley at the confluence of several tributaries of Red River. The Black Copper mine, which is usually considered in a separate district, is in this report grouped with the other mines of the Red River district.

It is said that prospecting for placer gold began in this district in 1869, about three years after the discovery of gold at Elizabethtown,[b] and some gold was obtained. Prospecting for lode ores probably went on simultaneously. In 1879 a smelter was erected near the site where the town now stands, but it was operated only a short time. The town was located in 1894 and was settled in a rush. Much prospecting was done then and has continued more or less actively to the present time. The total production is probably not large.

---

[a] By L. C. Graton.   [b] Jones, F. A., op. cit., p. 153.

GENERAL GEOLOGY.

The main range is made up of pre-Cambrian granites, gneisses, and schists, but in the vicinity of Red River igneous rocks of later age are present almost to the exclusion of the older series. These younger rocks are divisible into two groups. The older of the two, probably of pre-Tertiary age, is represented by large intrusive masses of monzonite porphyry, usually containing considerable quartz. The more recent group comprises intrusive and probably also effusive dark-colored andesites and coarse andesitic breccias, some of which contain fragments of the monzonite porphyry; there are also flows of rhyolite and beds of rhyolite breccia. The evidence regarding the relative age of the andesite and the rhyolite is not wholly satisfactory, but there is little doubt that the andesite is the older.

That these various rocks in a general way overlie the pre-Cambrian core is shown a few miles below the town of Red River, where the stream cuts through them to the granite and schist. The upper or eastern part of the gorge of Red River is cut in these rocks, which, being much altered, assume striking red and yellow colors on the faces of the great cliffs and readily explain the reason for so naming the stream. This alteration, which is widespread and affects both the monzonite porphyries and the andesite and rhyolite, is not connected with surface decomposition but rather appears to be closely related to the solutions that formed the mineral deposits.

The presence of so much coarse breccia and the extensive alteration of the rocks may be an indication that this is a volcanic center. The fact that bad air is encountered in many of the mines during damp weather may be a further indication of volcanism.

MINERAL DEPOSITS.

GENERAL FEATURES.

The ore deposits, although occurring more commonly in the monzonite porphyry than in the rhyolite, are nevertheless believed to be genetically related to the rhyolite. They certainly cut the rhyolite or andesite in places, and hence fall in the late Tertiary.

The veins are commonly not very wide, ranging from a few inches to 6 feet. Many of them are variable in width and direction, and most of them have not been traced very far. The general trend is within a few degrees of north and south and the dip is not far from vertical, but this is not true of all the veins. Brecciation appears to have taken place along the fissures before they were filled, for many of the veins hold abundant fragments of the country rock around which the vein materials were deposited. Replacement by silica and pyrite of the walls and of these included fragments is very marked and not uncommonly the inclusions are converted almost wholly into silica, which is bluish because of the presence of finely divided sulphides and tiny turbid specks, probably the last remnants of the original rock. Close to the main vein and usually parallel with it are in many places fractures along which replacement but little or no filling has occurred. From these altered inclusions and from the walls quartz, the predominant gangue mineral, radiates and fills the spaces intervening, and in these spaces minute, hairlike prisms of quartz and beautiful little crystals of sulphides are sometimes found. A striking feature is the relatively small amount of sulphides in the filled portion of the veins as compared with the replaced country rock. Whether or not the values are similarly distributed was not ascertained, but the apparent absence of native gold in the unoxidized ores makes it reasonable to suppose that the replaced rock is richer than the deposited quartz. Along with the quartz a carbonate, probably dolomite, is deposited in small amount in some of the fissures. Fluorite constitutes in certain veins an important portion of the gangue and is commonly associated with calcite. In such veins quartz is not generally plentiful. Adularia was not observed. Of the ore minerals pyrite is most abundant and usually occurs in grains and crystals of small size. Galena is present in numerous places and in some veins is the most important sulphide. Sphalerite, while apparently not of general distribution, is in certain veins comparatively plentiful. Chalcopyrite is rarely abundant, but in many places occurs in numerous small grains and in crystals in the quartz druses. Molybdenite has been found as thin coatings along the walls of narrow fractures. A telluride, possibly petzite, has been found at one place. On the whole these veins are characterized by rather sparing amounts and disseminated distribution of the sulphides.

Gold is the predominant metal, but silver, copper, and lead raise the value of the ore in some places and in one or two deposits lead is the principal product. The values of the ores vary between wide limits. Some of the ore is simply amalgamated and yields from $4 or $5 to the ton upward. Many of the ores are so rich that they warrant the expensive trip to the smelter without concentration. In the summer of 1905, at the Independence mine, on Bitter Creek, ore said to assay $400 to $500 a ton was being sacked ready for shipment.

The rocks which have been altered by solutions believed to be related to the mineral deposits are very susceptible to decomposition at the surface, and this change proceeds at so rapid a rate that the rock may crumble and decompose before the larger grains of pyrite are oxidized. On the other hand, the rapid erosion to which this district is subjected, as a result of the heavy precipitation and the steep gradient of its streams, removes the products of decomposition almost as fast as they are formed, leaving only a shallow zone of oxidation overlying the undecomposed rocks; and hence nearly all the openings reach solid, unoxidized sulphide ores.

Mining developments extend over a considerable area. The Black Copper mine is situated about 6 miles south of the town of Red River, a mile east of the river. Much work has been done on the slopes of Gold Hill, one of the high peaks of the main range, and particularly on Black Mountain, a spur of Gold Hill about 3 miles southwest of the town. Another cluster of prospects is along the river near Red River. Some work has been done 12 miles downstream, near a tributary known as Alum Gulch, and considerable activity manifested near the head of Bitter Creek, some 6 miles northeast of the town of Red River. In the summer of 1905 little aside from assessment work was being done in the district except at the Independence mine on Bitter Creek.

BLACK COPPER MINE.

The Black Copper mine, situated about 6 miles south of Red River town, has been closed for several years. It is reported that the shaft is about 250 feet deep and that considerable development work has been done to that depth. A 5-stamp amalgamating mill, with Bartlett concentrating table, constitutes the surface developments. The mine is situated at the contact of granite and monzonite porphyry. The parallel veins are reported as lying mostly in the granite, but also cut dikes or tongues of the porphyry which penetrate the granite. As seen on the dump the ore presents the characteristic features of radiating quartz surrounding pyritized and silicified country rock. Dolomite occurs with some of the quartz. Pyrite, galena, greenish-yellow sphalerite, and a little chalcopyrite are present. The porphyry wall rock is considerably impregnated with pyrite in places. Some very rich gold ore is said to have been taken out when the mine was working.

ANACONDA GROUP.

The Anaconda group, consisting of the Copper King and Copper Queen claims, is situated on the bank of Red River opposite the town. The development consists of a tunnel 1,030 feet long, mostly if not wholly in monzonite porphyry. This tunnel is said to cut an 85-foot zone showing sulphides and copper carbonates. The material on the dump shows that the porphyry is much silicified and impregnated with many tiny stringers of pyrite containing some intergrown chalcopyrite. In addition to this, narrow quartz veinlets carrying molybdenite and chalcopyrite are present in abundance. Chlorite and sericite are also plentiful along with the quartz. These altered portions appear to represent sheeted zones in the porphyry. It is said that the ore thus far encountered is of too low grade to be profitable. Some prospecting by means of well-drilling machines has been done on the hill above the tunnel.

OTHER PROPERTIES NEAR RED RIVER.

The Ione tunnel, just back of the town, follows a quartz vein in brecciated porphyry, which in places is heavily impregnated with pyrite. The quartz is drusy and much of it is dark, probably owing to the presence of finely disseminated sulphides.

At the Ajax tunnel, on Pioneer Creek, a drusy quartz vein cuts altered quartz monzonite porphyry. It is apparently not rich. The Tomboy tunnel, also on Pioneer Creek, is 400 feet long and at the breast cuts a drusy quartz vein carrying pyrite and a little chalcopyrite between silicified walls.

The Sampson mine is about one-fourth mile south of the river, a mile below town. A 65-foot incline follows a quartz vein considerably wider than the average, being fully 6 feet wide. On one side of the vein the white quartz that occupies the original fissure is clearly defined from the darker silicified wall rock, which, however, is pyritized and is said to average $4 in gold to the ton. On the other side there is no sharp line of demarcation between the vein filling and the silicified country rock, which as usual is quartz monzonite porphyry. Fragments of the porphyry held in the fissure have been pyritized and silicified and have acted as nuclei around which the quartz has deposited radially. Some of the dark mineral in these inclusions is reported to be a telluride, but this statement has not been verified.

On the Victor No. 2 claim, 3 miles down Red River, galena ore has been struck. It is said to run 60 per cent of lead and 1 per cent of copper as chalcopyrite, with 8 ounces of silver to the ton and a trace of gold. This property was not visited, but some specimens were seen. The galena is fine grained and might be expected to carry more silver than is stated above.

Specimens of ore from the Whip-Poor-Will claim show purple fluorite in quartz-carbonate vein matter holding pyrite and a little chalcopyrite.

HIGHLAND CHIEF.

The Highland Chief claim, formerly known as the Bueno, is about 8 miles downstream from the town of Red River, and 2 miles north of the river up Alum Gulch. This deposit is in rhyolite, which is in many places brecciated and altered. A quartz vein 1 foot wide, striking southwestward, and dipping about 50° SE., cuts the rhyolite. It is fairly well defined, but the walls are much altered. Greenish fluorite occurring along with the quartz holds a little pyrite. The quartz contains partly oxidized pyrite, also chalcopyrite and galena in places. The gold is said to be contained in the pyrite. The developments are slight.

JAY HAWK MINE.

The Jay Hawk mine is situated on Black Mountain about 3 miles southwest of the town, though by the road it is probably 5 miles. Considerable development work in the way of tunnels and shafts has been done. The deepest shaft is 100 feet deep. The prevailing rock is monzonite porphyry, which is cut by dikes of andesite. So far as could be seen these dikes are not related to the ore deposits. The ore consists of drusy and fibrous quartz, inclosing mineralized fragments of both monzonite porphyry and andesite. A considerable amount of ore has been hauled down the mountain and treated in the June Bug mill. It is said that good returns were received.

GOLDEN TREASURE MINE.

The Golden Treasure mine, on Black Mountain, has similar ore to that of the Jay Hawk. The country rock is brecciated andesite much decomposed. Some ore was extracted a few years ago.

OTHER PROSPECTS ON BLACK MOUNTAIN.

Prospects situated near the head of the south fork of Placer Creek are said to show veins with good assays. At the Bunker Hill, near the divide between Placer and Pioneer creeks, the workings are said to penetrate granite and schist and to encounter good ore. At the Deadwood claim two arrastres were working in Placer Creek when this district was visited. At the Blue Rock claim, still farther down Placer Creek, a tunnel was being driven into decomposed gneiss rock, containing quartz with pyrite along little seams; the pyrite carries gold, some of which is visible to the naked eye.

In Long Canyon, which lies southwest of Black Mountain and not far from the Twining camp, some good prospects are said to have been discovered.

BITTER CREEK AND INDEPENDENCE MINE.

Many places along Bitter Creek have been prospected by tunnels. Where veins have been struck the ore appears to be similar to that already described—radiating, drusy quartz inclosing fragments of country rock and holding pyrite. The Memphis mine, about 4½ miles from Red River, has received considerable development by shaft and some good ore from it is said to have been put through the June Bug mill. It is said that some of the gold was present as a telluride.

A rich strike at the Independence mine was made in 1904. This property, on a branch of Bitter Creek about 6 miles northeast of Red River and 12 miles northwest of Elizabethtown, has been acquired by a Philadelphia company. In August, 1905, a force of men were at work putting up buildings and bringing in mill machinery from an abandoned Elizabethtown mine. The country rock is a porphyry, probably related to diorite or monzonite. A narrow stringer, containing quartz in some places and in others only kaolin, cuts the decomposed porphyry, striking S. 12° E. and approximately vertical. This was followed by an adit for 350 feet without obtaining an encouraging assay and then hard lumps of quartz containing pyritic ore were struck. The ore proved to be a mixture of pyrite and a telluride and to carry high values in gold. From this point the vein has been exposed for 80 feet farther and at the breast was 2 feet 8 inches wide. This width comprised the rich streak and included not only a narrow quartz-filled fissure, showing beautiful comb structure and numerous druses, but the surrounding blue, silicified wall rock, which is impregnated with pyrite and the telluride, and which carries better values than the quartz itself. Outside of this zone the altered porphyry, less silicified than the "blue ore," also carries low values. In the druses of the vein perfect little sphenoids of chalcopyrite are found. The telluride is present in very small lustrous grains of dark-gray color. It is so intimately associated with pyrite that a separation of the mineral for the purpose of analysis would be difficult. A qualitative test proves the presence of tellurium, and the mineral may be petzite, as it has been called.

PLACERS.

As was stated at the beginning of this section, placers were the source of the first gold obtained in the Red River district. Washing of the streams has gone on intermittently ever since. It is probable, however, that the total production of placer gold in this district is not large.

Recent operations with a giant have been conducted at "the narrows" about 4½ miles up the river from town. Some rocking and sluicing has also been done on Placer Creek, which drains the prospected region of Black Mountain. It is said that several attempts have been made to recover fine gold from the river bed below town. No placer work is being done at present.

ANCHOR DISTRICT.

The district variously known as the Anchor, Midnight, or Keystone lies in northeastern Taos County, about 8 miles north-northeast of Red River. It is situated on the west slope of the Cimarron Range at an elevation of about 9,700 feet. A small camp known as Anchor was established in the midst of a group of prospects, but it is now almost deserted. At the time of visit a little assessment work was being done and an arrastre was handling such ore as could be dug out here and there on the Little Gem claim.

The prevailing rock of this district is a porphyry, probably of monzonitic composition. It is rather badly decomposed and few of the workings expose fresh rock. The mineral deposits consist of narrow quartz-pyrite veins in this porphyry, wholly similar to the veins in porphyry at Elizabethtown. These veins range from a fraction of an inch up to several inches in width. The wall rock is silicified and sericitized for an inch or two from the vein, and beyond that is bleached and softened. So far as known the only mineral besides quartz which the veins contain is pyrite. In most places, however, this sulphide is oxidized to pseudomorphic masses of hematite or limonite which inclose the gold originally contained in the pyrite. Pyrite also commonly impregnates the surrounding rock for a slight distance, is likewise auriferous though probably not so rich, and is oxidized along with the pyrite in the quartz. The ores of the camp are therefore to a large extent free-milling, although it is doubtful if all the gold can be recovered by simple amalgamation from thoroughly oxidized ores. The attrition in the arrastre probably increases the percentage of extraction.

On the Edison claim an east-west lode is made up of these quartz veins. The quartz holds hematite in which are contained particles of free gold, and is said to average $14 a ton. A 10-stamp amalgamation mill was erected, but was operated only eighteen days.

The Rosita claim is on the continuation of the Edison vein. It is developed by a 65-foot shaft and is said to have encountered good values.

The Little Gem claim is near the Rosita and has a short tunnel and a 40-foot shaft on a small vein parallel to the Edison vein. A 3-inch streak of the quartz carrying oxidized pyrite is said to assay from $120 to more than $200 a ton. A little of this ore is being put through an arrastre.

The Lillian or Keystone tunnel, 1,900 feet long, attains a maximum depth of 160 feet near the breast. It is all in porphyry and is said to have cut a few ore stringers.

The Cashier group is prospected by means of a shaft 85 feet deep. It is said that the lode is 27 feet wide and carries free-milling gold with values ranging from $4 to $20 a ton. An arrastre was used to mill the ore. According to reports operations were suspended because of too much water in the shaft.

The Midnight mine, situated three-fourths of a mile up the gulch from the Edison mine, is reported to have been operated profitably for three years on ore of good grade similar to that of the Edison. The ore was treated in a 10-ton Huntington mill. The shaft, 185 feet deep, extended below the zone of oxidation to sulphides. Bad managment is given as the cause of closing the mine.

LA BELLE DISTRICT.[a]

The La Belle district is situated at the extreme eastern edge of Taos County, just over the divide from the Anchor district and about 12 miles due north of Elizabethtown. A few miles to the northeast rises the beautiful Costilla Peak. The district experienced a boom in 1895 and a town was built. The inhabitants now number about half a dozen.

Mica diorite and diorite porphyry are the rocks most commonly seen. In some places these rocks are considerably brecciated. The Aztec mine, close to the town, is said to have had some rich ore, as quartz veinlets in porphyry. The mine has been abandoned for a number of years. The Criterion claim, higher up on the mountain, was developed a little in 1904, but was not working when visited. Quartz with decomposed pyrite was seen on the dump and is said to carry good gold values. In 1905 the Snyder tunnel in Gold Gulch, north of La Belle, was being continued. It is a crosscut tunnel 800 feet long, being driven to cut the supposed vein which furnished the rich float found on the hill above. A few quartz-bearing seams have been cut. The rock is diorite and diorite porphyry breccia.

It is probable that the veins of this district are of the same type as the quartz-pyrite veins of Elizabethtown.

COPPER HILL OR PICURIS DISTRICT.[a]

The Copper Hill district is situated in southern Taos County, between the Rio Grande and U. S. Mountain. It lies a little north of east from Embudo station, on the Denver and Rio Grande Railroad, and about northwest of the Indian pueblo of Picuris, by which name the district is often designated.

Although the presence of copper-stained quartz veins in the region has been known for many years, little attempt was made to develop them until about 1900. In that year the Copper Hill Mining Company began development operations and erected a concentrating mill, which burned soon after completion. The company failed and the property is now held by New York capitalists. About two years later the Green Mountain Copper Company prose-

---
[a] By L. C. Graton.

cuted development work in a small way. A little prospecting has been done by other parties. The district is credited with practically no production, and in 1905 no work was being done in it.

The district lies near the north end of the southwest terminal lobe of the Santa Fe Range. U. S. Mountain, which lies a little north of east of the district and reaches an elevation of over 10,700 feet, is a somewhat isolated mass at the north end of this range. On the west the rough highland which flanks the peak proper breaks off sharply to the deep canyonlike valley of the Rio Grande, which here has an elevation of about 5,500 feet. The Copper Hill district is situated on this highland close to the valley rim and has an average elevation of about 8,000 feet.

The geology of this region is not known in detail. The pre-Cambrian rocks, which in most places in the Territory consist of granite and granitic gneiss with more or less basic intrusives, here comprise in addition ancient sediments which have been profoundly metamorphosed, presumably by the intrusive pre-Cambrian granites. East and south of U. S. Mountain coarse alluvial-fan conglomerates of Tertiary age rest both on Carboniferous grits and limestones and on the pre-Cambrian gneissic granite, and in places have been capped by flows of basalt, which probably is also Tertiary.

The ancient metamorphic rocks prevail in the Copper Hill district. They consist of quartzites, in many places conglomeratic, and of knotty schists which contain quartz, sericite, tourmaline, garnet, corundum, sillimanite, andalusite, staurolite, and perhaps cordierite. The schists give way toward the west to garnetiferous slates, which probably are less metamorphosed equivalents of the schists. A slate deposit a little southwest of the district is said to have been quarried to some extent and to have furnished slate of good quality.

The mineral deposits consist of veins of glassy quartz carrying copper, silver, and gold. Quartz veins to the east of Copper Hill carry abundant black tourmaline prisms, but this mineral seems to be absent in the ore-bearing veins. Chalcocite, cuprite, malachite, and chrysocolla are present in the veins and it is said that argentite and tetrahedrite also occur. A little limonite is present in places and has probably resulted from the oxidation of pyrite. In such material carrying iron stains gold values are sometimes encountered. Developments have not passed below the zone of partial oxidation and it is consequently impossible to decide what the exact character of the unaltered ore may be. In places, however, the extent of oxidation is so slight as to make it seem doubtful that the solid lumps of chalcocite there occurring can have been produced by enrichment of a previously existing ore. It appears more probable that the chalcocite is an original constituent of the vein. In fact, the resemblance is very close to pre-Cambrian quartz veins that carry bornite and chalcocite in the Virgilina district of Virginia, where the primary nature of these sulphides is plain.

There is reason to believe that these deposits are of pre-Cambrian age, and it seems probable that they were formed immediately after the metamorphism of the inclosing rocks and were genetically dependent on the same agents.

The property of the Copper Hill Mining Company is developed by a 180-foot and a 60-foot shaft and a 350-foot adit. The country rock, of alternating quartzite and poorly fissile schists, strikes slightly north of west and dips very steeply to the south. On the Champion claim an approximately vertical vein of northerly strike has been followed by an adit for 350 feet with a maximum attained depth of about 70 feet. The vein ranges in width from 8 inches to 3 feet. It splits and forks considerably, but produces no apparent alteration of the glassy quartzite wall rock. Chalcocite and derived cuprite in massive form are carried in the green-stained quartz, but the average value appears to be low. Near the breast of the adit, where a bunch of ore considerably better than the average was encountered, silver values were said to be very good and were attributed to argentite, but this mineral was not observed by the writer. Practically no stoping has been done on the vein. On the Oxide King claim, a short distance to the south, a 180-foot shaft has been sunk to explore a northward-striking vein that dips about 50° W. Some ore similar to that in the Champion claim was found, but no important development was done.

The Wilson mine lies just east of the property of the Copper Hill Company. Several shallow shafts have been sunk on a vein of northerly strike, which lies mostly in quartzite but cuts knotty schist also. The vein shows copper and iron stains and is said to carry good gold values in places. The dark, glassy quartzite wall rock also contains iron-stained streaks and some free gold. The average value, however, is doubtless low.

About 4 miles northeast of the Copper Hill and Wilson mines oxidized copper ores with horn silver and gold and a little copper glance have been found. On Copper Mountain veins similar to those of the Copper Hill mine were encountered. Chalcopyrite, gold, and silver have been reported 1½ miles west of the Wilson mine. These localities were not visited by the writer.

GLENWOODY DISTRICT.[a]

A camp called Glenwoody was established in 1902 on the Rio Grande almost west of the Copper Hill district and a few miles above Rinconada. A wide band of quartzite, intercalated with other greatly metamorphosed pre-Cambrian sediments, was said to carry $1.40 to $3 a ton in gold and to yield satisfactorily to cyanide treatment. A water-power plant was installed and a mill built, but the amount of gold actually recovered was far too little to pay. The most favorable account is that the mill return was 40 cents a ton, although some people in the region have never been convinced that the quartzite contains any gold.

The general geology at this point on the river is interesting. Heavy pre-Cambrian schistose rocks rise abruptly on the east side 900 feet or more above the river, but on the west side these rocks, where present, are but little above the water level. On the west side is a thick series of basaltic flows, whose top forms a plain about 900 feet above the river level. Patches of basalt remain on the east side, resting against the side of the valley. The probable conclusion is that the Rio Grande follows approximately the line of an old fault, and that the floods of basalt coming from the west were stopped by the wall of schist on the east of this fault. The schists contain quartz, biotite, paragonite (?), magnetite, garnet, epidote, sillimanite, staurolite, corundum (?), and a number of other minerals. The grains are so crowded with inclusions of the groundmass that satisfactory optical determinations can not be made, but it is probable that andalusite, cyanite, and cordierite are present. The schists are generally similar to those at Copper Hill and undoubtedly belong in the same series.

COLFAX COUNTY.

By L. C. GRATON.

GENERAL FEATURES.

Adjoining Taos County on the east is Colfax County, which contains 3,897 square miles. On the south it is bounded by Mora County, on the east by Union County, and on the north by the State of Colorado. The western boundary of the county lies along the summits of the lofty Mora, Taos, and Culebra ranges, these being really separate names for the main chain of the Rocky Mountains extending southward from Colorado, where it is known as the Sangre de Cristo Range. From peaks reaching over 13,000 feet in height these ranges slope rather steeply to the east; from their foot, at elevations of 7,000 to 8,000 feet, the vast plateau drained by Canadian River slopes gently eastward to elevations of 5,500 to 6,000 feet in the eastern part of the county. Almost the whole of Colfax County is drained by the numerous branches of the Canadian.

The main range as far north as Elizabethtown is composed mainly of flatly arching Carboniferous strata, but north of this place the pre-Cambrian granites and schists predominate. Near the Colorado line the Culebra Range, here about 15 miles wide, is sharply defined, with a narrow band of upturned sediments at its eastern foot, similar to the front ranges of Colorado. A short distance north of the Colorado boundary line at Culebra Peak an elevation of 14,049 feet is attained. Farther south this typical structure is missing; the narrow longitudinal valley of Moreno and Cieneguilla creeks extends for 30 miles on the east of the main divide. East of this valley rises the Cimarron Range, which has almost the same length and attains its greatest height of 12,491 feet in Baldy Peak, Elizabethtown. The Cimarron Range is narrow in the

---

[a] By L. C. Graton.

north, but at the south end widens to nearly 10 miles. It is composed very largely of an intrusive rock, a monzonite porphyry, which has domed and disturbed the strata up to and including the latest Cretaceous. To the south the Cimarron Range is continued by the great basalt flows of Ocate Creek, in the southwestern part of Colfax County, which rest on Dakota (?) sandstone, on shales of Montana or Colorado age, or on rocks which have been thought to belong to the Laramie (?) formation. The same formations extend eastward to the county line east of Raton, and here, in the northeastern part of the county, the Cretaceous is again covered by extensive basalt flows, which in places attain elevations of over 9,000 feet.

The Cretaceous formations of the county contain beds of excellent coal, which is mined on a large scale near Raton and at Dawson.[a] The coal is of a coking quality and of great importance for the western industries. The production in 1905 amounted to 1,031,829 short tons. In 1907 the greatest output, of 1,844,550 tons, was reached. In 1908, 1,781,635 tons were produced.

Near Springer cement rock and gypsum beds occur.

The metal-mining districts center around Elizabethtown, in the western part of the county, and are chiefly gold placers from which a considerable production is annually obtained. Since their discovery in 1866 these placers are estimated to have yielded between two and three million dollars in gold. Northeast of Elizabethtown are the Poñil placers; east of it the Baldy or Ute Creek subdistrict; southeast of it the Cimarroncito district. All these localities are grouped along the Cimarron Range and the ore-bearing area seems to coincide in extent with the monzonite porphyry.

ELIZABETHTOWN DISTRICT.

LOCATION AND HISTORY.

The Elizabethtown district, one of the best known and most important mining centers in New Mexico, is situated in the western part of Colfax County, in the heart of the Cimarron Range. Mining developments are limited to the slopes of Baldy Peak, an imposing mountain mass which rises to an elevation of nearly 12,500 feet. Elizabethtown lies 4,000 feet lower, in the Moreno Valley, at the western foot of the mountain. Baldy, a camp high on the east side of the mountain, is near the head of Ute Creek, which empties into Cimarron River. Poñil Creek, which drains the northeastern part of the district, also flows into the Cimarron. On the west side several small, steep creeks flow down the mountain side into Moreno River. Most of these valleys carry water during only a portion of the year. The more important ones, named from south to north, are Willow Creek, Mexican Gulch, Anniseta Gulch, Grouse Gulch, Humbug Gulch, Big Nigger Gulch, and Pine Gulch. A few miles south of Elizabethtown, Moreno River, uniting with the northward-flowing Cieneguilla Creek, turns eastward and becomes known as Cimarron River.

Elizabethtown is 35 miles northeast of Taos and 53 miles west-northwest from Springer, on the Santa Fe Railway. At present Dawson, 40 miles to the east, is the nearest railroad station.

According to the interesting narrative in the book by F. A. Jones,[b] gold was discovered in this district in 1866. It seems that an Indian found rich copper float on the upper slope of Baldy Peak, probably at what is now known as Copper Park, on Poñil Creek. He exhibited his find at Fort Union, early in the sixties, and some of the men stationed at the fort located claims where this float had been found. In October, 1866, three men, including a Mr. Kelly, were sent by the owners to do assessment work on these claims. Kelly attempted panning in Willow Creek and found the ground rich. In the spring of 1867 a great rush took place and the boom which resulted lasted for years. Elizabethtown was built and was for some time the county seat of Colfax County. Although some lode locations were made, including the famous Aztec mine, most of the early work was placer mining and required an abundance of water—more than was furnished by the drainage from Baldy Peak. It was found that a large

---

[a] No reports on these coal fields have yet been issued by the Geological Survey.
[b] Op. cit., pp. 141-144.

supply of water at sufficient elevation could be obtained near the headwaters of Red River, about 11 miles west of Elizabethtown. A company was formed and a ditch to convey this water to the placers was completed in 1868. It has a length of over 41 miles and was designed to carry 600 miner's inches. Built at a cost of over $200,000, the "big ditch," as it was called, has stood as a monument to the energy of those early prospectors in the Elizabethtown district. The cost of maintenance proved so great that the ownership changed hands two or three times and in 1875 the ditch was bought by Mathew Lynch, who with his brothers operated it successfully for several years in working placers opposite Elizabethtown.

This region was embraced in an old Spanish grant, known as the Beaubien and Miranda grant. It is now the property of the Maxwell Land Grant Company, with headquarters at Raton. Until recent sales of coal lands in the eastern part were made, the grant comprised nearly 1,750,000 acres. Within the last ten years the company has adopted a liberal policy toward prospectors. The regulations concerning the location and tenure of claims are more favorable than those of the United States mining laws, and over 500 claims have been located on the grant in this way.

In point of total production this is one of the leading mining districts in the Territory. It is estimated that the placers have produced about $2,500,000, and the lode mines probably over $2,000,000, making the total nearly $5,000,000.

Very little work was being done in the district at the time it was visited. A dredge was working in Moreno River just below Elizabethtown during the open season; a little hydraulic

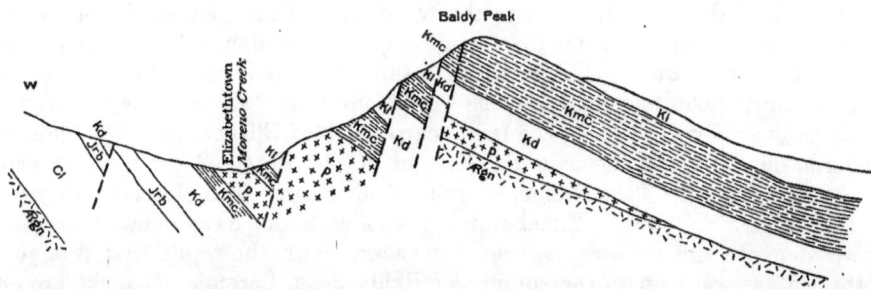

FIGURE 4.—Diagrammatic section through Cimarron Range at Elizabethtown. Only most important porphyry masses shown. Kl, Laramie (?) formation; Kmc, shales of Montana and Colorado age; Kd, Dakota (?) sandstone; Jrb, "Red Beds;" Cl, Carboniferous limestone and grit; Ægn, pre-Cambrian granite, gneiss, and schist; p, porphyry. Dashed lines are probable lines of faulting.

work was done in Humbug Gulch, when there was enough water, and here and there on the hills a few men were exploring prospects.

GENERAL GEOLOGY.

The geology of the Elizabethtown district is rather complex. Folding, faulting, intrusion, and contact metamorphism unite to render unsatisfactory the results of any but a careful study of the region. The broader features, however, stand out with fair distinctness and may here be described. (See fig. 4.)

The district comprises a narrow strip of country extending east and west across the Cimarron Range, with Baldy Peak at about its center. The Cimarron Range, like the other main ranges of the central mountain region, owes its elevation to faulting with accompanying upbending of the strata along its flanks. Unlike most of the high ranges in the northern part of the Territory, its crest, instead of being composed of a core of pre-Cambrian rocks, is made up of sedimentary strata. The pre-Cambrian basement is exposed in places, however. It consists of coarse feldspathic granites, granitic and dioritic gneisses, sheared diabase, and various schists. The sedimentary rocks belong chiefly to the Cretaceous system, and are broadly divisible into Dakota (?) sandstone, Colorado, Montana, and Laramie (?). West of Elizabethtown, near Comanche Canyon, rocks which have been called "Jura-Trias" are said to rest directly on the pre-Cambrian rocks and to be conglomeratic at the base. Farther south, along the west side of the valley, Carboniferous strata come in between these red beds and the pre-Cambrian. It has also been stated that the Dakota (?) sandstone rests directly on the pre-Cambrian in the main Cimarron Range. It is at least certain that near the head of Cimarron Canyon neither Carboniferous, Triassic, nor Jurassic rocks intervene between the Dakota (?) sandstone and the pre-Cambrian rocks, but a thick sill of intrusive rock separates these two and makes absolute proof regarding stratigraphic sequence impossible. If the Carboniferous and Jurassic or Triassic were once present they can have been removed from under the Dakota (?) sandstone only by a faulting almost parallel to the stratification, and the character of the faulting which is known to have occurred seems certainly to preclude this supposition. It is probable, therefore, that a portion of this region was an island of pre-Cambrian rocks during Jurassic and Triassic time, and that a still larger area had been exposed during Carboniferous time. Faulting and intrusion render an estimate of the thickness of the Cretaceous strata unreliable, but it is probable that the Dakota (?), Colorado, Montana, and Laramie (?) rocks are each several hundred feet thick. On looking up at the west face of Baldy Peak from Elizabethtown the edge of the shales of Colorado and Montana age is seen. The shales extend from the summit down for probably more than a thousand feet. Whether or not faulting has increased the apparent thickness, it is impossible to decide.

The major faulting which has affected this region is of the same general character as that prevalent throughout the central mountain province. It consists of upbending of the strata on one side of the fault, then downthrow on the other side. In this region the beds on the east of the fault are tilted upward toward it, so that all the beds dip eastward, and the blocks on the west side of the faults are always thrown down. This is shown in the section (fig. 4). Minor faulting has occurred in some places and may be discordant with the greater movements, but it is of little importance. The principal result of the faulting is to heave the Cretaceous sediments up fully 4,000 feet to form the Cimarron Range, so that the underlying pre-Carboniferous rocks are exposed along the fault scarp south of Elizabethtown, while the same Cretaceous formations are left down in the bottom of the Moreno Valley, which separates the Cimarron Range from the Taos Range, a similar fault ridge. The Moreno Valley may thus be termed a "graben." Opposite Elizabethtown smaller faults have allowed narrow blocks to slide down stepwise and veneer the face of the scarp, with the result that in a journey from Elizabethtown eastward up to the summit of Baldy Peak, Laramie (?) rocks are encountered no less than three times, besides the main body of the formation, which laps up on the east side of the range almost to the summit. On the east side of the mountain the dips are fairly uniform, gradually increasing as the summit is reached and ranging from 20° to over 45°, mostly to the east, but in the northern part north of east. On the west side of the mountain the fault blocks seem to be somewhat broken and disturbed, but there, too, the dip is generally rather steep and to the east. At Iron Mountain, on the southwest slope of Baldy Peak, the local dip is 50° NW.

Cutting these Cretaceous rocks are numerous large and small intrusive masses of monzonite porphyry, which in many places is quartz bearing. This porphyry, which may almost with certainty be correlated with the pre-Tertiary or early Tertiary porphyries in many other parts of the Territory, is nevertheless later than the faulting; in one place 2 or 3 miles above Ute Creek, in the Cimarron Canyon, a large unsheared dike of this porphyry occupies a fault cleft. The faulting, then, doubtless took place at the close of the Cretaceous and was accompanied or closely followed by the intrusion of porphyry. The rock is light gray in color and of distinctly porphyritic texture, containing prominent feldspar crystals in a finer-grained base of somewhat darker color. Both hornblende and biotite are present in rather sparing amounts. In its typical development the rock is a monzonite porphyry. Quartz phenocrysts are not uncommon; they are usually much rounded by corrosion and are locally designated as "pebbles." In some places the proportion of quartz is so large that the rock must be called a quartz monzonite porphyry.

The principal porphyry mass is a sheet or sill several hundred feet thick intruded between the Dakota (?) sandstone and the pre-Cambrian rocks. This body is best shown in the Cimarron Canyon, where the river has cut completely through it and exposes it for several miles.

In some places a rudely columnar structure is developed and at certain points unequal weathering has produced fantastic pinnacled forms. From this main sheet dikes cut upward into the Dakota (?) sandstone. Sills of much less thickness are plentiful in shales of Colorado and Montana age. A large intrusive mass, apparently of irregular form, extends from a point just south of Elizabethtown along the south side of Grouse Gulch and over into Willow Creek. Another elongated mass occurs on the east slope of South Baldy.

The igneous rock has caused a pronounced metamorphism of the shales near the contact. The purer aluminum silicate rocks seem not to have been greatly altered in composition, although decidedly baked or indurated. The shales which appear to have been calcareous or limy, however, have been in many places extremely metamorphosed. Diopside, hornblende, epidote, garnet, scapolite, magnetite, and specularite are the principal contact minerals. The vicinity of the Ajax mine on Willow Creek and Iron Mountain are the localities where contact metamorphism is most pronounced.

MINERAL DEPOSITS.

GENERAL FEATURES.

The Elizabethtown district is essentially a producer of gold, although of course some silver occurs along with it and iron and copper are known in quantities of possible economic importance. The primary gold deposits are of two types, quartz veins and contact-metamorphic deposits in calcareous sedimentary rocks. The veins have produced much the greater portion of the gold. The secondary or placer deposits, which have to their credit a sum probably greater than that extracted from the lodes, have unquestionably been derived from these primary deposits and probably most largely from the veins.

The quartz veins belong in the group which is closely associated with intrusions of porphyry—a group plentifully represented in New Mexico. Owing to the presence of sedimentary rocks in all portions of the district, however, many of the veins reach beyond the limits of the porphyry mass out into the adjacent strata, and several veins are, so far as known, in the sedimentary rocks entirely. What seems to be a favorite position for these veins is at the contact of sills of porphyry with shales of Colorado and Montana age. Several such veins are known on the Baldy side of the mountain, notably in the Rebel Chief group. These veins and those like the Aztec vein, which occurs between the shales of Montana age and the Laramie (?) formation, are locally spoken of as contact veins. While they do actually lie at the contact of two different kinds of rock, their position is probably due solely to the fact that fractures formed more readily at such places; these veins apparently differ not at all from those which do not occur at contacts, and it is probable that the character of the vein is not influenced by the fact that its walls are of different materials.

The veins might be further subdivided into single fissure veins and lodes in which several narrow parallel veinlets constitute an individual ore body. The latter are more common in the porphyry, occurring along the fractures of sheeted zones. All the veins are narrow, ranging from tiny threads scarcely discernable except by their superior resistance to decomposition to quartz sheets a few inches wide. The strike is in general fairly constant; in places, however, following a gentle curve. Most of the veins are not traceable for any great distance, but certain of them, particularly on the east or Baldy side of the mountain, have been followed on the surface for over half a mile. The developments thus far prosecuted have not as a rule been sufficient to settle the question of persistency in depth. It seems reasonable to suppose, however, that the vertical extent is at least of the same order of magnitude as their length on the surface.

In composition these are essentially quartz-pyrite veins. Quartz is the only gangue mineral of the actual fissures, except calcite in one or two places, and pyrite is by far the most important ore mineral. Chalcopyrite and galena are present here and there, usually in small amount, and in the shale the pyrite is locally intergrown with a little pyrrhotite. Magnetite accompanies pyrite in at least one place, at the Denver mine, an occurrence especially interesting in view of the abundance of magnetite developed in the contact-metamorphic deposits.

Neither quartz nor pyrite is confined wholly to the space between the fissure walls, but they invade and partly replace the adjoining rock. This alteration of the country rock is most pronounced in the porphyry, which close to the fissures is bleached, partly stained with iron, and commonly softened. It is probable that sericite was abundantly developed by this alteration, but it is now largely decomposed to kaolin. The sheeted zones, made up of rock probably somewhat shattered and crushed and exposed to the action of the vein solutions along numerous channels, appear to have suffered greatest alteration. There is little evident alteration of the shale along most of these fissures, but in a few where the shale has been shattered, leaving the fissure walls irregular and giving fragments of the shale in the fissure, considerable silicification has taken place immediately adjoining the original spaces, while a little farther away the shale is much sericitized.

So far as can be learned the gold was originally contained wholly in the pyrite and has been liberated in its present free state only by the oxidation of that sulphide. Oxidation has extended nearly everywhere to the depth reached by development—340 feet at the Black Horse mine—and this is what would be expected when it is known that few, if any, of the mines require pumping. The veins are therefore all iron stained, lumps of limonite occur in the quartz and the altered wall rock, and a large proportion of the gold occurs free. This surface decomposition also obscures the nature of the original vein alteration.

The veins have unquestionably some relationship to the porphyry and probably represent products derived at a slightly later time from the same source as that of the porphyry. In other words, the ore-bearing solutions are believed to be the final product of a magmatic differentiation whose most abundant and characteristic product was the monzonite porphyry.

The value of many of the veins themselves is probably high, for much of the ore as broken, including large quantities of barren or low-grade rock, has averaged $10 to $20 a ton, and in some places much richer ore has been found. At the Aztec mine ore running several hundred dollars to the ton was not unusual.

The contact-metamorphic deposits are fewer in number, so far as known, and have produced less than the veins. They occur in sedimentary rocks at or near the contact with the quartz monzonite porphyry. The metamorphic effects are different in different deposits and are probably dependent on the chemical character of the sedimentary rocks. In the Ajax deposit the rock was probably a calcareous shale originally. It is now a dark crystalline rock composed largely of diopside and scapolite, with smaller amounts of calcite, epidote, and dark amphibole and plentiful magnetite in small grains. A little pyrite is present and in some places chalcopyrite also, generally intergrown with the magnetite. Gold values in this material are decidedly variable, ranging from a dollar or two to $80 a ton, but on the whole of low grade. The gold may possibly be contained in the magnetite.

On Iron Mountain the sedimentary rocks were probably richer in lime and may have been argillaceous limestones. The minerals resulting from the contact metamorphism are epidote, garnet, quartz, calcite, magnetite, and hematite. Certain bands representing original strata are very rich in iron and have been prospected to some extent for that metal. There is some gold in this ore; on an average probably less than $2 a ton.

In Copper Park, near the east summit of Baldy Peak, rich copper float has been known for years. This was probably what first attracted attention to the district. The writer has recently learned from Mr. L. S. Preston, engineer for the Maxwell Land Grant Company, that the source of this copper float is a contact-metamorphic deposit. Work is now in progress to explore it.

AZTEC MINE.

The Aztec, the richest and one of the oldest mines in the district, is situated near the camp of Baldy, on a spur from the main range. It was discovered in 1868 by Lynch and Foley, whose names are closely associated with the history of the district. The deposit was very rich and changed hands rapidly, several times having been held by force. At one time it belonged to Maxwell, the former owner of the grant, and he probably made the most money out of it. At present it is under the control of the Maxwell Land Grant Company, and is virtually idle. The production is estimated as between $1,250,000 and $1,500,000, of which about $1,000,000 was taken out in the first four years.

The development consists of several tunnels and shafts, the greatest depth attained being 220 feet, at which some water is encountered. One adit is 1,000 feet long. A 15-stamp mill was built in 1868 and was run for four years. A modern 40-stamp mill with Wilfley concentrators replaced it a few years ago. A 1,600-foot inclined tramway connects the main tunnel with the mill. Two or three men are now at work on a lease from the Maxwell Land Grant Company. The main underground workings are near the contact of the shales of Montana age with the overlying coal measures, which are commonly regarded as Laramie. The base of this series is here a grit or coarse sandstone ranging to a fine-grained conglomerate. In places it is almost a quartzite, possibly as a result of porphyry intrusions. On the Ben Hur claim, a few hundred feet east of the principal workings, a sill of porphyry showing a tendency toward columnar jointing occurs either at this contact or just above it in the Laramie (?) formation. Toward the west this sill pinches from a thickness of 30 or 40 feet till it disappears. It was not seen by the writer in the mine workings, but it is said that bunches of the porphyry occurred with much of the rich ore.

The ore deposit is in some respects unusual. It occurs mainly in the coarse sandstone just above the contact with the shales and it consists of a number of veins of which a few are at the actual contact of the shales and nearly all are parallel to the bedding of the rocks. The lode thus comprised strikes about west-northwest and has a variable dip, which averages probably 25° to 30° NE. A few veins, which are apparently branches or offshoots from the main lode, cut upward almost vertically through the sandstone. The walls of the narrow fractures in the sandstone, which constitute the veins, are covered with a thin layer of dark material, presumably limonite, and this is said to have been fairly bristling with free gold in the richest ore. It seems probable that this was once rich auriferous pyrite, which was later oxidized. The mill concentrates contain some pyrite, a fact supporting this probability.

The stopes from which the ore of the main lode was extracted are disconnected, and as not all of them are on the same fracture they overlap in places. Many are but small gouges. The largest stope, known as the Bridal Chamber, is two and three sets high and probably 50 feet wide by 100 feet long. Some very rich ore was taken from it, but authentic figures as to its actual value could not be obtained. Most of the ore from the mine, in fact, has been rich but pockety. The ordinary range of values is said to have been $5 to $70 a ton. Above and west of the Bridal Chamber is another stope of good size. Several of the stopes on the main flat lode extend out to the surface on the steep hillside. North of the locality where most of the development has been carried on the lode apparently flattens, so that it is nearly horizontal, and the statement has often been made that this horizontal portion is barren. In view of the irregular distribution of the valuable ore, the amount of development done in this neighborhood appears too small to warrant such a generalization. West of most of the important stopes a vertical vein a few inches wide running in a northerly direction carries kaolin, bunches of copper carbonate, and some chalcopyrite, with reported good values in gold. A little ore has been taken from it. Another cross vein about parallel to the strike of the rocks is nearly vertical, and is sometimes called the vertical vein. It has been regarded by some as a bent-up portion of the main lode, but it seems to be instead a transverse fissure which breaks off abruptly from the flat lode. It has supplied some good ore similar to that from the main stopes.

Much of the placer gold that has been found in Ute and Poñil creeks probably came from this deposit. The coarse placer gold encountered in Ute Creek is generally believed to have been derived from the Aztec lode, as that is the only lode deposit known in the district which carries coarse gold.

BLACK HORSE MINE.

South of the divide known as Aztec Ridge, which separates Ute and Poñil creeks and which is capped with Laramie (?) rocks, the eastern slope of Baldy Peak is made up mainly of shales of Colorado and Montana age. Penetrating these sedimentary rocks are many small tabular and irregular bodies and a few larger masses of porphyry. The Black Horse mine, which lies about a mile southwest of Baldy camp and about the same distance east of the saddle between Baldy Peak proper and South Baldy, is located in one of these larger masses of porphyry. The group of claims known as the Black Horse group is extended nearly parallel to the long dimension of this irregular porphyry dike—about west-northwest. The lode which has been explored has a similar strike, but bends slightly more toward the north in its western part. The developments consist of a shaft 340 feet deep and a 1,700-foot tunnel, which intersects the shaft at a depth of about 120 feet. Other workings on the Paragon claim, which adjoins on the northwest, make the total known length of the lode about 3,000 feet. An old mill connected with the mine by a 1,200-foot gravity tramway is now pretty much dismantled.

The lode ranges from almost nothing to 4 feet in width. It consists of rusty seams in the porphyry, which seem commonly to be little quartz veinlets once holding pyrite but now thoroughly oxidized. At one place a narrow cross seam appears to come into the main vein, and at this place malachite and a little copper glance are present. The values are said to have been good in spots, reaching $100 a ton, but were exceedingly irregular in distribution, making the average ore of low grade. It is stated that during a period of fifteen years several thousand tons of ore was milled with a product of only $27,000. The Paragon claim, lying farther up the hill on the same lode, has produced probably almost as much.

MONTEZUMA MINE.

The Montezuma claim lies just southwest of Baldy camp, on the eastern slope of Baldy Mountain. The amount of development is small, consisting of several shallow shafts and pits on a nearly vertical quartz vein. Black shale of Colorado age is the prevailing rock in the immediate vicinity of the mine, but porphyry in small dikes and sills and in masses apparently of irregular form is exposed at a number of points. The vein, which is 2 to 4 feet wide and strikes nearly west, cuts across both the porphyry and the bedding of the shales. Narrow stringers make off from the main vein along the shale bedding planes, along the porphyry and shale contact and across the shale structure. The shale is considerably baked and silicified. Work has opened the vein more or less continuously for about 2,200 feet—almost the full length of the Montezuma claim and in the adjoining property on the west, the Bull of the Woods claim. Gold was apparently present in pyrite, which is now decomposed to limonite. The sulphide appears to have been finely disseminated through the quartz, for no large cavities nor bunches of limonite are observable. The ore yielded $10 to $20 a ton on the plates. It is said that about $300,000 has been taken out of the small workings on the Montezuma claim and that a considerable additional amount was yielded by the same vein in the Bull of the Woods claim.

REBEL CHIEF MINE.

The Rebel Chief group of claims is about one-third mile south of Baldy camp. The development consists of many small openings and a number of tunnels, most of them rather short, at different elevations on the hillside. Reliable figures of production were not obtained, but the output probably compares favorably with that of any of the mines on the east slope except the Aztec and the Montezuma. The property is underlain by the dark Cretaceous shales, which here strike about N. 25° W. and dip 30° to 35° E., the steepest dip being at the west. Intruding the shales, parallel to their bedding, are at least three regular and distinct porphyry sills. Most of the ore lies at the contact of these porphyry masses with the shale. Although considerably altered and baked by the porphyry, fossils, especially *Inoceramus*, can be recognized and determine approximately the geologic horizon of the formation. Most of the tunnels run a little south of west to crosscut the stratification. The Mountain Queen tunnel, after passing through some slide, reaches porphyry and at a distance of 45 feet from the portal encounters the contact with shale. Along this contact is a zone from 2 to 12 inches wide, consisting of decomposed porphyry impregnated with more or less vein quartz. A short drift and a little stope have been made on this streak. About 8 feet below, measured at right angles to the bedding, is a vein 4 to 12 inches wide, consisting of quartz and decomposed shale. It is known as the slate vein and has been developed to a small extent. Ore of good value has been taken from both these veins.

Pyrite was the principal mineral and was abundant, especially in the slate vein, but is almost wholly oxidized and obscures any other effects of alteration by the vein solutions.

The Chief tunnel, next above, goes 60 feet through porphyry and there strikes a tight contact with the shales, but in a drift to the north-northwest this contact opens and a little ore was found. This is said to be the same as the vein at the contact in the Mountain Queen tunnel, but no corresponding slate vein occurs here. A crosscut from this point goes westward 80 feet through shale and then 85 feet through porphyry, reaching the third contact between porphyry and shale. Here, resting on the shale foot wall, is a quartz vein 1 to 2½ feet wide, made up of decomposed porphyry and shale, with many quartz stringers, and carrying much limonite that was derived from the oxidation of pyrite. The dip is a little steeper than in the Mountain Queen tunnel, but the strike is practically the same. Most of the stoping has been done here, the stope reaching at several points all the way to the surface, 100 or more feet above. Unverified report as to the value was that the ore averaged over $20 a ton. The sketch (fig. 5), though not reliable as to scale, suggests the probable relations of the veins and the masses of porphyry and shale as indicated by the mine workings.

Still farther up the hill a narrow zone between porphyry above and slate below was struck in the Virginia Hudson tunnel and some ore was taken from it.

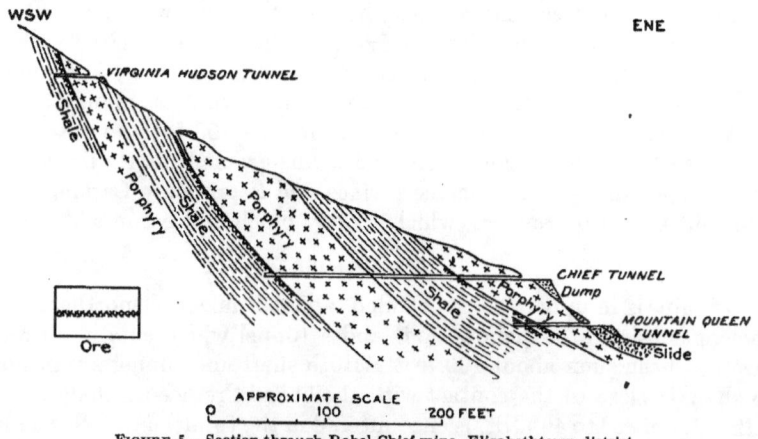

FIGURE 5.—Section through Rebel Chief mine, Elizabethtown district.

FRENCH HENRY MOUNTAIN.

The spur north-northeast of Baldy Peak and north of Poñil Creek, which is known as French Henry Mountain, has received considerable attention from prospectors. This part of the district was not visited by the writer, who is indebted for the following notes to Mr. L. S. Preston. French Henry Mountain, like the Aztec Ridge, is capped by sandstone (Laramie?), which is underlain by shales of Montana age. Porphyry sills are numerous, especially at and near the contact of the two sedimentary formations. To the similarity of these geologic conditions and those existing near the rich Aztec mine is undoubtedly due much of the attention which this locality has received. The application of this analogy seems to have been reasonably successful, for though no mines comparable in richness to the Aztec have been discovered in this portion of the district the most encouraging results have in almost all cases been obtained at approximately the same horizon as that of the Aztec mine—the contact of the Laramie(?) formation and the shales of Montana age.

Among the properties in which good ore has been encountered are the French Henry, Harry Lyons, American, and Harry Bluff. The first three are said to be at the contact of Laramie(?) with masses of porphyry, and in the last the ore lies at the contact of the Laramie(?) and the underlying Cretaceous shales, no porphyry being present so far as known. The French Henry has a 15-stamp mill on South Poñil Creek, with a 2,700-foot aerial tramway. The ore averaged somewhat under $25 to the ton, with a little much-richer ore that was shipped to smelters at Pueblo and Denver.

MORENO CENTENNIAL MINE.

The Moreno Centennial property lies in Grouse Gulch, on the west side of Baldy Peak, about a third of the way up the mountain from Elizabethtown and just south of the Empire. The mine is opened by a shaft about 200 feet deep. Porphyry here cuts across the bedding of limy Cretaceous shales, which are much baked and metamorphosed. The strike of the contact is about S. 85° E. and the dip approximately vertical. A 12-inch streak along this contact constitutes the vein. It is mainly altered porphyry with quartz stringers through it, is pretty thoroughly oxidized, and is said to have averaged $30 a ton by mill runs. Ore was also taken from a near-by parallel vein in the porphyry. The total production has amounted to several thousand dollars.

RED BANDANA MINE.

The Red Bandana property lies just north of the Moreno Centennial. A shaft 100 feet deep and 600 feet of drifts constitute the development work. A Huntington mill was operated on ore from this mine. The lode, striking slightly north of east with nearly vertical dip, is a zone along which the porphyry has been somewhat fractured and sheeted and ore-bearing solutions have penetrated. This lode, which has an average width of about 12 feet, consists of silicified seams with abundant limonite, from pyrite. Pyrite also penetrated the porphyry adjoining and between these seams, but has there likewise been oxidized to limonite. As at most other places in the district, the gold is directly associated with this iron oxide. Low-grade or barren horses were encountered at places in the lode, but some of the veinlets were rather rich; the average value of the main zone, however, was about $5 to the ton. Some of the better portions of the vein were stoped from the 100-foot level to the surface. It is said that on the east the vein abruptly entered a fine-grained quartzitic conglomerate, which appears at a corresponding position on the surface and is probably Laramie. In this rock the vein is said to split into two branches, which shortly pinch and become of very low grade.

DENVER MINE.

The Denver mine is in upper Anniseta Gulch, about 2 miles east-northeast of Elizabethtown. The development consists of a shallow shaft and a tunnel which cuts it. From the tunnel an inclined winze has been sunk about 100 feet. Both shaft and tunnel are in porphyry, but the collar of the shaft is close to the contact with the baked Cretaceous shales. A narrow vein in porphyry with a dip of 35° to 40° NE. is encountered in the tunnel and followed by the winze. It consists of calcite and quartz carrying plentiful magnetite, some pyrite, and spotted values in gold. A little stoping has been done at the tunnel level.

LEGAL TENDER MINE.

The Legal Tender property lies on the east side of Willow Creek, about west of South Baldy. The vein is reached in about 75 feet by a crosscut tunnel. It strikes about N. 10° E. and dips approximately 60° W. Below the tunnel level the dip increases to 90° and the vein is reached by crosscuts, 50 and 100 feet below the tunnel, from a vertical winze. A Huntington mill treated the ore. The vein is a narrow quartz seam and a bordering alteration zone in decomposed porphyry. In the upper workings it is 5 to 6 feet wide, but decreases to 4 feet in the winze and pinches near the bottom to about 1½ feet. Pyrite was rather plentiful in both quartz and the adjoining porphyry, but is now completely oxidized. The average value of the ore is reported as about $30 a ton. A stope of considerable apparent size, now somewhat caved, extends above the tunnel level and is said to have yielded about $15,000, which was recovered by an arrastre. About $10,000 more is said to have been recovered in a Huntington mill on ore from the winze workings. Cyaniding was attempted, but did not prove successful.

ALABAMA MINE.

Next southeast of the Legal Tender lies the Alabama claim. Here an 18-inch vein consisting of altered limonite-bearing porphyry with numerous narrow quartz stringers has been developed to a small extent. At one opening on this claim, southeast of the Legal Tender tunnel, a 1-inch dike of porphyry cuts across the Cretaceous shales, which here and at many other places are much altered. This dike, which is richer in quartz than the normal porphyry, is almost pegmatitic in appearance; it carries altered pyrite and some gold, and formed the vein explored at this place. The value of this material was not learned. The occurrence affords an interesting illustration of the close relation of the vein deposits to the porphyry.

OTHER VEIN DEPOSITS.

On very many other claims veins similar to those already described have been opened, with varying success. In nearly every place the vein either cuts porphyry or lies near some body of that rock. Among the few properties visited were those mentioned in the following paragraphs:

The Galena claim in Grouse Gulch is located on a typical vein in porphyry carrying particles of partly decomposed galena with completely oxidized pyrite.

The Confidence mine, close to the Denver, has a similar vein in porphyry, but of low grade.

The Hidden Treasure, on the west slope of Willow Creek above the Legal Tender, opened a narrow vein of $20 ore in porphyry, a little of which was treated in an arrastre.

The Only Chance and Gold Dollar claims, just west of the Legal Tender, expose quartz veinlets in the baked and much-epidotized shales. Good values were encountered erratically.

On the south side of Ute Creek, just above the Aztec mill, a narrow quartz vein is present in baked shale close to the porphyry contact. Owing probably to the impervious character of the shale, oxidation has scarcely affected this vein, though it is only a short distance below the surface. Pyrite is plentiful, with small grains of galena and a little intergrown chalcopyrite and pyrrhotite, the latter especially in the much-silicified walls of the vein. The values are said to be rather low.

In Cimarron Canyon, between Moreno Valley and Ute Creek, are a number of deposits in pre-Cambrian rocks, which are probably unrelated in origin to most of the veins of the district. About 4 miles up Cimarron River from the Ute Creek stage station the upper portion of the pre-Cambrian rocks emerges from beneath the thick porphyry sill that underlies the Dakota(?) sandstone. The pre-Cambrian at this point is a dark diabase, somewhat altered. This rock gives low assays for gold, and the values appear to be concentrated in narrow red oxidized seams which traverse it. There is no sign of quartz veinlets. Farther west or upstream this diabase gives way to amphibolitic gneiss and schists, with patches of much-sheared granite. In the schist are a number of quartz stringers which carry oxidized copper minerals and some gold values. One zone containing many of these stringers is exposed on the south side of the canyon. It strikes northwestward, is practically vertical, and is about 25 feet wide. It is said to give low-grade assays across its entire width.

It seems probable that these deposits in pre-Cambrian rocks are not connected with the porphyry intrusion, but are much older—in fact, pre-Cambrian.

AJAX MINE.

The Ajax claim lies across Willow Creek just south of the Legal Tender. It was located along a fissure in the porphyry having an east-northeast trend, but it was later found that the best ore was in a zone that runs across the claim practically in a north-south direction and dips steeply to the west. The principal workings consist of a shallow open cut, a short crosscut tunnel, and a shallow winze from the tunnel. Between 1898 and 1905 considerable ore was put through the 10-stamp mill at odd intervals; but the figures of total production were not ascertained.

The gold is irregularly distributed through a dark, heavy, finely granular rock which has generally been regarded as igneous and called diorite. This rock mass lies between porphyry and much-altered shale. In reality the rock is an extreme phase of contact-metamorphosed shale. Careful examination enables the bedding planes of the readily recognizable shale to be traced into the more altered mass, and the transition and gradation in degree of alteration toward the porphyry contact can be seen at various exposures on the surface. At the north end of the open cut the much-epidotized and indurated shales are seen in contact with the porphyry. The contact is minutely irregular, strikes nearly north, and dips 50° to 60° W. Where the ore occurs practically all trace of the original structure and appearance of the shale has been removed. The principal minerals developed are a nearly colorless pyroxene, a dark and a light amphibole, plentiful epidote and magnetite, a little zoisite and specularite, some quartz, oligoclase, and calcite, possibly resulting from recrystallization of smaller particles of the same minerals in the original shale. Perhaps the most characteristic mineral is a scapolite of rather low double refraction which is present in abundant individuals with prismatic cleavage. Another mineral of similar occurrence corresponds in optical properties to apatite and may be that mineral. The magnetite is developed in spongy skeleton crystals and surrounds grains of pyrite and a little chalcopyrite. Magnetite and dark amphibole are in places especially concentrated along narrow parallel bands that probably represent original bedding, but this was not demonstrated with certainty. However, the gradation from undoubted shale to a dark, basic-looking granular rock is very apparent.

The average yield on the plates is reported as $20 a ton, but the ore is very erratic in value and is said to range from $2 to $75 without any noticeable difference in appearance, character, or position. It is possible that the gold is present mainly or wholly in pyrite and that this mineral, which is nowhere more than sparingly present, may show a range in abundance corresponding to the gold content. But the maximum amount of pyrite present seems hardly sufficient to contain $80 gold to the ton of ore if any ordinary gold content is assumed for the pyrite. It seems more reasonable that the gold is in some way intimately associated with the magnetite. It would be interesting to learn what results concentration of this ore would give, but so far as known experiments of this sort have never been performed.

The principal ore body lies just below the tunnel level and must be close to the porphyry contact. A chamber more than 20 feet in each dimension has been stoped out here.

IRON MOUNTAIN.

The elevation known as Iron Mountain lies at the head of Mexican and Anniseta gulches and west of the divide which separates them from the Willow Creek valley. It is about west of the Ajax mine. A small amount of development work in the shape of tunnels and pits has been done, mostly on the upper west slope, to explore deposits of iron ore.

The hill consists principally of inclined strata of Cretaceous shales penetrated by numerous dikes and sills of porphyry, and plentiful porphyry float on top of the hill probably indicates the presence of a considerable mass of that rock. The shales have been metamorphosed as a result of the intrusion, and beds that probably were especially rich in lime have been much altered. Owing perhaps to the predominance of silica over lime in the shales, garnet is less plentiful than epidote, though in certain places garnet is abundant. Seams of epidote are present in the porphyry near the contact. Their presence may indicate that the marginal portions of the magma solidified sufficiently to permit fracture before the process of contact metamorphism was completed.

In some places the rock has been converted into almost solid magnetite, forming large masses that assay 70 per cent or more in iron. In other places the product of metamorphism is a garnet-specularite rock, the micaceous ferric oxide being either irregularly disseminated through the rocks or arranged in rude narrow layers that may correspond to original bedding. With the garnet and specularite is associated in some localities a little pyrite and chalcopyrite, and more or less epidote is universally present. Specular hematite lies along some of the seams or veinlets in the rock. There seems to be a tendency for the rock rich in epidote to carry magnetite and for the garnet-rich rock to predominate in hematite.

The main iron-bearing band is opened by several pits and tunnels. A 60-foot tunnel west of the crest of the hill is the principal working. The strike of the bedding is here N. 50° E. and the dip 50° NW., but owing to the steepness of the hill slope the strike of the outcrop is southeasterly. The accompanying sketch (fig. 6), indicates the attitude of the ore-bearing zone and the relation of beds within it. Little irregularities in the boundaries of the extremely metamorphosed portions show that the metamorphism has not been absolutely restricted to the confines of certain beds, but has encroached to a slight extent on beds which in the main show little chemical alteration. This circumstance would seem to prove that the marked difference in composition now present in adjoining beds is not an original difference due to sedimentation, and that therefore the iron oxide is not a bog ore deposit, as has been suggested, but that it has been brought about almost wholly by the agencies attending the porphyry intrusion.

Nearer the top of the hill and over on the east side other openings, which encounter similar material, indicate by their position the probable presence of a second horizon of intense metamorphism. It seems doubtful, however, if the locality can be made productive of iron ore on an economic scale. The ore persistently carries gold, but the values are almost invariably low, only in places reaching as much as $2 a ton. At the breast of the 60-foot tunnel above referred to a cross seam of quartz in much-decomposed material carries a little pyrite. It suggests the probability that the contact deposits were formed in advance of the lode deposits, a sequence naturally to be expected.

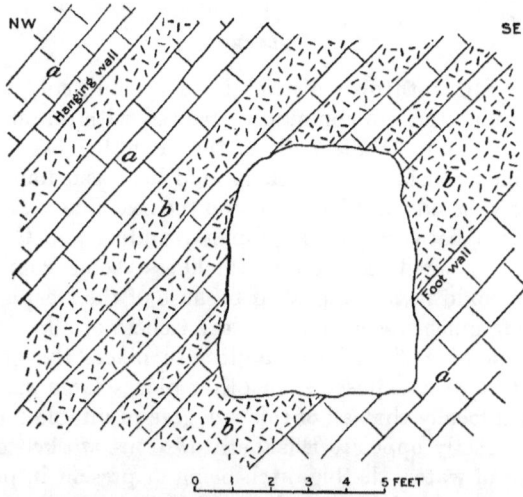

FIGURE 6.—Section at mouth of main tunnel, Iron Mountain, Elizabethtown district. *a*, Partly metamorphosed limy shale; *b*, garnet-epidote bands with magnetite and specularite.

BALDY TUNNEL.

The Baldy tunnel site lies high on the slope of Baldy Peak, at the head of Big Nigger Gulch, about one-half to 1 mile west of the summit of the mountain. It is owned by the Gold and Copper Deep Tunnel Mining and Milling Company. The principal development is a long tunnel directed S. 86° E., approximately toward the summit of Baldy Peak. At the time of visit this was about 1,670 feet long and was being extended. A mill is situated near the tunnel mouth.

The prevailing rock on the upper western slope of Baldy Peak is Cretaceous shale and the dominant attitude of the beds is a northerly strike and decided eastward dip. Minor faulting and local intrusions of porphyry have in the neighborhood of the Baldy tunnel somewhat disturbed this common structure. The shale is everywhere much baked and is easily recognizable as having been affected by a near-by intrusive rock, but it was not observed that metamorphism was generally more intense at the actual contact of the intrusive rock than at a considerable distance away.

Some fairly good ore, supposed to be but not certainly in place, was found some years ago at or near the porphyry-shale contact on the Chester claim, which lies higher than and to the north of the Baldy tunnel site. The Baldy tunnel was started in the hope of cutting the continuation of this ore or some similar body. Near the breast of the tunnel some particularly metamorphosed shale was encountered. This consisted of epidote, oligoclase, diopside needles, calcite, and quartz, with stringers and intergrown particles of chalcopyrite, pyrrhotite, and pyrite. The amount of copper is low.

COPPER PARK.

Near the headwaters of South Poñil Creek is an amphitheater or park in which abundant copper float is found and which is therefore known as Copper Park. It is supposed that it was here that the Indian who brought the first news of this district to Fort Union obtained the copper ore which he exhibited. It is said that thousands of tons of float of fairly good grade are present. The occurrence has always attracted interest and some small shipments have been made, but little work of importance has ever been done because it is maintained that the title to the land is in doubt. The locality was not visited by the writer, but from a pile of the float hauled to the Aztec mine some specimens were taken. This lot of ore, comprising several tons, contained probably 5 per cent of copper and, according to report, some gold. The copper is mostly in the form of malachite and cuprite in irregular veinlets all through a mass of limonite. A little chalcopyrite is present and was probably the original copper mineral. For many years the source of this float was unknown, but in 1906 the writer was informed that the deposit in place was located near the float and that it was of contact-metamorphic origin, lying in the shales close to porphyry.

PLACERS.

The placers of the Elizabethtown district have undoubtedly been the most productive source of gold in the region, and are commonly credited with an output of at least $2,500,000 up to 1905. The west or Moreno Valley side of the range has been most productive, but the east side has also supplied considerable. Little gold has been found in the gulches north of Big Nigger Gulch on the west side and Ute Creek on the east side. Except in the Moreno Valley and in a considerable area between and adjoining Humbug and Grouse gulches, most of the gold-bearing gravels are of small extent and thickness, being confined to narrow valleys in which little aggradation could be accomplished because of the steep stream gradients.

Grouse and Humbug gulches were perhaps the richest of all. It was to work the gravels of these gulches that the "big ditch" was built. In places the gravel has been completely sluiced away, leaving bed rock exposed. In other places large thicknesses of gravel remain unworked. It is claimed locally that all of this will give profitable returns. The Lynch placers along Humbug Gulch, directly opposite Elizabethtown, are worked only about one day out of three on account of lack of water, the big ditch being at present in poor repair. Other ditches bringing small supplies of water from less distant sources are allowed to fill reservoirs on the hillside, from which the water is drawn intermittently. Hydraulic mining is employed here. It is said that a shaft on the Reservation claim of the Red Bandana property was sunk 440 feet wholly in gravel, and that a crosscut to the east at that depth encountered bed rock dipping steeply to the west. Such an occurrence would indicate either the existence of an old drainage independent of the present system or else that faulting has taken place at a date so recent that the drainage has not yet adjusted itself, for bed rock appears to the south (downstream) in Grouse Gulch at a higher elevation than the bottom of this shaft is said to be. If 440 feet or more of workable gravel is present here it ought to be worth further exploration. In the lower part of Grouse Gulch porphyry bed rock has been exposed. The rock is considerably decomposed and is traversed by many small iron-stained bands, which are tiny veinlets carrying some gold. The prevalence of great numbers of these small but comparatively rich stringers in the easily disintegrated porphyry offers a very favorable combination of conditions for the supply of gold for placer concentration.

Spanish Bar, just below the mouth of Grouse Gulch and on the opposite side of the Moreno, was also a rich piece of ground which probably derived its gold mainly from the Grouse Gulch drainage. The Moreno River valley has recently been the scene of dredging operations by the Oro Dredging Company, of Chicago. The dredge has buckets of 5 cubic feet capacity working at the rate of about 19 per minute for twenty hours a day. The average daily amount handled is a little over 3,000 cubic yards. All the material elevated goes to a revolving screen with 3-inch openings and by far the larger amount of the material passes through. It is said that several hard clay layers carried rich gold on their upper surfaces and that rich streaks were found in the open gravel. Crevices in the decomposed bed rock also contained gold and a foot of bed rock was often taken up. Rich ground was found at the mouth of Grouse Gulch and near Spanish Bar. By August 1, 1905, the time of the writer's visit, the dredge had worked upstream about 1 mile and had covered about half the ground of the company. Later in the same year the company was forced to suspend operations because the gravels above the mouth of Grouse Gulch were too poor to be profitably worked. It is stated that the company went into bankruptcy.[a]

Placers have been worked by ground sluicing along the upper slopes of Willow Creek. The Last Chance placers are near the head of the northwest fork of Willow Creek; the exposed porphyry bed rock here shows countless rusty seams similar to those in Grouse Gulch. The Grub Flat placers are higher up, on another branch of Willow Creek. These placers on the southwest flank of Baldy Peak have not been so productive as those on the west.

In Ute Creek considerable ground sluicing was done in the early days and it has been resumed at irregular intervals since. The gold here differs from that found in the other creeks in being decidedly coarser. In the nineties one nugget weighing nearly 12 ounces was found in this stream. The source of this coarse gold is commonly ascribed to the Aztec lode, and in consideration of the character and richness of the ore of that mine, it seems probable that the Ute Creek placer gold did come from that lode and perhaps from other similar lodes either undiscovered or possibly now entirely eroded away.

CIMARRONCITO DISTRICT.

LOCATION AND GENERAL GEOLOGY.

The Cimarroncito district is situated near the head of the middle fork of Cimarroncito Creek, which flows eastward and unites with Cimarron River just below the town of Cimarron. The district is about 14 miles east-southeast of Elizabethtown and lies on the east side near the crest of the extension of the Cimarron Range, of which Baldy Peak is a part. The mines may be reached either by traveling directly west from Cimarron about 12 miles or by going south from Ute Creek to the Cimarroncito and then ascending the middle fork of that stream. Access is thus rather difficult and the roads are now in poor repair.

Prospecting and development have been carried on in this region with more or less regularity for many years, but as yet no extensive deposits of ore of sufficient value to be worked under the conditions existing have been found, and the production has been very slight.

The geologic features of the region are considerably diversified. In approaching the district from Ute Creek stage station the observer leaves the Cretaceous shales in which the Cimarron River and Ute Creek valleys are cut at this point and climbs southward over a ridge of Laramie (?) which divides the Cimarron and Cimarroncito drainage basins. On the south of this ridge the shales reappear and, except for small intrusive masses of porphyry, continue southward beyond the north fork of the Cimarroncito. Turning about westward up this stream one soon enters a steep-sided canyon which continues for one-half to three-quarters of a mile and is cut wholly in monzonite porphyry, locally quartz bearing and identical in every respect with that occurring in the Elizabethtown district. This porphyry mass is a broad dike which strikes nearly north and can be seen projecting prominently above the general surface for several miles at least. It appears to be nearly vertical and may possibly mark the line of an earlier fault. West of this porphyry wall is a flat divide that separates the north and middle forks of the Cimarroncito; the rocks are concealed for the most part, but are probably of Colorado and Montana age, mainly if not wholly. Not far west of the great porphyry dike on the middle fork, Dakota (?) sandstone appears. It is locally gypsiferous and in places is converted practically into a quartzite. It is intruded by many dikes and sills of monzonite porphyry, especially in its lower part, and these separate it from the underlying sediments. With local variations the general attitude of these Cretaceous strata is almost horizontal, the gentle dip being eastward and increasing toward the west. From the Dakota (?) outcrop westward the country rises to the crest of the main range

---

[a] No dredging operations have been carried on in the district from 1906 to 1909.

and the eastward inclination of the strata becomes increasingly marked toward the base of the sedimentary series. Underneath the Dakota (?) is a series of red beds which have generally been ascribed to the "Jura-Trias." They consist of thinly bedded, fine-grained shales and sandstones underlain by heavier bedded sandstones and conglomerates. As at Glorieta, in Santa Fe County, where what is undoubtedly the same series is exposed, the red color is seen to be the result of oxidation rather than the original color of the sediments. The rocks are mottled gray, drab, olive, and green, with spots and bands of red, reddish brown, and purple. The red colors extend away from joints, seams, and bedding planes, and are sufficiently widespread to impress their tint on the general mass of the rocks. This red-bed series is repeated several times by faulting. In passing westward over the upturned edges of the strata, the same ground, geologically, is traversed time after time. In one or two places narrow strips of sandstone, probably Dakota, are also brought down. Northward-trending and approximately vertical dikes of quartz monzonite porphyry in many places occupy what are undoubtedly fault planes, and by this occurrence confirm the observation made in the Elizabethtown district that the porphyry intrusion was subsequent to the major faulting. Sills and other dikes also cut the red beds. Porphyry is especially abundant near the point where the middle fork of the Cimarroncito itself splits into north and south branches, and float of this rock largely obscures the base of the red beds and their contact with the Carboniferous. A good exposure of this contact, however, is seen on the north wall of the north branch of the stream. The Carboniferous with conformable dip consists both of limestones and of the characteristic arkose or grit which forms so large a part of its strata in the northern part of the Territory.

Just east of Cimarroncito camp the road cut shows the contact of porphyry with the Carboniferous limestone. The contact is irregular and the limestone is greatly metamorphosed, consisting of epidote, garnet, specularite, and crystalline calcite. West of the district the pre-Cambrian granite and gneiss are said to appear and to form the crest of the range. Bowlders of these rocks are found in the stream beds at Cimarroncito camp.

MINERAL DEPOSITS.

GENERAL FEATURES.

The explored mineral deposits of the Cimarroncito district consist almost wholly of contact-metamorphic deposits in limestone near masses of intrusive rock. Quartz veins have in one or two places been developed to a slight extent; these veins, as will be shown, are probably also genetically related to the intrusive rock.

The characteristic intrusive in the Cimarroncito region is a porphyry related to monzonite, but carrying quartz in varying proportions, and on the whole much more plentifully than in the vicinity of Elizabethtown. In the immediate neighborhood of the mineral deposits the rock is very rich in quartz, microcline takes the place of part of the orthoclase, and the texture is rather finely granular. The rock is in fact a soda granite. An extreme phase has a faint pink color and an aspect suggestive of pegmatitic texture. It contains plentiful orthoclase and microcline, little plagioclase, and some magnetite; the smaller grains of quartz penetrate the feldspars poikilitically and also as regular micropegmatitic intergrowths; the larger quartz individuals interspersed among the feldspars have a flamboyant or rudely radial structure almost similar to that of quartz deposited from aqueous solution. The general tendency toward pegmatite[a] is undoubted and presumably indicates an early stage of that differentiation of water and silica which finally resulted in quartz-vein formation. Quartz veins are not uncommon in the intrusive rock, and the pegmatitic phase just described is in many places injected with countless narrow seams of quartz which have a sort of opalescent, semitransparent appearance. These range in width from one-half inch down to microscopic size. Most of them have rather indefinite boundaries, as if some replacement had taken place along the walls of the tiny fractures which they occupy. Under the microscope this replacement is seen

---

[a] Float found abundantly at this locality is a coarse, graphic pegmatite, whose composition is rather close to the soda granite described above. But it may have come from the pre-Cambrian axis to the west.

to have actually gone on. In several places where tiny quartz veins cross grains of original quartz of the rock something takes place as suggested in the sketch (fig. 7). This shows a permeation of the rock to some extent by the vein solutions and taken in connection with the flamboyant character of the original quartz of the rock would seem to indicate that vein formation began before the crystallization of the rock was completed. The quartz of the rock contains many minute inclusions, mostly of irregular shape, that contain a liquid, a gas bubble, and a transparent cube of some salt. Similar but still smaller inclusions are contained in the quartz of the veinlets. If these small quartz veins are related to the intrusive magma, as seems probable, it is reasonable to assume that the larger ones bear the same relation. Thus both the contact deposits in limestone and the quartz veins are probably referable to the same initial source, the monzonite and soda granite magma.

The contact-metamorphic ores are typical of the general class and have few distinctive features. They are characterized by andradite garnet, epidote, quartz, calcite, specularite, magnetite, pyrite, and chalcopyrite. The gold values are rather better than ore of this class

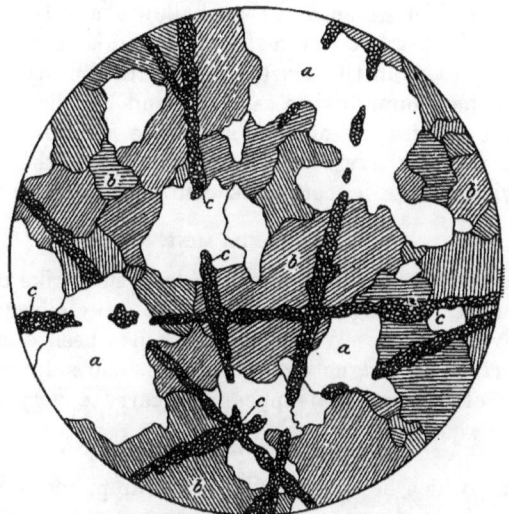

FIGURE 7.—Sketch of thin section of granitic phase of porphyry at Cimarroncito, showing the behavior of quartz veinlets. Note shadowy places in the quartz individuals, where veinlets fail to penetrate; magnified 46 diameters. a, Quartz; b, feldspar and micropegmatite; c, veinlets of microgranular quartz.

generally carries. Specific descriptions will be given under each mine. The properties visited comprised most of the important ones in the district. A brief description of each follows.

THUNDER MINE.

The Thunder mine is situated a short distance south of camp. The development consists of several short tunnels, and one 800 feet long from which a shallow inclined winze has been sunk. The contact of limestone with soda granite appears to be irregular and the metamorphic zone where ore has been found is also irregular. The gangue of most of the ore is dense, massive garnet rock, with veinlets of calcite, usually associated with more or less epidote in small prisms. Magnetite is abundant in many places, locally constituting probably three-fourths of the rock. Clusters of beautiful dark-red specularite plates occur here and there, especially with calcite, epidote, and some quartz. Of the sulphides, pyrite and chalcopyrite, the latter usually predominates in amount. They are intergrown with the other minerals, especially the iron oxides. Some of the pyrite is in fairly well formed octahedral crystals that are embedded in granular chalcopyrite, and chalcopyrite occurs commonly along narrow streaks or incipient fractures in the main mass, indicating that its period of deposition was, at least in part, a little later than that of the other minerals. Some of the ore contains but scanty amounts of garnet and epidote, consisting mainly of crystalline calcite, from the limestone, impregnated by grains of chalcopyrite and magnetite. The precious-metal values range up to $10 in gold and 3 or 4 ounces of silver to the ton. Copper runs as high as 8 to 10 per cent, but none of these values are averages, and but very little ore of such grade has been encountered. A little ore was taken from near the portal of the 800-foot tunnel. Another body east of this, struck near the breast of the tunnel, appears to strike west-northwest and to dip northward. A 40-foot inclined winze goes down on this body, but it held water at the time of visit. A raise extends to workings in a short tunnel above, where a considerable stope has been made. This is close to the contact with soda granite, in which are narrow seams that carry crystalline epidote and specularite plates. Between the portals of these two tunnels fine-grained soda granite is exposed, cut by veins of quartz with oxidized pyrite. Assays of this rock are said never to have been made.

GARST MINE.

The Garst claim adjoins the Thunder and is apparently on the same metamorphosed zone of limestone. The workings are of small extent, comprising a short tunnel near the Thunder workings, with a 35-foot winze, and two shallow shafts farther west along the general strike of the contact. The best ore occurs in the winze and is very similar to that found in the Thunder mine. Water is present in the winze. The shafts are dry and the adjacent rocks are considerably oxidized, so that copper exists as stains and bunches of malachite; the limonitic material derived from the oxidization of both andradite and sulphides is said to carry some gold. The contact with the intrusive rock is struck in one of the shafts at a shallow depth. As at the Thunder, crevices in the igneous rock contain epidote and specularite, also in some places quartz.

ANACONDA MINE.

The Anaconda property lies still farther west, along the strike of the contact of the limestone and the eruptive rock. The igneous rock is here especially granular and grades to the pegmatitic facies already described. What little work has been done in the way of trenches shows garnet-epidote rock, with magnetite, specularite, and sulphides, but the sulphides are mainly oxidized. Some of the material appears to carry a very respectable percentage of copper.

CONTENTION MINE.

The Contention mine, which lies just north of the camp, has been developed by two short tunnels. Limestone, fine-grained grit, and porphyry occur in the vicinity of the workings. The grit has been altered to quartzite and the limestone to garnet-epidote rock or to a baked hornfels. The lower tunnel follows a fracture of northeasterly trend, which cuts both altered limestone and quartzite. Bands of specularite occur in both rocks, but little copper is to be seen. Some of the material is said to give good gold assays. The upper tunnel runs a little east of north and cuts an eastward-trending vein, carrying much limonite and some pyrite and quartz, which occurs at or close to the contact of limestone and quartz. That porphyry is present is shown by the dump. A drift has been carried along this vein for some distance and a small stope made. The gold values are said to be fairly good.

MORA AND SAN MIGUEL COUNTIES.

By L. C. GRATON and WALDEMAR LINDGREN.

GENERAL FEATURES.

Mora is one of the smaller counties of the northwestern part of the Territory. It contains 2,512 square miles and lies south of Colfax County and north of San Miguel County.

The larger and western part is occupied by the high plains of Canadian River and its tributaries, principal among which is Mora River. Along its western border lies the Mora Range, the summits of which are followed by the county line. Mora Peak, at the head of the river of the same name, attains 12,020 feet in elevation, and Truchas Peak, on the southwestern boundary line, 13,150 feet. The geologic features are mentioned below, together with those of San Miguel County. No important mining districts are contained in the county. At Rociada, near the

San Miguel County line, the pre-Cambrian rocks contain some copper veins. Near Guadalupita, in the Coyote district, the occurrence of copper-bearing shales is reported.

San Miguel County lies south of Mora County and east of Santa Fe County; it has an area of 4,893 square miles. The eastern half of the county is drained by Canadian River and its tributaries; the western half is within the basin of the Pecos. The whole county, with the exception of the northwest corner, lies in the Great Plains province. The elevations range from 5,000 to 7,000 feet; near the mountains the plains are dissected into high mesas. Cretaceous formations cover almost the whole of the area. Immediately north of the boundary line, in Mora County, the great basalt flows which emerged from the Ocate crater flooded the Cretaceous strata at the foot of the Mora range.

That part of the county which lies northwest of the Santa Fe Railway is of mountainous character. The main chain of the Rocky Mountains continues southward from the Sangre de Cristo Range in Colorado into New Mexico and as far south as Taos is fairly narrow, its width here being about 15 miles, exclusive of the eastern front range, called the Cimarron. South of Taos this great uplift of Carboniferous and pre-Cambrian rocks widens to the broad Mora Range, attaining a width from east to west of 30 or 40 miles. Mora River on the east side and the Pecos on the south have cut deep north-south canyons into this great mass. North of the town of Mora and up as far as Taos the whole is known as the Mora Range. In the latitude of Mora, at the head of the Pecos Canyon, the range splits in two, the westerly, higher, and rougher ridge being called the Santa Fe Range and the easterly, broader, and more level summits the Las Vegas Range. At the head of the Pecos the range culminates in a cluster of high peaks, mostly in the Santa Fe Range, known as Jicarilla (12,944 feet), Truchas (13,150 feet), Baldy (12,661 feet), and Lake (12,405 feet). The highest point in the Las Vegas Range is formed by the broad shoulder of the Solitario (10,258 feet). Between Pecos town and Las Vegas the mountains sink into high mesas of Carboniferous strata and "Red Beds." The Santa Fe Range drops more abruptly underneath the gently southward dipping "Red Beds" and Cretaceous sandstones at Apache Canyon, near Glorieta.

The geology of this broad uplift has been mapped in a preliminary way and described by J. J. Stevenson.[a]

The ranges consist in the main of flat or gently dipping Carboniferous strata, but they also contain two areas of pre-Cambrian rocks. One forming the core and main mass of the Santa Fe Range extends from a point near Taos to Apache Canyon, a distance of 50 miles; the other belt, more narrow, lies in the eastern part of the mountains and follows their whole extent down to the latitude of Las Vegas.

This system of ranges, like the Rocky Mountain system in Colorado, has the appearance of having been created by vertical uplift. There is little or no corrugation, but many monoclinal folds and a number of profound north-south faults. Such a profound break separates the Carboniferous at Santa Fe from the granites and schists of the main range and the latter again from the Carboniferous shallow syncline of upper Pecos River and the Las Vegas Range. In the latter repeated folding brings the pre-Cambrian up to the level of the Carboniferous. During earlier and possibly later Paleozoic time land areas existed in these ranges, as attested by the littoral character of the sediments. The mountains undoubtedly owe their present form to post-Cretaceous uplift and faulting.

In spite of a complicated structural history, there are few deposits of useful minerals in the county. Coal beds up to 3 or 4 feet in thickness but not proved extent are found in the upper Carboniferous in the Pecos Valley, one locality being at the Hamilton mine and another a few miles above Valley ranch. The metal deposits contain chiefly copper. In the Tecolote district, some miles southwest of Las Vegas, copper ores occur in Carboniferous and Cretaceous strata and a small production is recorded. At Rociada, 20 miles northwest of Las Vegas, copper-bearing veins in granites and schists have been prospected. At the Hamilton mine, on the upper Pecos, a copper deposit of some extent, occurring in schist, has been prospected during the last ten years.

---

[a] U. S. Geog. Surveys W. 100th Mer., vol. 3, supplement, 1881; also maps 77 B, 78 A, 69 D, 70 C.

COPPER DEPOSITS OF THE SANTA FE RANGE AND UPPER PECOS RIVER.[a]

GEOLOGY.

A good view of the structure of the Santa Fe and Las Vegas ranges is obtained along the road and trail from Santa Fe due east to Macho Creek on the upper Pecos, a distance of 14 miles. The old city of Santa Fe lies at the western foot of the range of the same name, at an elevation of 7,000 feet. Sloping gravel mesas, deeply dissected by recent gulches, extend westward for 12 miles to the Rio Grande. Half a mile above the city the first outcrops of the range appear. They consist of a down-faulted block of Pennsylvanian ("Upper Carboniferous") beds, possibly 3,000 feet wide; 2 miles east of the city is another small block of this kind.[b] The dip is 35° W.; the strata consist of heavy-bedded gray limestone rich in typical fossils and covered by about 100 feet of yellowish calcareous shale containing in places some coaly material. The conditions indicate a fault along the front of the Santa Fe Range of probably not less than 2,000 feet. From this fault to the crest, a distance of 9 miles, pre-Cambrian rocks constitute the bulk of the range; the divide is crossed at an elevation of 9,300 feet.

The first outcrops of the pre-Cambrian consist of mica schists whose appearance strongly suggests sedimentary origin; they dip west at gentle angles and are thoroughly injected by granite and pegmatite. Above these first hills the range consists in general of coarse reddish granite with numerous dikes of pegmatite. A few miles above the city, in the canyon of Santa Fe Creek, which is followed by the road, are some outcrops of a schistose amphibolite striking N. 20° W. and dipping steeply west-southwest. This amphibolite is intruded by pegmatite dikes, parallel to the schistosity. At the contacts epidote, garnet, and quartz have developed. About 7 miles west of Santa Fe some biotite schists striking N. 60° E. appear, as well as aplitic granites, but the normal coarse granite prevails. Near the divide a rough north-south sheeting is noted.

About 3 miles east of the divide, on the bluffs overlooking Macho Creek, the Carboniferous limestone is encountered and the aspect of the country changes. More abundant vegetation covers the ground than on the sandy hills of the granitic range; far up and down the Pecos Canyon, 2,000 feet lower, the sedimentary, almost horizontal rocks are to be seen; and across the Pecos rises Las Vegas Range, with its heavily forested broad plateaus and gently sloping sky line.

There is here undoubtedly an important fault line, along which there has been a downthrow of the limestone on the east side of approximately 2,500 feet. This fault line will probably be found to continue south-southwestward on the east side of Thompson and Penacho peaks of the Santa Fe Range. About 1 mile northwest of Canyoncito, on the road to Santa Fe, a fault contact between Carboniferous limestone and pre-Cambrian rocks is crossed; it may prove to be the same dislocation that crosses the upper part of Macho Canyon.

For 2 miles below and 3½ miles above the mouth of Macho Canyon Pecos River has exposed the pre-Cambrian granites and also some diorite. In places these older rocks appear to rise 600 or 700 feet above the bottom of the canyon. From this area southward no pre-Cambrian rocks are exposed in the canyon; neither are there any such exposures above, except a small area at the Hamilton copper mine, 7 miles above the mouth of Macho Canyon. The sedimentary rocks continue up the headwaters of the river far north of the Hamilton mine.

The Carboniferous strata are splendidly exposed in the canyon of the Pecos. The thickness has not been accurately measured along the river, but is probably not less than 3,000 feet. J. J. Stevenson measured 3,276 feet of Carboniferous sediments in the Mora Creek section,[c] which appears to be rather similar to the Pecos section; the latter is not described. The dip along the Pecos is very gentle, about 5° S.

---

[a] By Waldemar Lindgren. In order not to separate the description of closely related localities, a few small deposits in the Santa Fe Range and near Glorieta are here included, though situated over the line in Santa Fe County.

[b] First described by J. S. Newberry (Report of the exploring expedition from Santa Fe, N. Mex., to the junction of the Grand and Green rivers, Washington, 1876, p. 43.)

[c] U. S. Geog. Surveys W. 100th Mer., vol. 3, supplement 1881, p. 77.

On the whole, the series consists of sandstones, shales, and some limestones, in rapidly alternating benches. Near the Hamilton mine at least 1,500 feet of this sequence is exposed. There are here about twenty prominent benches of yellowish sandstone, separated by shale, weathering light gray, and a few limestone beds. The basal portion presents the following section:

Section of basal portion of Carboniferous rocks near Hamilton mine.

	Feet.
Fine-grained sandstone with mica scales	30
Coarse sandstone with angular fragments	8
Fine-grained sandstone with mica scales	10
Sandy limestone	45
Dense blue limestone with crinoid stems	60
Brownish sandstone, locally with fine-grained conglomerate	6
Granite and schist.	159

Well-preserved fossils indicating a Pennsylvanian age are found 250 feet above the basement in a gulch opposite the mine, on the west side of the river. Near by is a small coal seam. Few dislocations break the continuity of the series. Three miles south of the Hamilton mine is a monoclinal fold carrying up the granite to a height of 600 feet or more above the stream; 2 miles farther south a fault brings the contact down again nearly to the water level. Three miles below the mouth of Macho Creek a coal bed has been opened in the beds on the west side; the section here represented is probably in the lowest part of the series, about the same as opposite the Hamilton mine. At the river limestone is exposed, overlain by 110 feet of light yellowish-gray shales, at the top of which there is a seam, 2 to 3½ feet thick, of coal of apparently good quality, capped with yellow sandstone; the strata dip 6° to 8° SW. and the coal seam has been opened by an incline 300 feet long. Several hundred tons of coal has been mined.

A sample was taken in the Gilmore coal mine over a thickness of 3½ feet. An analysis of the air-dried material made by the United States Geological Survey runs as follows:

Analysis of coal from Gilmore mine.

Moisture	1.90
Volatile matter	38.50
Fixed carbon	48.60
Ash	11.00
Sulphur	5.00
British thermal units	12,620

South of the coal mine the beds are almost level for 4 miles to the Williams Valley ranch, where gentle southward dips begin again. A short distance south of Pecos town, 3 miles farther south, the Pennsylvanian series finally disappears below the "Red Beds," which continue for 3 miles until the edge of the Glorieta Mesa, followed by the Santa Fe Railway, is reached. The thickness of these "Red Beds" is somewhat doubtful; probably it is about 1,000 feet.[a] They are best exposed in their upper part along the Glorieta Mesa; they consist of dark-red clayey shales with some cross-bedded sandstones; in the Glorieta railroad cut the red color of the beds appears to be due to the rapid oxidations of glauconitic greenish material. At the top of the Glorieta Mesa rests 100 feet of a yellow sandstone believed by Stevenson to be the Dakota. Probably the larger part at least of the "Red Beds" at this place belongs to the Manzano group of Gordon and Lee, representing the upper part of the Pennsylvanian series. No beds of gypsum were observed.

On La Cueva Creek, between Glorieta and Pecos, an interesting series of beds is exposed, but their relations to the beds at Pecos and Glorieta are not altogether clear. They consist

---

[a] J. S. Newberry (op. cit., p. 48) gives the following section of the beds in Pecos Valley:

	Feet.
Yellow and brown foliated sandstone with strata of red, purple, and gray shale	200
Yellow massive sandstone	150
Red, white, and green soft calcareous sandstone	800
Red and green sandstone and conglomerate, separated by thicker beds of green, blue, and purple shales, with oxides of iron and copper and with ferns, *Walchia* (?), and *Calamites*	200
Limestone and sandstone (Carboniferous) to bed of stream.	

of a thick, gently southward-dipping succession of white or reddish, extremely coarse arkose and beds of angular granitic detritus, with a few thin beds of limestone showing crinoid stems. These extremely coarse sediments, indicating proximity of shore and doubtless derived from the slopes of Thompson and Penacho peaks, suggest that granitic bed rock was exposed somewhere in this vicinity at the close of the Carboniferous. This again suggests an unconformity somewhere between the light-colored beds of lower Pennsylvanian age at Pecos and the typical red shales of the Glorieta Mesa. (See below under "Copper deposits in Paleozoic sandstones.")

IRON DEPOSITS.

The Kennedy iron mine is situated on top of the Glorieta Mesa, 2 miles south of Fox siding and 5 miles south-southeast of Glorieta. After ascending 400 feet over the "Red Beds" and the Dakota(?) sandstone, which caps the mesa, the road continues for 1 mile southward over the sandstone, which dips 10° S. The iron deposit outcrops in the bottom of an open little valley in the sandstone and forms a bed of dark-red, earthy hematite about 3 feet thick. It is covered by several feet of yellow clay mixed with concretions of hematite or limonite.

Probably several thousand tons of good-quality ore have been mined from this locality and shipped as flux for the lead smelters at Socorro and El Paso. Much of this ore has been taken from open cuts, but there are also underground workings where the seam dips underneath the side slopes. They are, however, caved, and it is thus difficult to judge of the extent of the deposit. The horizon is probably in the uppermost "Red Beds," of which the iron ore would seem to form an integral part. No work was being done in 1905.

COPPER DEPOSITS IN PALEOZOIC SANDSTONES.

A number of copper deposits known as the Kunklin prospects are located 2½ miles northeast of Glorieta, in the sandstones of La Cueva Creek, which have been mentioned in the preceding pages. The horizon is probably in the uppermost Carboniferous. In a bed of limestone with crinoid stems exposed in the lower part of the creek 100 feet above Long's ranch grains of chalcocite may be found. The arkose beds 1¾ miles farther up, on the west side, contain many cuts and short tunnels exposing cupriferous strata. The copper ores appear in the form of chalcocite, azurite, and malachite and seem to be most abundant where the sandstones contain most organic remains. Ore has been found at two horizons at least, 50 feet apart on the south side of the creek and about 250 feet above it, but it appears to be of too low grade for profitable working. The beds dip about 4° S. and appear to follow the creek, so that the cupriferous sandstones would be at a somewhat higher horizon than the limestones at Long's ranch. The relations between these coarse grits and the strata of the Pecos section are not clear.

COPPER AND GOLD DEPOSITS IN PRE-CAMBRIAN ROCKS.

GENERAL FEATURES.

The Santa Fe Range contains no mining districts of importance. A little flaky gold is said to occur in Santa Fe Creek. Five miles from Santa Fe a pegmatite dike breaks up through amphibolite, and this dike is reported to carry a little gold. Seven miles east of the city, on the south side of the canyon, are some prospects, one of which is called the Mikado. It shows a vertical fissured zone in aplitic gneiss, parallel to the schistosity and striking N. 54° E. The best ore is stated to assay $6 in gold and 6 ounces in silver to the ton and contains also some galena and zinc blende. Specimens show a fine-grained mosaic of quartz, microcline, and some other feldspars; also small foils of green biotite and muscovite. Grains of galena, zinc blende, and chalcopyrite are directly embedded in the rock-forming minerals, suggesting that the ore was formed under deep-seated metamorphic conditions.

Other prospects of copper and gold ores are reported to occur 6 or 7 miles farther south, on the slopes of Penacho Peak, to the northwest of Glorieta. Among these, the Bradley mine has the most extensive development. Deposits of molybdenite are reputed to have been found recently in the Santa Fe Range.

On the east side of the range, due east of Santa Fe, are several copper prospects, among them the Dalton, on the creek of the same name, and the Jones and Mailleuchet prospects on Macho Creek, 5 and 4 miles, respectively, above its confluence with Pecos River. At the Mailleuchet prospect a fresh pre-Cambrian granite outcrops in the creek; it contains microcline, quartz, and oligoclase, with biotite and some large grains of titanite; in places this granite shows a dioritic facies. Beginning at the tunnel and continuing downstream for about 1 mile is a mass of dark-green amphibolite, schistose in places, and there containing garnets. The granite is probably intrusive into this rock. Quartzose streaks in this amphibolite contain the ore, which under the microscope is seen to be a quartzitic schist of irregular quartz grains, intergrown with aggregates of light-green mica and grains of pyrrhotite, chalcopyrite, and much zinc blende. As in the Mikado prospect, previously described, the metallization appears to be directly connected with the metamorphism.

HAMILTON MINE.

The Hamilton copper mine of the Pecos Copper Company is situated in the canyon of Pecos River at the confluence of Willow Creek, 13 miles north of Pecos town and 17 miles east-northeast of Santa Fe. The elevation is 8,000 feet. The deposit was opened about 1882, but the more extensive development has been undertaken only during the last few years. It is also known as the Cowles mine, from the name of the principal owner in the company. In 1905 the developments consisted in a tunnel at the level of Willow Creek, just east of the river, and a vertical shaft 180 feet deep, its collar being 70 feet above the creek. Since then developments have been continued and the shaft sunk to the 400-foot level. A few carloads of selected ore have been shipped.

At the mouth of Willow Creek is exposed a small area of the basement rock upon which the prevailing Carboniferous beds rest. The dip of the sediments is about 7° W., so that at the west bank of Pecos River they reach down to the water level. A small area of pre-Carboniferous (and also pre-Cambrian) rocks is exposed on both sides of Willow Creek at its mouth, and was followed for 1,000 feet upstream. The croppings of the deposit are contained in this area of old rocks. The prevailing rock is an amphibolite, locally with foils of biotite and showing more or less perfect schistose structure. The microscope reveals small bluish-green prisms of hornblende extending in all directions; there are also imperfect crystals of a colorless epidote. Between these grains lies a mosaic of feldspar grains, in part certainly andesine or labradorite. There is some pyrite in fine division, also veinlets of quartz, pyrrhotite, and pyrite, as well as serpentinoid aggregates, products of decomposition of the hornblende.

The deposit forms a zone or belt 60 feet wide in this amphibolite parallel to its strike, or N. 50° E.; in this belt the amphibolite is in places changed to chloritic schist; it contains large scales of biotite and needles of tourmaline, as well as irregular masses of pyrite, chalcopyrite, and zinc blende. The best ore, of which some has been shipped, is said to contain 17 per cent of copper; the average of the whole mass would doubtless be of low grade. The ore in places carries gold to the extent of a few dollars to the ton, especially where it is more quartzose, as in the croppings. Encouraging gold values are said to have been obtained on the fourth level. The silver content is said to average 5 ounces a ton.

The actual croppings form a bluff on the north side of Willow Creek, at the tunnels; they have a highly variable appearance, being gray or brown, in part cherty, with transitions to amphibolite; in other places they are honeycombed, with clear evidence of oxidation and dissolution of sulphides. The croppings are said to contain no zinc and very little copper.

The tunnel level penetrates the ledge obliquely in a north-northeast direction for 75 feet, the actual width perpendicular to the schistosity being at least 60 feet. The tunnel shows much zinc blende but little chalcopyrite. In 1905 the first level of the shaft extended 180 feet N. 50° E.; largely in fair-grade copper ore, and about parallel to the schistosity. A crosscut to the northwest about 40 feet long reached the barren amphibolite, but a crosscut 40 feet long on the southeast was still in ore. The water level in the shaft is 20 feet below collar. Some of the ore consists of coarse dark-green hornblende of varying grains, intergrown with pyrite and chalcopyrite. Quartzose streaks of unoxidized ore contain a mosaic of quartz grains with cubes of pyrite, the latter intergrown with a pale-green mica, probably a biotite, identical with that from the Mailleuchet prospect. Other specimens contain, in a fine-grained amphibolite, well-developed prisms of bluish-gray tourmaline, inclosing grains of pyrite and hornblende. A zinc blende of very dark color accompanies the hornblende and chalcopyrite, and adds a difficult question to the problem of treating this ore. Galena occurs sparingly. Small seams of fluorite and yellow zinc blende form apparently secondary segregations.

The intergrowth of the ore minerals with the hornblende of the amphibolite and the absence of well-defined fissures make it evident that the metallization was almost contemporaneous with the metamorphism which produced the amphibolite from some dioritic or diabasic rock, and in this respect the deposit is similar to the prospect previously described. The appearance of biotite and tourmaline as gangue minerals indicates, moreover, the presence of conditions of high temperature and pressure during ore formation. It is suggested that the metals may originally have been contained in the basic magma and that their concentration was effected during metamorphism.

The most interesting feature of this deposit is its relation to the Carboniferous strata. Its croppings are immediately overlain by these rocks, and the contact is well exposed on the surface and in several cuts and short tunnels. The lowest sandstone stratum rests directly on the level surface of the oxidized ore. This sandstone is made up chiefly of quartz grains, but the microscope shows the presence of chloritic and serpentinoid cement, as well as of many rounded fragments of a brown cherty material which looks exactly like the oxidized ore. Nearer to the river the sandstone is replaced by a fine-grained conglomerate, with partly rounded pebbles reaching at most 1 inch in diameter. Most of these pebbles are of quartz, but some consist of a brown chert identical with that of the croppings. In places the cherty brown croppings are brecciated near the contact with sandstone and indistinctly show partly rounded fragments, conveying the impression that the uppermost part of the deposit consisted of subangular fragments at the time of the deposition of the Carboniferous strata. The Carboniferous in this vicinity contains no ores of any kind.

It is thought that these exposures prove conclusively the pre-Carboniferous age of this deposit. As the early Paleozoic formations, both in southern New Mexico and southern Colorado, are conformable with the Carboniferous and no epoch of mountain building and regional metamorphism intervened, it follows that the Hamilton copper deposit must be of pre-Cambrian age.

ROCIADA DISTRICT.[a]

GENERAL FEATURES.

The Rociada district takes its name from the town of Rociada, which is situated in the northwestern part of San Miguel County, close to the Mora County line. The Las Vegas topographic sheet of the United States Geological Survey shows the town, formerly known as Rincon, to be about 20 miles northwest of Las Vegas, but the mail and stage route, via Los Alamos and Sapello, is 32 miles. Mineral deposits are said to have been discovered in the region in 1900. Only a small amount of mining work has been done. All properties were idle when visited in August, 1905.

The geology of this region, although rather simple, is nevertheless interesting as throwing much light on the structure of the ranges of northern New Mexico. Las Vegas, situated at the western edge of the deeply dissected plains country, is built upon the Cretaceous shales. Immediately to the west rises the prominent north-south ridge formed by the upturned and steeply dipping strata of the Dakota (?) sandstone. West of this the red sandstones and sandy shales which have commonly been referred to as the "Jura-Trias" lie at a gentle angle, in places overlain by caps of the Dakota (?) rock to form prominent buttes. Still farther west the Carboniferous strata are exposed and with increasing easterly dips give way finally to the pre-Cambrian core of the Las Vegas and Mora ranges. These disturbances of the strata are in several places

---

[a] By L. C. Graton.

accompanied by faulting which causes a repetition of the strata, and which, probably more than the folding and tilting, accounts for the elevation of the range.

Rociada lies near the eastern foot of the range, in the region where the Carboniferous rocks alternate by faulting with those of the pre-Cambrian core. The town is situated in a beautiful and fertile valley, surrounded by low hills composed mostly of pre-Cambrian gneisses and schists; a few miles to the west is seen the remarkably even crest of the lofty Mora Range. The hill to the north of Rociada is made up of gneisses and schists, and some of the latter are of sedimentary origin, for bands of crystalline limestone and of micaceous schist alternate with quartzitic bands. The strike of these rocks at the Good Hope mine is N. 48° W. and the dip 60° SW. The Carboniferous rocks east of Rociada contain much limestone, but west of the town only the lower portion of the Carboniferous section appears and this consists of sandstones and coarse arkoses. Between the town and the axis of the range Carboniferous rocks are brought down into the pre-Cambrian rocks three times by faulting.

The mineral deposits are mostly referable to the pre-Cambrian. They consist of fissure veins in the ancient gneisses and schists. Copper is present in Carboniferous strata, however, in some places. A small smelter was erected near the town, but was operated for only a short time. There are two localities in which prospecting has been done—one close to Rociada, north and west of the town, the other at Hadley post-office, 5 miles west of Rociada, on the steep slope of the Mora Range.

MINERAL DEPOSITS.

PROSPECTS NEAR ROCIADA.

On the hill north of Rociada the Sam Adams claim is located on a group of quartz stringers in a schistose granite. The veinlets, which are parallel to the foliation of the country rock, strike N. 45° W. and dip 40° SW. A few of the quartz seams cut across the schistosity. An inclined shaft, whose depth was not ascertained but is probably not more than 100 feet, follows one of these quartz veins. The ore seen on the dump is much oxidized. Pyrite was evidently present and has left limonite in its place. Streaks and bands of galena only partly oxidized are present in the quartz and are associated with what is probably oxidized zinc blende. The adjoining schistose wall rock is much altered, being silicified, sericitized, and somewhat impregnated by pyrite. Copper stain is seen in places. It is said that assays of total values as high as $60 a ton have been obtained, gold and silver being most important, but the zinc content lowers the commercial value of the ore.

At the Loring mine, a little farther south on the same ridge, an incline shaft has been sunk 140 feet on a quartz vein carrying chalcopyrite and also some galena.

The Good Hope mine is northwest of the claims just mentioned, a little north of the village of Upper Rociada. A 90-foot incline has been sunk in crystalline limestone along its bedding, which strikes N. 48° W. and dips 60° SW. Silicified streaks through this limestone parallel to the bedding carry disseminated chalcopyrite, which near the surface is changed to azurite. The hanging wall of this ore-bearing zone appears to be fairly pure calcite; the foot-wall limestone is impure and micaceous. A drift from the bottom of the shaft is said to strike a body of ore containing galena and good values in gold and silver. Work was temporarily suspended during the summer of 1905.

The Joe and Jenny mine lies about one-third mile west of Rociada. The country rock is a sheared and much-jointed granite traversed by many quartz veinlets. A quartz vein 30 inches wide, trending N. 15° W. and dipping 35° E., has been explored by an incline said to be 240 feet long. The vein shows some cuprite and malachite near the surface and contains chalcopyrite at greater depth. It is said that gold accompanied the sulphide. Some good assays have been obtained, but from the small lot of ore shipped unsatisfactory returns were obtained.

A short distance southwest of the Joe and Jenny prospect several openings have been made in the Carboniferous grit or coarse sandstone, which here rests flatly upon the gneissic granite. The grit contains disseminated copper—mostly as carbonate, although a little glance is present. Considerable excitement prevailed at the time these developments were made, but no production has resulted. The grade of the ore is probably too low to be profitable. Underneath the sediments a shallow trench discloses a quartz vein in the much-sheared granite. The vein, which is vertical and strikes N. 20° W., is parallel to the sheeting of the granite and incloses a horse of that rock. The quartz holds masses of limonite which retain the form of the pyrite crystals from which they were derived. Whether or not this vein contains any gold or copper is not known. The interesting feature concerning it is that it appears certainly to be cut off and covered by the Carboniferous sediments which overlie the granite; this fact affords another proof of the pre-Carboniferous age of the deposits, and as no period of mineralization is known to fall within the early Paleozoic in this region it is also undoubtedly pre-Cambrian.

Other pits in medium-grained Carboniferous sandstone have been made one-fourth mile to the west. About a mile farther west the sedimentary rocks are turned up rather sharply so that they dip strongly to the east, and within a short distance they disappear and are replaced by granite. It seems probable that a fault separates the sedimentary from the massive rock.

DEVELOPMENTS AT HADLEY.

At Hadley most of the work has been done by O. A. Hadley, ex-governor of Arkansas. Three claims have been developed to some extent. The prevailing rock is a gray micaceous schist which is composed chiefly of quartz and biotite and which may have once been a sedimentary rock. A series of northwesterly veins cuts this rock and produces very marked alteration close to the veins. The Azure shaft, 150 feet deep, explores one of these veins which dips steeply northeast. Specimens seen on the dump indicate that quartz and calcite are the principal constituents of the vein filling. Epidote, garnet, and amphibole have been developed in the wall rock, and the product in some places closely resembles that derived from the contact metamorphism of limestone. In many places these metamorphic minerals have been developed in the quartz as well as along the sides of the vein. On the 100-foot level a 50-foot drift along the vein, which is 2 feet wide, is said to have encountered good chalcocite ore. At a depth of 115 feet water was struck. It is said that on the 150-foot level the vein carried good values, but was too narrow (14 inches) to make work profitable. Bornite, locally surrounded by glance, is the principal copper mineral at this level. Molybdenite is present in the calcite. The content in gold is reported low. It was reported that a little ore had been shipped from this mine.

The Rising Sun shaft, a short distance northwest of the Azure, is 170 feet deep. It is in schist, with low-grade copper values. A little crosscutting was done at the 100-foot level, but the heavy flow of water caused work to stop.

The Sammock prospect is situated about one-third mile northeast of the Azure. A shaft in schist exposes numerous quartz stringers which develop abundant garnets in the surrounding schist. Some heavy, partly oxidized pyrite ore, apparently containing considerable zinc, is said to carry a few dollars to the ton in gold. Only a small amount of work has been done and so far as known no ore has been shipped.

It is said that quartz veins carrying much tourmaline occur higher in the range and some specimens from these veins were seen.

TECOLOTE DISTRICT.[a]

GENERAL FEATURES.

The Tecolote district comprises many prospects and groups of claims scattered over a considerable territory in western San Miguel County. It includes the Tecolote, or Las Vegas, Salitre, San Pablo, and San Miguel camps or subdistricts. These places lie to the west and southwest of a prominent butte some 6 miles southwest of Las Vegas, known as Tecolote Mountain. This butte is shown near the southern boundary of the Las Vegas topographic sheet of the

---
[a] By L. C. Graton.

United States Geological Survey, and the principal camps lie in the northern or northwestern part of the area shown on the Bernal sheet, which adjoins the Las Vegas quadrangle on the south. The nearest shipping point is Chapelle (Bernal) which lies 8 or 10 miles south of the district on the Santa Fe Railway.

The occurrence of copper widely disseminated in rocks of this region seems to have long been known but it was not till about the beginning of the present century that serious attempts were made to ascertain the importance of these deposits.

The geologic features of this region are similar to those of the Rociada district, already described. In the Tecolote district, however, erosion has not carried away so much of the younger rocks, and the Dakota(?) sandstone and the "Red Beds" are exposed over large areas, while the Carboniferous is in the main confined to a rather narrow belt of steeply tilted strata on the eastern slope of the southern continuation of the Mora Range. The Dakota(?) is a thick-bedded, white, fairly fine grained sandstone which forms elevated and little-dissected mesas, usually bounded by bold scarps. The "Red Beds," which lie conformably between the Dakota(?) and the known Carboniferous, may belong either to the Carboniferous, or to the Triassic, or to both; it is not known that fossils have ever been found in them in this portion of the Territory. Tecolote Mountain is composed mainly of these red sandstones and shales, which here must have a thickness of over 300 feet, and is capped by a practically horizontal mass of the Dakota(?) sandstone.

Faulting has been prominent in this locality. The Dakota(?) is repeated and much tilted by faulting, and near the contact of the "Red Beds" and the Carboniferous step faulting has repeated the beds several times and in one place brings the pre-Cambrian through the Carboniferous.

The mineral deposits consist mainly of copper disseminated through beds of the sedimentary rocks. The only locality visited was the Tecolote subdistrict, which lies to the south and southwest of Tecolote Mountain. It includes the Blake and Bonanza mines, and the deposits of Burro Hill and a few other scattered localities.

MINERAL DEPOSITS.

BLAKE MINE.

The Blake mine is situated close to Bernal Creek about 12 miles southwest of Las Vegas. It is but little developed, the workings consisting of a small open cut and a 25-foot incline. A leaching plant of 50 tons daily capacity has been built close to these workings, with which it is connected by a tramway. After coming from the crusher, the ore is passed through dry rolls which reduce it to about 16-mesh. It is then treated in vats with a solution of sulphuric acid containing a little hydrochloric acid. The filtered solution is then subjected to the action of iron in the shape of old "tin" cans which are held in a perforated drum that revolves in the solution. The copper is thus precipitated in the metallic state and settles in the vat along with sediment from the solution and impurities from the cans. Up to the middle of August, 1905, about 5,000 pounds of 70 per cent cement copper had been produced during trial runs. At that time the property was idle, lack of funds for further development being the reported cause.

Bernal Creek at this place is cut in the sandstones and shales of the "Red Beds." Its east bank is a sharp bluff showing a thickness of 200 feet or more of these rocks in horizontal position covered conformably by a hard cap of the Dakota (?) sandstone, whose upper surface forms a large mesa. The west bank of the creek rises gently to the crest of the Mora Range, the summits of which here attain a little over 7,000 feet in elevation. Bernal Creek must occupy the line of flexure of a distinct fold, for while the "Red Beds" are flat-lying east of the stream, they have a marked eastward dip on the west side, and toward the west this easterly dip continues increasingly to the very base of the sedimentary rocks, where the bedding diverges only slightly from the vertical. As a matter of fact, the inclination of the beds is a little to the north of east, for the prevalent strike is about N. 25° W. The lowest portion of the "Red Beds" is a soft clayey shale. Under this are the characteristic coarse arkoses of granite-like appearance which belong to the Carboniferous. These alternate with richly fossiliferous limestones. A narrow band of shale somewhere near the middle of this series carries a seam of coal having a maximum width of 2 feet. Limestone is present near the bottom of the series almost down to the pre-Cambrian contact. The pre-Cambrian consists of granitic gneiss, cut by pegmatite dikes and scattered quartz veins. It is somewhat sheared along steeply dipping planes parallel to the range, but is little decomposed.

Faulting has taken place along the strike at several places. Toward the west from Bernal Creek red shales and light-colored grits alternate repeatedly and appear to belong to the bottom of the "Red Beds" and the top of the Carboniferous, respectively, indicating faulting instead of alternate sedimentation. About one-fourth to one-third mile east of the main pre-Cambrian contact a narrow ridge of pre-Cambrian granite and pegmatite is faulted up through the Carboniferous, and at this place a large spring issues. A sketched section of the structure is shown in figure 8.

The ore of the Blake mine is the arkose grit immediately underlying the westernmost band of red shale. It is doubtless the top of the Carboniferous. This ore-bearing band dips about 30° ENE. and is about 8 feet thick at the mine openings. Its outcrop is said to be traceable for a long distance, but its value is unproved at most points. Copper occurs similarly but in apparently smaller amount in bands of grit, which alternate with red shale on the low ridge

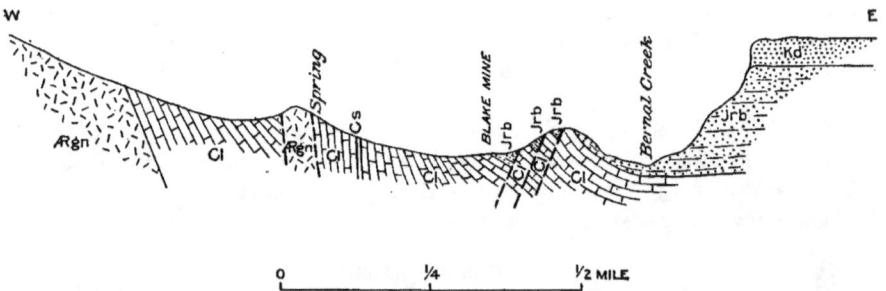

FIGURE 8.—Section through Blake mine, Tecolote district. Kd, Dakota (?) sandstone; Jrb, "Red Beds;" Cl, Carboniferous limestone and grit; Cs, Carboniferous shale with coal seam; Ægn, pre-Cambrian granite, pegmatite, and gneiss. Dashed lines are probable lines of faulting.

between the mine and Bernal Creek. No copper has been found in the seam of coal in the shale, which lies lower in the series than the ore-bearing zone, nor in any of the "Red Beds." The ore consists of small particles of copper minerals disseminated through the grit. The basic carbonate, malachite, is the predominant copper mineral. It is present in small tufts of tiny fibers, which in the aggregate make little bunches of the mineral and, where plentiful, give a greenish color to the rock. Azurite, the blue copper carbonate, is present in sparing amount. Malachite and azurite, however, are not the original copper minerals and can be seen to have been derived by oxidation from sulphides. The ore is said to range from 1.5 to 5 per cent of copper, and the quantity milled is said to have yielded about 2.8 per cent. No data were available in regard to the amount of silver present in the ore; it is doubtless small. Developments are not sufficiently extensive to permit much generalization as to the distribution of the copper, but the values seem to vary erratically. For example, in the midst of ore of average grade near the bottom of the incline an irregular mass or horse of almost barren grit was struck. Except for its much lower content of copper, this material, so far as could be seen, was identical with the ore.

Microscopic examination of this ore shows that the arkose or grit is made up of abundant quartz, orthoclase, microcline, calcite, and dolomite, with a little plagioclase and here and there a flake of muscovite. The two carbonates, calcite and dolomite, are present almost wholly as a cement between the other grains, and may have been introduced from solution since the deposition of the other minerals. They serve to close up the interstices which would otherwise exist, and though the resulting rock is not impervious, it is far less porous than it would be but for the carbonate cement. The grains of the rock consist for the most part of fragments of crystaline individuals from the parent granite; in other words, the disintegration of the granite effected the separation of the constituent grains. Rarely, however, single particles of the grit consist of fragments of two or more crystal individuals joined as in the original granite along an intimate boundary of crystallization. Since consolidation, the rock has suffered crushing or some other strain that has produced fractures which break across many of the mineral grains, indicating that the grains were probably already cemented at that time. These fractures, like the interstitial cavities, are generally filled with the carbonate minerals, calcite and dolomite. The rock is commonly even grained, and, so far as observed, there is a complete absence of those accessory minerals so common in granite, such as magnetite, apatite, zircon, etc., except where they may be included within individual grains of the grit. There is nothing to indicate that sulphides were present in the granite or in the grit at the time of its deposition. Neither is there evidence that carbon, carbonaceous matter, or other reducing material was present in the sediment. The feldspars are not greatly altered, and are, in fact, fresher than would naturally be expected in a water-laid rock. Part of the orthoclase has been somewhat kaolinized, though nowhere has this alteration progressed sufficiently to obscure the cleavage or twinning. In the most kaolinized portions some calcite has been introduced and a very little sericite has developed. Nearly all the microline is practically fresh. These several facts seem to indicate that erosion of the granite mass which furnished this material proceeded so rapidly that although thorough disintegration was permitted, the granite was not allowed to suffer any marked alteration; that assortment was carried on quickly but effectively; that deposition was rapid, and finally that long-continued circulation of superficial waters has not gone on throughout the mass of the rock.

Into this rock have been introduced the copper minerals, chalcocite, bornite, and chalcopyrite, named in decreasing order of abundance, and also a little pyrite. That these minerals were not present in the constituent grains, either when the grains were a part of the granite or when they were deposited as sediment, there is abundant proof. The sulphides are present mainly in the carbonate cement, and it is possible that part of the calcite and dolomite was introduced with them. With the exception of pyrite, which commonly occurs in tiny cubes, these sulphides are present in irregular grains and patches. In some places they occupy the full width of the space between the fragmental minerals of the rock; in others they are wholly surrounded by the carbonate cement. Pyrite occurs both as minute isolated cubes in the cement, especially hugging the boundaries of the fragmental minerals, and as grains of similar size and the same form embedded in chalcocite or bornite. Chalcopyrite is of similar occurrence except that it does not have crystal form, and many of its grains are larger than the pyrite crystals. Most commonly it lies within chalcocite or bornite. The two yellow sulphides, pyrite and chalcopyrite, appear to have been deposited slightly earlier than the darker, copper-rich sulphides. Chalcocite is much the most abundant sulphide. It is intimately and irregularly intergrown with bornite, which in some places incloses and in others is inclosed by it. These two minerals seem to be contemporaneous. There is no evidence whatever to indicate that either the chalcocite or the bornite has been derived from pyrite or chalcopyrite by deposition of copper from sulphate solutions according to the process known as secondary sulphide enrichment. All four minerals appear certainly to be primary so far as their deposition in this rock is concerned. The chalcocite especially, and the bornite to a small extent, are not confined to the calcite-filled space between the original grains. They enter many fractures and cleavage planes in the quartz and feldspars, and even invade the actual mass of these minerals by replacement. Feldspar has been most replaced, especially the more kaolinized individuals of orthoclase, but microcline and to a less extent quartz have been partly dissolved and chalcocite has been deposited in their stead. In some specimens early stages of the replacement of feldspars are to be seen, in which a spongy mass of chalcocite is included in the feldspar individual near to a calcite and chalcocite filled space or channel, and this spongy mass is metasomatically invading the feldspar, presenting an appearance somewhat analogous to the "skeleton crystals" of muscovite observed in some pegmatites. Elsewhere replacement by chalcocite of one part of a feldspar grain has been complete, but the lines of cleavage still present in the unreplaced or little-replaced portion can be traced into the solid chalcocite of the replaced portion. It is impossible to decide from certain examples whether chalcocite actually replaced feldspar or replaced calcite that had already been formed from feldspar, for calcite is present in unreplaced portions of feldspar grains that have been more or less replaced by the sulphide in another part of the same individual. It is perfectly clear, however, that in other individuals direct replacement of feldspar by chalcocite has occurred; for example, it can be observed that perfectly fresh grains of microcline have been partly replaced, especially along fractures.

In one specimen a grain of considerably kaolinized orthoclase has an irregular outer zone and many veinlike streaks which are brighter and more transparent, as if the kaolin had been for the most part removed. In these clearer portions are included minute particles of calcite and of chalcocite, but neither mineral is present in the turbid interior. Just at the border between the brighter and the turbid portions is a narrow, irregular, and locally interrupted band of chalcocite. These facts suggest that cupriferous solutions gradually penetrated the turbid feldspar grain from its periphery and from cracks; that these solutions attacked chiefly the kaolin and, as they advanced, deposited calcite and chalcocite in the clarified part of the feldspar in place of the removed kaolin; that deposition of chalcocite was greatest where the solution met unattacked kaolin; and that the band of chalcocite so deposited progressed inward with the front of the advancing solution.

The foregoing description of the microscopic character of the ore applies only to that portion which has escaped marked oxidation. Little of such ore has been encountered at the Blake mine, although it should be expected to prevail at no great depth provided the ore horizon is at all extensive. Most of the ore has undergone more or less oxidation and, as already stated, the most abundant ore mineral is malachite, with some azurite. Native copper, cuprite, and chrysocolla appear to be absent. Thin sections of the ore show well the gradual transformation of chalcocite into the carbonates. The change almost invariably begins at places where the cuprous sulphide is in contact with the calcite-dolomite cement. From such places the zone of copper carbonate gradually encircles and encroaches upon the chalcocite until in the final stage all the copper has been converted into carbonate. In some of the rock, azurite appears to form first and later to alter to malachite, so that a narrow band of the blue carbonate separates the black sulphide from the green carbonate. Generally, however, so far as can be seen, malachite is formed directly. The copper also extends out from the original chalcocite grains into the calcite cement as a pale-green stain, which even with a high power appears uniformly distributed and can not be resolved into definite particles of malachite. The appearance suggests that part of the calcium may have been replaced by the cupric ion, thus giving an isomorphous intergrowth of ordinary calcite with a small amount of the unknown anhydrous cupric carbonate. Except for the green color, the optical properties of this material can not be seen to differ from those of calcite, but it is possible that in masses larger and better individualized than those present some difference might be detected. Close to the sulphide cores and intimately associated with malachite, especially as inclusions or in little protected embayments in the malachite, are very small patches that consist of aggregates of minute particles of uncertain form and poorly determinable properties. The colors of polarization appear low and the refractive index is much below that of either malachite or calcite and apparently not greatly different from that of balsam. No certain identification of the mineral can be made, but the association suggests that it may be gypsum, as that mineral would be produced by the interaction of calcite with either the cuprous or the cupric sulphate that would almost certainly be formed sooner or later in the oxidation of chalcocite. The scanty optical observations, so far as they go, are in accord with the supposition that this is gypsum.

BONANZA MINE.

The Bonanza mine lies about 4 miles northeast of the Blake and about $2\frac{1}{2}$ miles south-southwest of Tecolote Mountain. It is situated on the northwestern slope of the mesa known as Burro Hill and about a mile southwest of Tecolote Creek, the main drainage line of the vicinity. But little underground development work has been done. A small leaching mill has been erected. No work was in progress and no one was on the property at the time of visit. It was learned in the district that the percentage of copper ranged about the same as at the Blake mine, but that the ore was more pockety.

Horizontal strata of the "Red Beds" underlie the valley bottoms and are exposed part way up the sides of the mesa divides, which are made up of conformably resting Dakota (?) sandstone. About 3 miles to the west are the upturned Carboniferous beds resting on the eastern flank of the pre-Cambrian core that constitutes the main range. The workings of the Bonanza mine are in the Dakota (?) sandstone just above the "Red Beds." The rock is nearly white and of rather fine grain. Like that of the Blake mine, it is an arkose, consisting of quartz and feldspar, cemented by calcium and magnesium carbonates. The proportion of quartz to feldspar is rather greater than in the Carboniferous grit at the Blake mine, and the feldspar comprises microcline with a somewhat smaller amount of plagioclase and a little orthoclase. Other minerals are not observed. The microcline and plagioclase are fairly fresh on the whole, but the orthoclase is very turbid owing to the development of kaolin. Individual layers or beds are of even grain and hence well assorted, but a variation in coarseness among the different beds is distinct though small, and indicates that changing influences were effective during deposition. The great part of the rock is of fine grain and the layers that are a little coarser are as a rule thin, commonly less than an inch, and relatively rather widely spaced. As the grains are smaller than those of the Carboniferous grit, the interstices are also smaller, and the carbonate cement which fills the interstices occurs therefore in thinner films and smaller particles than at the Blake mine.

The ore mineral is malachite chiefly; but, as at the Blake mine, this is the result of oxidation of chalcocite. The copper has been deposited principally in the calcite and dolomite. Replacement of the fragmental minerals of the rock appears to have taken place to a small extent but, possibly owing to the greater obscurity attending more advanced oxidation in the specimens examined, is not so well indicated as in the Blake ore. Material similar to that which was considered to be possibly gypsum in the Blake is present in the Bonanza ore in similar association with malachite and calcite. Instead of being fairly evenly distributed through what may be classed as ore, as at the Blake mine, the copper here is largely confined to narrow bands parallel to the bedding and gives the ore a somewhat striped appearance. Certain of these greener bands lie on both sides of one of the coarser layers of the rock, as if this coarser portion had offered relatively easy passageway for the copper-bearing solutions but had been less inducive to precipitation than the finer-grained material alongside. Owing to the small size of the interstitial masses of carbonate cement and the probably small amount of replacement of other minerals of the rock by chalcocite, the individual grains of copper minerals are on the average somewhat smaller than at the Blake mine. This size of grain would become an important matter if ore should be found containing sufficient chalcocite to make mechanical concentration possible.

OTHER DEPOSITS.

Near the bend of Tecolote Creek, about 1½ miles northeast of the Bonanza mine, two small custom leaching mills had been erected in 1905, although one, known as the Ute mill, had not yet had its machinery installed and the other was not in operation. These mills were designed to treat ores which have been found in the Dakota (?) sandstone near the east side of Burro Hill, not far from the little plaza of Tecolote. These deposits were not seen, but specimens from them indicate that the ore is of the same character as the Bonanza ore, except that the greater part of the copper is unoxidized and occurs as glance which is covered with thin films of malachite. The specimens seen show a more even distribution of the copper than in any material seen at the Bonanza mine and are richer than any ores seen at either the Blake or the Bonanza, carrying probably 8 per cent of copper at least. As to the extent of these deposits and their average value no important information was gained. It is said that they have been but little developed, though to judge from the specimens they would appear to afford as favorable an opportunity for exploitation as any deposits yet known, especially in view of the possibility of taking advantage of their sulphide character and working them by concentration.

Two miles south of the Blake mine, on property owned by the Blake Company, chalcocite and bornite are present in the Carboniferous grit. The constituents of the rock are very little altered, and the copper sulphides have undergone almost no oxidation. A little pyrite and chalcopyrite are present, and in every respect the ore is identical with the freshest found at the Blake mine. It is stated that some of the ore is of good grade, but that the values are irregularly distributed and pockety. From this point copper is said to have been found at intervals along a zone parallel to the range for at least 8 miles north-northwest of the Blake mine. Specimens from the San Pablo district, which lies about 6 to 7 miles in this direction from the Blake mine, are of the same Carboniferous grit containing partly oxidized chalcocite and bornite. Azurite appears to be more abundant than in ores from other portions of the region and occurs as blue patches or kernels in which lie particles of glance and bornite; malachite surrounds the whole.

Patches of Carboniferous sediments are said to occur at several points along the crest of the Mora Range, and some of them extend down on the west slope and connect with the main mass of strata which lap up on that side of the range just as the corresponding strata do on the east side. A little south of west of Tecolote Mountain copper occurs in patches of this Carboniferous grit. A specimen said to have come from one of these deposits on the very crest of the ridge is the characteristic, rather coarse arkose of these Carboniferous strata. The rock is considerably weathered and holds many small pits that contain yellow limonite. Little malachite is present, but unoxidized cores show that somewhat blackened chalcopyrite is rather plentiful and is distributed through the rock in the same manner that chalcocite and bornite are in most ores of the region. Examination under the microscope shows that a little pyrite is present with the chalcopyrite and that these minerals have partly replaced calcite, dolomite, and feldspar in the same way that chalcocite and bornite have done in the other ores described. It is not certain that quartz has been replaced. The chalcopyrite is in most places surrounded by a thin, irregular coating of chalcocite, in which is mixed considerable limonite. Bornite seems to be absent. It is probable that this chalcocite coating is the result of action of copper sulphate waters on the chalcopyrite. It is undoubtedly different in appearance and almost certainly different in origin from the glance which occurs in the ores that have already been described.

CONCLUSIONS.

The ores of the Tecolote district occur in two arkose formations furnished by rapidly degraded land areas. These formations are of Cretaceous and Carboniferous age, and are separated without any unconformity by a considerable thickness of the "Red Beds," which, so far as known, are barren. Except for the differences resulting from precipitation in rocks of somewhat different nature, the ore deposits at these two horizons are almost identical in character and doubtless are of the same age and derived their copper from the same source. They occur over a region more than 10 miles long and about 4 miles wide. The sediments in which the copper is found extend, however, without notable change of character, for long distances away from the deposits. Although, commercially, these deposits are, on the average, of low grade, the total quantity of copper which they hold is certainly very great, and if the metal, as has been suggested, was derived from preexisting deposits in the pre-Cambrian rocks, it must primarily have been distributed through an enormous mass of those rocks—probably a greater mass, to judge from the very low average copper content of the pre-Cambrian, than the relatively small bulk of these rocks that has been removed by erosion since their exposure under the bottom of the Carboniferous. Moreover, the presence of chalcopyrite and chalcocite in the Carboniferous directly above the highest part of the pre-Cambrian, which is almost unaltered, practically excludes the possibility that this copper could have been derived by leaching of copper deposits in the pre-Cambrian rocks. In fact, it may have been that at the time of copper deposition the pre-Cambrian core in this vicinity was completely covered with sedimentary rocks, and thus protected from degradation and decomposition. If, on the other hand, the present deposits represent concentrations of more disseminated syngenetic deposits, such previously existing deposits presumably were at least coextensive with the sea in which they were formed, and may have had an even greater areal expanse than that of the sea at any one time.

At any rate, such deposits must have been of great extent. The fact that copper is not known to be generally distributed through any bed in the geologic column in this vicinity probably indicates that the metal is present in very small amount, or that, though once present, it has been removed, or else that it was never present. But granting that such a previous dissemination did exist, the degree of concentration into the present deposits must have been extremely great, and it would be expected that numerous centers of concentration would result. As a matter of fact, however, these deposits are sporadic phenomena, and it seems probable that they are the result of localized influences.

The copper is confined to a belt which strikes about north-northwest and parallels the main range. This belt corresponds in direction and in position with a zone of fracturing and faulting, and most of the deposits lie fairly close to the faults. Copper has not been reported at a distance greater than 4 miles from the nearest known fault,[a] and sharp fractures, without relative displacement of the walls, may possibly be present much nearer. The occurrence near the faults of pyrite, chalcopyrite, and bornite, along with chalcocite, and the apparent absence of all but chalcocite at greater distances from the faults, may indicate that the character of the precipitate changed with increasing distance from the main channels of supply. On the whole, the relation of the copper deposit to the channels provided by rock fractures is extremely suggestive and makes it easy to believe that the solutions which deposited the copper may have circulated along these channels.

It is believed that the copper ores were deposited by hot waters ascending along fault lines, and that these waters contained copper sulphides.[b]

SAN JUAN AND McKINLEY COUNTIES.

By Waldemar Lindgren.

San Juan County, containing 5,598 square miles, and McKinley County, containing 5,377 square miles, are situated in the northwest corner of the Territory, San Juan County bordering against Colorado on the north, and both being bounded by Arizona on the west. A large part of both counties is occupied by the Navajo Indian Reservation. Topographically, the two counties form a part of the plateau province characterized by mesas, cliffs, and canyons, and geologically they are almost entirely occupied by the horizontal or gently dipping sandstones and shales of Cretaceous or later age. The elevations in San Juan County range from 4,500 to 6,000 feet. McKinley County contains many prominent buttes and ridges of higher elevation, Hosta Butte in the center of the county attaining 8,837 feet, and Powell Mountain 8,851 feet.

San Juan County lies wholly within the Pacific drainage area, but the Continental Divide crosses McKinley County diagonally from northeast to southwest. The monotony of the prevailing Cretaceous or later strata which cover this wide area is broken only by some areas of Tertiary volcanic rocks or a few very recent lava flows. In the Zuni Plateau, in southern McKinley County, a huge swell brings up the "Red Beds" and the Carboniferous strata underlying the Cretaceous.

Prospects containing copper ores are said to have been found in the southeastern part of McKinley County, but no metal production of any kind is reported in either county. Their principal mineral resources consist in extensive coal beds, mined chiefly near Gallup, in McKinley County. The tonnage mined in that county was 480,490 short tons in 1905, 629,821 tons in 1907, and 539,050 tons in 1908.[c]

---

[a] Copper ore of the same character is said to occur in Pintada Canyon, in Guadalupe County, about 45 miles south-southeast of the Tecolote district. The distance is too great to allow any correlation of the two occurrences, but it may be pointed out that the Pintada occurrence is in the direction of the general strike of the Tecolote zone.

[b] For further theoretical discussion concerning the origin of this class of deposits see pp. 76-79.

[c] Mineral Resources U.S. for 1908, U. S. Geol. Survey, 1909. Schrader, F. C., The Durango-Gallup coal field of Colorado and New Mexico: Bull. U. S. Geol. Survey No. 285, 1906, pp. 241-258. Shaler, M. K., A reconnaissance survey of the western part of the Durango-Gallup coal field of Colorado and New Mexico: Bull. U. S. Geol. Survey No. 316, 1907, pp. 376-426. Gardner, J. H., The coal field between Gallina and Raton Springs, N. Mex.: Bull. U. S. Geol. Survey No. 341, 1909, pp. 335-351. Gardner, J. H., The coal field between Gallup and San Mateo, N. Mex.: Bull. U. S. Geol. Survey No. 341, 1909, pp. 364-378.

RIO ARRIBA COUNTY.

By L. C. Graton.

GENERAL FEATURES.

Rio Arriba County, which embraces an area of about 5,800 square miles, is situated west of the Rio Grande and is adjoined by Colorado on the north and by Taos County on the east. Its western half belongs in the plateau province and is covered by Cretaceous and Tertiary flatlying sediments containing valuable coal beds, which have been described in other publications of the Survey.[a] This part may best be described as a dissected plateau with elevations ranging from 5,500 to 9,000 feet.

The eastern half of the county lies in the Rocky Mountain province. The Paleozoic and pre-Cambrian rocks are first brought to light in the Sierra Nacimiento in longitude 106° 50', reaching elevations of somewhat over 10,000 feet, and continuing southward into Sandoval County. East of this range rises the circular group of the Valles Mountains, in southeastern Rio Arriba and northeastern Sandoval County. This group appears to consist largely of heavy masses of rhyolitic lavas. No mineral deposits are reported from this vicinity.

The northeastern part of Rio Arriba County is covered to a considerable extent by the late Tertiary basalt flows of the Rio Grande valley, which almost surround the isolated group of the Hopewell Mountains, an outlier of pre-Cambrian granites and schists, containing the most important gold and silver deposits of the county. The elevations in this group attain 10,000 feet.

Eastern Rio Arriba County is drained by Chama River, which joins the Rio Grande a few miles above Espanola in the southeast corner of the county and heads near the boundary line between New Mexico and Colorado.

Of nonmetallic substances, coal is mined chiefly in the vicinity of Monero, near the Colorado line, and mica exists in commercial quantities at Petaca, about 15 miles south of the Hopewell mining district, where it occurs in pegmatite dikes contained in gneiss.[b]

Gold and copper are the most important among the metallic resources. Placer gold has been and is still won in the Hopewell district, being derived from the veins in this neighborhood. Placer gold is said to exist also on the Tierra Amarilla grant, near the headwaters of the Chama, and in paying quantities in the gravels of Chama River near Abiquiu, the gold here being probably derived from low-grade deposits contained in the pre-Cambrian rocks cut by the Chama in its deep canyon.

The chief gold and silver deposits are found in the Hopewell and Bromide districts, in the Hopewell Mountains. The former yields principally gold, the latter copper ores with some silver and a little gold. The small metal production occasionally reported from the county is derived from these districts.

Copper in sandstones is found in the Sierra Nacimiento and near Abiquiu; these occurrences are described under Sandoval County. No production of copper has lately been reported from the districts in Rio Arriba County.

HOPEWELL AND BROMIDE DISTRICTS.

LOCATION AND HISTORY.

The Hopewell and Bromide districts are in northeastern Rio Arriba County, about 15 miles from the Colorado line. Tusas, the post-office of the Bromide district, is about 2 miles north of the center of the developed portion of that camp, and is 12 miles west of the town of Tres Piedras, on the Denver and Rio Grande Railroad. Hopewell (formerly Good Hope), situated at about the center of the camp of that name, is 8 miles north of west from Tusas. The two regions adjoin and were originally part of a single district known as Headstone, but were after-

---

[a] Schrader, F. C., The Durango-Gallup coal field of Colorado and New Mexico: Bull. U. S. Geol. Survey No. 285, 1906, pp. 241-258. Gardner, J. H., The coal field between Durango, Colo., and Monero, N. Mex.: Bull. U. S. Geol. Survey No. 341, 1909, pp. 352-363.

[b] For the production of coal and mica see the annual reports on the Mineral Resources of the United States. For the occurrence of mica see a report by J. A. Holmes, Twentieth Ann. Rept. U. S. Geol. Survey, pt. 6, continued, 1899, pp. 706-707.

ward separated. Ore is said to have been first discovered in 1881[a] at what became known as the Bromide mine, from which the eastern portion of the Headstone district later took its name. Considerable activity followed this discovery, but it was only after the discovery of the Cripple Creek district in Colorado had stimulated the mining industry in this portion of the West that much real mining was done. Extensive work has never been done at any one place in these districts, although in the aggregate much money and labor have been expended; the development of the district consists mainly in prospects, of which the greater number are now idle.

The general geology of the Hopewell and Bromide districts will be described in a single section. The ore deposits of the two districts, however, although very similar in many respects, are sufficiently unlike to merit separate descriptions.

GEOLOGY.

The districts are situated in pre-Cambrian rocks, which protrude through coarse conglomerates and sandstones, probably of Tertiary age, and are surrounded on nearly all sides by later basaltic flows. On the east the pre-Cambrian rocks are first exposed at Tres Piedras, at an elevation of about 8,100 feet, where the "three rocks" are small abrupt knobs of red gneissic granite. Farther east the basalt begins and extends across the full width of the Rio Grande valley. From Tres Piedras westward toward Tusas the elevation gradually increases and culminates in the Tusas Peaks at an elevation of 9,500 feet. Rock exposures are not prominent, the hill slopes being covered by soil holding many bowlders of granite, schist, and rhyolite up to a foot in diameter. These detrital deposits are probably due to the disintegration of a very coarse conglomerate of Tertiary age. In a few places a rather coarse sandstone or arkose is exposed, imperfectly bedded and lying about horizontally. An exposure in a tunnel near Tusas shows that the sandstone is poorly consolidated and that it contains good-sized rounded bowlders of rhyolite. Tusas Creek has cut through these sedimentary formations to the underlying crystalline rocks. Dark, much-foliated dioritic gneiss is cut by reddish, very fine grained gneiss in masses of indefinite form and extent. In most places the dark rock is so much sheared that it constitutes an amphibolite schist. The strike of the foliation of this schist ranges from west to northwest, and the dip is usually very steep. The greater portion of the area in which mining has been done is composed of this schist, but more massive rocks, locally known as granite, are exposed in places. The three elevations known as the Tusas Peaks are composed of a somewhat gneissic reddish granite. Kiawa Mountain, a prominent eminence a few miles to the southeast of the Bromide district and a part of the same general province, is also said to be made up of granite. In other places, for instance in the vicinity of the Ivanhoe mine, dark gneissic granodiorite occurs. A reddish rock resembling granite but having the composition of granodiorite is encountered in the Santa Fe tunnel north of Hopewell.

Porphyritic rocks of at least two varieties cut these metamorphosed granular rocks and have been themselves been more or less sheared and altered. Dikelike masses of a grayish-green rock with yellow spots representing original feldspar phenocrysts are known in several places—at Iron Hill north of Hopewell, near the Red Jacket mine and the Rubicon prospect at Hopewell, and at the Snow Flake and War Eagle claims in the Bromide district. These various masses of porphyry, though closely related in composition and character, nevertheless exhibit sufficient variation to range from diorite porphyry through granodiorite porphyry to quartz monzonite porphyry. In some places the rocks show the effect of a smaller amount of shearing and are slightly schistose. The other variety is more constant in appearance and composition. It is a reddish, fine-grained, plainly schistose rock, containing small rounded or lenslike grains of bluish-white quartz. It is a much foliated granite porphyry. It occurs in dikes, one of which, probably several hundred feet in width, is said to extend from the center of the Bromide district slightly north of west for miles. Another mass is encountered in the workings of the Jaw Bone mine north of Hopewell.

---

[a] Jones, F. A., op. cit., p. 163.

Closely associated with these crystalline rocks of igneous origin are several exposures of grayish and bluish quartzite, very dense and very massive, forming great blocks and bowlders. In places this rock becomes conglomeratic and the form of the squeezed and flattened pebbles is readily seen, especially on somewhat weathered surfaces. This ancient sedimentary formation is confined to the western part of the district so far as known. It forms prominent hills north and west of Hopewell and probably extends in interrupted patches as far south as Ojo Caliente, where similar quartzite is observed. It is undoubtedly of pre-Cambrian age.

On the dump of a prospect on the Homestake claim, in the eastern part of the district, was found a rock consisting of streaks and lenses of impure crystalline limestone interfoliated with biotite-chlorite schist. This probably represents original limestones and shales which were doubtless of pre-Cambrian age and which were presumably members of the same series as the quartzites.

A fine-grained quartz rock holding many minute needles of hornblende is shown on a dump on the Red Fissure claim, a little farther east. It is doubtless a metamorphosed sandstone.

In the upper part of Eureka Gulch, west of Hopewell, a cemented gravel is encountered in the workings of the Dixie Queen mine. It is probably of Tertiary age, although it may be younger.

MINERAL DEPOSITS.

GENERAL FEATURES.

The primary ore deposits of these two districts occur wholly within the old metamorphic rocks and are all believed to be of pre-Cambrian age. In most places their relation to the metamorphosed sediments is not clear, although at the Jaw Bone and Freeport claims, in the Hopewell district, the quartzite lies close to the deposits, and on the Trenton claim, just east of the Freeport, a quartz vein occurs at the contact of the quartzite and the granite porphyry. Another evidence of the relative age of the sediments and the ore deposits is that some of the veins occur in granite porphyry and this cuts the quartzite. It is plain, therefore, that both eruptive and vein-forming agencies were active after the deposition of these ancient sediments.

A feature particularly worthy of mention is the association of ores with the old porphyritic rocks. Just how intimate this relation is it has not been possible to determine, but the distribution at least is significant. Diorite porphyry has been found in or near the Jaw Bone, Iron Mountain, Red Jacket, and Mineral Point mines, in the Hopewell district, and the War Eagle and Snow Flake mines, in the Bromide. Granite porphyry occurs at the Jaw Bone, Freeport, Arrastre, Ivanhoe, Danbury, and other mines. The location of a number of claims, as shown on a map of the district, marks the location of one of the principal granite porphyry dikes; but in all these claims the amount and value of the ores have been small, and it seems probable that they have been located not because of the occurrence of known values but because a contact or a dike is commonly regarded as a favorable place for prospecting.

The principal difference between the Bromide and Hopewell districts lies in the kind and relative amounts of metals present. The Hopewell district is essentially a gold producer, but little copper or silver having been obtained. The Bromide district, on the other hand, is a producer of copper ores, with good values in silver in certain deposits and generally only low values in gold.

HOPEWELL DISTRICT.

GENERAL GEOLOGY.

The deposits in the Hopewell district are of two types—quartz veins and fahlbands (disseminations of pyrite along certain zones of the schist). The quartz veins, which occur principally in the schists, are commonly irregular after the manner characteristic of the so-called schist veins; they are of rudely lenticular form, pinching and swelling, closing out, and then beginning again at some distance.

Usually the veins are narrow, only a few inches wide, but in many places two or more are parallel and within a workable width, collectively forming a lode. Locally the intervening schist carries values also. The predominant mineral of these veins is quartz, but other minerals are frequently found to constitute a portion of the fissure filling. Siderite is one of the most common of these and occurs in irregular aggregates of crystal grains, as a rule near the margin of the veins. It is very commonly accompanied by bunches of small chlorite flakes, which appear to have been derived from some of the wall-rock minerals. This association of chlorite and siderite seems rather characteristic of a number of these pre-Cambrian veins. Tiny flakes of muscovite accompany some of the chlorite and occur in minute fractures in the body of the quartz. Tourmaline is found in several places. It usually occurs as good-sized groups of tiny prisms of lustrous black color, either parallel, suggesting a crystal individual with cleavage, or in irregularly radiating arrangement. Some tourmaline is present in imperfect bands in the quartz; and here and there it appears to impregnate the wall rock for a slight distance from the fissure. Calcite is less plentiful than siderite. It is present mostly in narrow veinlets, many of which are interbanded with quartz. Good-sized masses of specular hematite occur in the quartz vein at the Jaw Bone mine. The quartz contains pyrite, in some places in abundance. Chalcopyrite is generally much less abundant, but in a few veins it is comparatively plentiful. The sulphides in the quartz are in the main poorly crystallized, distributed usually in bunches and patches of irregular grains. In the adjoining schist copper pyrite is less common than in the quartz; much of the pyrite is well crystallized, both as cubes and as pyritohedrons. In many places, however, particularly close to the fissures, the pyrite has no crystallographic outline. Galena is sparingly present here and there, intergrown with the pyritic sulphides. Zinc blende has been noted only in the fahlbands and there only sparingly.

So far as can be ascertained no native gold has ever been found in the unoxidized ores of this district. The gold is contained in the pyrite and perhaps to a small extent in the chalcopyrite, and the free gold which has in the past been recovered by panning or amalgamation has doubtless all been liberated from sulphides by oxidation. Not only does the pyrite in the quartz carry gold, but that replacing the schist is also auriferous, although probably to a less extent. It seems to be the general experience, also, that whereas pyrite may extend to a considerable distance from the veins, the gold values drop off more abruptly; there seems to be some kind of a filtering action or osmosis which causes the gold to be precipitated near the fissure and at the same time allows the pyrite to penetrate farther away. It is said that the well-crystallized pyrite is usually low in gold, and this statement seems to be borne out by the occurrence of good crystals mostly at some distance from the fissure.

Of the effects which the vein solutions have produced in the adjoining wall rocks the impregnation by pyrite already described is the most noticeable. Other changes have gone on along with this, however. Silicification is locally important. The microscope shows that the quartz has been introduced mainly in minute lenses parallel to the foliation of the schist. In certain parts of the rock the feldspars and perhaps other constituents of the schist have been converted into sericite. The bisilicate—generally amphibole—has in places been changed into a chestnut-brown biotite or into chlorite.

The fahlbands are similar in character to the altered rock adjoining the veins. In some localities narrow quartz veinlets are present, with the altered rock making up the greater part of the deposit. Silicification is perhaps a little more intense where there is no distinct filled fissure than in the wall rocks of the quartz veins.

Both the quartz veins and the fahlbands are characterized in general by two structural features—first, most of them conform to the foliation of the inclosing rock and hence a northwesterly strike and steep dip are most common; second, few of the veins or ore bodies have great continuity on the surface—instead of a few extensive veins there are many which can be traced only for short distances.

The zone of oxidation is not deep and reaches barely to the ground-water level, which ranges from a few to 50 or 60 feet below the surface or even to more than 100 feet in some of the higher places. The schists are not much affected except where they contain sulphides. In some of such places they are much decomposed into crumbling masses, but more commonly the pyrite is converted to limonite, which in certain localities remains as more or less perfect pseudomorphs after the pyrite and in others is wholly dissolved and removed. As a rule, neither the veins nor the pyritized zones in the schist outcrop noticeably, although here and there on very steep slopes croppings are to be seen.

The ores vary between wide limits of value, ranging from those which are far too low to permit profitable working up to those running as high as $800 a ton. The length of the haul to and from the railway, the distance from the shipping point to the smelter (Pueblo), the long winters, during which transportation is often impossible on account of the deep snows, and the consequent high price of supplies, all work to the disadvantage of the miner and render impossible the shipment of low-grade ores. It is doubtful if ore running under $25 a ton would pay to ship.

Milling and amalgamation have been done to some extent in the district and are said to have been remunerative on the thoroughly oxidized surface ores of the Croesus and Buckhorn mines, but when only partly oxidized sulphides were reached, recovery fell off below the point of profit. Concentration seems to have thus far been attended with but little success, although there is no obvious reason for such a result.

JAW BONE MINE.

The Jaw Bone mine is situated about 2 miles a little west of north from Hopewell. The shaft is a little more than 100 feet deep, and some cross-cutting and drifting have been done from it. In addition, two diamond-drill holes, 600 and 1,000 feet deep, have been sunk. The country rock is principally a dense gray schist of uncertain origin. Pink granite porphyry, rather schistose in structure, was encountered in one of the drill holes and probably came from a dike of that rock. Diorite—perhaps the granodiorite porphyry which occurs on Iron Mountain near by—is said to have been reached during the drilling. The shaft is situated near the foot of a large hill lying to the north and is just below the contact of the schist with the massive gray quartzite that caps this hill.

The vein, which does not appear prominently on the surface, is said to strike northwestward and to dip about 65° NE. Tourmaline is present in places; close to the quartz it completely replaces the schist with a felted aggregate of black needles. Siderite, associated with chlorite and muscovite derived from the wall rock, occurs at the borders of the vein in some places. Pyrite, chalcopyrite, and specular hematite are present in the vein. The ore body is said to have been a lens 8 feet wide, 30 feet long, and 45 feet deep. Beyond these limits the vein pinches. Some ore was shipped from this body. It is said that the total values ranged from $17 to $25 a ton, mostly in gold, but the copper content varied from 1 to 5 per cent. The diamond-drill holes are reported to have encountered no quartz nor ore, although sunk in the hanging wall of the vein. The property is now idle.

IRON MOUNTAIN.

On the small elevation known as Iron Mountain, a short distance south of the Jaw Bone mine, a moderate amount of work has been done on a deposit of iron ore. A slaty schist, striking somewhat north of west and dipping 65° N., about parallel to the Jaw Bone vein, holds an interfoliated band of iron ore. To judge from the pits sunk upon it, it is about 6 or 8 feet wide and, instead of having definite walls, grades off into the inclosing schist. The ore, which is mostly magnetite, is slaty in structure and fairly rich in iron. The microscope shows that it is a mixture of minute grains and cubes of magnetite mingled with quartz and perhaps a little feldspar. Sheared diorite porphyry is encountered in a 300-foot tunnel which runs in from the north but fails to cut the iron-rich zone. But little ore has been taken out. It is said that a slight amount of gold is contained in the ore.

DIXIE QUEEN MINE.

This mine is situated on the south side of Eureka Gulch, perhaps a third of a mile west of Hopewell. It is owned and operated by the Dixie Queen Mining Company. A shaft nearly 100 feet deep penetrates a conglomerate cemented by a rather soft gray material which is probably clay. This is evidently a former stream bed and gold placer and is presumably of Tertiary age. The material usually crumbles on reaching the surface and is screened and sluiced like placer material, the gold being caught in riffles. Broken crystals of partly oxidized pyrite also occur in the matrix. The gold content of this gravel is not known to the writer. Water is a very troublesome factor and adds greatly to the cost of working.

RED JACKET MINE.

The Red Jacket mine is just across Eureka Gulch from the town of Hopewell. It also is the property of the Dixie Queen Mining Company and was the only mine in the district in active operation in the summer of 1905. The country rocks are schist and sheared diorite prophyry, the latter occurring probably as a dike. From the bottom of the 120-foot shaft a crosscut to the north encounters small streaks of galena and limonite in the schist and parallel to its foliation, N. 60° W. and nearly vertical. The galena is said to carry good gold values and assays as high as $75 a ton have been obtained. It is possible that the gold is actually contained in the limonite, which resulted from the oxidation of pyrite. A zone of soft, much-foliated schist is next encountered. Beginning from this zone, a rather ill-defined band with streaks of limonite and carrying gold cuts across the schistosity in a northerly direction. A little pyrite is encountered in places. This ore is said to be of good grade.

A 5-stamp mill is located not far from the shaft. Both amalgamation and concentration are practiced. The ore is stamped to 60-mesh, which appears too fine, for in order to save all the pyrite too much sand is retained in the concentrates. It is possible, however, that amalgamation has been found to be more effective with fine crushing.

MINERAL POINT MINE.

The Mineral Point mine is situated just east of the Red Jacket, where Eureka Creek turns from an easterly to a southerly course. Considerable exploration work in the way of shafts, tunnels, and pits has been done at this place, but none of the workings are accessible at present. The schist is heavily impregnated along certain zones with pyrite and in places with galena and zinc blende also. In some of the workings there are quartz veins carrying pyrite, chalcopyrite, and a little galena. The deepest working, a tunnel just above the creek level, encounters the sulphides undecomposed, but in most of the upper openings the limonite and hematite, locally stained with malachite, are commonly found. It is said that some of the ore is of good grade. A little was milled years ago, but the production has been only trifling.

CROESUS MINE.

The Croesus mine has to its credit a larger production than any other in the district. It is situated on the west bank of Eureka Creek, only a short distance below the Mineral Point. The creek here cuts deeply into the pyritized schist, almost at right angles to the foliation. Along the steep valley side several tunnels and pits have been made on zones along which pyrite is especially abundant. Stringers and veinlets of quartz occur in the middle of these zones in some places. The principal tunnel is said to be several hundred feet long. It follows one of these fahlbands, which is parallel to the foliation of the schist and strikes N. 55° W. Stopes have been carried upward to the surface in several places, but most of the ore is said to have come from winzes following certain rich streaks below the tunnel level. At a comparatively shallow depth the ore is said to have pinched out. Some of the ore is said to have been very rich, reaching as high as $800 a ton. The average value is said to have been $50 a ton and the production is estimated locally at about $15,000. Most of the ore was treated in an old Chilian mill run by water power, and later in a stamp mill, the remains of both of which still exist. The richest sulphide ore was shipped.

OTHER PROSPECTS.

Just below the Croesus mine there is an exposure of the old quartzite, which is here seen to be a sheared conglomerate. On the south this is cut by a dike of the pink granite porphyry already described. This dike continues eastward or southeastward and numerous prospects have been located along it, including the Freeport, Trenton, Park View, Gold Dollar, and Arrastre.

The Hidden Treasure group lies about half a mile southeast of Hopewell. A 100-foot shaft is sunk in schist, with some granitic rock and narrow stringers of quartz and calcite carrying tourmaline, siderite, and auriferous pyrite. It is said that a little ore obtained near the surface was once shipped, but values decreased below.

The Emerald prospect, a short distance east of the town, is at the contact of granodiorite on the north with schist on the south. The contact is parallel to the foliation of the schist, striking a little north of west and dipping steeply. A 75-foot shaft is sunk on a 2-foot quartz vein in schist close to the granodiorite contact. Some pockets of exceedingly rich ore were found near the surface. All the ore was oxidized. A little cuprite was present. The quartz pinched out at a shallow depth, although the fissure and gangue continued. A little ore was once shipped. The shaft is now half full of water.

A shallow shaft on the Cash Entry mine is said to be on the same vein a few hundred feet farther west. Here the vein is actually at the contact of granodiorite and schist. Among the many other prospects may be mentioned the Duck, northwest of town, where 87 per cent copper ore has been reported, and the Buckhorn, located between the Croesus and Mineral Point mines and showing small stopes in pyritized schist similar to that at these two mines.

PLACERS.

The western portion of the original Headstone district, or what is now the Hopewell district, first received attention because of the discovery of rich placer ground in Eureka Creek. These placers were worked actively for several years after their discovery, and according to general report $175,000 were taken out during the first three years. The common estimate of the total production of these placers is about $300,000. It is said that one nugget worth $96 was found, and another worth $34. Jones[a] says that a $15 lump of gold was found in the lode workings of the Croesus mine, so the size of these nuggets is not surprising.

There are two placer areas in the district, both on Eureka Creek. One, known as the Fairview placer, bounds the town of Hopewell on the west. The Red Jacket and Mineral Point mines are situated on its southwest bank. The Buckhorn and Croesus mines are along the creek, just a little farther down. The gravel of the Dixie Queen is probably an early representative of this placer. Practically no valuable veins have been discovered higher up the creek. It is said that the ground was very rich just where the valley narrows from a broad flat to a steep-sided channel.

The other placer, known as the Lower Flat placer, is about a mile farther down the creek, at the junction of Eureka Creek with another branch of Vallecitos Creek. A flat of considerable acreage is being prospected here by the King William Company. A shaft has been put down exposing 35 to 40 feet of alluvium. A small storage tank for water has been erected and there are two giants on the ground. The outlet of the flat is blocked with large bowlders of quartzite, which at the time of visit were being blasted out. What steps other than the sinking of the shaft have been taken by the company to ascertain the value of this ground are not known. It appears doubtful whether enough water can be obtained for steady hydraulicking.

BROMIDE DISTRICT.

GENERAL GEOLOGY.

Much of what has been said regarding the Hopewell district applies equally well to the ore deposits of the Bromide district, which consists likewise of quartz veins and of impregnations of the schist. Regarding the structure of these deposits little need be added to the description given for the Hopewell district. The mineralogy also is the same in many deposits, but pyrite is less and chalcopyrite more abundant, the gold values are lower, and the silver values are higher. The minerals developed in the rocks immediately adjoining the vein are

---

[a] Op. cit., p. 166.

more strikingly characteristic of the pre-Cambrian veins than in the Hopewell district. Garnet, epidote, hornblende, and tourmaline appear in streaks and zones parallel and close to the vein at the Dewey mine. It is probable that the rock was a biotite schist similar to that found not far away near the Admiral mine. Epidote is also present close to the vein on the Ora and Boston claims. At the Strawberry mine biotite and crystalline hornblende occur with chalcopyrite, without quartz vein filling. Tetrahedrite occurs at the Bromide mine, molybdenite at the Tampa, and fluorite at the Joe D. prospect and the Elliott & Kennedy mine.

SIXTEEN TO ONE MINE.

The Sixteen to One mine is situated in the western part of the Bromide district. It is developed by a 153-foot shaft and short levels at depths of 80 and 115 feet. The country rock is schist, and the vein strikes slightly south of west and dips steeply north. The ore seen on the dump is almost wholly oxidized and consists of schist impregnated with magnetite and bright-red hematite (not cuprite, as commonly stated), with small stringers of quartz and a little copper. Assays are reported to average $11 to $15 a ton, of which about $2 is gold and the remainder about equally divided between copper and silver. It is said that the ore zone is 9 feet wide and that the values are slightly higher in the lower part of the mine. A parallel narrow streak encountered at the bottom of the shaft is said to run 4 per cent of copper and to assay $25 a ton in total value. A few assays near the surface were high, mostly in silver.

WHALE MINE.

The Whale claim is situated in the central part of the district. A 100-foot shaft is sunk on a quartz vein in schist. The vein averages about 2 feet wide, strikes N. 55° W., and dips steeply northeast. A drift at the bottom of the shaft followed the vein for some distance, but finally lost it. The vein is said to carry gold, silver, and copper, and the value of the ore is reported at various figures, from $35 to $75 a ton. The upper portion of the vein was wholly oxidized, but below 60 feet chalcopyrite is present. Several carloads of ore were once shipped. The property is idle at present.

PAY ROLL MINE.

The Pay Roll mine lies a couple of claims east of the Whale, about in the center of the district. The property has recently been acquired by the Keystone Bromide Mining Company, and the information given here was obtained from the manager, Mr. J. P. Rinker. The country rock is a sericite-chlorite schist, with foliation striking northwestward and dipping very steeply to the northeast. Along a zone of some width the rock is impregnated with chalcopyrite in stringers parallel to the foliation of the schist. Little lenses of calcite occur here and there along with the sulphide. A shaft 250 feet deep is sunk in the hanging wall of this sulphide zone or fahlband, and crosscuts are driven to it. The upper portion of the ore body is oxidized and consists of hematite and limonite streaks stained with copper, in the schist. At a depth of about 60 feet flakes of native copper are found, and below this depth sulphides appear. Chalcopyrite is the principal sulphide, pyrite being only sparingly present. A 50-foot crosscut to the south at the 150-foot level shows some ore in certain streaks. At the 250-foot level a crosscut to the south, after passing through a kaolin seam 4 feet from the shaft, reaches soft schist with stringers of chalcopyrite, said to average 4 per cent of copper and $6 to the ton in gold. It is said to be 20 feet wide, and the last 4 feet of the ore is harder, carrying 9 per cent of copper and $30 a ton in gold. Beyond this is a kaolin streak, then a width of 70 feet of 2 to 3 per cent copper ore, gradually decreasing in value beyond a point 90 feet south of the shaft. Some of the chalcopyrite along fractures appears to be secondary.

ADMIRAL GROUP.

The Admiral group lies in Cunningham Gulch a short distance northwest of Tusas Peak. The claims, which were discovered in 1902, have received considerable development, but appear not to have been remunerative.

In this part of the district there is a pronounced deviation from the general strike of the foliation of the rocks. Instead of the strike about west-northwest that is prevalent in other places, the trend of foliation here is a little east of north.

Numerous veins and stringers of quartz, carrying calcite and chlorite penetrate what was probably an amphibolite, but is now, especially where exposed near these veins, a biotite schist, with local development of hornblende and red garnet. The Admiral incline on the Sampson claim is sunk on a quartz vein in this biotite rock. It is said that sulphide ore, carrying $12 to $20 a ton, mostly in gold, was encountered in this incline. The amount of work done was not ascertained. At the Dewey shaft to the south, quartz stringers occur in a rock composed of biotite, epidote, hornblende, and garnet. This is the mineral composition only close to the vein; farther away the rock is a biotite schist. This alteration of the wall rock is very striking and corresponds with that of many pre-Cambrian veins.

TAMPA MINE.

The Tampa mine is situated just north of the middle knob of Tusas Peak. It is owned by the Tusas Peak Gold and Copper Mining Company. The shaft is 400 feet deep, the deepest in the district, and is connected with 800 or 1,000 feet of drifting on five levels at depths of 45, 100, 200, 300, and 400 feet. Eight men were working at the mine in the summer of 1905. Water is held just below the 300-foot level, no attempt being made at present to work the 400-foot level. From 4,000 to 5,000 gallons are bailed every twenty-four hours. Most of the water was struck below 300 feet. Only the 300-foot level, which is the most extensive level of the mine, was accessible at the time of visit. For information regarding the other levels the writer is indebted to the superintendent.

The vein strikes about N. 30° W. and dips steeply to the northeast. On the 45-foot level good ore was reached just north of the shaft. It consisted mainly of malachite, copper glance, and a little chalcopyrite. In a few places azurite, cuprite, and a little native copper were found. Some free gold was present in hematite, which probably resulted from pyrite. Little scales of molybdenite were occasionally found. On the 100-foot level a strong body of partly oxidized ore was reached, but the grade was lower than that of the ore above. The best ore is said to have been found on the 200-foot level. A crosscut extending 25 feet northward from the shaft reached the ore body, which was 14 feet wide. Specimens from this level show several narrow and coarsely spaced quartz veinlets, carrying along their borders a mixture of chlorite and chalcopyrite, and in some places considerable molybdenite. Two carloads shipped from this mine are said to have averaged about $15 a ton—one-eighth ounce of gold, 6 ounces of silver, and the rest copper. Some of the heavy chalcopyrite ore from this level is said to carry platinum.[a] On the 300-foot level the vein cut on the levels above has not been sought, but a 300-foot crosscut has been run a little north of east to reach an ore body which is known on the surface. In this crosscut the country rock is granite with a few streaks of schist. Chalcopyrite and malachite are encountered in small bunches here and there, but no ore body has yet been reached. A vein holding low-grade ore is said to have been cut on the 400-foot level, but whether it is the same as the one known on the 200-foot level is not certain.

BROMIDE MINE.

The Bromide mine, the oldest location in this region, is about 2 miles southwest from the Tusas Peaks. It is reported that a large part of the production was yielded by a carload of very rich silver chloride ore from the surface. Some silver-bearing material was of a deep-blue color and was thought to be bromyrite (silver bromide), and on this account the mine was so named. Copper carbonates were also encountered in the upper portion. Sulphides were first reached at a depth of about 50 or 60 feet. The rock is a fine-grained amphibolite, somewhat altered by the ore-bearing solutions. A quartz vein containing siderite and chalcopyrite is said

---

[a] A copper ore, consisting mainly of pyrite and chalcopyrite with some covellite, from the Tampa mine was tested for platinum by Ledoux & Co., of New York. None was found, but the ore contained 0.03 ounce of gold and 7.80 ounces of silver to the ton.

to strike about N. 70° W., with almost vertical dip. Parallel to and near it the schist is impregnated with stringers and lenses of tetrahedrite, associated with calcite. This copper-antimony sulphide, which has erroneously been called stephanite, is in some places intergrown with chalcopyrite and associated with quartz, and it seems probable that it is a primary mineral. This mineral is probably the one which carries the silver. The ore is said to have followed a shoot in the vein. In the bottom of the shaft, at a depth of 140 feet, a 22-inch streak, running well in copper and silver, was left when work was stopped. It is stated that the gold values are low, ranging from $1 to $4 a ton.

A parallel vein about 20 feet north of the Bromide vein furnished some good copper ore to lessees when water interfered with working on the main vein. The production of this mine is reported at $27,000.

DILLON TUNNEL.

The Dillon tunnel, a development enterprise, is being driven by the Keystone Bromide Mining Company. It starts at the lowest point on the Bromide claim, renamed the Hutchison, and runs parallel to the schistosity, N. 67° W., for almost 800 feet, being then about 190 feet below the surface and 50 feet below the bottom of the Bromide workings. At this point it turns to a course about N. 30° W., heading for the Pay Roll mine some 2 miles distant, which it is expected to cut at a depth of about 640 feet. In July, 1905, the tunnel was a little over 800 feet long, and was being continued. Near the breast a drift was run along what was thought to be the north vein of the old Bromide workings. A 65-foot raise from this drift cut the Bromide vein, but at the time of visit values had not been encountered.

STRAWBERRY MINE.

The Strawberry mine lies about half a mile south of Tusas Peak and near the head of Cow Gulch. It is the property of the New Mexico Gold and Copper Mining Company. The country rock is a chlorite-amphibole schist, probably once an amphibolite. The foliation of this rock strikes N. 76° W. and dips 80° S. Parallel to this schistosity is a zone along which chalcopyrite has impregnated the rock. Near the sulphide coarsely crystalline hornblende has been developed in the chlorite. An inclined shaft, following ore constantly, had been sunk a little more than 100 feet. At the surface copper occurred as malachite. At a depth of 60 feet was a brownish-red oxide, giving high assays in copper and good gold values. A little native copper was found at this level. Below it chalcopyrite was the principal copper mineral. At the 80-foot level a crosscut has been run 200 feet to the north and has cut several streaks of ore. One of the zones is 4 or 5 feet wide but of low grade.

OTHER PROSPECTS.

Among numerous other claims in this district that have received some development are the Ivanhoe and the Danbury, carrying copper minerals at the contact of gneissic granodiorite with the large granite porphyry dike that continues eastward from the Hopewell district; the Ora and the Boston, also carrying copper minerals in epidotized schist; the Sardine, near the Pay Roll mine, where gold and copper have been found in quartz lenses in the schist; the Snowflake, showing pyrite with a little zinc blende and galena in amphibolite at the contact with sheared granodiorite porphyry; the Continental, between the Tampa and War Eagle, on which has been opened a quartz vein cutting granitic gneiss and carrying galena; the King Richard, near the Strawberry, containing oxidized gold ore from which high assays were reported; and the Elliott & Kennedy, in Cow Gulch, where cuprite, with smaller amounts of other oxidized copper minerals and of chalcopyrite, has been found, together with fluorite and fine, scaly muscovite, in gneissic granite. On Kiawa Mountain, a few miles southeast of the Bromide district, gold ore holding much tourmaline is reported as occurring in an impure quartzite at the contact with granite.

VALENCIA COUNTY.

By F. C. Schrader.

GENERAL FEATURES.

Valencia County, situated in the west-central part of the Territory, occupies an area of 5,712 square miles; it is bounded on the north by McKinley and Sandoval counties, on the east by Bernalillo and Torrance counties, on the south by Socorro County, and on the west by Arizona. In general the county belongs to the plateau province and most of its large area is covered by flat or gently inclined beds of Cretaceous and Tertiary age, cut by many water courses, which expose the geologic structure. The "Red Beds," of late Carboniferous or early Mesozoic age, are also exposed over large areas, especially in the west, south of the Zuni Reservation. The region may be broadly characterized as a land of mesas, terraces, cliffs, and canyons. The average elevation is probably about 6,000 feet.

In the northwestern part of the county and extending into McKinley County is the broad uplift of the Zuni Plateau, in which the Carboniferous rocks have been brought up to the surface and surround a central area of pre-Cambrian granites and schists. This area of old rocks, about 15 by 8 miles in extent, culminates in Mount Sedgwick with an elevation of about 9,200 feet. The highest point in the county is, however, Mount Taylor, in the volcanic plateau of the same name. It attains 11,389 feet.

Basalt flows, in part of very recent age, occupy large areas in the central part of the county, and similar flows followed San Jose River for a long distance. In the northeast the earlier eruptions of the Mount Taylor Plateau form prominent features of the landscape.

The extreme eastern part of the county is crossed by the Rio Grande and includes a part of the Manzano Mountains, which are mentioned with more detail in the description of Bernalillo and Torrance counties. The westernmost part of Valencia County is in the drainage basin of Colorado River; the waters of the larger eastern part flow into San Jose River and thence into the Puerco, which ultimately empties into the Rio Grande.

The geology of the northern part of the county is admirably described by Clarence Dutton in his report on Mount Taylor and the Zuni Plateau;[a] the southern part has been discussed by C. L. Herrick.[b]

Of nonmetallic mineral products the county contains coal seams, as yet little developed, along Puerco River. Salt and gypsum also occur in large quantities, generally in connection with the "Red Beds,"[c] but are not as yet utilized. Metalliferous deposits containing copper and gold are found in the Zuni Mountains and are described in detail on the following pages. The copper occurs both in the sedimentary beds and in the pre-Cambrian rocks; the gold in the latter exclusively. The only other mining district is in the Manzano Mountains, in the extreme southeastern part of the county. These deposits, which have been but little developed and seem to be of small importance, were not visited. Some of them are mentioned by F. A. Jones.[d]

COPPER DEPOSITS OF THE ZUNI MOUNTAINS.

The observations on which this paper is based were made during a hasty visit to the Zuni Mountain field late in November, 1905. For courtesies and aid received there the writer is indebted to Mr. W. J. Skeed, of Copperton. The assays and chemical tests of the ores were made in the chemical laboratory of the Geological Survey, and the microscopic determinations of the pre-Cambrian rocks by Mr. Waldemar Lindgren.

---

[a] Sixth Ann. Rept. U. S. Geol. Survey, 1885, pp. 105-198.
[b] Report of a geological reconnaissance in western Socorro and Valencia counties: Bull. Hadley Laboratory, Univ. New Mexico, 1900, p. 13.
[c] Herrick, C. L., Salt and gypsum in New Mexico: Bull. Hadley Laboratory, Univ. New Mexico, vol. 2, pt. 1, pp. 12. Darton, N. H., The Zuni Salt Lake: Jour. Geology, vol. 13, 1905, pp. 185-193.
[d] New Mexico mines and minerals, 1904, pp. 190-192.

VALENCIA COUNTY.

DESCRIPTION OF THE REGION.

The deposits occur in the northwestern part of New Mexico, about 85 miles west of Albuquerque, in the northern part of Valencia County, south of the Santa Fe Railway. The nearest stations are Grant and Bluewater; from these points Copperton, the principal mining camp, is about 20 miles west and southwest, respectively, and is easily reached in a few hours' drive.

The deposits center chiefly about Copperton, in the heart of the Zuni Mountains. Mount Sedgwick, near the center of the district, is the culminating peak; it has an elevatian of 9,200 feet and rises 2,000 feet above its base and the surrounding plateau. The Zuni Mountains are mapped on the Wingate reconnaissance sheet of the Geological Survey. They trend northwest and southeast and have in this direction an extent of about 50 miles. Their northwestern extremity lies near Fort Wingate and their southeastern near Agua Fria. Their mean height is about 8,500 feet. Toward the southeast, between San Lorenzo and Bluewater, in the vicinity of Copperton and Mount Sedgwick, they have a maximum width of about 20 miles. The belt which they occupy has been studied, mapped, and admirably described by Dutton as the Zuni Plateau.[a] Edwin Howell characterizes the uplift as an elongated quaquaversal from which all the rocks above the Carboniferous have been denuded. Like the Sierra Nacimiento, the mountains are due to the regional uplift, which has here raised above the general surface of the

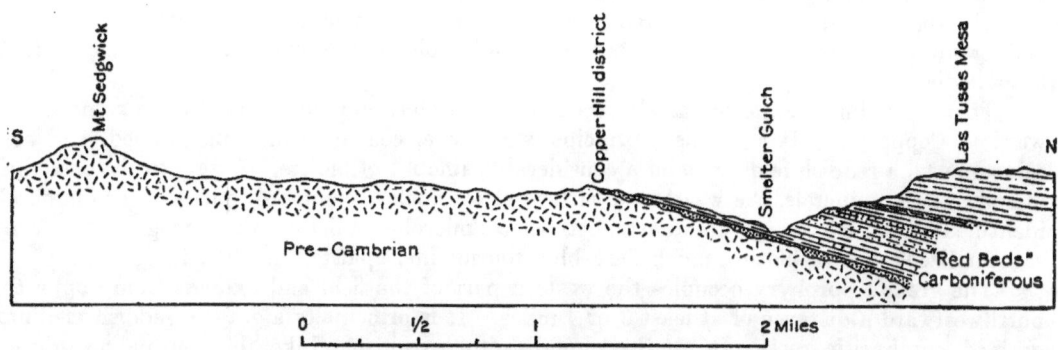

FIGURE 9.—Section from Las Tusas Mesa to Mount Sedgwick, Zuni Mountains.

plateau the vast expanse of sedimentary rocks ranging in age from Carboniferous to Cretaceous, with a thickness of 8,000 to 10,000 feet, and brought the underlying pre-Cambrian complex to view.

In the vicinity of Copperton these pre-Cambrian rocks occupy a belt about 8 miles wide and form the crest and core of the mountains, with remnants of the overlying upturned sedimentary rocks resting upon their flanks and gradually merging into the plateau. To this type of geologic structure Dutton has applied the term "swell."

By reason of their color certain parts of the sedimentary rocks, from Pennsylvanian ("Upper Carboniferous") to Cretaceous, are commonly known in the Southwest as the "Red Beds." About the region here discussed they form an almost encircling inward-facing scarp, about 1,000 feet high, with the Pennsylvanian resting directly on the pre-Cambrian (fig. 9), the Mississippian ("Lower Carboniferous"), Devonian, Silurian, Ordovician, and Cambrian here being absent. The Zuni Mountains seem to have been a land area, at least during the early Paleozoic. This scarp forms the edge of Las Tusas Mesa on the north (Pl. VI, C) and the face of Bear Mountain on the southwest; the latter here represents the Continental Divide.

The deposits are comprised in a rectangular field of about 70 square miles trending nearly north-northwest and south-southeast diagonally across the crest of the mountains; in this direction it has an extent of about 12 miles. The middle and most of the southern part of the field are known as the Copperton district, the northern part as the Copper Hill district, and the extreme southeastern part as the Montezuma district.

---

[a] Mount Taylor and the Zuni Plateau: Sixth Ann. Rept. U. S. Geol. Survey, 1885, pp. 113-198.

The topography is hilly and mountainous, but except at Mount Sedgwick it is not rugged. The relief forms are of the rounded type, produced by the weathering down of granitic rocks. Almost any portion of the field can be reached by light wagons. The rougher country is in the northern part, where from the summit of Mount Sedgwick the surface descends nearly 2,000 feet in a horizontal distance of about 3 miles. The drainage is into the Rio Grande, principally by way of Bluewater and San Jose creeks on the north and the Agua Fria on the south.

THE ROCKS.

PRE-CAMBRIAN ROCKS.

The pre-Cambrian consists essentially of granitic rocks, but include also gneisses, schists, and other undetermined metamorphic rocks. By Dutton they were assigned to the Archean. They correspond closely with the pre-Cambrian rocks at the Hamilton mine, on Pecos River east of Santa Fe, described by Mr. Lindgren (p. 113), and with similar rocks of the Sierra Nacimiento, described in a preceding chapter (Mora and San Miguel counties). They exhibit great variation in character and are locally schistose; a typical coarse-grained granite, for instance, being changed beyond recognition to a schist almost indistinguishable from a similar rock of sedimentary origin. The schistosity trends on the whole a little north of west. The prevailing dips are southerly and are usually steep, at angles of 70° to 80°.

The rocks consist chiefly of three principal types—tourmaline granite, granite porphyry, and granitoid gneiss. Locally they contain considerable barren white quartz, much of it in large bodies.

The tourmaline granite practically occupies the eastern part of the field and extends westward to Copperton. It is reddish, principally massive, coarse to medium grained, and contains quartz, a reddish feldspar, and a considerable amount of biotite. Owing to the corroding of some of the minerals, the weathered surface of the rock is commonly pitted. The minerals identified by the microscope are quartz, orthoclase, microline, microperthite, plagioclase (partly saussuritized), brown biotite, much dark-blue tourmaline, apatite, and titanite.

The granite porphyry occupies the western part of the field and extends from Copperton northwestward a distance of at least 3 or 4 miles. It is principally a gray to reddish medium-grained porphyritic rock. It weathers reddish to brownish. Like the tourmaline granite, it occurs chiefly in massive form and seems to be the youngest of the three rocks here described. It exhibits two different phases, in one of which it seems to have been a tuff or pyroclastic rock, and in the other the rock is reddish and free from dark minerals, and the quartz phenocrysts are surrounded by flow lines. It consists essentially of phenocrysts of orthoclase or perthite and corroded quartz, with some green and brown mica in small foils. The groundmass is a microcrystalline aggregate of quartz and orthoclase.

The granitoid gneiss occurs in the northwestern part of the field, in the Copper Hill district, where it underlies the fringe of overlapping copper-bearing sedimentary rocks to be described later. It ranges from a gneiss to a schist, is very micaceous, weathers reddish, and is evidently a medium-grained pressed granite. It consists principally of crushed quartz and orthoclase in long-drawn lenses and of straight foils of muscovite and biotite and contains a small amount of plagioclase and magnetite. Besides possessing the structure common to the other two rock masses it has also an older, nearly north-south schistosity, which dips west at angles of about 60°. It is inferred to be the oldest rock in the field.

The less important of the pre-Cambrian rocks in areal extent are amphibolites, gabbros, quartz-sericite schists, probably sedimentary, and diorites, granites, and gneisses. Nearly all of these rocks are more or less altered and schistose, and contain copper along certain shear zones; some of them probably represent ancient dikes.

The pre-Cambrian rocks are sparingly cut by relatively young dikes of a reddish, fine-grained, somewhat pressed biotite and chlorite bearing microperthite granite, whose intrusion took place subsequent to the disturbance that produced the shear zones in the rocks. These dikes were observed chiefly in the tourmaline granite. They rarely exceed 2 feet in thickness and they closely resemble the red granitic dikes cutting the pre-Cambrian granites and schists at the San Miguel mine, in the Sierra Nacimiento, except that the latter bear more mica.

"RED BEDS."

Resting upon the lower slope of the pre-Cambrian masses about the margin of the field on the north is a remnant of the tilted and eroded sedimentary rocks, usually less than 100 feet in thickness, and consisting of reddish conglomerate, grit, sandstone, and shale, whose coarser members have been derived directly from the pre-Cambrian mass (fig. 9). These beds seem to be the base of the Pennsylvanian ("Upper Carboniferous"), which forms the lower part of the encircling scarp toward which they dip, and they lie within the area mapped as Carboniferous by Dutton.[a] The rocks composing this encircling scarp, represented in the section in fig. 9, between the Archean at the base and Las Tusas Mesa at the summit, aggregate nearly 3,700 feet in thickness and are composed in ascending order mostly of Pennsylvanian ("Upper Carboniferous") and Triassic sandstones and shales which Dutton[b] has subdivided as follows:

"Upper Carboniferous."—This portion of the section is composed of two parts. The lower part consists of about 800 feet of bright-red, usually heavy-bedded, fine-grained sandstone, which Dutton calls "Lower Aubrey." The upper part (Dutton's "Upper Aubrey") consists of 400 feet of mostly light-yellowish, gray and brown, very hard, more or less coarse-bedded sandstone that occurs in several bands with which are intercalated thick beds of pure fossiliferous limestone.

Permian.—Above the "Upper Carboniferous" is about 450 feet of variegated, mostly sandy shale, with gypsum and thin beds of limestone, and with conglomerate and silicified tree trunks in its upper part.

Triassic.—The Permian is succeeded by the Triassic, whose lower part (Dutton's "Lower Triassic") consists of 650 feet of dark, strongly colored and variegated, sandy shales abounding in gypsum and silicified wood, above which occur 800 to 900 feet of lighter-colored, dull-red, easily weathered shales, which are in turn succeeded by a conspicuous massive bright-red sandstone (the Wingate sandstone of Dutton), 450 feet in thickness, reaching to the top of the section and the summit of the mesa, and thought by Dutton to be the equivalent of the Vermilion Cliff sandstone in southern Utah.

LAVAS.

Capping Copperton Mountain, a low hill on the divide at an elevation of about 9,000 feet, near the middle of the eastern edge of the field, on the new Copperton-Grant road, is a red, scoriaceous medium to coarse grained rock, which appears to be an andesite of uncertain, probably Tertiary, age.

Three miles north of this locality, near the north edge of the pre-Cambrian mass, stands a reddish hill rising several hundred feet above the surrounding surface. It is locally known as the Sierra Colorada, and is reported to contain at its summit a typical crater nearly filled with congealed lava. It seems undoubtedly to be the source of the dark basalt flows which are encountered near by on the above-named road and which for several miles along it floor the canyon cut in the Carboniferous rocks of the encircling scarp, down which the lava flowed. This lava appears to be the same as that which occurs in the Zuni Canyon farther east, followed by the old Copperton-Grant road, and which also forms the rough surface of the lowlands along the railroad about Grant. Its freshness betokens great recency of eruption.

---

[a] Op. cit., p. 128.   [b] Op. cit., pp. 132-137.

COPPER DEPOSITS.

GENERAL OCCURRENCE.

The copper occurs in two distinct classes of deposits—(a) in the pre-Cambrian rocks and (b) in the overlying "Red Beds," which are upper Carboniferous or younger. Of these, the former constitute the more important class, have the wider distribution, and seem to give the more promise for the field as a whole. These pre-Cambrian deposits occur chiefly in the middle and southern parts of the field, in the Copperton and Montezuma districts.

DEPOSITS IN THE PRE-CAMBRIAN ROCKS.

In the pre-Cambrian rocks the ore is found mainly in lodes which are shear zones or lines along which rock movement and sheeting have taken place. Locally they are known as veins. These lodes are particularly characteristic of the Copperton district, where they lie chiefly in the tourmaline granite and the granite porphyry. They generally occur independent of the distribution of the rocks, but are in places found on contacts or courses of ancient dikes. Here and there the shearing along these lodes develops into schistosity. Many of the croppings present the appearance of a belt of jagged upturned schist, which contrasts strongly with the massive country rock on either side. Where a zone is narrow and its walls well defined, it may locally partake of the nature of a fissure vein, but such occurrences are so few in number and so small in extent that it seems best to characterize the copper ores of the pre-Cambrian rocks as a whole as shear-zone deposits.

The most important and youngest of these shear zones is known as the Compromise vein. It has a width of 60 to 800 feet and extends across the field, cutting everything that lies in its path. It trends about N. 70° W. and dips at an angle of about 80° S., from which, however, the dip may vary locally to vertical or steeply north. In its hanging or south wall there usually appears a more or less continuous body of impure brownish to reddish iron-stained quartz, from 1 to 10 or 12 feet in thickness. This quartz seems to be barren, but contains veinlets, seams, and stringers of buff or light-colored clayey material. It gradually merges into the country rock on the one hand and into the shear zone on the other, without sharp lines of demarcation on either side. Its hardness gives prominence to the croppings of the zone, which extend with a height of 10 to 20 feet for several miles along the Bluewater in the western part of the field.

Next in importance to the Compromise lode is a series of less persistent, older shear zones, which lie at nearly right angles to the Compromise on the south and extend to or toward it. Apart from their position, however, they bear no relation to the Compromise lode, nor does their junction or contact with that lode in any wise affect it. There are five or six of these "spur" zones in the Copperton district. They trend a little west of north and range from a few hundred yards to nearly a mile in extent and from 1 foot to 60 feet in width, probably with an average width of 5 or 6 feet. They dip eastward toward the Compromise lode at angles ranging from 30° nearly to 90°.

In the above-described shear zones the copper ore is somewhat irregularly distributed in areas, patches, or bodies, some of which have an extent along the lode of several hundred feet. It occurs principally as the green and blue carbonates, malachite and azurite, and to some extent as chalcocite or glance. The ore is in places associated with quartz; here and there it is finely disseminated in the rock and in the walls of the zone, the carbonate frequently replacing the minerals of the rock, particularly along the planes of schistosity and sheeting. Chalcopyrite, which probably was the primary mineral, now occurs very sparingly, having been altered to carbonates. In one place a small amount of native copper is reported. Another reported type of occurrence of the copper is as kidneys of the carbonates in the lodes of limonite described below.

The croppings most commonly consist of conspicuous green, greenish, or bluish-green stained patches or areas along the zones, at some places in the jagged schist or quartz protruding a few feet above the surface, and at others in the surface detritus of the weathered rock.

In a few localities the covering is an iron capping; at the Matthews-Whiteside mine this consists of ferruginous black pulverulent material from 6 inches to 3 feet in thickness.

Assay samples collected by the writer from the shear-zone deposits yielded from 3.16 to 6 per cent of copper. The sample that yielded 3.16 per cent was taken at a depth of 40 feet (the present bottom of the workings) across the whole Matthews-Whiteside ore body, in the Copper Hill district. The ore consists essentially of malachite in the altered granitoid gneiss and has a known extent of 120 feet along the croppings and a width of 30 feet. The malachite also occurs finely disseminated in the inclosing wall or country rock, which has been tunneled laterally to the extent of about 100 feet from the ore body and is reported to yield about 2 per cent of copper.

The sample which yielded 6 per cent of copper was collected from croppings and several shallow openings on the Emperor group of claims, 3 miles east of Copperton, on the Compromise lode. Here the ore-bearing rock seems to be mostly a dark schistose phase of the tourmaline granite; the ore is chiefly a mixture of malachite and azurite and has in association considerable quartz.

The Montezuma district, in the southeastern part of the field, was not visited by the writer, but reports indicate that its deposits form a continuation of the Copperton deposits just described. It comprises a group of 10 or 12 claims, on some of which shafts have been sunk to a depth of 100 feet. The ore from the district is reported to yield principally copper, but carries also the usual gold and silver values. The recognized minerals associated with the copper ores are quartz, hematite, limonite, pyrite, galena, gold, and silver. The amount of gold present is reported to range from $1 to $10 and of silver from 5 to 10 ounces a ton.

DEPOSITS IN THE "RED BEDS."

In the northwestern part of the field, in the Copper Hill district, between Smelter Gulch on the east and Copper Gulch on the west, a fringe of "Red Beds," resting upon the edge of the pre-Cambrian mass at a mean elevation of about 8,000 feet, constitutes a copper-bearing group, which consists in ascending order, so far as made out, of reddish conglomeratic quartz grit, sandstone, or marl, and shale, aggregating a thickness of 30 to 60 feet. The underlying rock is the granitoid gneiss, on which the "Red Beds" seem to rest unconformably, as shown in figure 9.

The sandstone and grit are medium to heavy bedded and show cross-bedding, some of which is also present in the shale. The shale is greenish buff, soft, fine, argillaceous, and somewhat talcose, and is reported to be underlain to the northeast by leaden-gray limestone. The rocks dip west of north, away from the mountains, at angles up to 25°, and are copper bearing more or less throughout an area of about half a square mile. About 20 or more claims are located in this area and they constitute the portion of the field known as the Copper Hill district.

The copper-bearing beds and the occurrence of the copper in them are similar to those in the better-known Sierra Nacimiento and other "Red Beds" deposits in the Southwest. The copper occurs principally as blue and green carbonates and as glance, replacing wood, as in the Sierra Nacimiento, or forming thin layers or stringers interbedded and disseminated in the rocks. Some of it, especially that in the conglomeratic grit, has the appearance of having been deposited synchronously with the inclosing sediments and to have been derived like them from the pre-Cambrian mass.

The best ore horizon seems to be that of the shale, which occurs in a stratum about 20 feet thick. The upper 7 feet is exposed in a shallow drift on the Copper Queen claim, in the western part of the district. A 2,645-pound sample of picked ore reported to be from this shale, probably consisting mostly of glance, treated by the American Smelting and Refining Company, of Pueblo, Colo., yielded 36.5 per cent of copper and 3.6 ounces to the ton of silver, and an assay sample of the deposit taken by the writer vertically across the strata, 7 feet in thickness, in the face of the drift, including everything, yielded 4.72 per cent of copper. It is reported that a boring 12 feet deep made in the bottom of the drift showed no decrease in the ore value of the deposit down to that depth.

As pre-Cambrian rocks now outcrop both to the east of Smelter Gulch and to the west of Copper Gulch, the original lateral extent of this copper-bearing group can not be ascertained. To the north, however, it probably underlies Las Tusas Mesa, where one would expect to find its edges exposed in the head of the north-south canyon cut through the mesa by Bluewater Creek.

DEVELOPMENTS.

According to oral accounts of the Mexicans, the Indians from time immemorial have been collecting copper ore from this field to adorn their persons and barter as gems or curios, and in Smelter Gulch, on the barite vein, described as occurring in the Copper Hill district, stands a couple of adobe smokestacks whose construction antedates the memory of the oldest settlers now living in the region. It is reported that they stood there eighty-five years ago just as they are to-day. They were evidently built for the purpose of treating ore, possibly silver found in the barite vein, or in the copper deposits, of which some are present near by. They bear no evidence of having been used.

No mining on a larger scale has yet been undertaken. The recent developments, mostly in the last five years, consist principally of pits, cuts, shafts, drifts, and tunnels, from 10 to 200 feet in depth or extent. In all there are about 200 copper claims on which assessment work is being kept up. These are owned or controlled by about 75 men.

A diamond core-drill test hole put down near the middle of the field a few years ago to a depth of 385 feet is reported to have shown copper ore with some gold and silver for a depth of 113 feet from the surface down.

OTHER DEPOSITS.

In the pre-Cambrian granitoid gneiss occur deposits of gold, barite, and limonite. The barite forms a persistent vein which is 10 feet or more in maximum thickness and trends N. 70° W. The vein, to judge from prospect holes and a shaft 30 or 40 feet deep, seems to have been prospected, probably for silver. The barite is reddish and coarsely crystalline and is associated with much greenish fluorspar.

The auriferous gneiss, locally known as the "iron dike" from the iron pyrites it contains, consists of a 20-foot belt of coarse gneiss trending N. 60° E. across the field and having a nearly vertical dip. It has a typical gneissoid structure and is composed essentially of large quartz and feldspar grains. It contains much biotite and sericite, with pyrite in cubical crystals and a small amount of chalcopyrite. It has been prospected at a number of points and is reported to assay from $2 to $4 in gold to the ton, the gold presumably being in association with the pyrite.

The limonite occurs in workable bodies or lodes, one of which, where prospected at the head of Smelter Gulch, has a width of 50 feet and extends interruptedly for a distance of about a mile. It trends northeast and southwest, nearly parallel with the "iron dike," from which in certain localities it lies less than one-eighth mile distant, but with whose occurrence it has not been observed to have any connection. It is heavy, weakly magnetic, and very uniform in occurrence, except that kidneys of copper carbonate are reported to be scattered through it. In texture it is uniformly dense, and under the microscope appears like a comminuted chert breccia cemented by hydrated iron oxide. These and similar iron-bearing deposits widely distributed in the pre-Cambrian rocks are probably the chief source of the coloring matter in the overlying "Red Beds" of the Carboniferous and lower Mesozoic.

SANDOVAL COUNTY.

By F. C. SCHRADER and L. C. GRATON.

GENERAL FEATURES.

Sandoval County, in northwestern New Mexico, lies chiefly on the west side of the Rio Grande, but includes a small triangular area on the opposite side, a short distance north of Albuquerque. It covers 3,833 square miles and is adjoined on the north by Rio Arriba, on the west by McKinley and San Juan, on the east by Santa Fe, and on the south by Bernalillo County. The western and southern portions have the characteristics of the plateau province, with its gently dipping Cretaceous strata, in part covered by basaltic flows. It is drained from north to south by Rio Puerco, a tributary which farther south enters the Rio Grande. The eastern and larger part lies in the region of monoclinal ranges, the first of these encountered being the Sierra Nacimiento, which extends north and south for 30 miles with its steep westward-facing scarp. This range continues into Rio Arriba County. Its highest point attains 10,045 feet in elevation.

The Valles Mountains rise to elevations of about 12,000 feet in the northeast corner of the county and lie partly in Rio Arriba County. Unlike the monoclinal Sierra Nacimiento, with its scarp of pre-Cambrian granite and back slope of Paleozoic and later sediments, the Valles Mountains form a circular group of volcanic origin and, so far as known, are built up almost entirely of rhyolitic and andesitic flows. Along the Rio Grande are found the deposits of a late Tertiary lake, and these white lacustrine beds are at many places covered by strongly contrasting black basalt flows. The part of the county to the east of the Rio Grande includes the northern half of the Sandia Range, a westward-facing monocline even more characteristic than the Sierra Nacimiento.

The nonmetallic mineral resources of this county include the important Cretaceous coal field of Una del Gato.[a] Coal beds are also known to occur in the Cretaceous Mesaverde formation at the western foot of the Nacimiento Range, where gypsum likewise occurs in abundance. A deposit of sulphur occurs at Jemes, in the country adjacent to the Valles Mountains, and has been commercially utilized on a small scale. The mineral appears to have been formed by the hot springs.

The metal deposits comprise principally the copper-bearing sandstones of the Sierra Nacimiento, which continue northward into Rio Arriba County, and the gold-silver veins of the Cochiti district, in the foothills of the Valles Mountains. At Placitos, near the north end of the Sandia Range, are some copper deposits said to occur as flat sheets in Carboniferous limestone. This district was not visited.

COPPER DEPOSITS OF THE SIERRA NACIMIENTO.[b]

INTRODUCTION.

The observations of F. C. Schrader in the Sierra Nacimiento, which form the basis of this paper, were made in September, 1905, while carrying on a reconnaissance along the eastern border of the Durango-Gallup coal field, in which work he was assisted by M. K. Shaler. For courtesies and information received acknowledgments are due to the Juratrias Copper Company; to Mrs. M. A. Gorman, proprietor of the San Miguel mine; and to numerous men who have operated or worked in the Nacimiento district.

DESCRIPTION OF THE REGION.

The copper deposits of the Sierra Nacimiento lie in the northwestern part of New Mexico, west of the Rio Grande, in the eastern part of the Colorado Plateau. The sedimentary rocks that underlie this extensive plateau have an aggregate thickness of 12,000 to 13,000 feet and range from Carboniferous to Tertiary in age. They have here been broken by upheaval that has produced the Sierra Nacimiento, which is practically a single-ridged mountain that trends nearly north and south, with a length of about 50 miles, a uniform elevation of about 10,000 feet, and a height of about 3,000 feet above the surrounding plateau. Somewhat more than its southern half is shown on the Jemes reconnaissance sheet of the Geological Survey; the northern portion is shown on the Gallina sheet. The larger number of the deposits lie near the middle of the length of the mountain, on its west slope. They are situated about 60 miles west-northwest of Santa Fe, about the same distance in a nearly northerly direction from Albuquerque, about 70 miles south of the Colorado state line, and about 45 miles south of Elvado, the nearest railroad station, the terminus of the Burns-Biggs Lumber Company's branch road extending southward from the Denver and Rio Grande Railroad at Lumberton.

---

[a] Campbell, Marius R., The Una del Gato coal field, Sandoval County, N. Mex.: Bull. U. S. Geol. Survey, No. 316, 1907, pp. 427-430.
[b] By F. C. Schrader.

On the west the range presents a nearly straight front throughout its extent and the surface descends to the plateau at a nearly uniform rate of about 500 feet to the mile. On the east the limits are less regular, the range here being flanked by mountains of intermediate height. Still farther east are the Valles Mountains and adjoining high plateaus which extend eastward a distance of 20 to 30 miles before the surface begins to decline into the valley of the Rio Grande.

At both ends the Sierra Nacimiento gradually falls off into the Colorado Plateau, on the south into barren desert, and on the north into timbered slopes and parks. The drainage finds its way into the Rio Grande, that of the east slope by way of Chama River and Jemes Creek and that of the west slope by way of Rio Puerco, a small desert stream, the upper portion of whose valley occupies the acute reentrant angle between the Sierra Nacimiento and the low Continental Divide lying but a few miles distant on the plateau to the west. The Puerco drains to the south, nearly parallel with the range, which in the zone of the upturned sedimentary rocks is usually fronted on the west by narrow longitudinal hogback ridges and valleys connected with the Puerco Valley by cross canyons and arroyos that give exit to the drainage. The plateau is mainly arid, but the mountain slopes are well timbered.

THE ROCKS.

Along the summit and upper slopes of the range the upraised axial core of underlying pre-Cambrian rocks, consisting mostly of massive red granite, but including also gneiss, schist, and other undetermined metamorphic rocks, is exposed at intervals throughout its extent; the flanks of the mountain, notably on the west, present admirable sections of the overlying upturned sedimentary rocks. These sedimentary formations trend parallel with the mountain and are conformable one with another.

The geologic section may be roughly stated, in ascending order, as follows: An axial core of pre-Cambrian granites and metamorphic rocks, forming the great mass of the range; on this rests the Pennsylvanian, a series of mostly massive gray limestone and grit of variable thickness, averaging about 800 feet; this is in turn succeeded by a series of nodular concretionary limestone, variegated marls, and soft red and white sandstone, aggregating about 400 feet in thickness, which also may belong to the Pennsylvanian; above this are about 3,000 feet of mostly reddish sandstones and marls, containing at the base a reddish-white copper-bearing Triassic (?) sandstone and conglomerate 100 or more feet in thickness; next follow the Cretaceous formations, composed of light-brown sandstones, shales, and marls 7,000 or more feet in thickness; and above these, at a few miles distant from the range, Tertiary marls and sandstones about 3,000 feet thick, forming the top of the section and the summit of the plateau. The inclination of these sediments ranges from nearly vertical in the Carboniferous and Triassic, next to the granite core, to nearly horizontal 2 or 3 miles distant from the base of the range. A large portion of the section from Pennsylvanian to Lower Cretaceous, inclusive, by reason of its color is commonly known in the Southwest as the "Red Beds."

As seen near the San Miguel mine, the pre-Cambrian granites seem to consist of two types, or perhaps two phases of the same rock, a pronounced red and a dark gray. Both are medium to coarse grained and in part schistose, but the gray rock seems to be more generally schistose than the red. Both are cut by dikes of a younger, fine-grained red granitic rock, whose intrusion took place at a date later than that of the shearing of the granite. Nowhere, however, was any dike or igneous rock found to cut any of the overlying sedimentary rocks.

The red granite seems to be the dominant pre-Cambrian rock throughout the Sierra Nacimiento, and megascopically corresponds closely with a similarly prominent pre-Cambrian rock of the Zuni Mountains, which was identified as a tourmaline granite. The above-described red granitic dike rock seems to be the same as that which similarly cuts the tourmaline granite in the Zuni Mountains and which has been found to be biotite and chlorite bearing microperthite granite.

COPPER DEPOSITS IN THE "RED BEDS."

GENERAL STATEMENT.

The copper deposits occur essentially in the "Red Beds" and belong to a series of similar deposits comprised in the somewhat irregular belt extending from north-central Utah southeastward through the southwestern part of Colorado, the northwestern part of New Mexico, and adjacent portions of Texas.[a]

They were originally discovered and located by Indians and Mexicans, who about the middle of the nineteenth century brought out specimens of rich ore, but the first published account of them is that given by Newberry, who as geologist of the Macomb expedition visited the region in 1859. In the report of that expedition published in 1876 Newberry states: "In the coarse yellow sandstones and conglomerates overlying the Red Beds of the Trias I find large quantities of the sulphide of copper replacing trunks and branches of trees."[b]

As early as 1868 the Nacimiento Mining Company is reported to have been an owner of the deposits, but no systematic effort seems to have been put forth until about 1880, when claims covering what now constitute the principal part of the deposits in the northern part of the district were located; a little later those in the southern part were discovered. The first company that operated in the district was the New Mexico-Illinois Company, but with what success has not been learned. It finally withdrew from the field, and the Providence Company, subsequently the owner of the San Miguel mine, was organized. In 1900 the Juratrias Copper Company was organized, with headquarters at Albuquerque. This company now has a 25-ton smelter at Senorito, near the mines, and owns or controls a score of claims, including the most important deposits.

In the year 1880 an article on the occurrence and probable economic value of the deposits was published by F. M. F. Cazin,[c] and further comments were made on their economic value by the same author in two later papers.[d] In 1899 the deposits were discussed by William Jenks.[e] More recently the deposits have been referred to by S. F. Emmons in a paper entitled "Copper in the Red Beds of the Colorado Plateau."[f]

NACIMIENTO DISTRICT.

GENERAL FEATURES.

The most important of the deposits are comprised in the Nacimiento mining district,[g] an area of somewhat indefinite limits, but having a north and south extent of about 12 miles and a width of about 3 miles. It lies near the middle of the range, on its west slope, partly in Rio Arriba County, but mainly in Sandoval County. The principal mining camp and post-office is Senorito, in the northern part of the district.

The accompanying sketch map (fig. 10), adapted in part from the Jemes sheet of the Geological Survey and in part from the reconnaissance traverse made during the season of 1905, shows the topography of the district and the relative location of the principal deposits and mines. The topography is mountainous, but not unusually rugged. The surface descends from elevations of 9,500 feet at the crest of the range to 7,000 feet at the edge of the plateau on the west, at the rate of about 500 feet to the mile.

The district is drained chiefly by Nacimiento and Senorito creeks, San Miguel Arroyo, and one or two similar small streams flowing westward across the upturned sedimentary rocks into Rio Puerco. Of these streams the most important is Senorito Creek, in the mountainous portion of whose valley all the important deposits in the northern part of the district occur. This

---

[a] Jenks, William, Juratrias copper: New Mexico Mining Record, vol. 1, No. 4, 1899, p. 1. Emmons, S. F., Bull. U. S. Geol. Survey No. 260, 1905, p. 221.
[b] Newberry, J. S., Geological report, exploring expedition from Santa Fe to junction of Grand and Green rivers in 1859, 1876, p. 117.
[c] New Mexico vs. Lake Superior as a copper producer: Eng. and Min. Jour., vol. 30, 1880, p. 87.
[d] Eng. and Min. Jour., vol. 30, 1880, pp. 108, 381.
[e] New Mexico Mining Record, vol. 1, No. 4, 1899, pp. 1-2.
[f] Bull. U. S. Geol. Survey No. 260, 1905, p. 225.
[g] Rept. Governor New Mexico to Sec. Interior, 1902, p. 472.

creek has here trenched its bed to a depth of about 1,000 feet below the normal surface of the mountain slope, and its valley is bordered on the west by a more or less continuous scarp several hundred feet high, along which some prospects are located.

The most important of the deposits occur in two main localities about 10 miles apart, one near the site of Copper City, the former mining camp in the northern part of the district, the other at the San Miguel mine, in the southern part. They all occur in practically the same "Red Beds" formation, consisting of sandstones, marls, conglomerates, and shales, which so far as known are of Triassic age.

The Copper City deposits are chiefly found at the Copper Glance and Eureka mines and the Bluebird claim. Numerous other prospects have been opened, mostly in the Senorito

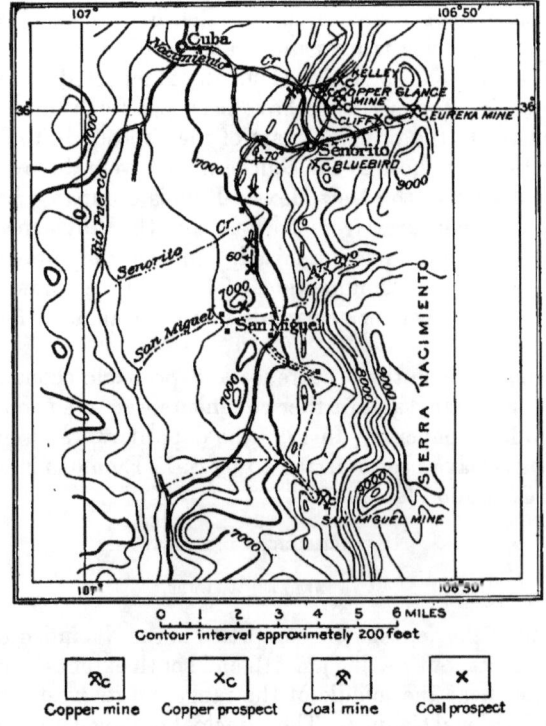

FIGURE 10.—Reconnaissance map of Nacimiento district.

Creek valley. The Juratrias Copper Company, now virtually in possession of the northern part of the district, holds many of them.

COPPER GLANCE MINE.

The Copper Glance mine is situated about a mile above Copper City and about 4 miles nearly east of the old Mexican town of Cuba, on the Puerco. It lies at an elevation of about 8,000 feet, at the head of a gulch that drains southward into Senorito Creek. The ore consists of sulphides, carbonates, and silicate of copper, contained in whitish, yellowish, or reddish sandstone or grit, near the base of the Triassic, within a few hundred feet of the underlying pre-Cambrian granite. These ore-bearing rocks vary chemically from siliceous to calcareous, and texturally from fine-grained sandstone to fine conglomeratic grit. The beds are from 2 to 4 feet thick and locally exhibit cross-bedding. They aggregate about 100 feet in thickness, and dip northeastward, toward the mountain, at an angle of about 45°. Areally they occupy a portion of three contiguous claims, aggregating an extent of about 4,500 feet along the strike, across which they have a known width of about 250 feet. The richer part of the deposit, however, to judge from the development work that has been done, seems to be contained in a section of the strata about 30 or 40 feet in thickness and extends about 400 feet along the strike. At and near the surface of most of this area several thousand tons of ore has been mined by crude methods, largely by quarrying. The area now presents the appearance of a vast dump.

None of the rocks require blasting and the ore is easily separated from the rock. Some underground work has also been done by means of slopes, drifts, tunnels, and shafts, mainly within the area last noted. Although these works had mostly caved since the mine was shut down a few years ago, and the observations were confined almost wholly to the worked-over surface croppings and the resulting dump débris, sufficient information was obtained to form an adequate idea of the occurrence.

The principal ore is chalcocite, but considerable quantities of bornite, malachite, azurite, and chrysocolla are also found. The ore is in part finely disseminated through the rocks, but by far the most of it occurs in connection with wood and plants which it has wholly or partly replaced. In the main the deposits are conformably interbedded with the strata or inclosed in an individual stratum, but they are also developed along joints and lines of weakness and variation in rock texture, especially where ferruginous or organic material is present. In places the coarse granitic grit or arkose is favorable for ore deposition. Locally associated with the deposits is some crystalline calcite, mostly in short, discontinuous veins, or irregular thin masses and lenticular bodies.

The part played by wood in the formation and concentration of the deposit is most important. The wood is present throughout the rock in the form of fragments of trunks, stems, or branches of trees, much of it exhibiting the cellular structure, and now converted into coal, siliceous matter, or ore. A cross section of a piece about 3 inches thick, which seems to be from near the periphery of a large tree, consists mainly of chalcocite in banded concentric layers and lenses, ranging from one-sixteenth to one-half inch in thickness, alternating with seams of coal or lignite about one-sixteenth of an inch thick. The chalcocite is incrusted by a thin layer of malachite, surmounted by beads of blue azurite. With the chalcocite is associated a little hematite and with the carbonate some crystalline calcite.

It is reported by the operators of the mine that, of the ore produced, about 60 per cent was chalcocite and 40 per cent carbonate; that the bulk of the copper-bearing sandstone worked averaged about 5 per cent of metallic copper; that many of the chalcocite nodules, of which a large number have been smelted from the surface work, ran as high as 65 per cent; and that 100 tons of the sorted ore yielded 15 per cent. During the season of 1886–87, 920 tons of the ore was smelted, which, according to the statements of the company yielded 677,120 pounds of metallic copper, an average of 36.87 per cent. The returns of still later work are said to have been 572,666 pounds of copper from 606 tons of ore. These figures undoubtedly refer to picked ores. Little is known regarding the tenor of the average ore, but it will probably be found to be of low grade. It seems probable that the deposits may extend down the dip slope of the rocks to a considerable depth. If so they are likely to contain a large amount of low-grade ore. The ore carries 1 to 3 ounces of silver to the ton, but no silver ores proper have been found.

BLUEBIRD CLAIM.

The Bluebird claim lies about 1½ miles nearly south from the Copper Glance, on the opposite or south side of Senorito Creek, near Senorito. Its deposit seems to be in the same strata as the Copper Glance, though situated at a somewhat lower elevation—about 7,800 feet. It was not visited by the writers. The appearance of the dump and oral reports of the men on the ground indicate that the rock and ore are essentially the same as at the Copper Glance, except that the ore is more exclusively chalcocite. But little development work has been done.

EUREKA MINE.

The Eureka mine is situated at the head of Senorito Creek, near the crest of the mountain, on its west side, at an elevation of about 9,000 feet. It is about 2 miles east of the Copper Glance and 3 miles northeast of the Bluebird. It seems to be in the same formation as the Copper

Glance and Bluebird deposits; at least, fossils reported to have been collected at the Eureka mine seem to be Triassic, and the occurrence of the ore is likewise much the same as at the Copper Glance mine. At the Eureka, however, the rock is coarse and a considerable portion of the ore occurs in conglomerate. The two principal claims are known as Copper Conglomerate No. 1 and No. 2; both have produced considerable ore. The strata intervening between the deposits and the underlying pre-Cambrian granite are said to be about 300 feet thick and to consist principally of red sandstone. The dips are 30° to 35° SW.

According to the reports of the mine operators the deposits are distributed through a thickness of about 25 feet of sandstone and conglomerate, but the richer and most important part is contained in a bed of the copper-bearing fine conglomerate, 3 to 4 feet thick, that is said to average several per cent of metallic copper. Above this lie "low-grade" ore-bearing sandstone and conglomerate about 12 feet thick, which are thought to carry about 5 per cent of metallic copper. Considerable glance, much of it in the form of nodules, is present, especially in the lower conglomerate.

As seen in the ore bin at the Senorito smelter, the ore-bearing conglomerate is very siliceous. Its pebbles are essentially quartz, dense quartzite, some of which is schistose, some light-colored or buff aphanitic igneous rock, and yellowish-brown and slate-colored flint. The largest pebbles are those of quartz, some of them reaching $1\frac{1}{2}$ inches in diameter. The quartz resembles that seen in granitic areas in different localities on the mountains. The cement is composed of detritus of igneous rocks, thoroughly impregnated with copper ore, chiefly the green carbonate. This has led mining engineers who have visited the deposit to report the cementing material to be copper ore.

The ore has mostly been quarried, but there has also been some underground development, consisting of tunnels or drifts aggregating about 600 feet in length, with crosscuts and raises. The Eureka tunnel, 500 feet long, extends through beneath the crest of the mountain to open air on the other side. These works and the croppings show the ore deposit to have a linear extent of about 400 feet on claim No. 1 and of more than 100 feet on claim No. 2. There is said to be about 1,000 tons of ore on the dump.

In the vicinity of the mine there are also half a dozen or more additional claims, showing good prospects, on the most important of which considerable development work has been done.

SAN MIGUEL MINE.

The San Miguel mine is located in the southern part of the district, on the middle-western slope of the range, at an elevation of about 7,700 feet. It is reached by a road ascending a narrow ridge from the north. The ore-bearing formation is the same as that of the Copper Glance mine at Copper City, and consists of medium to heavy bedded white to reddish fine-grained sandstone and grit, with less coarse grit and calcareous material and more shale than at the previously described locality. Much of it is micaceous and some contains hornblende. Some thin lenses of conglomeratic material are also present.

The rocks dip about 40° S. 15° W. They are directly underlain by several hundred feet of "Red Beds," which are probably of Pennsylvanian or Permian age and which are separated from the axial pre-Cambrian granite by 700 or 800 feet of mostly coarse semicrystalline Carboniferous limestone, with some variegated marls and sandstones. (See Pl. VI, A.)

The San Miguel deposit was discovered about 1880. It was first opened as a mine in 1884 and was closed in 1900. The surface workings are comprised within an area of about 250 yards in width, extending 300 yards along the strike, and having a vertical range, due to the slope of the mountain, of about 200 feet. Five or six drifts or tunnels have been driven, nearly all along the strike, but as the mine has been abandoned for six years, owing to cave ins and infilling of wash, these were not examined far beyond the mouth. The drifts and tunnels are all situated at different elevations, but do not seem to represent definitely spaced levels. The purpose of each seems to have been to follow an immediate exposure or lead of ore. The shorter ones are from 300 to 400 feet long and one of the larger 700 or 800 feet. Some of them pass entirely through the ore-bearing strata into the underlying barren "Red Beds."

The ore occurs mainly as chalcocite in fossil remains of wood or trees constituting the so-called "petrified forest." It is of a much higher grade than the Copper City ore. Hand specimens seen by the senior writer are almost pure glance, and according to Mr. Jenks it is not rare to find trees 60 feet or more in length and 2½ feet in basal diameter almost completely replaced by ore. Where the remains of wood are absent, much of the ore occurs in concretionary nodules of glance. There is also present, however, a considerable amount of chrysocolla, impregnating the sandstone. The carbonates seem to be present as secondary products only, where the ore is oxidized or weathered. The ore begins in the white sandstone and extends downward into the reddish sandstone. It carries 2½ to 3 ounces of silver to the ton, but no gold.

Some of the ore that was smelted ran 64 to 65 per cent of copper, much of it 48 to 50 per cent, and very little of it less than 35 or 40 per cent. While the mine was in operation a 30-ton smelter at Copper City produced daily a carload of copper from the ore, but as the ore had to be hauled 10 miles by wagon to the smelter, the lowest grade, forming a relatively small percentage of the whole, was not included. In consequence of this fact a considerable quantity of low-grade ore, running from 10 to 20 per cent of metallic copper, now lies on the dump.

By far the largest part of the production of the Nacimiento district, amounting to about 2,500 tons of metallic copper, has come from the San Miguel mine. The copper was hauled by wagon to Bernalillo, 60 miles distant, whence it was shipped by rail to New York.

Some faulting has taken place at the San Miguel mine and the most important fault, according to Mr. Jenks, has cut off the ore-bearing beds south of the present workings. This fault is normal and hades 25° S. As the downthrow is probably not great, the faulted-off portion of the deposit can doubtless be recovered at no great depth by a shaft. Some ore, which is probably for the most part drag, occurs along the fault plane.

So far as the present workings of the San Miguel mine extend the high-grade ore seems to be practically worked out, but it is said that as only the high-grade ore was sought at the time the mine was operated pillars and other large bodies of easily accessible good-grade ore still remain standing. Little prospecting has been done, and it is reasonable to infer that ore in paying quantities may be developed.

FACILITIES.

The natural facilities of the Nacimiento district for mining are exceptionally favorable. Coal, gypsum, lime, and iron are plentifully present and easily obtained in the upturned Mesozoic rocks along the base of the range and were utilized with satisfactory results in operating the mines and smelters during the latter part of the last century. Coal was obtained from the Mesaverde formation of the Cretaceous, and fluxing iron from the adjacent shales. Timber is abundant on the ground and water can nearly everywhere be made available. A few miles to the northwest, on the Puerco and its tributaries, La Jara and other creeks, lies a fairly productive agricultural district susceptible of great development under irrigation.

GALLINA DISTRICT.

The term "Gallina district," as here used, refers to a large field of indefinite limits in which several prospects occur on and about the north end of the Sierra Nacimiento. This part of the range is about 9,200 feet in elevation, and its width is 8 or 10 miles. On the north it gradually falls off through timbered slopes and parks, and in the latitude of the Mexican village Gallina, about 20 miles north of the Nacimiento district, is lost in the neighboring plateau. On the east the general slope is also gentle and timbered, but the surface is more dissected. On the west the slope is steep but not precipitous.

The drainage for the most part converges at the northwest base of the range, in the neighborhood of Gallina, forming Rio Gallina, a small stream that flows northward in a broad arroyo for about 12 miles, turns eastward and breaks through the mesa, which is 6 to 7 miles across, and then flows into the Chama, whose course here is southeastward, nearly opposite to that of the main part of the Gallina. The portion of the mesa lying between the Gallina and the Chama, a crudely quadrangular timbered area of about 200 square miles, is known as Mesa Blanca Capulin. It seems to have been raised by differential uplift probably 1,000 feet above the portion of the plateau adjoining it on the south, immediately skirting the base of the range (Pl. VI, B). The sedimentary rocks, particularly the lower Mesozoic rocks resting upon the flanks of the range, ascend well up the middle or upper slopes, above which, as seen from a distance the underlying pre-Cambrian red granite seems to form the summit and core of the range.

The copper deposits of this district, so far as observed and according to oral reports, are all in the same class of sedimentary rocks as those at Copper City and the San Miguel mine, but appear to be of less economic importance. As yet they are mere prospects; none of them has produced ore in commercial amount.

The prospect known as the Sefrina is about 2½ miles southwest of Gallina, near the head of the creek that flows through the village, well up in Gallina Mountain, at an elevation of about 8,000 feet. The deposit occurs in massive reddish to brownish sandstone, and an open cut shows it to consist of a mixture of a small amount of impure malachite and azurite and reddish oxidized iron ore, disseminated in brownish-stained, impure, calcareous, soft and shaly sandstone, some associated light and some dark barite in tabular crystals, and some blackish and brownish coaly or lignitic material and impressions of plants, all constituting a bed about a foot thick, dipping about 80° SE.

The Jarossa prospect, named from the ranch on which it occurs, lies about 20 miles southeast of Gallina and about the same distance northeast of Copper City, on the east slope of the Sierra Nacimiento. It is on the Abiquiu-Cuba road, in a well timbered and watered region, and is accessible by wagon. The prospect lies at an elevation of about 8,400 feet, on the side slope of an open ravine. Three exposures occur here within a few hundred yards of one another. The country rock is massive or heavy-bedded reddish micaceous sandstone and conglomeratic grit, consisting essentially of sediments eroded from the granite that forms the core of the range or some similar granite. It contains almost perfect red feldspar crystals one-half inch or more in length. From fossil plant remains found in the beds and determined by F. H. Knowlton, of the United States National Museum, the beds are referred to the Triassic. The dips are low, 3° to 4° SE. The ore deposits consist principally of copper carbonates scattered sporadically through a rock section about 40 feet in thickness, and they are reported to have, with interruptions, a horizontal extent of about 2 miles. They extend northwestward up the ravine for about half a mile, and also northeastward for about 1½ miles, where an exposure occurs on the Valdes ranch. The ore shows fairly well in the croppings at various points, but presents a fresher appearance in the open cuts, which represent the only development work that has yet been done. The principal cut extends across the strike and is about 20 feet long and 12 feet deep at the face. In the sides and face of this cut most of the ore is contained in reddish sandstone and conglomeratic grit, in a vertical thickness of about 8 feet. It consists of segregations of principally malachite and azurite, with a few seams of bornite from one-sixteenth to about one-fourth inch thick. It occurs mostly in more or less impure bodies, thin sheets, or stringers, ranging from about a foot down to less than an inch in thickness, with no very marked persistency exhibited in any. It lies mostly parallel with the bedding planes of the rocks, but much of it is irregularly disseminated through the rock layers. Some of the ore dips northward at angles of about 20° or 30° with the rock bedding, and marks disturbances or breaks in the rocks before the ore, which now fills the cracks and breaks, was deposited. Plant remains are present in small amount. They occur mostly in the fine-grained sandstone and do not seem to have any intimate association with the ore, as they do in the Nacimiento district. Though the Jarossa deposits seem to have considerable geographic distribution and vertical range in the geologic section and are preeminently the best visited outside of the Nacimiento district, without further development their economic value is hard to estimate.

The Capulin prospect is in the southern part of Mesa Blanca Capulin. It was not visited by the senior writer, but was described to him by Pedro Valdes, a local resident, as being similar to the Jarossa prospect.

U. S. GEOLOGICAL SURVEY PROFESSIONAL PAPER 68 PLATE VI

A. VIEW FROM SAN MIGUEL MINE, SANDOVAL COUNTY.

Upturned beds of Mesozoic age along western base of Sierra Nacimiento. Portion of the Colorado Plateau of flat-lying Cretaceous and Tertiary in left background. The white rocks are beds of gypsum and limestone. Elevation, 7,700 feet. Looking northeast.

B. SOUTH EDGE OF MESA BLANCA CAPULIN, SANDOVAL COUNTY; ELEVATION 8,000 FEET.

Showing sedimentary rocks extending from the Dakota (?) sandstone down to the copper-bearing Triassic, below white band of limestone and gypsum. Looking north from Agua Sarca road, eastern part of Gallina Valley.

C. PORTION OF LAS TUSAS MESA, VALENCIA COUNTY.

Showing 1,000 feet of Triassic and Carboniferous "Red Beds," some of which are copper bearing. Looking north from Copper Hill district.

ABIQUIU "MINES."

The Abiquiu "mines" lie beyond the Gallina district, about 20 miles east of the Capulin prospect, and east of Chama River, in the Conejos Mountains. According to Newberry, they are situated about 9 miles north of Abiquiu. They were formerly worked to obtain copper, probably by the early Spanish explorers. They were not visited by the writer, and for a description of them the reader is referred to Newberry's report,[a] the substance of which follows:

The mines are situated in the face of the cliffs bordering the eroded valley (Arroyo Cobre) that drains into the Chama below Abiquiu. They comprise a series of galleries aggregating a length of perhaps 100 yards, and the work exhibits considerable skill in the use of tools and a familiarity with mining. The cliffs are composed, at the base, of the saliferous sandstone and interstratified marls, about 250 feet in thickness; above these, blood-red marls and calcareous sandstone 200 feet thick; the whole crowned by coarse yellow sandstone having a thickness of about 150 feet. The copper occurs in the base of the coarse yellow sandstone just above the marls, much of it in a conglomerate, with beds of light-gray clay. The ore is distributed with considerable uniformity through a layer 4 or 5 feet in thickness. "It occurs in the form of sulphide of copper and iron (erubescite) and green carbonate, replacing trunks of trees and fragments of wood, and in concretions and botryoidal masses scattered among the pebbles of quartz, or as minute points of carbonate specking the shales. It has evidently been deposited from solution, investing and replacing the wood precisely as the sulphide of iron is prone to do."

ASSOCIATED MINERALS.

As has been indicated in the descriptions of the several deposits, the minerals observed in association with the copper in the "Red Beds" are calcite, barite, pyrites, iron, and coal. In this connection it should be noted that Jenks[b] reports that there is one deposit in the Sierra Nacimiento in which carbonate of lead (cerusite) is associated with the copper in the sandstone, and that in nearly all the deposits there is some sulphate of gypsum.

ORIGIN OF THE DEPOSITS.

The presence of abundant sediments derived from the pre-Cambrian rocks in the copper-bearing strata of the "Red Beds" in both the Nacimiento and the Gallina districts has been indicated. The same relations are even more clearly shown in the Copper Hill district of the Zuni Mountain field, where sediments derived from pre-Cambrian mass largely compose the cupriferous "Red Beds." In both the Sierra Nacimiento and the Zuni Mountain fields copper deposits, as will be presently shown, occur in these pre-Cambrian rocks. It is believed that the deposits in the pre-Cambrian rocks represent the source of the deposits in the "Red Beds" and that the copper, in one form or another, probably for the most part as cupriferous carbonate-bearing sediments eroded from the pre-Cambrian mass, was deposited essentially at the same time as the inclosing "Red Beds" or adjacent sediments. This view finds support in the geologic and topographic localization of the deposits, but, as has been pointed out by S. F. Emmons,[c] it does not follow that the deposits as they exist to-day are therefore syngenetic in origin with the inclosing sediments.

COPPER DEPOSITS IN THE PRE-CAMBRIAN ROCKS.

The ore deposits thus far described are all in the "Red Beds," which alone, so far as present knowledge extends, seem to contain workable bodies. This probably accounts for the fact that the underlying pre-Cambrian rocks seem to have received but little attention from the prospector. At a single locality, however, in the Nacimiento district, near Copper City, on the south fork of Nacimiento Creek, about three-quarters of a mile north of the Copper Glance mine, a copper prospect called the Chalcocite occurs in the pre-Cambrian rocks. The deposit lies chiefly in a zone or band of schist about 35 feet wide contained in the granite,

---

[a] Geological report, exploring expedition from Santa Fe, N. Mex., to the junction of the Grand and Green rivers of the great Colorado of the West, in 1859, under the command of Capt. J. N. Macomb, 1876, pp. 67-69.
[b] Jenks, William, Juratrias copper: New Mexico Mining Record, vol. 1, No. 4, 1899, p. 1.
[c] Bull. U. S. Geol. Survey No. 260, 1905, p. 225.

which is here the dominant country rock. The ore is malachite and glance. It occurs in moderate amount in the schist, and seems to attain its best development near the contact of the schist with the granite. The schist and ore croppings trend in a northeast-southwest direction, and when the locality was visited a tunnel in the canyon at 150 feet below the croppings, then 60 feet in, was being driven in the granite to crosscut the "lead" at about right angles. The jointing in the granite dips about 80° NE., and the tunnel follows a 2-foot diabase dike which conforms to the southeasterly trend of this jointing. The dike carries some iron pyrites and possibly a trace of copper pyrites, but does not seem to have any connection with the deposits in the schist.

In connection with the above-described occurrence, it may be also noted that Jenks[a] reports having seen at Jemes Pueblo what he believes to be copper glance in quartz from a fissure vein in the granite at the south end of Jemes Mountain, in the southern part of the Sierra Nacimiento. These two localities are the only ones known or reported to the senior writer in which ore occurs in the pre-Cambrian rocks of this range, but as its manner of occurrence in these rocks is necessarily very different from what it is in the "Red Beds," and as exploitation has been slight, it is reasonable to infer that other deposits, possibly productive, may yet be found in the pre-Cambrian rocks. The mere occurrence of copper in these rocks is of special importance, for, as has been shown, these deposits seem to be either directly or indirectly the source of those contained in the "Red Beds."

COCHITI OR BLAND DISTRICT.[b]

LOCATION AND HISTORY.

The Cochiti or Bland district is situated in the northeastern part of Sandoval County, about 9 miles northwest of the Rio Grande, about 30 miles almost due west from Santa Fe, and about 45 miles slightly east of north from Albuquerque. The town of Bland, the principal camp of the district, lies 25 miles north-northwest from Thornton, a water station on the Santa Fe Railway. A daily mail stage connecting these two points passes through the little plaza of Peña Blanca, close to the ford across the Rio Grande, and on the west side of the river through the old Indian pueblo of Cochiti, from which the mining district is about 10 miles distant. From the broad river valley the road ascends to a mesa which borders on the southeast the outskirting elevations of the Valles Mountains. The mining district lies up in these hills, about 6 miles northwest of the locality where they join the mesa. Bland lies at an elevation of about 7,400 feet, in Pino Canyon, a steep-sided trench 700 to more than 1,000 feet deep, which heads 2 or 3 miles farther northwest. Colla Canyon, a similar and nearly parallel valley, lies 1½ miles to the southwest, but turns and unites with Pino Canyon a couple of miles below Bland. Medio Dia Canyon, also similar in character and direction, lies on the northeast, separated from Pino Canyon by a sharp, crestlike divide; it unites with the Pino-Colla trench a few miles below Bland. Most of the mines are situated along the walls of Pino and Colla canyons.

Together with much of the mountainous portion of the Territory, this region received attention from the prospector in the seventies or eighties. The opposition of Mexicans, who claimed possession of the region under private grant, impeded development, and it was not till 1889 that prospecting began in earnest. In 1893 much activity was manifested in the district and many claims were located. Early in 1894 the Albemarle group was located. This proved to be the most important group in the district. In 1896 a mill, known as the Woodbury mill, was built in the valley about 7 miles below Bland, for the purpose of treating the ores of the Iron King mine, but it was never a success. Late in 1899 the Albemarle cyanide mill was completed. This was closed in the spring of 1902 and was later dismantled. In the eleven years from the opening of 1894, when production began, to the close of 1904 the district had produced slightly over a million dollars. The greatest production in any one year was $359,135, in 1900. No production was reported in 1905. At present very little mining is being done in the district, and the prospects for resumption of work in the immediate future are not bright.

---

[a] Jenks, William, New Mexico Mining Record, vol. 1, No. 4, 1899, p. 2.   [b] By L. C. Graton.

GENERAL GEOLOGY.

The general geology of the Cochiti district is, in the main, rather simple. The lowest rocks exposed are monzonite, monzonite porphyry, and diorite porphyry. Apparently overlying these in some places is an impure feldspathic sandstone which may be Cretaceous. The underlying igneous rocks are younger than this sandstone, however, having domed up and in some places cut across the strata, and they correspond lithologically with types which appeared at the end of the Cretaceous and which are represented abundantly in the Territory. Over the surface of these pre-Tertiary rocks, made uneven by erosion, there was extruded a great flow of rhyolite. About 6 miles southeast of Bland the rhyolite area ends, appearing to overlie coarse gravels which are believed to be of early Tertiary age. It seems probable that the rhyolite is of Miocene age. This rhyolite, which at Bland is probably 500 to 800 feet thick, came from the northwest and undoubtedly had its source at a vent somewhere in the vicinity of Pelado, a conical pile over 11,200 feet high, which lies about 12 miles north-northwest of Bland. Pelado Peak, together with the other peaks near by, which are almost as high or higher, constitute the central core of the Valles Mountains and are probably parts of a considerably dissected volcanic cone, which had a diameter of about 15 miles and was made up of ash, cinder, and other ejecta. Surrounding this relatively steep cone was a broad, out-sloping border of lower gradient averaging about 10 miles in width. This bordering apron, which doubtless represents the accumulation of flows of lava similar in composition, has suffered much less by erosion than the higher and probably less compact central cone, and is simply divided into many narrow segments by sharp, deep canyons cut by radiating streams that head against the higher core; except where the streams are close together, the divides that separate them are gently inclined planes which doubtless closely correspond with the original flow surfaces of the rhyolite. On the east this rhyolite flow is locally veneered with basalt.

The physiographic expression here described is well brought out by the Santa Clara and Jemes sheets of the topographic map of the United States.

Pino, Colla, and Medio Dia canyons are three of these deep incisions in the low border cone of rhyolite. Bland is situated in Pino Canyon, about midway between the inner, upper margin and the lower, outer margin of the slope. These canyons have been cut down through the rhyolite into the older porphyritic and granular rocks that lie underneath. From the exposures it can be seen that the surface covered by the rhyolite was an uneven one. In some places the rhyolite extends almost or quite down to the bottom of the canyon; in others, the bottom of the rhyolite is several hundred feet up on the canyon walls.

The petrology of the rocks in this region presents little of particular interest. The scanty exposures of the rocks older than the rhyolite make it impossible to draw generalizations with certainty, but it appears probable that the monzonite, monzonite porphyry, and diorite porphyry are all varying types of the same magma, or at least belong to the same general period of intrusion.

The monzonite is well exposed in Jenks Draw, along the road connecting Pino and Colla canyons, on the Colla slope of the divide. It is a grayish-green granular rock which gives the impression of being porphyritic. It is fairly fresh, and good specimens can be obtained where a little blasting has been done along the road. When examined under the microscope, plagioclase feldspar is seen to be the predominant co nstituent. It comprises both albite and oligoclase, and many of the individuals are surrounded by narrow turbid borders, probably of orthoclase. Large individuals of orthoclase are also present, and quartz occurs in irregular, interstitial grains. Biotite was sparingly present in irregular plates, but it is now wholly altered into yellowish chlorite and epidote. Epidote has resulted also from the decomposition of the plagioclase. Apatite, a little zircon, and plentiful magnetite are especially associated with the decomposed biotite. The hypidiomorphic granular texture and the mineral composition class the rock as a quartz-bearing monzonite.

Monzonite porphyry is the country rock of the Albemarle deposit and is exposed both on the dump of that property and in cuts on the surface. The freshest specimens obtained were somewhat altered; the color ranges from rather dark green in fairly fresh, noticeably porphyritic specimens to light greenish gray in those of denser texture, in which more silicification has taken place. Under the microscope these porphyritic rocks appear rather more acidic than the granular type just described. Orthoclase predominates over plagioclase, but no original quartz is to be seen. Greenish chlorite, which is plentiful, was derived from some ferromagnesian constituent. This may have been an amphibole, as aggregates of chlorite suggesting the amphibole outline are present in a few specimens. Besides this mineral the phenocrysts are mostly oligoclase; a few are orthoclase, the potash feldspar being mainly confined to the trachytic groundmass. Apatite and magnetite are present as accessory constituents. Silicification has caused the development of mosaics of minute quartz grains in patches in the groundmass. In some of the specimens the rock seems to have been subjected to a sheeting, and along the extremely narrow shear zones silicification has been most marked. Locally a very little pyrite accompanies this introduction of silica.

The diorite porphyry, which can be found in reasonably fresh exposures at several points in Pino Canyon above Bland, ranges from fine grained to dense and from greenish gray to almost black. A good exposure is near the mouth of the Navajo tunnel.

Thin sections of these specimens show augite and plagioclase of medium composition to be the chief components. A little orthoclase may be present, but can not be proved such with certainty, because of its turbidity. Feldspar and augite are present both as phenocrysts and in the groundmass. The feldspar, which is in large part a sodic oligoclase, furnishes on decomposition kaolin in some places and epidote in others. Augite is decomposed into epidote, serpentine, and chlorite. Some of the decomposed material may have been derived from biotite. A little magnetite is present in small grains. Similar rocks, but finer grained and darker, are exposed near the lowest tunnel of the Lone Star mine. The augite, which is rather more abundant, is in places decomposed to a nearly colorless, micaceous material on the interior, surrounded by small grains of black iron ore. An approach toward intersertal texture and the finer grain may indicate that this is near a contact.

The rhyolite of the region is a reddish rock which ranges from somewhat porous to almost pumiceous; in places it is tuffaceous or brecciated, but generally it is massive. A noticeable feature is the partial or complete filling of the vesicles with iridescent opal. Angular fragments of white quartz several inches across are present in some brecciated portions of the rhyolite. Examination of thin sections reveals abundant sodic orthoclase, usually in well-formed individuals, some of which are twinned. There is a little albite, and biotite was sparingly present, but is now decomposed. Blebs of opal and quartz, both in crystals and grains, with curving faces, are plentiful in the vitrophyric, partly devitrified groundmass. In some specimens a few grains of pyroxene are seen, and small masses, which probably represent resorbed olivine, are surrounded by grains of magnetite. The groundmass in some of the fresher specimens shows spherulitic texture. Gas and fluid inclusions are abundant in the quartz grains.

The only other rock observed in the region is the impure sandstone already mentioned. Small exposures of this rock, which has been called granite, were seen near the creek just below the Casina claim and in Jenks Draw, on the slope toward Colla Canyon. It ranges from brownish to greenish gray in color and from medium to fine in grain. Bedding planes are preserved, though rather poorly, but the structure over the small areas exposed is not significant. Quartz is predominant in small rounded grains, but plagioclase, microcline, and orthoclase are also present. These grains are held by a cement, which may consist of any or all of the following: Calcite, quartz, chlorite, serpentine, kaolin, and sericite. A few grains of secondary epidote are scattered through the rock.

Specimens can be procured at probably all the mines which do not correspond with the foregoing descriptions. They represent the result of the action of ore-bearing solutions on the rocks just described, for the most part on the monzonite and related porphyries.

MINERAL DEPOSITS.

GENERAL FEATURES.

The ores of the Cochiti district contain both gold and silver, gold being subordinate in quantity but predominant in value. The deposits belong to one general type, the mineralogy and, in a broad way, the structure being similar in all. The ore bodies or lodes are closely connected with fissures. These fissures, as is common where relief of stress takes place at comparatively small depths, are not the simple, sharp, clean-cut breaks which the term fissure vein first brings to mind and which would result from the same forces acting at greater depth, where the rocks would be under greater load. They are, instead, complex fractures, due to shattering that extended mainly along planes or zones, although these exhibit a rather wide range in width and regularity. It is usually not possible to decide whether or not actual faulting took place along these fracture zones. There was, however, before the deposition of the ores, considerable displacement of the smaller fracture fragments from their former relative positions, so that much space existed between them, while what might be regarded as the walls were traversed by many small irregular cracks extending away from the main fissure. One vein, the Little Casina, occupies a fracture along which, since the formation of the ore, the rhyolite has been faulted down against the porphyry.

Through these zones of fracturing and brecciation the ore-bearing solutions found easy channels. The shattered and crushed country rock, offering much surface for attack, relative to its mass, was vigorously acted upon and part of its constituent elements were dissolved. These elements entering into solution doubtless disturbed the equilibrium that had previously existed in the solutions and thereby caused the precipitation of the ores, along with other materials which united with the remaining constituents of the country rock. The ore minerals were deposited partly in the spaces between the rock fragments and partly within these fragments and in the surrounding shattered walls. The deposits are therefore due in part to the filling of open spaces and in part to replacement.

Veins formed wholly by the filling of fissures are usually characterized by reasonable uniformity in width and direction. Replacement deposits, on the other hand, are likely to be irregular in width and outline and to possess boundaries which are not sharp, the ore gradually passing, by decreasing amount of alteration, into the normal country rock. Replacements along fissures commonly partake somewhat of the character of each component type; their form is determined in a general way by the fissure, but their width is largely dependent on the distance from the fissure walls to which replacement has gone, and is in many places variable. The veins or lodes of the Cochiti district exemplify this composite type. They are known to range from 1 foot to more than 150 feet in width, though no single vein exhibits such variations. The narrower of the veins which have been explored appear to be rather more regular in width than the larger ones, although marked variations in width are not uncommon; the wider veins may pinch down to a few feet and within a short distance widen out to 50 or 100 feet or more. These variations are not due wholly to differences in the distance to which the replacing solutions have penetrated, but partly to the irregular width of the fracture zones. Although alteration of the country rock, caused by the ore-forming solutions, may extend a considerable distance from the open channels through which they ascended, the distance through which ore minerals in noteworthy amounts penetrated and were deposited by replacement is doubtless less than 10 feet from the open channels, and perhaps not over 5 feet in the extreme, averaging probably 2 or 3 feet. The uncertainty in determining the exact width of the replacement ore is caused by the extreme alteration of the country rock near the fractures, so that the filling and the replacement material are locally indistinguishable. In the majority of deposits the boundaries of the lodes are not at all sharply defined, though this is not true of some of the narrower veins. Beyond the limits of extreme replacement of the wall rock the cracks which traverse it are in many places filled with vein material. With one exception, there is no evidence that faulting or movement, either along or across the veins, has taken place since the deposition of the ores. On the Little Casina claim a fault on the hanging-wall side of a vein has brought the rhyolite down opposite the porphyry.

In strike the veins differ widely, ranging at least from N. 35° E. to N. 60° W., but veins having a strike close to N. 10° W. are noticeably common. Fissures having this prevalent strike are directed toward the vent through which the rhyolite found egress, and may be due to radial cracking, corresponding thus to the radial dikes which accompany many large masses of igneous rock. The veins usually dip at steep angles, ranging from 60° to 90°, but no uniformity in direction of dip is apparent.

These fissures cut the igneous rocks which underlie the rhyolite. It was stated that a vein on the Posey claim, close to Bland, goes up into the "cap rock," but there forks and pinches to stringers and within a short distance gives out. Search for this occurrence at the bottom of the rhyolite sheet, along the canyon wall, was not successful. A short adit tunnel on a vein on the Posey claim, close to the rhyolite cap, did not show such a condition, but the junior writer was not able to learn whether this is the vein that was said to penetrate the rhyolite. Other veins nowhere extend into the rhyolite.

The mineralogy of the veins is comparatively simple. Quartz greatly preponderates in the fissure filling. Sphalerite, or zinc blende, and pyrite are the most plentiful sulphides, the former being rather more abundant, but neither of them constituting a large proportion of the ore, probably not more than 3 per cent, except in unusually rich ore. Chalcopyrite is widespread in small scattered grains. The mineral that appears to contain the principal value is a dark sulphide, probably argentite. A tiny grain of galena is occasionally seen. Barite, calcite, and adularia (valencianite) have not been observed.

The quartz, which is usually white, is commonly of somewhat fine grain and distinct from the massive, vitreous quartz of many veins in other localities. It consists of a mosaic of rudely polygonal grains, most of which have similar but not identical optical orientation. Owing to the irregular shape of the spaces filled by the quartz, drusy cavities are rather abundant and are most common where many small fragments of the country rock are included in the vein. In these places the individual grains are usually less than 1 millimeter across, and the prismatic crystals are, as a rule, not more than 5 millimeters long by 1½ millimeters across. Elsewhere, usually where the grains are larger, polarized light discloses a radial structure in what appear to be definite individuals. This produces the "flamboyant" quartz so frequently encountered in fissure fillings. Inclusions of minute size, many of them fluid-filled cavities, are common in the quartz and are in many places arranged along certain crystallographic directions, especially parallel to the vertical axis. Inclusions, both fluid and solid, are likely to be most abundant close to the periphery of the grains. Some quartz from the Iron King mine, showing comb structure, is colored faintly violet, approaching amethyst.

An occurrence of quartz worthy of mention is found in the Little Casina vein. The quartz is not compact but, in addition to small cavities left from the solution of sulphides, contains many tabular gashes up to 2 inches in diameter. The quartz occurs as thin, broad plates, in some places parallel, in some radiating, and in others irregularly arranged. Between these plates are the tabular cavities or spaces. This "hackly quartz," which is rather common in the veins of this character, has been well described and pictured by Lindgren in a report on the veins of Silver City and De Lamar, Idaho.[a] It is undoubtedly due to the solution of some mineral with which the quartz was intergrown, or, less probably, which was replaced by quartz along certain planes. The two minerals that come to mind in this connection are barite and calcite. Of these calcite seems more likely to be the one, and the plates of quartz in some localities apparently correspond well with the cleavage planes of calcite. It is impossible to decide as to the conditions under which this calcite, if such the mineral was, in the Little Casina vein was dissolved. All of the vein which is exposed is thoroughly oxidized, and it may be that surface waters have dissolved the mineral. Lindgren states that surface waters have caused the solution of calcite, leaving the same hackly quartz, at the Waihi mine in New Zealand, and he believes that a second ascension of hot waters has produced similar results in the veins

---

[a] Lindgren, Waldemar, Twentieth Ann. Rept. U. S. Geol. Survey, pt. 3, 1900, pp. 65-256.

at Silver City, Idaho. This structure of the quartz is seen in the Mogollon veins, in western Socorro County, N. Mex.

Sphalerite, the sulphide of zinc, is in a few veins present in individual grains or compact aggregates which are as much as 5 millimeters across. Grains of this size occur in parts of the Iron King vein. Generally, however, the sphalerite occurs in groups of tiny crystalline grains, making dark patches in the quartz. This mineral is in most places closely associated with the other sulphides. Pyrite varies in abundance in different veins and in different parts of the same vein. It is not always closely associated with the zinc blende, some of it occurring alone in groups of small grains. The crystal form commonly assumed is the cube, but irregular veins are more plentiful. Most of the individuals are less than 1 millimeter across. Chalcopyrite occurs very sparingly, either in tiny shapeless grains or in small tetrahedrons. A black mineral present in small crystals, most commonly associated with zinc blende, proves on chemical examination to contain silver. It is the mineral which carries the silver content of the ore. The absence of antimony shows that it is not polybasite, which it resembles, and makes it probable that the mineral is argentite. The crystalline grains are too small to allow determinations of the crystal form. This mineral is contemporaneous with the other minerals, being intergrown promiscuously with pyrite, chalcopyrite, and blende. It is not plain how the gold occurs. Free gold is said to have occurred in rich ore, though it could not be ascertained whether or not this was oxidized ore. It is reasonable to suppose that the gold is intimately associated with the pyrite, though there is no positive evidence of it.

In the replacement zones along the fractures the mineralogy is a little different but no more complex. Replacement has naturally gone on both in the walls of the fissures, which in most places are more or less shattered, and in the fragments which partly filled up the fissure and around which the filling was deposited. To outward appearance the alteration of the country rock results in a hardening and a change in color from dark greenish to light greenish gray. Addition of silica is evident and under the microscope the alteration is seen to have been principally a silicification. The original minerals of the rock have been almost completely replaced by an extremely fine grained aggregate of irregular particles of quartz. Under crossed nicols this aggregate resembles somewhat the microgranular groundmass of certain porphyries, but has a more uneven, ragged appearance. It would seem that the farther from the open channels the finer grained is the siliceous aggregate, whereas close to the fissures the minute particles have been recrystallized into grains of larger size. Very commonly there is a gradual transition in size of grain from what in unquestionably unaltered wall rock right up to the quartz prisms lining a druse, and in such places it is not always possible to determine at what point the original rock adjoined the fissure. As a rule, however, there is a fairly well marked though not always absolutely sharp change in size of quartz grains in passing from the fissure to the silicified wall. Furthermore, wherever the wall rock has been altered it is common to find a narrow cloudy strip between the quartz grains. These strips are caused by swarms of minute dark particles which are too small for identification, but which in the aggregate give a green color and, from the occurrence of the mineral in less silicified portions of the rock, are assumed to be chlorite. The same sulphides that are present in the fissure filling also penetrate the rock adjoining. Some small, altered inclusions of the country rock are favorable places for the deposition of these heavy elements, even where the surrounding vein quartz holds little of them. In general, however, pyrite appears to be relatively more abundant in the replaced rock, the other sulphides are decidedly less plentiful, and there can be little doubt that in general the portions of these lodes which were deposited in open spaces are richer than the replacement portions. In this respect the Cochiti ores differ from those of Red River, in Taos County, where the altered inclusions of country rock appear to be the richest portions of the veins.

Farther away from the solution channels the country rocks are less altered. Quartz is added in places where permeation seems to have been most easy, chlorite is rather plentifully developed from the ferromagnesian constituents, and the feldspars are in part converted into sericite. Pyrite is practically the only sulphide which can penetrate to this distance, and such rock carries at best very low values.

What has already been said regarding the druses, the hackly quartz, the irregularity of the fissures and the replacement zone, and the numerous altered inclusions covers the principal structural features of these lodes.

The decomposition of these lodes under the action of agencies at the surface does not afford much of especial interest. Oxidizing waters attack and decompose the sulphides, taking the iron, zinc, and copper, as well as the sulphur, into solution. If the gold is contained in pyrite, it is liberated, but in any event it is not transported by the oxidizing solutions. What becomes of the silver is not known definitely, but inasmuch as oxidized ores near the surface have been worked, and no marked variations in silver content have been reported, it seems probable that the argentite is converted into some oxidized and not easily soluble silver mineral, which remains close to the place where the argentite was deposited. The iron taken into solution is partly deposited again as limonite, which here and there stains the vein stuff with rust. In certain lodes siderite is formed from the pyrite. It may be that the solution of the mineral, probably calcite, which was associated with the quartz in such veins as the Little Casina was brought about by surface waters. Tiny cubes of pyrite deposited on the quartz that lines druses may have been deposited by descending waters which, oxygen bearing near the surface, took up iron and sulphur, and then as they passed downward were reduced by more sulphide, and finally were able to give up their load as pyrite. It is possible, however, that this secondary pyrite was deposited by a second generation of ascending solutions. In many places the oxidizing waters have penetrated the granular and drusy portions of the lode, due to simple deposition in the fissures, while the replacement portions, dense and in places hornstone-like, have been impermeable and have protected their inclosed sulphides from oxidation. The depth to which oxidation has proceeded was not ascertained, partly because none of the deeper workings were accessible. The unequal permeability of different portions of the veins, as just described, would doubtless permit evidence of oxidation to be found in the drusy portions at a considerable distance below undecomposed sulphides in the tighter portions. The best data obtainable are from the Albemarle mine. Sulphides are present in plenty on and below the sixth or 425-foot level, although practically no water was encountered. Much of the ore above is said to have been oxidized, but it seems probable that the dark patches containing zinc blende and argentite, without visible pyrite, were not recognized as sulphides. Sulphides are plentiful on the dump of the Iron King shaft, whose deepest point is only 136 feet below the surface. It is said that water was encountered at that depth.

The resistance to decomposition which these highly quartzose masses possess, as compared with the surrounding less silicified porphyries, has led to the development in certain localities of outcrops that project boldly above the general surface. The outcrop of the Albemarle lode is one of the most notable.

Evidence furnished by these deposits in themselves does not permit the drawing of conclusions regarding their genesis, or at least regarding the source of their materials. Such inferences as can be drawn are based mainly on comparison with similar deposits in other regions where data bearing on genesis are more plentiful. It is believed that the Cochiti lodes were deposited by ascending hot waters which were given off by the rhyolite magma as crystallization of the molten rock took place. It is doubtful if the molten material that reached the surface and was poured out as flows contributed much, if any, to these solutions; but the lower portions of the magma, which consolidated below the surface, were probably the first source of the water and dissolved materials constituting these solutions and were the cause of their high temperature.

The mineralogy and structure of these ores correspond closely to those of ores occurring in rhyolites, andesites, and dacites in many other places. The deposits of the Comstock lode, Tonopah, and Goldfield, Nev., and of Silver City and De Lamar, Idaho, have many features in common with the Cochiti ores, and in New Mexico the Red River deposits and the ores of the Mogollon Mountains are very similar.

It is difficult to account for the absence of veins in the rhyolite at Cochiti. In the occurrences at Red River most of the veins, as here, are found only in the older porphyries, although this may be because the rhyolite cap has in most places been eroded away. Some veins cut the rhyolite in the Red River district, however, and there are evidences of alteration in the rhyolite, caused by the vein solutions. At Cochiti there is no reason to believe that the solutions which deposited the ores entered and dissipated themselves in the rhyolite. The absence of deposits in the rhyolite can be explained by the assumptions that the veins were formed before the flows extended so far from their vent, and that the veins were soon covered by the advancing lava. In such case it would not be reasonable to expect much erosion in the interval between the formation of the veins and the covering of their outcrops by the rhyolite. It would therefore seem that there should remain, between the earlier porphyries and the overlying rhyolite, some evidence, such as sinter deposits, of the surface outpouring of the vein solutions. No such evidence is to be found, however. It seems more probable that the rhyolite which covers the veins was poured out during the same general period of eruption, but was of somewhat later appearance than the rhyolite to which the veins are considered intimately related. The earlier rhyolite may not have covered and probably never did cover the Cochiti district, being restricted to the more immediate vicinity of the vent, while the vein solutions, taking a diagonally upward course along channels unchoked by the rhyolite, reached the surface at this place. The interval which elapsed between the appearance of this earlier rhyolite and that seen in the district to-day would permit erosion to carry away the hot-spring deposits that were probably formed at the surface of the vein channels. Such conditions apparently existed in the Mogollon district, where the veins, occurring in andesites and dacites, are cut off and covered by slightly later flows of similar rocks.

If the Cochiti ores were deposited just before the rhyolite covered the veins, as seems most probable, the tops of the veins as known at present can not have been far from the surface at the time of their formation. It is believed that this distance nowhere reached 500 feet. The conditions close to the surface may have differed from those at greater depth, and may have caused the deposition of the ores only within a few hundred feet of the surface. The characteristics of the Cochiti veins correspond with those of veins elsewhere which have been formed relatively near the surface.

The hot springs near Jemes Pueblo, about 20 miles southwest of Bland, and the "Sulphur Springs" just northwest of Pelado Peak, may possibly be related to and survivors of the hot solutions which deposited the Cochiti ores. It is to be regretted that time did not permit visits to these localities.

The ore bodies of the Cochiti district which have been exploited have yielded ores of rather low grade. Most of the mines are closed and their operators have departed, so that it is difficult to learn the value of the ores mined. The information regarding the values given here was gained mainly by hearsay, except for the Albemarle mine. The value of the ore as broken from the stopes averaged probably about $6 a ton and ranged from $2 to $100, or higher, in small bunches. It is stated that a small lot from the Iron King mine ran between $2,000 and $3,000 a ton and that some assays were as high as $9,000. By careful sorting the value of the average ore can be raised to $30 to $75 for shipping. Just how much is lost in the discarded waste was not ascertained. In sorting, the dark portions, carrying the sulphides, are saved, and as a rule the plain white quartz goes over the dump.

The principal metal recovered is gold. The returns to the Director of the Mint and the Geological Survey indicate that the total production of the district to the end of 1904 was about $695,000 in gold and $345,000 in silver, or approximately $2 of gold to $1 of silver. The bullion from cyanide recovery at the Albemarle mine was said to average about $1.85 a ton in 1902, when silver sold for about 50 cents. If it is assumed that this bullion contained only silver and gold, the ratio in value of the recovery was about gold 3 to silver 1, the quantity ratio being about 0.067 ounce of gold to 0.933 ounce of silver.

The precious metals are unevenly distributed through the lodes. Small rich pockets, surrounded by lower-grade ore or separated by practically barren vein stuff, are not uncommon. Few if any lodes are of workable grade throughout, the values being concentrated in pay shoots. These richer portions are usually irregular, but in the plane of the vein are more commonly equidimensional than elongated. Many of them appear to occur where the lode is widest, and

158   ORE DEPOSITS OF NEW MEXICO.

perhaps the richest portions are in the widest places. Beyond this, present development has not revealed any general rule that can be applied to their shape, extent, or location.

It was stated to the writer by Mr. L. B. Smith, who was superintendent of the Albemarle mine, that paying ores seem to pinch out at depths of a few hundred feet; the experience at the Albemarle, Lone Star, and Washington mines was cited as evidence. The deepest ore known in the district was in the Albemarle mine, and gave out at a depth of about 600 feet below the outcrop.

Considerable development work has been done on the ore deposits of the Cochiti district. The workings range in importance from small prospect pits and assessment holes up to those of the Albemarle mine, which is the most highly developed mine in the district, with workings 750 feet below the outcrop. The Lone Star tunnels and the Navajo crosscut tunnel explore considerable ground. The Washington, Iron King, Crown Point, and Puzzle are among the claims on which several hundred feet of work has been done.

Extraction of the ores has never been successfully worked out. Three or four mills, including the Woodbury stamp and cyanide plant, have been built in or near the district, but have not been successful. Amalgamation was the process used in most of them. The Albemarle mill, which gave better results, depended on cyanidation of dry-crushed ores. The tailing, however, from ore averaging from $4 to $8 a ton contained $2 or more, so that this can hardly be called a success. The absence of a plentiful supply of water in the district is a serious drawback in milling. High-grade sorted ore was hauled to the railroad and shipped to Pueblo or El Paso, but owing to high freight and treatment charges this was usually not profitable.

ALBEMARLE MINE.

The Albemarle mine, situated on South Fork of Colla Canyon, is owned by the Navajo Gold Mining Company, of Boston. Ore is said to have been discovered on this property early in 1894 as the result of prospecting along the prominent outcrop. The mine, which was sometimes called the Altoona, passed into the hands of the Cochiti Gold Mining Company, which afterward became the Navajo Company. From the time of its discovery the mine was conceded to be the most important in the district, and from 1899, when its mill was put into operation, till the closing of the mill in the spring of 1902, it is said to have produced $667,500.[a] Since the latter date the property has not produced.

The mine is developed by an incline shaft 625 feet deep, connecting with eight levels. From the bottom level a 90-foot winze has been sunk. The drifting amounts to perhaps 5,000 feet. The first level is really an adit starting at the shaft collar. From 1899 the ore was treated in a mill in which the gold and silver were extracted by cyanidation. The capacity was originally 125 tons daily, but was doubled later, the actual quantity of ore treated monthly then ranging from 5,000 to 9,000 tons. Power was derived from a current generated at the company's power house at the Madrid coal mines, in Santa Fe County. The ore was dry crushed by rolls to pass a 34-mesh screen. The fines were blown out by air and were leached by air agitation. The sands went to other vats and were leached by percolation. There were ten 130-ton circular vats and eight 250-ton square vats. The strength of solution and time of extraction were not learned. It is stated that the tailing contained $2 to $2.15 a ton, and this loss was attributed to incomplete percolation. It seems probable that the presence of part of the values in the form of sulphides may have also had something to do with this loss. At the time of visit the machinery and framework of the mill had been sold and were being hauled away.

The country rock is monzonite porphyry, which, where reasonably fresh, is a greenish rock with plainly visible crystals of feldspar in a fine-grained groundmass. In the immediate vicinity of the vein the rock is silicified, becoming harder and of lighter color. In some places the rock has been somewhat sheeted and silicification has gone on, especially along the planes of foliation.

The vein, whose cropping projects as a prominent ridge, strikes N. 30° to 35° E. and dips 62° NW. Most of the following notes were kindly supplied by Mr. L. B. Smith, the superintendent of the mine.

---

[a] Jones, F. A., op. cit., p. 186.

The ore is characteristic of the whole district and needs no special description. Much of it is said to be oxidized, but sulphides are present on the sixth or 425-foot level and on lower levels. It appears, however, that the dark patches in the ore have not been regarded as sulphide bearing unless pyrite could be seen, and it is therefore probable that unoxidized sulphides were present above the sixth level. Some of the richest ore contains much zinc blende in tiny grains, closely associated with argentite. Pyrite is not plentiful and chalcopyrite is rare.

With some exceptions the values were fairly constant but gradually decreased with depth. The average for the levels above No. 4—that is, about 230 feet deep—was $6.50 to $7.50 a ton; from the sixth or 425-foot level down the average was $4 to $4.50 a ton, and Jones[a] states that the level at the very bottom carried only $3.50 a ton at best. This was deemed unprofitable ore and the mine was closed.

The principal pay shoots of the mine are northeast of the shaft. The highest ore body is developed by the first or adit level, which has the same elevation as the shaft collar and starts close to the shaft. This ore body, right under the outcrop, is of small size but is separated by only a short stretch of low-grade ground from the one below it. The main ore body was opened from the third level, which is 150 feet below the collar. A stope was carried up 130 feet, or nearly to the adit level. This stope in its greatest dimensions was 50 feet wide and 100 feet long. On the fourth level, 80 feet below the third, the stope on this main pay shoot reaches its maximum width of 60 feet and is about 200 feet long. The fifth level, at a depth of 325 feet, is practically at the bottom of this ore shoot. The stope up to the fourth level averages 15 feet wide and 50 feet long, the ore body pinching and shortening steadily down to the fifth level. On this level, 100 feet northeast of the main ore body, another shoot of smaller size was struck. It continued down past the sixth and seventh levels, over 200 feet, but was not found on the eighth. A third small ore body, 75 feet southwest of the main ore body on the fifth level, was found on the sixth and stoped out; it was not encountered on the seventh level. Small bunches of ore were found northeast of the main ore body on levels 3 and 4. In the southwestern portion of the workings a small amount of ore was stoped from the sixth level 150 feet from the shaft. It was prospected for on the seventh and eighth levels, but was not found. A shoot of small horizontal dimensions was struck in the shaft near the sixth level and continued down to a point within 40 feet of the eighth level; that is, to a depth of 585 feet, where it gave out. This is the deepest ore known in the mine or in the district.

Practically no water of consequence was ever encountered in the mine.

WASHINGTON MINE.

The Washington mine, situated on the mountain side just west of Bland, was once one of the important producers of the district and is credited with a production of about $75,000. No work was being done in 1905 and the mine was not visited. It is said to have been considerably developed, the workings extending to a depth of 500 feet. The writer was informed that a 75-foot winze below the 500-foot or bottom level went below good ore and caused cessation of work. So far as known the ores were similar in character and value to those of other mines in the district.

LONE STAR MINE.

The Lone Star mine is situated a little north of the Washington, on the southeast slope of a side gulch which joins Pino Canyon just above Bland. It is on the same vein as and lies just south of the Iron King, the first discovery in the district. The present owner is the Navajo Gold Mining Company. A considerable amount of ore was shipped in the early history of the mine. Not very much ore has been taken out since the last sale of the property. Actual figures of production were not obtainable, but it seems probable that between $50,000 and $100,000 has been taken out.

The mine is developed principally by short crosscut tunnels at different elevations, which start on the side of the gulch and intercept the vein. Tunnel A is the lowest one and is about 200 feet below the outcrop. It is run eastward for about 400 feet till it cuts the vein, then

---

[a] Jones, F. A., op. cit., p. 186.

follows along the vein for 1,200 to 1,500 feet. There are four other tunnels almost directly above this, tunnel E being the highest and just below the outcrop. Each of these is shorter than the one below it. The drifts from these crosscut levels are very irregular. In all, the workings aggregate probably 4,500 feet. From these crosscuts very irregular workings explore the vein for comparatively short distances. A long crosscut tunnel, known as the Navajo tunnel, has its portal near the creek level in Pino Canyon. It runs southwestward, diagonally to the vein, which it finally intercepts. Some of the ores were treated in a mill about half a mile down the canyon from Bland. Much of the ore, however, was shipped.

The country rock is diorite porphyry. Most of the rock where not greatly altered by the vein solutions is greenish gray, with plainly visible porphyritic texture. Near the mouth of tunnel A, however, the rock is very fine grained and dense and is nearly black in color; it resembles in appearance a baked shale and may perhaps owe its difference in texture and general aspect to the nearness of a contact with sandstone which quickly cooled the igneous rock at the time of its intrusion.

The Lone Star vein strikes about N. 10° W. and dips in the aggregate almost vertically, although near the surface it is steeply inclined to the east. The vein is very irregular in width. The siliceous outcrop is 40 feet wide and on level E the lode is almost as wide, and on level A much wider. At the north end of the property, however, the vein becomes narrower and more regular, and in the Iron King property it averages 12 to 15 feet in width.

The ore is similar in character and occurrence to that at the Albemarle but is perhaps more drusy and porous. Oxidation has destroyed a large part of the sulphides, but in the replacement portions, particularly the silicified inclusions of wall rock, dark patches which are unoxidized remain. The values were not uniformly distributed throughout the ore. The average value of the ore extracted was probably about the same as in the Iron King mine, namely, about $6 a ton, but shipments of sorted ore gave returns of nearly $40 a ton.

The principal ore body of the mine began at the surface and continued below level D, probably about 100 feet. The open cut averages probably 30 feet in depth and 75 feet in length, and the ore was probably 25 feet wide. The southward continuation of this pay shoot was extracted underground on level E from a stope 50 feet long, 25 feet wide, and 25 feet high. This stope continues to level D. Bunches of ore were found on the lower levels, but nothing of importance was encountered by considerable crosscutting of the lode. On level A a crosscut extending 156 feet eastward across the lode was all in vein stuff and did not reach the eastern limit, but disclosed nothing of paying value. Little water is present in the mine.

IRON KING MINE.

The Iron King, the first mine discovered in the district, lies just north of the Lone Star, close to Pino Canyon. It is reported that the ore extracted amounted to $50,000. The development includes a small shaft and a short adit from which a winze was sunk, the greatest depth attained being 136 feet. An amalgamation mill was built in 1896, 7 miles below Bland. In 1902 it was replaced by a 10-stamp mill, with ten cyanide tanks, but this was operated only a short time. The vein is the continuation of the Lone Star vein. It strikes N. 10° W. and near the surface, where exposed in a stope above the tunnel, dips about 55° E., but is said to dip westward below. In this stope the vein is 12 to 14 feet wide but widens somewhat with depth. Where visible from the tunnel it appears considerably more regular and the walls are more distinct than in the Lone Star workings. A very little galena is present and in places the quartz approaches amethyst in color; otherwise the ore does not differ from that occurring elsewhere. From wall to wall the ore is said to have averaged $6 a ton, but sorted shipments ranged mostly from $40 to $65, with some up to $2,000 or $3,000 and occasional assays up to $9,000 a ton. Gold is said to have predominated in value at the surface, but silver became most important below. The ore body was short and began at the surface. It is stated by some that the ore gave out in the bottom, at 136 feet, and by others that the presence of water at this depth caused the suspension of work.

CROWN POINT MINE.

The Crown Point mine is situated on the northeast side of Pino Canyon, about three-quarters of a mile above Bland. It is one of the early locations in the district and is commonly credited with a production of about $50,000, being said to have sent out 1,500 tons of shipping ore. The workings are on a vein which strikes N. 10° W. and dips 70° W. but steepens in the bottom. The vein has been regarded as the same as the Lone Star vein, or at least a spur from it. This may be the case, as it is on the strike of the Lone Star vein and has the same trend, but there seem to be no other reasons for this assumption. The walls are not well defined and the lode ranges from 6 to 20 feet in width.

The ore in places shows cubical cavities resulting from the oxidation of pyrite; commonly these cavities are lined with quartz crystals, deposited since the removal of the pyrite, but in another part of the same specimen some of the cavities may hold the rusty limonite that resulted from the oxidation of the pyrite. Much of the ore is honeycombed, owing to the removal of the sulphides by oxidation and consequent solution. Oxidation seems rather more complete here than at other mines of the district. Part of the ore was taken from an open cut on the outcrop. A short crosscut tunnel strikes the vein just below the open cut, and from this crosscut short drifts run in both directions, making a total length of 250 feet on the vein. South of the crosscut a 250-foot inclined winze follows the ore body, and at its bottom a drift along the vein is run for more than 200 feet. Between the tunnel level and this winze level some stoping has been done. No water is encountered in the workings. A small stope has been made on the same or, more probably, a parallel spur vein, reached by a shallow tunnel a short distance to the south.

OTHER PROPERTIES.

The Tip Top mine, on the hill above the Crown Point, has shipped several carloads of ore. It is on a vein which strikes N. 60° W. and dips very steeply to the southwest. The vein is fairly well defined and is narrow, but pinches and swells rather abruptly. The ore differs from that of other mines mainly in the greater quantity of pyrite which it contains. A small stope has been driven overhand from a short adit tunnel.

The Laura S. claim is close to the Tip Top and the Crown Point and is stated to have had a small production. A little work was being done here, but no ore was seen at the time of visit.

The Little Casina claim is situated on the northeast side of Pino Canyon, in the upper part of the town of Bland. A vein striking N. 10° W. and dipping 60° E. has been developed by a rather shallow incline. It is said that the ore averages $8 to $12 a ton across the vein, and that a little ore sorted up to $40 grade has been shipped. The vein is rather sharply defined, better from the hanging wall than from the porphyry foot wall. On the hanging wall is a hardened clayey seam 1 inch or more wide, which separates the quartz of the vein from rhyolite. This is doubtless a fault plane parallel to the vein along which, after the formation of the vein and the extrusion of the rhyolite, the east side was downthrown, bringing the rhyolite opposite the vein. The throw of the fault can not well have been over 100 feet, to judge from exposures of the lower surface of the rhyolite on the two sides of the fault. This vein shows more evidence of banding than most others in the district, containing many narrow druses parallel to the plane of the vein. It is in this vein that the hackly quartz noted on page 154 occurs. The quartz is porous also from the removal of sulphides.

The Black Girl, Hopewell, Good Hope, Allerton, Posey, Union, and others, all on the southwest side of the canyon, are said to lie on the southward extension of the Little Casina vein. From the evidence obtainable such an assertion seems not well grounded. The Union is said to have shipped a little. The Posey shipped some float ore, of which considerable more remains on the hillside. These properties as at present developed have little importance.

The Little Mollie claim, near the head of Jenks Draw, is located on a wide vein of rather massive quartz, but the values are low, only a small quantity of pay ore having ever been found.

The Puzzle claim, just below the Albermarle, has a vein striking nearly north and dipping about 65° W. The vein is rather well defined and averages only 14 inches in width. It is explored by an adit at the level of the creek, which extends for 148 feet on the vein. A winze from the adit goes down the dip of the vein for 165 feet. Most of the ore is unoxidized, and the values are said to be rather better than the average for the district. This property was being developed on a small scale at the time of visit.

Many other claims have received some attention, but are not of present importance and probably few ever will be.

COPPER PROSPECT.

At the junction of Medio Dia and Cochiti canyons some work was being done on a copper prospect by Benham & Sayer. A tunnel penetrated the rhyolite country rock, which here is denser and browner than in the immediate vicinity of Bland. Faulting has evidently taken place along crushed zones, which are fairly well defined and most of which strike northwestward. Some of these zones are 30 feet wide or more. Some of them are present in the tunnel, but they are more easily discernible on the surface. In these zones of crushing copper occurs mainly as malachite, with a very little of the oxide, cuprite. The malachite forms a coating on the fragments of the crushed and recemented rhyolite. Assays are said to indicate an average content of 2½ per cent of copper and small quantities of gold and silver, the aggregate precious-metal value reaching $2 to $3 a ton. This ore is almost certainly of later formation than the gold-silver veins of the Cochiti district. There may be a similarity in its origin to that of the copper ores at Tecolote, Nacimiento, Estey, etc.

F. A. Jones [a] notes the presence in Peralta Canyon of oxides of uranium and vanadium "filling the interstices of a silicified volcanic breccia on the property of the Peralta Gold Mining and Milling Company." The occurrence may correspond with that noted above in which copper is present. If so, the association is of interest in view of the similarity of occurrence of copper and vanadium in southwestern Colorado and southeastern Utah.[b]

BERNALILLO AND TORRANCE COUNTIES.

By Waldemar Lindgren.

Bernalillo County, in which Albuquerque, the capital of the Territory, is located, is a small county in the central part, occupying only 1,240 square miles. It is bounded on the north by Sandoval County, on the east by Santa Fe and Torrance counties, on the south by Torrance and Valencia counties, and on the west by Valencia County. The Rio Grande traverses its central part from north to south. Its western part lies at the eastern limits of the plateau province and is mainly underlain by Cretaceous strata.[c] The eastern part contains the north-south trending Sandia Range rising to 10,400 feet, and continued southward by the Manzano Range which extends farther down into Valencia and Torrance counties.

Torrance County adjoins Bernalillo County on the southeast and covers 3,330 square miles. Santa Fe and San Miguel counties lie to the north of it, Valencia County to the west, Lincoln and Socorro counties to the south. The greater part of the county east of the Manzano Mountains belongs to the province of the Great Plains; Cretaceous rocks occupy much of the surface, but several mesas and small mountain groups rise above the general level; among them are the Gallinas, Animas, Pedernal, and Chameleon hills. The Manzano Range attains elevations of about 10,000 feet.

In this region the line between the Great Plains and the Plateau province is supposed to be marked by the great line of dislocation east of the Rio Grande. In fact, however, the characteristic features of the Plateau province extend far east of the Manzano Mountains. In eastern Bernalillo County the crust has broken along a great north-south dislocation and a correspondingly long monoclinal block, the Sandia Range, has been lifted up so that it presents an abrupt westward-facing scarp in which the pre-Cambrian granites and schists have been brought up, and a much gentler eastern slope covered by an evenly sloping plate of limestones and sandstones

---

[a] Op. cit., p. 186.

[b] Compare Hillebrand, W. F., and Ransome, F. L., Am. Jour. Sci., vol. 10, 1900, pp. 120–144; and Boutwell, J. M., Bull. U. S. Geol. Survey No. 260, 1905, pp. 200–210.

[c] Herrick, C. L., and Johnson, D. W., The geology of the Albuquerque sheet: Bull. Univ. New Mexico, vol. 2, pt. 1, 1900, pp. 1–67.

of Pennsylvanian ("Upper Carboniferous") age. That the dislocation along the west side exists admits of no doubt, the total throw being not less than 6,000 feet, but there seems to be evidence that this has been accomplished to some extent by distributed faults, dividing the pre-Cambrian into a great number of long, narrow blocks. The Manzano Range is less well known, but here too the Carboniferous has been brought up by faulting, and an excellent section of the strata is afforded at Abo Pass,[a] through which the Belen cut-off of the Santa Fe Railway has been carried.

In spite of the great dislocations and complicated faulting which mark the Sandia and Manzano mountains, there are few mining districts, and practically no production of metals, in the two counties. East of the Sandia Range F. A. Jones[b] reports copper ores in "Red Beds," similar to the deposits of the Oscura Range.

In Tijeras Canyon, which separates the Sandia and Manzano ranges, there are many prospects, mainly containing copper, in granite and allied rocks, but no important deposits.

In the Manzano Range, near Abo Pass, copper ores in sandstone are reported to exist, as well as prospects containing gold, silver, and copper in crystalline rocks.

Among the nonmetallic resources the salt lakes of Estancia, in central Torrance County, should find mention. These are situated between the Manzano Mountains and Trinchera Mesa. A considerable quantity of salt is obtained from the Big Salt Lake to supply local demand. The salt beds also contain crystallized sulphate like glauberite and bloedite. According to F. A. Jones, the lake lies in a basin within the "Red Beds," from which the saline materials are undoubtedly derived.

SANTA FE COUNTY.

By Waldemar Lindgren.

GENERAL FEATURES.

Santa Fe County in northern-central New Mexico, occupies a rectangular area of 1,980 square miles. San Miguel County adjoins on the east, Sandoval and Bernalillo counties on the west, Rio Arriba County on the north, and Torrance County on the south. It belongs for the most part in the Great Plains province, lying in the portion where it merges into the plateau province, and is covered by slightly inclined strata of Cretaceous sandstones and shales, or in places by "Red Beds." The Rio Grande cuts across its northwest corner. In the most northeasterly part rises the high Santa Fe Range, the southernmost of the Rocky Mountain chains, which near the corner of Mora and San Miguel counties attains its highest elevation of 12,623 feet in Baldy Peak. This part of the range is described under San Miguel County. Along the western boundary line rise four circular groups of eminences—the Cerrillos Hills, the Ortiz Mountains, the San Pedro Mountains, and the South Mountains—all of them formed by igneous intrusions uplifting Carboniferous and Cretaceous strata. The highest elevation in the Ortiz Mountains is 8,928 feet.

The principal nonmetallic mineral resources of Santa Fe County consist in the coal beds at Madrid,[c] south of Cerrillos, where a Cretaceous anthracite, now almost exhausted, and bituminous coal of the same age is mined. The production of coal in Santa Fe County in 1905 was 69,832 short tons and in 1908 about 55,000 tons. At the north end of the Cerrillos Hills is the Tiffany turquoise mine, which has produced considerable quantities of this gem. Gypsum is found near Lamy; brick clays at Santa Fe; mica near Nambe, in the Santa Fe Range.

The metallic resources consist of copper prospects in the Santa Fe Range; of lead-silver veins in the Cerrillos district; of gold placers and veins in the Ortiz district; and of gold placers, gold veins, and copper deposits in the San Pedro district, all three of them occurring in the early Tertiary intrusions of porphyry which formed these mountains.

---

[a] Richardson, G. B., oral communication.
[b] Op. cit., p. 191.
[c] Johnson, D. W., Geology of the Cerrillos Hills, New Mexico: School of Mines Quart., vol. 24, 1903, pp., 303-350. Stevenson, J. J., The Cerrillos coal fields near Santa Fe, N. Mex.: Trans. New York Acad. Sci.,vol. 15, 1896, pp. 105-122.

CERRILLOS DISTRICT.

GENERAL FEATURES AND GEOLOGY.

The Cerrillos mining district is situated a few miles to the north of the station of the same name on the Santa Fe Railway and about 20 miles south-southwest of the old city of Santa Fe. (See fig. 11.) The deposits are found in a prominent group of hills after which the district has been named; these hills rise to an elevation of 7,000 feet, 1,300 feet above the bed of Galisteo

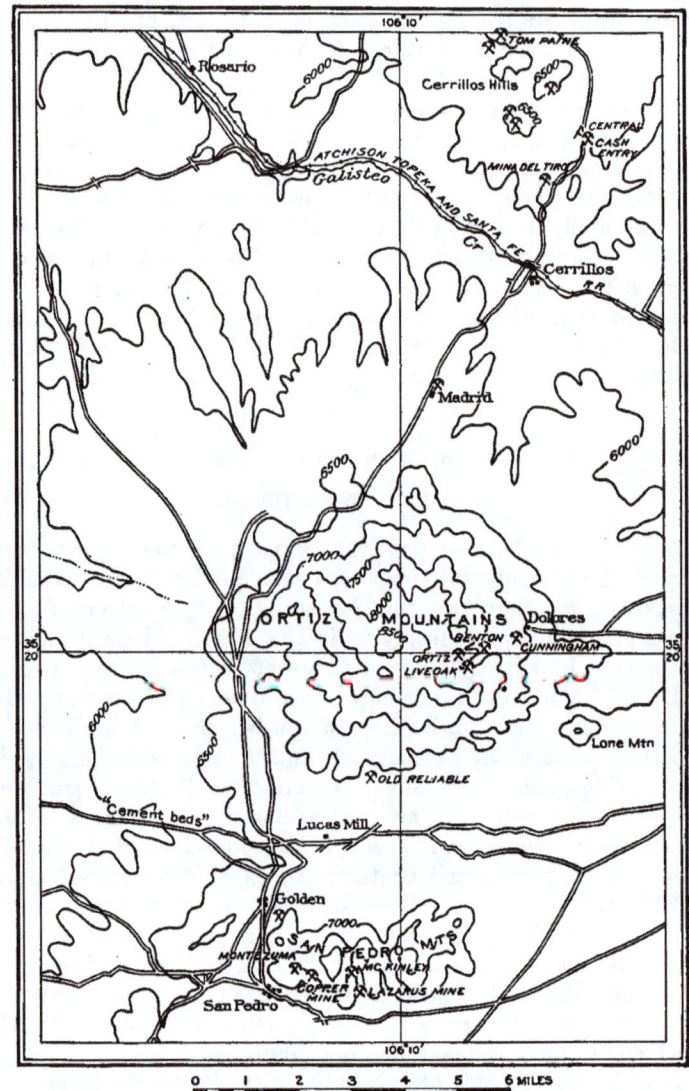

FIGURE 11.—Map of Cerrillos, Ortiz, and San Pedro districts. (From San Pedro sheet, United States Geological Survey.)

Creek, and their outlines are plainly visible from Santa Fe; the area occupied by them amounts to only a few square miles. The district was discovered in 1879, although there is some evidence of far older workings, both for turquoise and silver, by the Indians and Mexicans. Lead, silver, and zinc are the principal metals found.

The geology of the district has been described in great detail by D. W. Johnson.[a] Many other geologists have visited the region and a full list of the literature is given in this paper.

---
[a] School of Mines Quart., vol., 24, 1903 pp. 173-246 (paleontology), 303-350, 456-500 (general geology); vol. 25, 1903, pp. 69-98 (petrography).

According to Johnson's conclusions, with which the writer is in general accord, the six peaks of the hills represent an intrusion into Cretaceous strata, which consist of the Dakota sandstone and the Colorado and Montana groups. Johnson divides the formations above the so-called Dakota sandstone into a small thickness of Benton shale, 1,000 feet of Pierre shale, and from 1,500 to 2,000 feet of Fox Hills sandstone, which at Madrid, a few miles south of Cerrillos, contain workable coal beds, in part converted to anthracite by intrusive contact action. He considers the Galisteo sandstone (thickness 1,500 feet) to be the equivalent of the Laramie formation. Except where disturbed by the intrusive rocks this sedimentary series dips uniformly to the east at angles from 15° to 20°. Near the contacts the strata are turned up, are violently disturbed in some places, and generally dip away in all directions from the igneous rock. The conclusion is justified that the intrusive mass represents a laccolith, or more probably two. Intrusive sheets project into the adjoining strata, and on the southwest side radiating dikes extend out into the sediments. The age of the intrusion is probably on the border line between Cretaceous and Tertiary. The total thickness of the bent strata which covered it and prevented the magma from reaching the surface was at least 3,000 feet. During the whole of early Tertiary time erosion was active, dissecting the rocks and exposing the intruded rocks. Deposition asserted itself in the Miocene and on the surface of the peneplain the Santa Fe marl was laid down. Basalt flows followed the deposition of these beds, which are regarded as of subaerial origin.

According to Johnson the entire mass of the Cerrillos Hills proper is composed of rocks of the augite andesite series, and a portion of the lower country to the east is formed by hornblende andesite. A textural variation of augite andesite is recognized in the gabbro porphyry from Mount McKenzie. This nomenclature does not correspond with that in use during recent years. The rocks are intrusive and have consolidated in large masses far underneath the surface; they are decidedly holocrystalline and porphyritic. Such rocks are now generally designated as porphyries, with appropriate prefix of the name of the granular group to which they belong. The only analysis given is that of a "gabbro porphyry" (No. 1 in the table below), which contains 18 per cent of potassium feldspar and probably considerably more alkali feldspar when account is taken of the admixed soda molecule. The rock has a decidedly "intermediate" tendency and does not differ greatly from some of Brögger's typical monzonites.

The prevailing rock at the Cash Entry mine is dark gray and is rendered porphyritic by closely crowded feldspar crystals, at most 3 millimeters long. The microscope shows abundant andesine crystals, wholly uralitized or chloritized augite crystals, also some epidote; grains of magnetite abound. The groundmass is microcrystalline, is rather coarse, and consists chiefly of a closely massed mosaic of orthoclase grains. The results of a partial analysis are recorded under No. 2.

Analyses of rocks from Cerrillos district.

	1.	2.	3.	4.
SiO_2	48.21	53.99	56.68	60.82
Al_2O_3	17.96		16.62	
Fe_2O_3	5.18		6.50	
FeO	4.47			
MgO	4.11		.79	
CaO	9.72	4.89	.59	.25
Na_2O	3.68	4.08	1.03	5.86
K_2O	2.99	4.90	11.18	4.94
H_2O-	1.41		3.28	
H_2O+	.21			
TiO_2	.84			
P_2O_5	.58		.73	
MnO	.31		1.02	
BaO	.07			
FeS_2			2.21	
	99.74		100.63	

1. "Gabbro porphyry," Mount McKenzie, Cerillos Hills; George Steiger, analyst. Measured by the quantative system this rock is an andose.
2. Monzonite porphyry, Cash Entry mine, Cerrillos Hills; E. C. Sullivan, analyst.
3. Altered rock from turquoise quarry; F. W. Clarke, analyst.
4. Altered rock from turquoise quarry; E. C. Sullivan, analyst.

A medium-grained granular syenite rock of light color from the 600-foot level, Cash Entry mine, shows under the microscope a fabric of large orthoclase crystals, in part with microperthite, in which are embedded prisms of andesine. There are also long prisms of colorless augite, partly altered to greenish-brown hornblende. Much titanite and some magnetite are present.

Still another variety of rock from the Cash Entry mine is fine granular and consists chiefly of orthoclase with some andesine. This rock is partly altered and contains secondary muscovite, calcite, and tourmaline, the last in long prisms, intergrown with pyrite.

These notes confirm the statement that rocks of intermediate monzonitic composition are present in the hills and the writer believes that they entirely predominate.

The alteration of the turquoise-bearing rocks in the Cerrillos Hills has been the subject of considerable discussion. F. W. Clarke and J. S. Diller [a] consider these rocks to be trachytes, but D. W. Johnson [b] believes them to be somewhat altered augite andesites and thinks the analysis of the former authors (No. 3 in the table) very difficult to reconcile with the facts. In view of this statement, sections of the turquoise-bearing rock were examined and E. C. Sullivan, of the Geological Survey, made a partial analysis of a typical sample (No. 4).

In many discussions of the origin of turquoise the fact is not clearly appreciated that the mineral is a secondary hydrous phosphate, similar to many such compounds which easily form in the upper levels of ore deposits. The writer believes it safe to assert that turquoise is not a primary vein mineral, but is formed by surface waters of ordinary temperature descending through altered and mineralized rocks. The matrix from the Cerrillos Hills is a very much altered white rock with earthy fracture containing small seams of the blue or greenish gem mineral. The microscope indicates that probably it was originally a monzonite porphyry, like the normal rock of the hills. The feldspar phenocrysts are, however, filled with microscopic sericite, the groundmass to less extent. There are also some kaolin, small grains of tourmaline, very little epidote, and probably some fluorite. The turquoise-bearing area was evidently greatly altered by mineralizing solutions. Sericite was formed in large amounts, lime, magnesia, and soda were leached, and thus a great concentration of potash was effected. It is the same type of alteration so commonly found along mineral veins. Extreme sericitization undoubtedly accounts for the high potash in Clarke's analysis. In the specimen collected by the writer and partly analyzed (No. 4) similar conditions prevailed, but the sericitization had not proceeded so far.

When erosion had brought this altered rock within the reach of surface waters kaolin began to develop by the action of sulphuric acid from the pyrite on the sericite and various phosphates, and other hydrous minerals characteristic of the oxidized zone found congenial conditions to form.

MINERAL DEPOSITS.

The intrusive rocks in the Cerrillos Hills contain a great number of mineral veins, carrying silver, lead, and zinc. That the ores were known at an early day is shown by an old shaft, called the Mina del Tiro, which was found by the earliest American prospectors in the southern part of the hills. F. A. Jones [c] believes that Spanish priests had this work performed by the Indians prior to 1680.

After the rediscovery of the district in 1879 much activity followed and a considerable amount of silver-lead ore was extracted. The total value of the production to date is not likely to exceed a few hundred thousand dollars. In 1902 a small lead smelter was built at the town of Cerrillos, but it was never operated steadily and has been closed since 1904. Occasional shipments have been made by several mines during the last few years. In 1909 preparations were made to reopen the Cash Entry and Grand Central mines in the Cerrillos district. A concentrating plant was installed and the mine had been unwatered down to the 300-foot level.

The veins are very abundant, but mostly small. They are as a rule contained in the central monzonite porphyry, but also cut across the contact into the surrounding Cretaceous shales, which at the contact are somewhat baked but have suffered no extensive metamorphism.

---

[a] Bull. U. S. Geol. Survey No. 42, 1887, p. 39.   [b] School of Mines Quart., vol. 25, 1903, pp. 86–89.   [c] Op. cit., p. 30.

The strike is northeast, ranging to the north-northeast in the northern part of the hills. The dip is generally steep or the veins are vertical. Most of the deposits appear to be formed along sheared zones in the intrusive rock.

As stated, a great number of veins have been located and in the aggregate a large amount of development work has been performed. Silver-bearing galena, zinc blende, and a little chalcopyrite are the principal ores; the prevailing gangue is quartz with some calcite.

The Cash Entry is the deepest mine, situated 3 miles north of the town of Cerrillos. A considerable amount of ore has been shipped, but the mine has been idle and filled with water since 1904. The shaft is 650 feet deep and is said to have 5,000 feet of workings; water stands at 150 feet below the surface. The country rock is the monzonite porphyry described in a previous paragraph, but the vein is said to be followed by a dike of coarse syenite rock, also mentioned in the same place. The strike is N. 70° E., the dip steep to the northwest. The deposit is a well-defined shear-zone vein with several good walls distributed within a width of about 15 feet; evidently it is of the type of replacement veins. Down to water level the ore was oxidized with residual masses of galena and much zinc carbonate. Sulphide ore was entered at water level. The fresh ore occurs in a gangue of ankerite and quartz and contains galena, yellow zinc blende, and a little chalcopyrite. It is only moderately rich in silver, containing about 20 or 30 ounces to the ton. In depth zinc ores are said to prevail. Near the vein the country rock is partly converted to sericite; tourmaline is also present in some of the specimens.

The Central mine is situated close by the Cash Entry, on the west side. It has had considerable production. The shaft is 500 feet deep, and here, too, water stands at 150 feet below the surface. The ore is similar to that just described. In 1905 attempts were made to sort and ship oxidized zinc ores from the upper levels.

On the north side of the hills the Tom Paine is the most prominent mine, with a shaft 176 feet deep and a record of some ore shipped. There are many other small properties which were not examined.

The Cerrillos Hills contain two deposits of turquoise, both of which are believed to have been worked by the Indians before the Spanish occupation. The locality now worked by the American Turquoise Company, of New York, is in a low hill about a mile northeast of the main group of mountains, near the road from Santa Fe to Cerrillos. No opportunity was offered to examine this locality. The second turquoise mine is near the center of the hills, east of Grand Central Mountain, a short distance from the Cash Entry mine. The large excavations at this place, made by the Indians, have often been described. At present the place is abandoned, but close search will usually result in the finding of small turquoise-filled cracks. The white, earthy rock in which the mineral occurs is an altered phase of the monzonite porphyry, as described in a preceding paragraph, and occupies an area of about 20 acres.

The mineralization of the Cerrillos Hills in all probability took place in the earliest part of the Tertiary period and followed the intrusion of a laccolith of monzonite porphyry. The thickness of superjacent rocks at this time is not likely to have exceeded 4,000 feet. In spite of this, the conditions appear to have been those of rather high temperature and pressure, as indicated by the presence here and there of tourmaline. The solutions probably found difficulty in reaching the surface.

DOLORES OR OLD PLACER MINING DISTRICT.

GENERAL FEATURES AND GEOLOGY.

The Ortiz Mountains form an almost circular group 6½ miles in circumference, rising above old débris fans that spread in all directions and culminating in a central peak at an elevation of 8,928 feet. The base of the mountains follows the contours of 6,500 to 6,800 feet. Sparse forests cover the slopes. The group is in plain view from many directions, even from Santa Fe, and its center lies 7 miles south of Cerrillos, on the Santa Fe Railway. (See fig. 11.) In the southern and eastern parts of the group placers and lode mines are found, both being discovered early in the nineteenth century and successfully worked by the Mexicans. At the present time there is little activity, partly owing to the fact that the best mining ground is within the limits of an old Spanish grant, the owners of which do not appear to encourage this industry. The region is in the San Pedro quadrangle as mapped by the Geological Survey.

The district has been studied by Yung and McCaffery,[a] who published a geologic sketch map of the Ortiz and San Pedro mountains.

In brief, the central part of the group is formed of an intrusive mass of light-gray granular rock of medium grain, composed chiefly of white feldspar and black grains and prisms of hornblende. The hornblende is of the normal brownish-green type, the feldspar in part andesine, in part orthoclase, the former predominating. The rock is probably a diorite approaching a monzonite. In intrusive sheets at the northwest margin of the group of mountains at the Madrid coal camp, the rock becomes more porphyritic and very similar to the Cerrillos type of monzonite, described elsewhere. This rock contains phenocrysts of augite, hornblende, oligoclase, andesine, and orthoclase. The groundmass is microcrystalline and consists of stout prisms of triclinic feldspar embedded in orthoclase. D. W. Johnson[b] terms it a trachy-andesite, a name which seems inappropriate. A large dike just below the Old Reliable mine on the south side of the mountains is a light-colored, almost granular monzonite porphyry, which consists of phenocrysts of andesine and light-green augite in a microcrystalline groundmass of orthoclase and quartz. At the Ortiz mine, on the eastern slope, the country rock is a coarse breccia of granular rocks looking like diorite with darker or lighter porphyries of various kinds.[c]

The sedimentary rocks outcrop chiefly in the northern and southern foothills, the intrusive rock bordering elsewhere against the débris fans. On approaching the hills from Madrid on the north side the intrusive rock is found to adjoin coal-bearing sandstones and shales regarded of Fox Hills age, without exerting much contact metamorphism in the shales. It has, however, converted the bituminous coal of a seam with which it is in contact to an excellent grade of anthracite. On the northwestern spurs a contact is shown with striped and noticeably metamorphosed Cretaceous shales dipping 10° NE. At the Lucas mill, in the southern foothills, Cretaceous shales also appear, and a couple of miles farther north a series of shales and highly metamorphosed garnet rocks are exposed, dipping gently east; these rocks probably also belong in the Cretaceous. To judge from the observations of D. W. Johnson in the Cerrillos district, all these shales are probably of Pierre age. Some distorted casts of *Inoceramus* were found on the northwestern spurs. The plains to the southeast of the mountain group are probably underlain by the "Red Beds," which outcrop in the gulch 2 miles south of Lone Mountain. The contact metamorphism of the sediments is distinctly stronger than in the Cerrillos Hills.

So far as can be seen the igneous mass is probably a laccolithic intrusion. Here, as well as in the Cerrillos Hills, it is unlikely that the covering roof of the laccolith can have been more than 2,000 or 3,000 feet in thickness.

MINERAL DEPOSITS.

PLACERS.

The Ortiz Mountains contain a number of gold deposits, in part placers, in part lodes, but the auriferous areas seem to be confined to the southern and northeastern parts. The placers at the mouth of Cunningham Canyon, known as the Old Placers, were discovered by Mexicans in 1828 and were worked with good results, chiefly from 1832 to 1835, during which period, according to Prince,[d] from $30,000 to $80,000 was taken out annually. During the following decade the yield diminished somewhat. The total production is not ascertainable, but is probably nearly $2,000,000. In recent years, under the control of the present owners of the Ortiz grant, little placer mining has been done.

The placers are situated at the mouth of Cunningham Canyon, at the almost deserted little town of Dolores, the elevation of which is about 7,000 feet. The gravels form a mesa, the upper part of the old débris fan from the gulch, and this is dotted with shafts and pits. A considerable

---

[a] Yung, M. B., and McCaffery, R. S., The ore deposits of the San Pedro district: Trans. Am. Inst. Min. Eng., vol. 33, 1903, pp. 350-362.
[b] School of Mines Quart., vol. 25, 1903, p. 79.
[c] For analyses of the rocks of the Ortiz Mountains see p. 40 in the general part of this report.
[d] Prince, L. B., Historical sketches of New Mexico, New York, and Kansas City, 1883, p. 242.

area is occupied by the gravels and it is stated that much profitable ground remains. The depth of the angular wash increases with the distance from the mountains and in some places has been found to be more than 100 feet. Some twenty years ago Mr. Edison built an extensive plant at this place to extract the gold by a method of static electricity. The plan failed because the requisite dry gravels could not be obtained. Dolores Gulch, which lies to the west of Cunningham Gulch, was less rich. Arroyo Viejo, between the two, also carried good values in gold. The scarcity of water has at all times been a drawback to the effective treatment of these gravels. The fineness of the gold is said to be 918; in part it is coarse, many nuggets being reported. Placer gold is found also at the south side of the mountains, but the gravels are less rich:

QUARTZ VEINS.

As already noted, the principal lode mining has been carried on in Cunningham Canyon above the Old Placers.

The Ortiz mine is situated half a mile above Dolores, at an elevation of 7,600 feet. After its discovery it was worked in a desultory manner, but the largest shoot is said to have been taken out between 1854 and 1864. The depth then attained was 150 feet. Work was resumed in 1868 and the shaft was sunk to the 200-foot level by the New Mexico Mining Company. In 1878 it was reopened by F. Stecks. From 1895 to 1900 the Ortiz Mining and Milling Company leased it from the owners of the grant and sunk the shaft to the 400-foot level. The old mill of 15 stamps was erected at Dolores in 1854. In 1869 a 20-stamp mill was built in the same place, and its ruins still stand. In 1895 a plant of 10 stamps, 2 Huntington mills, and 3 vanners was erected at the mine. No reliable estimate of production was obtained. The mine was idle in 1905.

The Ortiz vein, which is traceable for half a mile, strikes north-northeast and dips 75° WNW. It is contained in a breccia of diorite or monzonite and porphyry, which in places contains much epidote and near the vein is bleached with development of sericite, pyrite, carbonates, and specularite.

Water stands at the 150-foot level in the workings. So far as could be seen from surface exposures the deposit is a vein about 4 feet wide and made up of several stringers of quartz in porphyry breccia. The gold above water level is free, but in depth is probably contained mostly in the sulphides.

The ore material on the dump shows numerous seams of chalcopyrite and crusts of small quartz crystals in the above-mentioned altered rock. They also contain specularite and magnetite, both unusual minerals in veins. Some arsenopyrite and galena are reported to occur.

The rich shoots appear to have been found in the oxidized zone above the 200-foot level; the exploration below that depth apparently has yielded little. F. A. Jones[a] gives a sketch of the stopes extracted prior to 1850, which appear as irregular pipes or channels for a distance of 700 feet along the vein. Measured along the plane of the vein these shoots are from 10 to 50 feet wide, the largest mass of ore being extracted from the foot of the old incline from the 200-foot level near to the surface.

The Benton claim, 1 mile above Dolores, is located on a well-defined fissure in porphyry, striking N. 50° E., with steep dip; it is continued southward on the Live Oak claim. The veins are stringer zones, with quartz as the principal gangue mineral. Some molybdenite was found in the Live Oak vein.

Yung and McCaffery, in the article cited above, describe without giving locality a deposit contained in hard white sandstone, cut by two dikes of porphyry which formed offshoots from the main mass of the mountains. Near the dike the sandstones are brecciated and gold is contained in these oxidized joints and cracks.

CONTACT-METAMORPHIC DEPOSITS.

A small 5-stamp mill belonging to J. M. Lucas is located in the Tuerto Arroyo 1½ miles from Golden. The ore treated in this mill comes from a deposit on the Old Reliable or Pat Collins claim, situated 1½ miles farther north, at the south base of the mountains, at an eleva-

---
[a] Op. cit., p. 23.

tion of 7,300 feet. The deposit has yielded some gold, exact figures not being obtainable. Cretaceous rocks outcrop near the mill. On going north over the gravel mesa spreading at the foot of the mountains one finds the first outcrops at the base of the mountains. They consist of a dike of porphyry of considerable width which crosses the little gulch where the mine is located. Immediately above and continuing at least for 1,200 feet up the gulch is a series of gently eastward dipping shales, which alternate with beds of yellowish-gray garnet rock that evidently is contact-metamorphosed limestone. Still farther up begins the diorite or monzonite, which forms the bulk of the mountains.

The workings are contained in the principal bed of garnet rock; they extend along the east side of the gulch for about 75 feet, and in one place continue into the hill for the same distance. The garnet includes scattered grains of chalcopyrite, which appears to contain gold; when oxidized this gold can be saved by the free-milling process. At one place vertical seams of oxidized pyrite striking north and south cross the garnet bed and here the values are reported to be exceptionally good, but there are also, Mr. Lucas states, good values in the solid garnet rock. This is a rather unusual type of deposit, but finds a close parallel in the San Lazarus mine, in the San Pedro Hills.

SAN PEDRO OR NEW PLACER DISTRICT.

GENERAL FEATURES AND GEOLOGY.

The small group of the San Pedro Mountains (see fig. 11) rises 3 miles south of the Ortiz Mountains to elevations of 8,325 and 8,375 feet, attained in two points situated on an east-west line. In this direction the San Pedro Mountains are 4 miles long; the width from north to south is at most 2 miles. The height above their basement is about 1,500 feet. Like the other groups, they stand above gently sloping detritus fans, and from the scantily timbered summits splendid views are obtained, reaching to the Mount Taylor volcanic region on the west and far out into the plains on the east.

South of the San Pedro Mountains are the South Mountains, similar in structure to the others of the laccolithic group. They contain some prospects, but no deposits of importance. Most of the mining developments in the San Pedro Mountains are in their western part. The Mexican town of Golden is situated at the northwest base of the mountains, and San Pedro, near the copper mine of the same name, lies at their southwest base. Madrid, the nearest railroad station, is about 15 miles to the north. The geology of the San Pedro Mountains has been well described by Yung and McCaffery,[a] whose paper also contains a somewhat generalized geologic map of the region and a detailed map of the vicinity of the San Pedro copper mine. The recent mining developments have been ably recorded by R. B. Brinsmade.[b]

The topography is mapped on the San Pedro sheet of the United States Geological Survey. In a general way the mountains are the result of the erosion of a laccolith intruded in Carboniferous strata. The western, northern, and eastern slopes are made up of porphyry; in the central part a considerable portion of the laccolithic roof of Carboniferous beds is preserved and has a thickness of about 700 feet.

A flat 3 miles wide intervenes on the north between the San Pedro and the Ortiz mountains. North of the former group exposures of flat Cretaceous shales with ammonites were noted at Lucas Mill; "Red Beds" dipping 10° NE. were observed to the northeast of the mountains; and flat-lying Carboniferous limestones outcrop at Golden on the northwest side. Three or four miles west of Golden, at the "cement beds" in Tuerto Creek, the "Red Beds" are again exposed, dipping 15° NNE. From San Pedro the road ascends 200 feet to cross a low divide south of the mountains. Carboniferous limestones in heavy flat-lying benches are here exposed. The road then turns northward to the copper mine. A wedge of the main mass of porphyry is crossed between the Lincoln mine and the copper smelter, and above the latter lies a moderately altered limestone. The whole hill, 700 feet high, above the copper

---

[a] Yung, M. B., and McCaffery, R. S., Trans. Am. Inst. Min. Eng., vol. 33, 1903, pp. 350-362.
[b] Mining World, Chicago, June 27, 1908.

smelter is greatly metamorphosed and consists first of beds of limestone in which (at the mouth of the Old mine) specimens of crinoid stems and a *Productus* were discovered; the horizon is probably in the Madera limestone. Above this is a succession of sandstones, calcareous in places, and black shales which probably also belong to the Carboniferous and to the Manzano group. The whole series dips 10° to 15° E.

Intrusive porphyries form sheets and dikes in these rocks, but the main intrusive mass is exposed on the western slopes of the mountains southeast of Golden. Its contact with the limestone at Golden is covered by débris fans. From the base up to the lower edge of the laccolithic roof, say, at Montezuma Point, west of the copper mine, the vertical distance is 800 or 1,000 feet. This intrusive sheet diminishes materially in thickness to the southeast, and at the copper smelter it is probably only 200 feet thick, measured in a vertical direction. The conditions indicate with certainty a laccolithic intrusion in the lower part of the Pennsylvanian ("Upper Carboniferous") and it is also evident that the uplift has been at least 800 feet.

The main mass of the igneous rock is a granodiorite porphyry of fairly uniform character. Specimens taken below Montezuma Point, at an elevation of about 7,670 feet, show white, closely crowded crystals of andesine feldspar averaging 2 to 3 millimeters in diameter and fewer black prisms of hornblende, both embedded in a dark-gray groundmass, which, under the microscope, is found to be microcystalline and composed of quartz and orthoclase. Incipient decomposition of the hornblende has yielded some chlorite and epidote. A partial analysis by E. C. Sullivan of this rock (28 New Mexico) gave 62.08 per cent SiO_2, 4.62 per cent CaO, 4.76 per cent NaO_2, and 2.84 per cent K_2O.

A dike in the roof of sedimentary rocks in the same vicinity is a grayish-yellow porphyritic rock without dark silicates; it contains large phenocrysts of andesine in a microcrystalline groundmass composed chiefly of orthoclase. Another dike near by is a white fine-grained porphyry, with a few corroded small phenocrysts of quartz, and also some of andesine. The groundmass consists of a micropoikilitic intergrowth of quartz and orthoclase. The rock contains no dark silicates.

The main ore zone in the San Pedro copper mine follows a sheet of white quartzose porphyry, which in the paper by Yung and McCaffery seems to be referred to as an andesite. Under the microscope this rock is in all respects similar to the one just described.

No dark lamprophyric dike rocks were observed.

Contact metamorphism is strongly developed, especially in the roof of the laccolith. The limestones at Golden, half a mile from the nearest outcrops of porphyry, show no alteration; neither do the limestones on the hill east of San Pedro, which are at a distance of about 2,000 feet from the contact. The whole sedimentary series, however, which overlies the porphyry and forms the western summit of the mountains, is greatly altered; the limestones nearest to the igneous rocks are irregularly recrystalized and garnetized and contain chalcopyrite. The shales and sandstones above show less alteration, but the shales have a black baked appearance and to some degree have lost their fissility. Where calcareous, they have been altered to hornfels containing diopside and lime feldspar. This alteration extends throughout the thickness of the beds overlying the porphyry, say, 500 feet; westward the altered sediments extend at least 1 mile to Lazarus Gulch. At intervals they contain sheets and dikes of a quartzose porphyry.

MINERAL DEPOSITS.

GENERAL FEATURES.

The San Pedro Mountains present an unusually interesting case of metallization attending igneous intrusion. Narrow gold-bearing veins are very abundant, both in porphyry and in altered sediments. At the contact, copper deposits with garnet gangue, of the contact-metamorphic type, have been developed. In the limestone farther away from the contact are replacement deposits of galena. Finally, as a result of degradation and erosion, gold-bearing gravels have accumulated at the foot of the group, especially to the north, west, and south.

GOLD-BEARING VEINS.

In the western and southern parts of the mountains the porphyry contains many small veins with steep dip; some of these trend to the northeast. They are filled with quartz, calcite, and pyrite and contain also free gold, in many of them coarse and abundant. Unfortunately, these veins are very narrow and the gold is capriciously distributed, so that although they have been prospected at a great number of places no permanent mine of decided value has yet been developed. According to Yung and McCaffery, these veins appear both as filled fissures and as crushed and fissured zones, but the gold values are invariably contained directly in the seams and veins and do not enter the country rock. The oxidation rarely extends deeper than 100 feet. In Lazarus Gulch there are several of these deposits. The McKinley vein lies 1 mile from the mouth of the gulch and about 500 feet higher up; it is said to consist of stringers of quartz carrying free gold in porphyry. The mine has yielded some production, and at the time of visit ore from it had been treated at the Lazarus mill. Other seams in porphyry, for instance, at the Brooks mine, consist below water level of calcite and pyrite with free gold.

The Lazarus mine, near the mouth of the gulch of the same name, was producing in 1905 at the time of visit. The ore was treated in two small Huntington mills. This deposit is contained in a series of contact-metamorphosed shales and sandstones, dipping gently east. Near the entrance a sill of porphyry 2 feet wide was noted in the sandstone, the cement of which had been metamorphosed to epidote for a distance of a foot from the dike. Between the shale and sandstone are seams of auriferous pyrite containing, it is said, $10 to $20 a ton in gold. Some vertical seams of similar nature cut across both shale and sandstone. The workings have followed these flat veins for several hundred feet. The fresh ore scarcely pays to mine, but within the zone of oxidation the gold is set free and the ore can be profitably treated by amalgamation.

LEAD DEPOSITS.

The Lincoln-Lucky mine, situated in unaltered limestone half a mile below the San Pedro copper smelter, has a considerable production of argentiferous galena to its credit. No work has been done in it for many years. According to Yung and McCaffery the deposit consisted of a chimney of ore about 60 feet in cross section, in a fractured zone following the dip of the strata. The ore is said to have consisted of galena, zinc blende, pyrite, alabandite, and a little chalcopyrite in quartz and calcite gangue. There were also present near the surface cerusite, limonite, and oxidized manganese ores.

SAN PEDRO COPPER MINE.

The copper deposits of contact-metamorphic type are confined to the lower part of the laccolithic roof, on the south slope of the western flat-topped mass of the mountain. Practically all the important copper properties are owned by the Santa Fe Gold and Copper Company, controlled by the Lewisohn Brothers, of New York. From 1889 to 1892 a production of several million pounds is recorded. After various vicissitudes the company was reorganized in 1899, and a smelter built just below the mine. It was blown late in 1900, but the run lasted only for a year, when declining copper prices caused the works to be shut down. At the time of visit in 1905 the mine and smelter were idle, but work was resumed late in 1906, only to be abandoned in 1908, when copper prices again declined.

The problem of matte smelting of poor ore at a considerable distance from the railroad is naturally a difficult one. However, the company has the advantage of owning a coal mine near Madrid, 15 miles to the north. According to Brinsmade[a] the cost of the coal is $3.50 a ton, delivered. The coke comes from the Raton field, and is stated to cost $10 a ton at the mine. The smelting plant lies on the southern slope of the mountains, 1 mile above the town of San Pedro. The mine is situated immediately above it. The smelter is equipped with a 10-tuyere water-jacketed blast furnace with iron settler. Brinsmade states that it was aimed, during the last period of operation, to produce matte with 50 to 60 per cent of copper.

---

[a] Loc. cit.

The ore is developed by the two-compartment Richman shaft at an elevation of about 7,550 feet, sunk to a depth of 300 feet. It is located in a little gully on the hillside 200 feet above the smelter. From the 180-foot level a raise, inclined 13° E., follows the dip of the ore bed to the so-called old mine, situated at the same elevation as the shaft. As the smelting ore is contained only in certain layers of the bed, drifts are turned from the incline at 30-foot intervals along the foot wall. No timber is needed to support the strong limestone hanging wall. In 1906 and 1907 about 160 tons was hoisted every twenty-four hours; by hand sorting this was reduced to 115 tons of first-class ore, assaying 5 per cent of copper and $2 a ton in gold and silver. The ore mined is largely chalcopyrite, and the sorted ore was smelted directly with limestone flux. The average of the gangue contains, according to Brinsmade, 60 per cent of insoluble matter, 10 per cent of iron oxide, 10 per cent of alumina, and 10 per cent of lime.

There is a large amount of low-grade ore, averaging about 2 per cent of copper and 80 cents a ton in gold and silver. It is proposed, according to the same author, to crush it, sort it on a belt, reduce it to 10-mesh size, roast it slightly, and concentrate it on Wetherill electromagnetic machines, which will deliver a product of chalcopyrite and garnet carrying 15 per cent of copper; this mixture is self-fluxing.

The copper-bearing zone containing the greatest developments extends from the Richman shaft northwestward for about 1,500 feet in a straight line to Montezuma Point, rising in this distance 250 feet. Along the contours of the hill the distance is greater. The main contact between the porphyry and the sediments lies about 200 feet lower. This interval is occupied mostly by a shaly limestone, which in some places is altered to hornfels and in others contains some coarser silicates, such as tremolite and diopside. No ore is found at this horizon.

The ore-bearing zone begins as heavily garnetized benches at the Old mine, extending underground to the Richman shaft and northwestward by the Copper Belle claim up to Montezuma. Above this ore zone, which is from 50 to 100 feet thick and which on the Copper Belle and Montezuma resolves itself into two garnet beds, lies an intrusive sheet of white quartzose porphyry, which has been described above. It is a facies of the main igneous rock, without dark silicates, probably also containing more quartz and orthoclase, thus approaching a granite porphyry. This thin sheet or sill has been traced for long distances on both sides of Montezuma Point. Toward the east, beyond the shaft, it thickens to about 100 feet and appears to cut down to lower stratigraphic horizons. It seems to exert little contact-metamorphic influence. At Montezuma Point there is a well-defined dike of granodiorite porphyry, which starts from the main mass and follows the spur up, eastward, for many hundred feet. It is believed that this dike has carried the contact metamorphism into the ore zone. The garnetized limestone begins at the Old mine, though there is here also some entirely unaltered rock. The altered stratum continues northeastward to Copper Belle, where a bed of massive garnet 20 feet thick underlies coarsely crystalline limestone. A little farther east the same stratum of garnet overlies a quartzitic sandstone. At Montezuma Point the above-mentioned dike of granodiorite porphyry is followed by another of the quartz porphyry, and between them lies a mass of garnet 12 feet thick.

The garnet bed in the Old mine is said to be 50 feet thick and has been worked at three horizons not far apart. The distribution of the ore in the bed is extremely irregular. Some of the stopes shown in the old incline are 8 to 12 feet high. The roof stands extremely well, but some low-grade pillars are left for support. The extent of the ore-bearing bed is in doubt, and no ore is said to have been found in the deeper part of the shaft. There was much low-grade ore in sight at the time of visit in 1905, and the great horizontal extent of the metamorphosed stratum indicates good possibilities.

The ore mineral is mainly chalcopyrite, though near the surface some oxidized ore has been extracted. No chalcocite or bornite seems to occur, and the zone of oxidation is shallow. The mine contains no water. The chalcopyrite is invariably associated with a yellowish garnet, which E. C. Sullivan determined as andradite, low in alumina, and containing 21 per cent of Fe_2O_3. Brinsmade, in the article cited, states that the gangue contains 10 per cent of alumina, but it is probable that much of this comes from an amphibole which is also contained in the ore. It is, of course, possible that garnets of varying composition occur in the ore bed.

The gangue forms a coarsely crystalline mass of granular calcite and doubly refracting, partly crystallized andradite, much of it in skeleton crystals. The garnet is altered in part to a chloritic mineral in radial bunches, which would tend to show that part of the mineral is really grossularite; there are further a green mica, some tremolite, and a little quartz. The ore contains some wollastonite, possibly also some vesuvianite, but these minerals are not prominent. The metallic minerals consist of chalcopyrite in irregular grains and specularite, the two at many places intergrown with each other and with the other constituents of the ore in such a manner as to prove contemporaneous deposition.

A dark hornfels found on the dump of the Richman shaft and probably taken from the lower levels contains grains of pyrrhotite intergrown with chalcopyrite. Ore minerals like magnetite, galena, and zinc blende, which are common in contact-metamorphic deposits elsewhere, were not found, and even pyrite is rare.

PLACERS.

The placers near Golden and San Pedro were discovered in 1839, before the American occupation. In the aggregate the production has been large and was derived chiefly from Lazarus Gulch and from the branches of Tuerto Creek near Golden. Mexican miners have done most of the work, and even at the present time there is an annual production of $2,000 to $4,000. Most of the placer work is being done on a small scale by dry washers. The gold is beyond doubt derived from the numerous narrow quartz seams in the porphyry. It is generally coarse and rough, and is said to have a fineness of 920.

As the erosion of the San Pedro Mountains has proceeded, apparently without check, since the middle or beginning of the Tertiary, it is not surprising to find great detritus fans extending from every gulch and gradually flattening out as they merge into the gravel-covered plains; the characteristic detritus of porphyry and lime extends for miles in every direction, particularly westward along Tuerto Creek. All of this subangular gravel contains gold, and every creek and gulch cutting into it has concentrated the gold farther along its course. Certain parts of this detritus may be of Tertiary age, and much of it certainly dates from the early Quaternary. It is stated that a fossil horse tooth was found in the gravels of Lazarus Creek; if so, it would indicate at least the age last mentioned.

The most successful operations have been undertaken between Golden and the first slopes of the mountains, perhaps half a mile from the town. The ground available here is believed to be large; Brinsmade mentions 1 square mile. The detritus fan at this place is covered with sand and subangular fragments and the depth is said to be from 10 to 40 feet. The yield is said to be variable, from 10 cents to the cubic yard up, but a layer on the bed rock is stated to average $1, and where some concentration has been effected by secondary streams the deposit is much richer. Previous to 1905 the Monte Cristo Company had been operating an old dipper dredge on this property, which consists of 480 acres. It was not a success. According to Brinsmade a new company, called the Gold Bullion, was organized and a modern dipper dredge with a capacity of $1\frac{3}{4}$ cubic yards and capable of handling 75 cubic yards an hour, was recently put at work. The dipper delivers the gravel to an elevator, the latter to a trommel, the undersize going to a 22-foot flume, 2 feet wide, with sheet-steel riffle frames, the first few of which contain quicksilver. The water is obtained from several deep wells owned by the company. Some placer ground is also available in the lower part of Lazarus Creek.

The greatest difficulty in working most of the gravels of the district is the great scarcity of water, for the mountains contain no perennial streams. Water under artesian pressure is, however, found in a stratum from 500 to 800 feet below the surface. The Gold Bullion Company has several wells near Golden and uses this source of supply for washing its gravel. When the supply bed is penetrated the water often rises several hundred feet in the hole. Some wells along Tuerto Creek are said to have overflowed. The yield of water for each well has been 25 gallons a minute, but as Brinsmade points out, it is doubtful if this would continue indefinitely independent of the number of wells. At the mouth of Lazarus Gulch is a well 400 feet deep from which water is pumped for the mines in the vicinity. The water stands 125 feet below the surface. It is somewhat alkaline, though potable.

Along Tuerto Creek, a few miles below Golden, about halfway between that place and the Una del Gato coal field, are the gold-bearing "cement beds," famous for a number of unsuccessful projects for recovering the gold. The road from Golden (elevation 6,750 feet) to the cement beds descends over a gravel-covered mesa, reaching creek level about 3 miles from town. Tuerto Creek here forms a gravel bottom several hundred feet wide at an elevation of about 6,400 feet; the low bluffs at the banks show red sandstone of the "Red Beds," dipping about 15° N. On the beveled edges of these strata rest 50 to 100 feet of roughly stratified, subangular, locally cemented gravel consisting largely of porphyry and limestone and evidently derived from the San Pedro Mountains. These gravels are the so-called "cement beds." It is stated that, as early as 1888, 11,000 acres of these gravels had been claimed. In 1905 there were several companies with various plans to work these gravels, which it is claimed contain paying values variously estimated up to $1 or $2 a cubic yard. The Interstate Mining and Milling Company has erected a 30-stamp mill, with amalgamation and cyanide plant, in order to treat these gravels, but no success seems to have been achieved. The consensus among mining engineers seems to be that, though the gravels contain gold, the tenor is far below workable basis, unless sluicing or hydraulic methods could be applied.

LINCOLN COUNTY.

By L. C. Graton.

GENERAL FEATURES.

Lincoln County forms a rectangular area in south-central New Mexico and includes about 4,950 square miles. Socorro County adjoins it on the west, Otero County on the south. In its southern and central portions the county contains the northern part of the great system of ranges which continues into New Mexico from the Guadalupe Range of northern Texas. The northern portion of the county is generally occupied by flat-lying strata of Paleozoic or Cretaceous age. In general the mountains have a short and steep westerly slope toward the northerly continuation of the undrained Sacramento Valley, which occupies the southwestern part of the county. The elevations in the valley range from 5,000 to 6,000 feet. Toward the east a network of long streams and creeks drain to Pecos River. The White Mountains, which reach elevations of almost 10,000 feet, lie in the southern part of the county and are adjoined on the northeast by the Capitan Mountains and on the north by the Jicarilla Mountains, which attain elevations of about 9,000 feet. Very little is known as to the geologic features of these ranges. They consist in general of Paleozoic limestones and Cretaceous shales, locally disturbed and intruded by large masses of monzonite. The long slope toward the Pecos is largely occupied by heavy beds of late Paleozoic limestone. A basalt flow, which is probably of late Quaternary age, occupies part of the valley west of the White Mountains.

The principal nonmetallic mineral resources of the county consist in coal and gypsum. Coal is mined near Capitan[a] and White Oaks and occurs in Cretaceous strata. At Capitan nearly 100,000 tons was mined in 1903, but since then the production has declined. Gypsum, mingled with carbonate of lime, occurs on a large scale at Ancho and is utilized in a local mill for the manufacture of cement plaster.[b] The mineral occurs at a number of other points in the county, for the most part probably in the form of strata in sedimentary rocks of late Carboniferous age.

Of metallic deposits those containing gold are of most importance. The Nogal district is situated in the White Mountains, in the southern part of the county, and embraces the subdistricts of Nogal proper, Vera Cruz, Bonito or Parsons, and Schelerville. Both placer and lode mines have been operated, but the production is not large. The White Oaks district lies in the

---

[a] Campbell, M. R., Coal in the vicinity of Fort Stanton Reservation, Lincoln County, N. Mex.: Bull. U. S. Geol. Survey No. 316, 1907, pp. 431-435.
[b] Jones, F. A., op. cit., p. 239.

central part of the county, in the northern continuation of the White Mountains. It contains the well-known Old Abe and many other mines, and has yielded a production estimated at nearly $3,000,000. The Jicarilla district is 11 miles northwest of White Oaks. Some placer mining has been done here and a number of lode mines have been prospected, principally for gold. Still farther north, near the northwestern corner of the county, is situated the Gallinas district, which contains deposits reported to carry gold, copper, lead, and iron. Iron ores are reported to exist also in the Jicarilla and White Oaks districts, but no information is available as regards the value of the deposits.

NOGAL DISTRICT.[a]

LOCATION.

The Nogal district is situated a little southwest of the center of Lincoln County. It centers about Nogal Peak, which is one of the main elevations of the Sierra Blanca, having a height of 9,983 feet, according to Wheeler. The town of Nogal is about 6 miles northeast of the peak and about 11 miles southeast of Carrizozo, on the El Paso and Southwestern Railroad. A branch line of this railroad, running from Carrizozo to Capitan, crosses the north end of the district about 2 miles north of Nogal.

According to local custom, the Nogal district is divided into the Vera Cruz subdistrict, at Vera Cruz Mountain, 4 miles north of Nogal; the Nogal subdistrict, comprising the country between Nogal Peak and the town of Nogal; the Bonito or Parsons subdistrict, lying south and southeast of Nogal Peak,[b] and the Schelerville subdistrict, west of Church Mountain and northwest of Nogal Peak. Owing, however, to their proximity and to the general similarity in their geology, they may here best be considered together.

Mining operations in this district date back many years, placer gold having been found in Dry Gulch, northeast of Nogal Peak, as early as 1865.[c] In 1868 the American lode mine was located, but not till after 1882, when this region was thrown out of the Mescalero Indian Reservation, did active prospecting begin. A few mines encountered good ore and produced for a time, but in recent years the production has been slight. The total output to date may amount to $250,000.

GEOLOGY.

The district lies mostly on the upper east side and at the north end of the Sierra Blanca. At this end the range consists principally of intrusive rocks that have broken up through the Cretaceous sediments which bound the mountains on the east, north, and west. On the east these sediments, consisting mainly of limestone and sandstone, form a broad mesa. At Capitan there are coal mines which have been worked for a number of years. In addition to the intrusive rocks, there are some surface flows. The rock most commonly seen in the vicinity of the mines is a gray or greenish-gray porphyry of medium grain, containing many phenocrysts of dull, light-colored feldspar in a groundmass mottled with slightly greenish or brownish aggregates of feldspar and green patches of chlorite or pyroxene. The microscope shows plagioclase feldspar predominant among the phenocrysts and constituting the only feldspar of the groundmass; it corresponds about to andesine. Orthoclase phenocrysts are of good size, but are not very abundant. An occasional grain of corroded quartz is seen. Augite is rather plentiful in clear individuals that were somewhat corroded by the magma; it is also present in irregular, interstitial grains in the groundmass. Apatite and magnetite are common in small grains. There is a marked tendency toward parallelism of the plagioclase laths of the groundmass, but not enough to stamp the texture as trachytic. The principal product of alteration of the feldspar is calcite, though epidote is also formed. Both these minerals, together with chlorite, are developed plentifully from the pyroxene. Close to the veins the feldspar phenocrysts are partly sericitized, the groundmass feldspars are more or less replaced by quartz, and the pyrox-

---

[a] The section dealing with the Nogal district is based in part on the observations made in 1905 by L. C. Graton and in part on a report by J. M. Hill, of the United States Geological Survey, as a result of examination of the district in 1908.
[b] Some prospects are known on Eagle Creek and Rio Ruidoso, southeast of Bonito post-office, but these were not visited.
[c] Jones, F. A., op. cit., p. 168.

ene is completely converted into the alteration products already mentioned. In view of the probable relation of this porphyry to the main intrusive of the White Oaks district, it may be called a monzonite porphyry, although the proportion of orthoclase is rather low.

The monzonite porphyry is cut by dikes of darker and somewhat finer grained diorite porphyry that resembles the monzonite porphyry in texture and mineral components but is somewhat more basic. It is probable that these two rocks appeared at the close of the Cretaceous period.

On the Ibex claim was found a light-colored porphyritic rock of fine grain, holding small rectangles of orthoclase. The finely trachytic groundmass of orthoclase laths holds countless small patches of calcite, many of which have rectangular shape and suggest a possible derivation from nepheline. The general aspect of the rock, moreover, suggests phonolite. It must be stated here, however, that phonolite is, in itself, no more an indication of the presence of gold than is granite or schist. The age and structural relations of this rock are not known.

Exposed at various places on the flanks of Nogal Peak is a dense dark-green or nearly black rock. Its relations to surrounding rocks were not ascertained, but microscopic study of specimens collected indicates that it solidified at or very close to the surface and that it has the composition of andesite. Sparse phenocrysts of somewhat altered plagioclase and pyroxene are surrounded by an extremely fine-grained groundmass consisting of the same minerals in pilotaxitic arrangement. Calcite, or dolomite, and chlorite are the chief products of alteration. Associated with this rock in places is a purplish tuff consisting principally of small fragments of andesite but holding also fragments of the monzonite and diorite porphyries, all somewhat iron stained. With the exception of a few basaltic dikes, the andesite and this andesitic tuff are doubtless the youngest rocks of the district, and the probability is that in point of age they belong with the Tertiary lavas.

ORE DEPOSITS.

The ore deposits may perhaps be best understood from the description of some of the typical occurrences.

AMERICAN, HELEN RAE, AND IOWA AND NEW MEXICO MINES.

The American and Helen Rae mines are situated in Dry Gulch, about $3\frac{1}{2}$ miles southwest of Nogal. They were located on the same vein, which strikes N. 15–35° W. and dips from 60° W. to vertical. The Helen Rae is about 1,000 feet north of the American. These were among the earliest locations in the district, and their combined production is probably more than half the district's total output. The American mine is developed by a shaft whose depth is variously stated at 300 to 400 feet. A mill near by is equipped with 15 stamps, plates, and bumping tables. The deepest shaft on the Helen Rae property is 265 feet deep and is cut at a depth of about 150 feet by a 980-foot tunnel. The mill, about half a mile from the tunnel portal, has a Huntington mill, plates, and vanners. Neither of these mines has been worked since the early part of the present decade. The American workings are flooded and admission to the Helen Rae was not obtained.

The wall rock of the vein is, for the most part at least, monzonite porphyry, though dikes of diorite porphyry are intersected locally and andesite is present not far from the American mine and may have been encountered in the workings. The lode consists of stringers of quartz and dolomite and of the shattered and much altered wall rock intervening and along the sides. The initial alteration of the rock results in the formation of a little sericite and much carbonate, but where alteration is more complete, quartz is chiefly formed. All the quartz has flamboyant, internally radial structure, and where it has replaced the wall rock it is in small grains that under the microscope give a ragged, shredded appearance to the field. Pyrite is the most abundant sulphide, mostly in small grains and crystals; zinc blende is next in importance and in places is the chief constituent of veinlets up to an inch or more in width; galena is not plentiful and is mostly intergrown with the pyrite; chalcopyrite is present in sparing amount. The lode averages 3 to 5 feet in width. It was stated that gold was found especially in pyrite-bearing branch stringers, close to the points where they left the main lode, in which zinc, lead, and copper sulphides were prominent. Such places were known as pockets and some were said to have yielded as much as $35,000 to $50,000 each. The intervening lode was in most places of too low grade to be profitable. In the American mine good bunches of ore were found at the intersection of the main lode with both the Vanderbilt vein, which strikes N. 83° E., and the Crosscut vein, which strikes N. 30° E. Both of these veins are similar to the lode described. The ore is said to have averaged about $20 a ton, mainly in gold. About 70 per cent of the total value recovered was obtained by amalgamation, the remainder in the concentrates.

The Iowa and New Mexico Mining and Milling Company was working in 1905 near the headwaters of Turkey Creek, directly south of Nogal Peak. Monzonite porphyry, dark-green andesite, and purple andesitic tuff are present. A quartz-calcite vein cutting andesite had been developed by a tunnel. The ore differed from that at the American and Helen Rae in holding chiefly pyrite and but little of the zinc, lead, and copper sulphides. The values as indicated by assays were said to be high, averaging $65 a ton in gold and from a very few dollars up to $85 a ton in silver. As little work has been done at the property since that time, it is probable that these values were not representative of the working average of the ore.

The internal structure of the veins above referred to suggests that they are genetically connected with the Tertiary lavas; and this suggestion is supported by their presence in the andesite, which is probably a Tertiary effusive. The other type of ore deposit in the district is represented by the Hopeful and Vera Cruz mines.

HOPEFUL AND VERA CRUZ MINES.

The Hopeful mine is situated in the Parsons subdistrict, a mile southeast of Nogal Peak. Operations in recent years have been conducted by the Eagle Mining and Improvement Company and some production has resulted. The mill, which has a capacity of 150 to 200 tons, is equipped with a crusher, six Huntington mills, and amalgamation plates. In 1908 a 200-ton cyanide plant was being erected and amalgamation was to be abandoned.

The ore body is simply altered rock, much bleached and kaolinized. It is said to occupy an area about 2,000 feet square and is of uncertain depth; in a well driven near the mill sulphides are said not to have been encountered till a depth of 124 feet was attained. The country rock appears to be mainly monzonite porphyry, but the other rocks of the district may also be present, disguised by the extreme alteration. The trend of most of the fractures through this body is about N. 20° E. The gold is said to occur free and to be present rather uniformly through the rock. In 1908 the ore was said to average $3.50 a ton in gold, with little silver. The ore was mined by open cuts. The company got into difficulty and in May, 1908, the property was advertised for sale by the county.

The Vera Cruz mine is located about 4 miles north of the town of Nogal, on the southwest side of Vera Cruz Mountain. The ore body is a mass of much-altered porphyry, brecciated and recemented. Fragments of limestone, sandstone, and shale are also present. The ore is traversed by fracture planes running about N. 20° E., and a plane of this direction, showing vertical slickensides, forms the east wall of the ore body. Zones stained with iron run parallel to these planes. The ore was formerly extracted from an open cut, but a 700-foot tunnel has been run and it is the plan to use this hereafter. The ore body is said to be 900 feet long and 120 feet wide and to be known to a depth of 260 feet. The average value is stated as $1.60 in gold and 48 cents in silver to the ton. The mill contained in 1908 a crusher, rolls, six Huntington mills, and amalgamation plates. It was said that the Huntington mills and plates were to be at once replaced by cyanide leaching tanks to treat the coarse product from the rolls.

It is probable that the ores of the Hopeful and Vera Cruz mines are related in origin to the veins of the other mines—that the mineral-bearing solutions spread through the porous rock at these places and made low-grade deposits.

No information was gained as to how careful and extensive the sampling has been on which are based the average values of the Hopeful and Vera Cruz ores here quoted.

A. WHITE OAKS, LINCOLN COUNTY.

B. STOPE IN SOUTH HOMESTAKE MINE, NOGAL DISTRICT, LINCOLN COUNTY.

OTHER PROSPECTS.

In addition to a number of other properties where gold has been sought, there are copper prospects consisting of veinlets carrying chalcopyrite at Schelerville, on the northeast side of Nogal Peak, and a lead-silver prospect on the Wiggins property, about a mile northeast of Bonito post-office.

WHITE OAKS DISTRICT.

The White Oaks district is situated in a group of hills called the White Oak Mountains, which really constitute the northern continuation of the Sierra Blanca. Baxter Mountain, one of these hills, is the locus of mining operations. The town of White Oaks (Pl. VII, A), situated at the northeastern base of Baxter Mountain, is about 12 miles distant from the El Paso and Southwestern Railroad at Carrizozo.

A small amount of placer gold had been produced intermittently in the fifties and sixties by the operations of Mexicans in Baxter Gulch at the foot of Baxter Mountain. In the late seventies this was one of the many regions explored by prospectors, and in 1879 gold was discovered at what became the North Homestake mine. Shortly afterward the Old Abe, South Homestake, and a number of other claims were located. The total production up to January 1, 1904, is stated by Jones[a] to have been $2,860,000. The production since that time has been small. Little was being done in the district at the time of visit in 1905, the only productive operation being that of lessees in the South Homestake mine. A brief description of the region has been given by Smith and Dominian.[b]

The region is occupied by Cretaceous sediments broken through by post-Cretaceous intrusive rocks. The stratified rock prevalent in and near the mines on Baxter Mountain is Cretaceous shale. It is cut by a large body of monzonite, of gray color and rather fine grain. Many of the individuals of feldspar and dark constituents are lath-shaped, and thus give the rock the appearance of a porphyry, similar to many of the monzonite porphyries of the Territory. The microscope, however, shows that, although there are porphyritic facies and in representative specimens certain individuals have better crystal outline than others, the main mass of the rock is too even grained to be classed as a porphyry. The chief constituent is plagioclase feldspar in long striated laths and in stouter prisms of zonal structure; in composition it ranges from andesine to albite, the more basic variety forming the core of the zoned crystals and albite, or in many crystals orthoclase, forming the exterior border. Distinct individuals of orthoclase are also present, as well as many grains of the intergrown orthoclase and plagioclase known as microperthite. Brownish-green hornblende, clear brown biotite, and pale-green pyroxene (augite?) are present in variable amount in large and small grains; each is predominant in certain phases of the rock. The hornblende and pyroxene are in many places poikilitically intergrown, whereas in some specimens individuals of hornblende occupy the center of large biotite flakes. Some of the biotite is partly faded and in places converted into muscovite, apparently as a result rather of internal instability than of ordinary alteration. Accessory constituents notably plentiful are titanite and black iron ore, and apatite is present in irregular grains of good size. Dark finer-grained segregations consisting of larger amounts of the colored constituents are here and there included in the normal monzonite. On alteration the colored constituents of the rock change mainly into carbonate, chlorite, and limonite. Kaolin and calcite are the chief products of the feldspars. Near the veins sericite forms sparingly, the more common alteration product being a cryptocrystalline mass of quartz.

The essential chemical features of the rock can be inferred from the following partial analysis, made by W. T. Schaller, of a specimen of a somewhat porphyritic phase in which hornblende and pyroxene are about equal in amount and biotite less plentiful. The specimen was collected near the North Homestake mine.

---

[a] Op. cit., p. 175. [b] Notes on a trip to White Oaks, N. Mex.: Eng. and Min. Jour., vol. 77, 1904, pp. 799-800.

Partial analysis of monzonite from White Oaks district.

SiO_2	56.56
CaO	4.77
K_2O	4.72
Na_2O	5.41

The monzonite and shale are cut at various points by dark dikes, commonly of fine grain and showing glistening flakes of the dark mica, biotite. The weathered rock is distinguished by its greenish rusty color. The microscope shows a division, not discernible in the hand specimen, into minette and kersantite, but except for the fact that the feldspar is orthoclase in the minette and plagioclase in the kersantite, the two rocks are alike and are probably connected by transitional types. Biotite, the essential bisilicate of the minette-kersantite series, is of deep reddish-brown color in thin section. In some sections it is practically the only ferromagnesian silicate present; in others it is equaled or considerably exceeded in amount by a nearly colorless pyroxene. Hornblende is rare. Magnetite and apatite are comparatively plentiful. Carbonate, epidote, and serpentinoid material are the chief products of alteration. The texture is holocrystalline, but ranges in both types from granular to porphyritic, in which laths of feldspar constitute the groundmass. Inasmuch as specimens were not collected from many of the dikes, and as proper classification can be made only by microscopic study, these minette and kersantite dikes will be referred to in the subsequent portion of this section simply as basic dikes.

Certain coarser portions of the shale have been somewhat epidotized, but in general a little induration by slight silification is the only effect produced by the intrusion of the monzonite and related dike rocks.

The ore deposits are of comparatively simple nature. They consist of quartz-pyrite veins that cut the monzonite, the dikes, and the shale. The veins themselves are for the most part mere narrow streaks or stringers, but in many places there are numbers of these veinlets which, in connection with the intervening impregnated wall rock, constitute workable lodes or irregular shoots and pockets. Cavities lined with quartz crystals are common, and coatings of calcite are found in places on the quartz, though possibly deposited subsequently to the original ore deposition. Hübnerite is inclosed in the quartz of many veins in single plates or intergrown clusters. It is lustrous black, shows good cleavage, and appears fairly resistant to alteration, but in places, together with the surrounding quartz, it is coated with a sooty film of manganese oxide, and here and there is completely decomposed, leaving small specks of a canary-yellow, glistening material, that may be scheelite. In reflected light under the microscope the mineral resembles clusters of specularite plates, but it proves to be opaque for the most part, with scattered plates or irregular patches not easily transparent and of a dull brown color. Hübnerite is especially noticeable in the South Homestake mine, where it is intergrown with radiating prisms of drusy quartz in veins that have no definite walls but gradually fade out into the surrounding monzonite. At this place purple fluorite is closely associated with hübnerite. In one vein, the Little Mack, dull black woolly sheets composed of parallel minute needles of tourmaline were found intergrown with the quartz and also penetrating the altered monzonite. Tiny needles of somewhat altered material supposed to be tourmaline were also found in brecciated shale immediately adjoining ore veinlets in the Old Abe mine. A closely striated feldspar, developed in minute plates or tablets, is sparingly embedded in the quartz. The twinning plane is parallel to the predominant crystal face. The mineral has practically the same index as Canada balsam and without doubt is albite.

It is commonly stated in the district that gypsum is found in the Old Abe mine, and occurrences of free gold included in that mineral have been reported. Material on the tenth level of the Old Abe mine, pointed out to the writer as gypsum, proved to be simply a soft, bleached, chalky phase of the much-altered monzonite. In the Little Mack vein, however, in the same extremely narrow seams in which tourmaline occurs, gypsum was found, both in crystalline plates and in bands with fibrous structure transverse to the walls. Interesting in this connection is the occurrence in the Cactus mine, at Newhouse, Utah, of felty tourmaline and anhydrite,[a]

---

[a] Lindgren, W., New occurrences of willemite and anhydrite: Science, n. s., vol. 27, Dec. 25, 1908.

along with quartz, sulphides, and other minerals, in veins and irregular impregnations in monzonite. The gypsum resulting from the alteration of the anhydrite in the Cactus mine is similar in appearance to that in the Little Mack vein, and the writer is inclined to believe that anhydrite was present in the latter deposit, but the material is pretty thoroughly oxidized and any anhydrite that might have been present would certainly have been converted into gypsum.

This mineral combination—quartz, albite, fluorite, tourmaline, pyrite, and hübnerite (possibly anhydrite also)—is of considerable interest, and so far as known has never before been recorded as present in veins. It probably indicates either deep-seated deposition or the practically equivalent conditions, probably more applicable in this locality, of deposition in close proximity to igneous rock just intruded. In other words, it seems probable that the White Oaks veins were formed so nearly contemporaneously with the intrusion of the monzonite magma that only the outer portion of this rock nearest the cooling shales had sufficiently consolidated to permit the development of fractures, into the very first of which was forced the molten material of the basic dikes and into those immediately subsequent the solutions from which the veins were deposited.

The effect of the vein-forming solutions on the monzonite has already been briefly referred to as slight sericitization and more common silicification. In connection with the presence of albite in the veins it is interesting to note that certain bands of the zoned plagioclases are but little affected, whereas the remainder of the individual is largely converted into a feathery mass of quartz. Likewise, in some of the microperthite, one of the component feldspars is silicified, while the other is almost unattacked. Although actual determination was not possible, it is probable that in both cases the unattacked feldspar is albite.

A great part of the production of the district has come from ore in which the gold was free. In most of the ore worked recently the gold is associated with limonite, or with partly oxidized pyrite, and doubtless was originally contained in the pyrite. The gold is bright and has a rough crystalline appearance, though definite crystal forms could not be made out. Especially rich specimens saved from early operations, however, show bright gold associated with clean white quartz, though usually in cavities. It is probable that gold existed in the veins both uncombined and contained in pyrite.

The mines are practically dry; even the Old Abe mine, at a depth of 1,380 feet at the time of visit, made very little water. The depth of oxidation is not coincident with the water level, for while oxidized material may be found in some of the deepest workings, sulphides occur in places close to the surface. On the whole, however, wholly oxidized ore was of course more abundant in the upper portion of the mines.

The greater part of the production has come from pockets and shoots of high-grade ore. From the North Homestake mine, for example, $35,000 is said to have been taken in two days.[a] The workings near the surface were richest, and some veins that held rich pockets near the surface became of low grade at comparatively shallow depth. The ore below the seventh level in the Old Abe mine is said to have been decidedly poorer than that above; it is stated that the average of all the ore milled from this mine is a little under $20 a ton. In the South Homestake mine the leasing company was sorting out about one-third of all the rock broken, and this third averaged $12 to $14 a ton; but the richest portions of the lode had previously been exhausted.

An idea of the distribution of values in the lodes may be gained from the following brief descriptions of the occurrence of ore shoots in the Old Abe and South Homestake mines, which were the only ones entered, but are said to be representative of the district.

The Old Abe vein strikes N. 10° W. and dips very steeply toward the west. It cuts both monzonite and shale and such dikes as are present. One especially micaceous dike is encountered just west of the vein on several of the lower levels. The vertical shaft reached the 1,300-foot level in 1905, and was the deepest in the Territory. Below the 1,300-foot level a winze had been sunk 80 feet. Ore was said to be present in this winze, but was not rich enough to stand hoisting twice. The main shaft has since been deepened, therefore, and some ore extracted from the bottom levels. The mine was practically dry to the thirteenth level. In driving

---

[a] Jones, F. A., op. cit., p. 173.

that level water was encountered; for some time it came in at the rate of about 20 gallons a minute, but later subsided. In November, 1905, though no unwatering had been done for a long time, there were only a few feet of water in the winze. The upper workings of the mine were reached through a shaft with levels at intervals of about 50 feet; this has been replaced by the new shaft, and the lower levels are mostly 100 feet apart. Only the lower half of the mine was visited. The ore was found in pockets and shoots in the lode. On the 700-foot or seventh level, for example, an ore shoot 50 feet long was struck about 250 feet north of the shaft. It was mostly in sheeted shale and the ore was much oxidized. As seen in the face of the drift 30 or 40 feet north of the stope the vein consists of a 4-inch seam of limonite and manganese oxide. Directly above, on the 650-foot level, the shoot was 80 feet long. Another good bunch of ore was found on the seventh level, where a flat seam crosses the main lode. This flat seam also shows that a little faulting has taken place along the main fracture, the west wall being thrown down 6 or 8 inches. On the eighth level the shoot extends from 50 or 60 feet south to 200 feet north of the shaft. It reaches from 45 feet above the eighth level to the 950-foot level. It was widest, attaining a width of 10 feet, just south of the shaft on the eighth level. The rich stopes known as the Fish Pond and the Duck Pond were also reached from the eighth level, or the fourteenth level of the old shaft. These stopes were close to but at one side of the main lode and consisted of brecciated masses of shale and monzonite, porous and somewhat silicified and stained with iron. The Fish Pond stope was 20 feet wide, 50 feet long, and 60 feet high; it is said to have yielded $80,000. On the tenth level a stope about 150 feet long begins at 100 to 150 feet north of the shaft. In going northward the ore was found to follow a branch stringer on the east, but the ore gave out where this stringer again united with the main vein farther north. At this level a considerable amount of pyrite is seen in the seams and the adjoining wall rock. On the twelfth level the wall rock was especially impregnated with pyrite, with values, though gradually decreasing away from the central fracture, sufficient to permit stoping to a width of 31 feet. On the 1,300-foot level sulphide ore is also found, but most of the ore is oxidized. About 150 feet north of the crosscut west from the shaft is a stope 35 feet in diameter and 15 feet high. The ore here was said to be of good value. Shale is present with the monzonite and both are more or less brecciated, as above. This ore body is followed down by the 80-foot winze and a small stope made at the bottom, but extraction was not profitable under the conditions then existing. Massive shale, little oxidized, is present 50 feet north of the ore body on the thirteenth level. It is stated that on the whole the ore was of a little better grade in the shale than in the monzonite.

At the South Homestake mine the lode strikes about N. 2° W. and dips about 76° E. Monzonite forms the walls. The ore shoot begins at the surface and a large stope 40 by 40 feet by 70 or 80 feet high was taken out here. Plate VII, B, gives a view of a face of this stope, known as the Devil's Kitchen, and shows the sheeted character of the rock. Most of the fractures are mineralized. Twenty feet west of the Devil's Kitchen is the Capitan stope, similar but not so large. This stope is continuous with what is known as the North stope in the adit level, more than 100 feet below. In the North stope the ore is found at the intersection of the main lode with an east-west seam of steep southerly dip. Below the Devil's Kitchen and almost connecting with it is a stope in the adit level 30 feet long, 20 feet wide, and 20 feet high. At this point a winze was put down. It is known as the old shaft and continues in ore to a depth of 150 feet. At the 50-foot level of this winze the stope enlarges where a flat southward-dipping seam intersects the main veinlets. The average width of the stope is 14 feet. At the 100-foot level a large stope 70 feet long, 40 feet wide, and 35 feet high was made at the intersection of the main lode with the same east-west seam that crosses the ore body in the North stope of the adit level. A new shaft cuts the ore shoot at a depth of 430 feet, on the 250-foot level of the winze. It is reported that this shaft has been deepened since 1905 and that the Wild Cat Leasing Company, which was operating in the South Homestake, is also working the North Homestake mine.

Among former producers not accessible at the time of visit were the Compromise, Little Nell, Lady Godiva, and Little Mack. The lodes of most of these mines cut monzonite and strike nearly north and south.

JICARILLA DISTRICT.

The Jicarilla district is situated in central Lincoln County, in a group of hills at the north end of the Sierra Blanca known as the Jicarilla Mountains. It is about 11 miles northeast of White Oaks; the nearest railroad station is Ancho, on the El Paso and Southwestern Railroad, about 8 miles to the northwest.

It is stated that certain of the streams in this district were worked for placer gold by the Mexicans about the middle of the last century. Thirty years later prospecting for lode deposits was begun, and work has continued intermittently up to the present time, but the production from lodes and placers combined has been trifling.

The geology is similar in certain features to that of the White Oaks district. An intrusive mass of quartz monzonite porphyry invaded and elevated the stratified rocks. Erosion has removed most of the sedimentary rocks, of which only limestone was seen,[a] and has exposed the porphyry. This is cut by dark basic dikes similar to those of the White Oaks district. The quartz monzonite porphyry is a gray rock with many pinkish or yellowish phenocrysts of feldspar. The microscope shows that these are of plagioclase, near oligoclase, and that numerous somewhat rounded crystals of yellowish-green amphibole are also among the phenocrysts. A few flakes of brown biotite are seen; both this mineral and the amphibole are bordered by scattered small grains of magnetite, and larger magnetite grains, together with apatite and titanite, are present here and there. The groundmass consists of abundant minute crystalline grains of quartz in a mesh of orthoclase grains, giving a microgranitic texture.

The porphyry is traversed by many joints, of which the most prominent group strikes northwestward, with a prevailing dip to the northeast. The ore deposits are lodes consisting of pyrite-quartz veinlets along one or more of these fracture planes. In some of them there is a principal pyrite streak with narrower ones alongside; the main veinlet may pinch down and another near by widen out so as to become the principal one. Quartz, of which some is drusy and some dense, is rather scanty in amount, but the adjoining rock is somewhat silicified, as well as impregnated with pyrite grains. Sericitization has taken place only slightly in most places. The pyrite in the fractures is commonly rather coarse grained. It contains the gold, and where it is oxidized gold can be panned out of it. The gold thus obtained is mostly in small, irregular, spongy grains. In a few mines, such as the Good Luck, Prince Albert, and Eureka, some copper and silver are present with the gold. Some of the lodes are a few inches wide and some are said to carry low-grade values over a width of 40 feet or more. At the Eureka mine the ore shoot occurs along a 5-foot basic dike. The values show a wide range, but depend mostly on the width of lode represented by the sample. A wide ore body in the Good Luck, for example, cut by a 235-foot tunnel, was said to average $1.20 in gold and about $4 in silver to the ton. A wide body in the Hawkeye, opened by shaft to a depth of 70 feet, was stated to carry $13.50 a ton in gold as the average of 52 assays, whereas a narrow streak encountered in a tunnel was reported to give assays of $200 to $400 a ton. In the Gold Stain mine, developed by a 35-foot shaft and a 140-foot tunnel, samples of sulphide ore were stated to range in value from $4 to $520 a ton. The water level and the limit of oxidation vary according to the topography, from close to the surface to a depth of 160 feet at the Eureka mine.

In most of the properties that have encountered any ore, in addition to those mentioned, the deposits are of the type just described. At the Honey Bee mine, however, in the southeastern part of the district, the country rock is blue limestone, much jointed, but in the mine it is in most places crystalline and leached. The strike appears to be about N. 60° E. and the dip 25° SE. The 150-foot shaft at a depth of about 60 feet cuts a mass of porphyry that may be a sill. In the vuggy limestone is a 14-foot zone containing cuprite, malachite, and glance mixed with crystalline calcite and limonite. Some azurite and chrysocolla are also said to be present. The ore body appears to be irregular and has been developed only a short distance along the strike. At the 60-foot level brecciated prophyry forms one wall of the streak, which is narrower and apparently leaner in copper than above. A drift at the bottom of the shaft had not encountered ore at the time of visit.

---

[a] Much of the surface was covered with snow at the time of visit, and other rocks may possibly be present.

The ores of this district are probably related genetically to the quartz monzonite porphyry which was doubtless intruded at the close of the Cretaceous. The placer gold has come from degradation of the lodes. It is asserted that the bottom 6 inches of the channel of Ancho Creek, which flows through the district, is good pay ground, but that the 6 to 20 feet of gravel above is valueless and is too deep to make operations profitable. This conclusion was reached after a large dredge had been installed in 1903 and failed to make a success.

Little work was being done in the district in November, 1905. The Wisconsin Milling and Smelting Company was erecting a mill to treat the ore from several claims, then practically undeveloped.

OTERO COUNTY.

By L. C. Graton.

GENERAL FEATURES.

Otero County is situated in south-central New Mexico and borders against Texas on the south; Dona Ana and Socorro counties adjoin it on the west; Eddy and Chaves counties on the east; Lincoln and Socorro counties on the north. It occupies an area of about 6,870 square miles.

The western part of the county extends for 95 miles along the Sacramento Valley, with elevations of 2,000 to 5,000 feet. Near the western boundary are salt marshes and large areas covered with gypsums and, the latter commonly known as "the white sands."[a] Smaller eminences like the Hueco Mountains and the Jarilla Hills rise above the valley, the former consisting mainly of Carboniferous limestones, the latter of limestones intruded by monzonite porphyries.

The eastern part of the county is occupied by a series of mountain ranges of considerable elevation. In the southeastern part and in the adjacent Eddy County rise the Guadalupe Mountains, with steep western scarps, to elevations of over 9,000 feet. They appear to be largely built up of gently dipping limestones of Carboniferous age, with a thickness aggregating over 5,000 feet.

Farther to the northwest extend the Sacramento Mountains, also with steep westward-facing scarps, reaching elevations of somewhat over 9,000 feet a short distance east of Alamogordo. This range also consists chiefly of Paleozoic limestones, with few rock masses of igneous origin. The Sacramento Mountains are continued northward by the White Mountains or Sierra Blanca, which reaches into Lincoln County and attains, in Nogal Peak, an elevation of 9,983 feet (Wheeler).

Otero County is poor in metallic ores, though such materials as limestone and gypsum are abundant. The Guadalupe Mountains are reported to contain prospects of copper ore southwest of Carlsbad, in Eddy County, and near the Texan boundary line. No metal mines are known in the Sacramento Mountains, although prospects of gold and copper are stated to have been found in the vicinity of Alamogordo.

The principal mining district of the county is located in the Jarilla Hills near the line of the El Paso and Northeastern Railway, about 50 miles north-northeast of El Paso. Copper and gold are mined here. Near Tularosa, in the northern part of the county, a producing copper mine has recently been developed. The metallic production of the county is small.

JARILLA DISTRICT.

The Jarilla district is situated in an isolated group of hills that rise to an elevation of about 600 feet above the great desert plain in western Otero County, about 50 miles north-northeast of El Paso, Tex. The post-office of Brice, in the center of the district, is about 2½ miles northwest of Jarilla Junction (now Oro Grande), on the El Paso and Southwestern system, and is connected with the main line of the railroad by a spur. The country is arid; vegetation is scanty and water has to be brought in by rail. Inconsequent attempts at mining were made about 1880. About the close of the last century attention was directed to the district by the

---

[a] Herrick, C. L., Geology of the white sands of New Mexico: Bull. Hadley Lab., Univ. New Mexico, vol. 2, pt. 1, 1900.

A. NANNIE BAIRD AND I MINES, JARILLA DISTRICT, OTERO COUNTY.

B. OPEN CUT, REPUBLIC MINE, HANOVER DISTRICT, GRANT COUNTY.

discovery of turquoise, and more extensive metal-mining operations were undertaken. After having produced approximately $100,000 in gold and copper, of which about $8,000 in gold is estimated to have come from dry placers, operations practically ceased. In 1905, however, the control of several properties was acquired by the Southwest Smelting and Refining Company. At the time of visit, in November of that year, exploratory work was being carried on at several points, and, although developments did not appear to justify such a procedure, ground was being broken near the railway for the erection of a smelting plant to turn out copper matte. This plant was blown in late in 1907.

The Jarilla Hills consist of limestone strata domed up by an irregular intrusive mass of fine-grained monzonite porphyry. Fossils found in the silicified limestone show clearly that it is Carboniferous; it is possible but not certain that both the Pennsylvanian and the Mississippian series are present. Near the base of the limestone is a little impure quartzite. Near the porphyry the limestone is much metamorphosed. Garnet, diopside, and quartz are formed in abundance and in certain of the more shaly layers much epidote has been developed, whereas other fine-grained bands have been converted into hornfels. It was in such silicified material that the fossils were found. The garnet is of the variety andradite, as shown by the rough partial analysis made by George Steiger of material from the Three Bears mine:

Partial analysis of garnet from Jarilla district.

SiO_2	35.3
(Al_2O_3) Fe_2O_3	32.8
CaO	30.0
CaO (soluble in acetic acid)	2.5

Inclosed with the other metamorphic minerals are radiating groups of acicular prisms, probably of tremolite. Pyrite is abundant locally and commonly associated with it is chalcopyrite. Rosettes of specularite are found, especially in narrow seams in the pyroxenized limestone, and some of them are intergrown with pyrite. The sulphides carry some gold and locally silver. Oxidation causes the production of much porous limonite, coatings and bunches of malachite, and in places chrysocolla with associated black specks of "copper pitch" ore (?). Little pyrite is present in the porphyry, even close to the limestone contact.

With the exception of the placers, all the deposits of the metals that had received attention up to the time of visit were associated with intensely metamorphosed limestone, but it has since been announced that copper ore is being mined from the porphyry where the turquoise was found. Of the deposits in metamorphosed limestone, some follow the bedding of the limestone but many probably represent original fracture zones across the limestone strata. All are either at or close to the contact with porphyry.

The Nannie Baird mine (Pl. VIII, A) was probably the largest producer up to 1905; the combined production of it and the Lucky mine is stated at $66,000, mostly gold, with some copper. A bed of metamorphosed limestone lies just above the contact with the monzonite porphyry. The strike of the contact and of the limestone bedding is about N. 60° W., and the dip averages about 20° NE., ranging from 12° to 30°. Practically at the outcrop of the metamorphosed stratum is a vertical fracture parallel to the strike, and above the intersection of this fracture with the ore-bearing stratum the metamorphic minerals have been formed along the fracture instead of in the upward continuation of the limestone bed. The upper portion of this zone is oxidized and limonite is plentiful, but the outcrop does not project and is not especially prominent. Garnet and pyroxene, with specularite, pyrite, and chalcopyrite, chiefly make up the ore-bearing bed, though quartz and calcite are also present. The overlying bed is shaly; the effect of metamorphism has been induration and the development of pyroxene and some epidote.

An incline 800 feet long follows the main shoot in the ore-bearing bed. This shoot, known as the "ore pipe," pitches a little more to the north than the dip of the bed. Many branches were run off from the main incline, following pockets and side streaks. Numerous veinlets of calcite and possibly dolomite cut through the limestone beds, and in some of these within the main ore-bearing bed some good ore has been found. It is said that a very good bunch of ore was found near the bottom of the incline at the intersection of a similar pipe of ore coming in from above on the northwest side. The ore bed outcrops near the top of a hill whose other slope is nearly parallel to the dip of the ore; a shaft was being sunk at the bottom of this slope which was expected to cut the ore body at a depth of 90 feet near the end of the incline.

The I mine (Pl. VIII, A) is situated farther west along the same bed, which curves in strike from northwest to west and finally toward the southwest. The porphyry seems to cut across the beds at a very slight angle and to be in contact at the I mine with a bed a little below the ore-bearing stratum.

The Lucky mine lies some distance west of the Nannie Baird. A 220-foot inclined shaft opened a lode striking about N. 7° E. and dipping 70° E. The ore is found in garnet rock at the very contact with porphyry, which lies on the east or hanging-wall side. The contact is tight and one of intrusion. The ore is found in bunches along this zone. In some places the gold values are good and in others copper is important. Most of the ore is oxidized, but at the north end of the mine large octahedrons of pyrite are present in the garnet rock and in places are inclosed by chalcopyrite. Some work was also being done on this lode in an adit tunnel south of the shaft. About 50 feet west of the shaft some good copper ore was found in a lode striking N. 15° E. and dipping about 75° E., with somewhat altered limestone on the east and garnet rock on the west wall. A second mass of porphyry lies 15 or 20 feet farther west. Other irregular shoots are found in the metamorphosed limestone.

The Three Bears mine is just south of the little camp of Brice. The mine is developed by a vertical shaft 525 feet deep and five levels. The production up to the end of 1905 was only a few thousand dollars. The mine was then idle. Limestone is exposed on the surface, cut by dikes and irregular intrusions of porphyry. The main mass of limestone strikes about N. 52° W. and dips 30°–35° SW. The ore deposit is similar in a general way to that at the Lucky mine, consisting of metamorphosed zones in limestone along what originally were probably fractures. The Three Bears lodes, however, contain more quartz than the Lucky ore and are essentially quartz-pyrite veins with indefinite boundaries at the walls, which consist of garnet with some pyroxene, calcite, specularite, and sulphides; in places porphyry forms one wall and nowhere is it far away. The pyrite in both the quartz and the garnet is associated with more or less chalcopyrite and, more especially in the quartz, it locally carries some gold. The sulphides have been pretty completely oxidized at many places. On the first or 65-foot level a lode was cut 20 feet south of the shaft. It strikes S. 70° W., dips about 80° N. and is about 10 feet wide. It is porous and completely oxidized; the values are rather low. A parallel zone a little farther south is separated from it by a dike of porphyry. Still farther south of the shaft a more definite vein was encountered, running approximately north and south and practically vertical. It consists of a few inches to 3 feet of quartz with much pyrite (or limonite). A narrow vertical shoot of ore said to average about $18 in gold to the ton was followed up for 130 feet, or nearly to the surface at that point. The ore of this shoot was thoroughly oxidized; it is said that at the ends of the shoot, where pyrite was encountered, the values dropped. A winze following down on this shoot got out of ore in a short distance and into limestone dipping about 20° W. The two lodes just south of the shaft on the first level appear to have united somewhere above the second level, 100 feet deeper, where they form a zone about 20 feet wide striking N. 78° W. and dipping about vertical. There is on the second level a lode striking about north and south, dipping 60° to 65° W. close to the contact with porphyry. A stope 50 to 60 feet long, 15 to 20 feet high, and averaging 4 feet wide has been taken out here. The values are said to be about $14 a ton in gold and 4 per cent of copper, mostly in the form of malachite and chrysocolla with some chalcopyrite. Limonite is abundant. Here, as on the level above, where less oxidized ore is encountered the values are lower. The lower levels were not visited; it was stated that little but exploratory work had been done on them. The copper values below the fourth level were said to be somewhat better.

A feature of interest in this Three Bears mine is the apparent association of oxidation with the richer grades of ore. If the variation in value applied only to copper the matter could be readily explained as the result of some enrichment by descending oxygenated waters. But as the gold values in particular are concerned and as the difference in grade between the oxidized material of the ore shoots and the immediately adjoining sulphides is too great to be explained either by removal of material from the oxidized ore or by secondary enrichment, the only explanation seems to be that, as appears to have been the case elsewhere,[a] the channels which evidently afforded the original ore-bearing solutions the easiest passageways, as indicated by the presence of the ore shoots, were also the most permeable places to the oxidizing waters descending from the surface.

The Garnet mine, about 2½ miles northwest of Brice, is credited with a production of $15,000 from ore found near the contact of porphyry with the limestone, and similar to that in the other mines.

The placer ground that has been worked lies on the southeastern slope of the Jarilla Hills, east of the Nannie Baird mine. Although the outcrop of the Nannie Baird deposit is on a slope drained in the opposite direction, some have believed that that deposit was the source of the placer gold. The writer was informed, however, that the gold of the placers is different in appearance and character from the free gold found in the Nannie Baird mine. The absence of a water supply has made necessary the operation of some form of dry washer. The productive ground is reported to average about $1 a cubic yard, and a total of about $8,000 is said to have been produced. It is reported that a nugget weighing over 6 ounces was found in 1904.

TULAROSA DISTRICT.

The Tularosa district is situated near the southwestern base of the Sierra Blanca in northern Otero County, about 13 miles east-northeast of Tularosa on the El Paso and Southwestern System and about 5 miles west of the Mescalero Indian Reservation. The only development of importance is the Virginia mine of the Tularosa Mining and Milling Company at the little camp of Bent. Development began in 1904, so that the production up to the time of visit in 1905 had been small; a greater production has been made since then. In 1909 the company now known as the Tularosa Copper Company was preparing to erect a 100-ton concentration mill. It is stated in the press that a body of low-grade ore 45 feet thick and averaging 2.6 per cent of copper had been developed.

In going up Tularosa Canyon one passes in descending order over tilted red grits and conglomerates. Near the Virginia mine these red beds become coarsely conglomeratic, containing bowlders of granite, syenite (?), and red, purple, bluish, and nearly white vitreous quartzite. Underlying these rocks is a fine-grained sugary sandstone of light color. Limestone, probably of Carboniferous age, is exposed on neighboring hills. A mass of dark, dense diorite porphyry is exposed in a gulch just southeast of the shaft, and arches up the overlying sandstone and red beds. Though the evidence on this point is perhaps not wholly conclusive, the rock is believed to be intrusive into the sediments. In the underground workings of the Virginia mine is encountered a body of unknown outline consisting of greenish-gray, finely crystalline calcite that probably represents a detached fragment from some limestone stratum through which the diorite porphyry was intruded. At the mine the strike of the sandstone is a little west of north and the dip about 25° W. Farther north the strike changes to northeast, with the dip about 25° NW. The ores are found in the sandstone and in the diorite porphyry and limestone. The sandstone is of ordinary character; the grains, which are almost wholly of quartz, are well rounded and fairly well assorted. The cementing material is calcite, with some dolomite, and is plentiful. The diorite porphyry consists of plagioclase feldspar, with green and brown hornblende and numerous grains of magnetite. The texture of the rock approaches the ophitic. Alteration near the ores makes the rock brownish gray in color and results in the formation of calcite, dolomite, and sericite from the feldspar and of a chloritic material from the hornblende. The only apparent effect produced on the limestone by the diorite porphyry is a slight marmorization.

---

[a] Cf. Bull. U. S. Geol. Survey No. 293, 1906, pp. 88, 92.

In the sandstone is found ore similar in general features to that of the Estey, Tecolote, Nacimiento, Zuni, and other districts. Malachite and azurite, with kernels of chalcocite in the larger masses, are disseminated through certain portions of beds and along minor cross fractures. There is no doubt that chalcocite was originally the chief mineral; but scattered grains of either pyrite or chalcopyrite are found, now pretty much oxidized. For the most part the copper has replaced only the carbonate cement between the quartz grains; replacement of the quartz itself, as in the Tecolote distirct, can not be proved with certainty, though there are places where this process appears to have gone on slightly. From the general character of the ore there is every reason to believe that the method of ore formation in this district was similar to that in the Estey, Tecolote, and other districts where the disseminated ores in sandstone have been found. Ore of this kind has been but slightly developed in the Tularosa district, and how extensive its distribution is through the sandstone formation is not known. The known occurrences are not far distant from the outcrop of diorite porphyry.

The ores in the limestone and the diorite porphyry have been developed only at the Virginia mine. These rocks are cut by veinlets and stringers of quartz, dolomite, and barite that carry more or less chalcocite, with a very little pyrite and chalcopyrite. Many of the veinlets are drusy and in them some of the minerals are well crystallized. The quartz develops in perfectly clear short hexagonal prisms terminated at both ends by pyramids; some of these crystals are more than half an inch long. Dolomite, which is the most plentiful gangue mineral, occurs mostly in small rhombohedrons that are almost cubes; in places the much curved crystal faces characteristic of the mineral are seen. Dolomite also occurs in the wholly filled veins in granular aggregates, showing a rude banding parallel to the walls. Barite does not assume regular external form, but is developed in slightly radiating plates, some of them an inch or more across; in places the mineral is pinkish. Chalcocite is present, ranging from minute grains sprinkled very sparingly through the gangue minerals to solid veinlets several inches wide, accompanied by insignificant quantities of gangue. In the druses the small grains observed appear to have a certain regularity of outline, but because of their small size definite crystal form can not be determined. A little chalcocite is disseminated as small grains in the diorite porphyry between near-by stringers. One small fragment of chalcopyrite was found on the dump partly inclosed in somewhat altered diorite porphyry, and rarely a tiny grain is seen in the glance. Pyrite is a little more common, but is likewise sparingly developed. It is present in small irregular grains and patches, mostly near the exterior of bunches of chalcocite. Here and there a small grain or crystal can be found in the hard diorite porphyry at some distance from the ore body. There is no evidence whatever that the chalcocite replaced corresponding masses of pyrite; the dolomite is evidently slightly ferruginous, and where exposed to oxidizing solution is covered with a brownish film; a little limonite stain is also formed from oxidizing waters that have coated the glance with malachite; and in the altered diorite porphyry adjoining the ore some limonite has been formed from the hornblende and possibly from the magnetite. Aside from these no other iron minerals appear to have been present.

There is another material present in the veins in addition to the minerals described. It is black, of pitchy or resinous luster and conchoidal fracture. Where surrounded by other minerals it forms irregular small grains and bunches, but in the druses the projecting portion is smooth, lustrous, and round, and when it is broken a concentric arrangement of layers can be seen. When held some distance above the flame of an alcohol lamp, the material quickly softens and then takes fire, giving a yellowish flame and a grayish smoke that smells like that from a kerosene lamp and leaving a residue of coke. The material is undoubtedly a hydrocarbon. It is found in the veins both in the limestone and in the diorite porphyry. In the limestone, at least, it occurs also in the wall rock for a short distance away from the veins.

All these minerals, including the hydrocarbon, are intergrown and were practically of contemporaneous formation. Chalcocite forms borders along the walls of many veins containing banded dolomite in the middle, and thus appears to be first in order of deposition, but it also occurs in the druses as one of the last-formed minerals. Chalcocite and the hydrocarbon occur both within and on the exterior of perfect crystals of the gangue minerals. In the center of one plate of barite grains of chalcocite and a bleb of the hydrocarbon were found, and in the dolomite these two black materials are commonly associated. The two seem to have been deposited at practically the same time. Some of the hydrocarbon has been replaced by chalcocite, however. Examination in thin section revealed these two materials apparently intergrown, but scrutiny of the minute rounded blebs of hydrocarbon in the druses showed that some of them, though of characteristic form, lacked the normal shiny surface and were instead a dull black or were covered with an exceedingly thin film of malachite. When these were broken open, they were found to consist in large part of chalcocite, and the pseudomorphous replacement of the hydrocarbon by the copper sulphide could not be questioned; even the original concentric structure was in some of them found preserved in solid glance.

Tiny crystals of calcite, somewhat stained with limonite, are present in small numbers in the druses deposited on the surface of crystals of the other minerals and are doubtless due to comparatively recent downward circulation of waters from the overlying rocks. These waters were oxidizing in character and have formed thin coatings of malachite on the glance that was exposed to them, and in one or two instances observed have caused the production of a little native copper and cuprite in druses.

Veinlets of dolomite carrying chalcocite outcrop in the diorite porphyry. The principal outcrop, that which became the scene of development work, is in the sandstone just where that formation feathers out. A mass of ore a few feet thick occurring here is similar in character to that already described as occurring in the sandstone, but is much richer. It is pretty thoroughly oxidized and somewhat iron stained. Directly underneath this in the diorite porphyry are the veinlets that have just been described. The workings are not such as to disclose exactly what happens to the veinlets when they reach the sandstone, but they were not found traversing that rock, although they are undoubtedly younger than it.

The copper deposits of these two kinds—disseminated deposits in sandstone and veins in diorite porphyry and limestone—are believed by the writer to be closely related in origin and through this relation to throw important light on the genesis of the puzzling type of copper deposit in sedimentary rocks. Many facts of the mineralogy and paragenesis suggest strongly that they were formed by rising, probably heated solutions that carried in some form the essential elements of the minerals now seen; in other words, that the solutions were mainly alkaline carbonates that held sulphides of the copper and iron. There is absence of evidence that the copper was brought from above in oxidized form in acid solution. It is believed that the solutions which filled the veins continued upward through the overlying sandstone, spread through its pores to considerable distances, and, dissipating their heat and pressure and perhaps mingling with foreign solutions held or circulating in the porous rock, precipitated part of their load mainly as chalcocite. The marked difference in conditions of deposition in the constricted fracture channels, which the copper-bearing solutions doubtless monopolized, and in the expanse of sandstone pores, which may have been occupied either by active reagents or by neutral diluents, might readily account for the fact that chalcocite is the only mineral common in important amount to both kinds of deposits. The uncommon richness of the sandstone ore directly above the richest veins known would seem to support the idea that the copper came from the veins into the sandstone.

In view of the relation observed in many places between carbonaceous material and copper deposition in sediments, the presence of a hydrocarbon in the veins is interesting. Careful examination of the sandstone specimens collected has not revealed its presence. No carbonaceous material of any kind was observed in the sandstone or the limestone, nor in fact in any of the rocks of the region. The metal-bearing solutions may have traversed at some lower depth some sedimentary rock which yielded the carbonaceous material, but this suggestion is wholly hypothetical. Whatever the source of the carbonaceous material, it was undoubtedly carried in the same solution from which the copper was deposited, and as the hydrocarbon was deposited, some of it, probably because of its reducing power, was attacked by the cupriferous solution and replaced by chalcocite. But just as some of the hydrocarbon was not thus attacked, so some of the chalcocite was not thus precipitated; in fact it appears probable that most of the glance was not precipitated by the hydrocarbon observed in the veins. Yet it may be that the precipitation of the chalcocite was the result of reduction by gaseous hydrocarbons from which the pitchy material observed is but the more inert or less easily oxidizable residue.

In this connection it may be noted that the most commonly observed and most important occurrences of chalcocite are the result of reducing action as a rule by pyrite, though in such places the reducing material is not carried in the cupriferous solutions. There may be significance also in the fact that most cupriferous veins contain no chalcocite as a primary constituent. May this possibly be due to the absence in most places of strongly reducing conditions? Finally, reference may be made to the discussion of similar deposits in the Tecolote district on pages 116–123.

From the observations above recorded it may be suggested that cupriferous solutions of alkaline character may react with carbonaceous material, with resulting deposition of the copper in the reduced or cuprous form as chalcocite.

The ultimate source of the copper of the Tularosa deposit is not known. A natural assumption would be that it was derived from the diorite porphyry. The deposit has certain features, however, suggesting relation to the Tertiary lavas, being similar in some respects, for example, to the deposit at the Cooney mine, in the Mogollon district. Tertiary volcanic rocks are known at various places in the Sierra Blanca; whether they occur in the vicinity of Bent is not known.

The Virginia mine was developed in 1905 by two 70-foot shafts, with levels at 35 and 70 feet. A stope of considerable size had been made at the 35-foot level, the ore consisting of diorite porphyry shot through with stringers containing glance. A important part of this ore was of shipping grade, said to run from 25 to nearly 45 per cent of copper, with 3 to 4 ounces in silver and 0.01 to 0.02 ounce in gold to the ton. It was the intention to make an open cut down at least to the 35-foot level, dropping the ore in chutes to the 70-foot level and hoisting it to the surface. On the 70-foot level ore similar to that above had been developed. A fairly well-defined flat vein was found in limestone in a crosscut from this level.

A concentration mill was practically completed at the time of visit. It was to be run by water power, for which purpose a 10-inch pipe line with 180 feet fall had been installed. It was stated that the milling ore was expected to carry 8 to 12 per cent of copper. A carload of screenings from the general mill-ore dump is said to have averaged 11 per cent.

SOCORRO COUNTY (EASTERN AND WESTERN PARTS).

By L. C. Graton.

GENERAL FEATURES.

Socorro, the largest county in the Territory, is situated in the west-central part and occupies an area of about 15,000 square miles. It is bounded on the north by Valencia County, on the east by Torrance, Lincoln, and Otero counties, on the south by Otero, Dona Ana, Sierra, and Grant counties, and on the west by Arizona. This large area exhibits a great variety of physiographic and geologic features; it contains a considerable number of mining districts with deposits of widely varying types.

The Rio Grande flows through the eastern part of the county from north to south. The region east of the river contains the narrow San Andreas Range, continued to the north by the Oscura Range. West of the river rise the Lemitar, Magdalena, and San Mateo ranges, which form a part of the central system of northward-trending ridges that follow the river through the central and southern parts of the Territory.

The western part of the county belongs to the plateau province, but the sedimentary rocks are mostly covered by a great thickness of Tertiary lava flows. The great basin of the Plains of San Augustin occupies the central part of the county and is bordered on the north and west by an arc of ranges built up of horizontal volcanic flows.[a] From east to southwest these are

---

[a] Herrick, C. L., Report of a geological reconnaissance in western Socorro and Valencia counties, N. Mex.: Am. Geologist, vol. 25, 1900, pp. 331–346. It is here stated that the Datil and Gallina mountains are composed of trachyte and rhyolite.

A. VIEW FROM POINT NEAR SUMMIT OF DIVIDE BETWEEN COONEY CANYON AND MOGOLLON, SOCORRO COUNTY.

Outcrop of Queen vein shows on two ridges.

B. LAST CHANCE SILVER-GOLD MINE, MOGOLLON, SOCORRO COUNTY.

the Gallina, Datil, Tularosa, San Francisco, and Mogollon ranges. Many of them reach elevations of 10,000 feet.

The mineral resources of the county are abundant and varied, but most of the mining districts are situated in the central part about Socorro. These are described under the heading "Central Socorro and Sierra counties" (pp. 241-260). Among the nonmetallic substances coal stands first. It is mined at Carthage, about 10 miles east of San Antonio, on the Rio Grande, and the beds are contained in strata thought to be of latest Cretaceous age. The production is not given separately in Mineral Resources of the United States. It is, however, small compared with the output of Colfax and McKinley counties.

There are salt lakes in the western part of the county. Guano, said to amount to 3,000 tons, has been shipped from bat caves near Lava station.

There are nine metal-mining districts in the county besides many places where indications of deposits have been found. The Mogollon district of silver-gold-copper veins in Tertiary lavas is situated in the southwest corner, not far from the Arizona line. No metal-mining districts are known in the extensive northwestern part of the county. Near Socorro, in the central part, are the Magdalena, Mill Canyon, Water Canyon, and Socorro districts. The Magdalena district is the most important, and has a large production of lead, zinc, and silver to its credit. The deposits form replacements in limestones. South of Magdalena, in the San Mateo Mountains, is the Rosedale district of gold-bearing veins in rhyolite. Along the eastern margin of the county are the Estey district of copper-bearing sandstones and the Jones district of contact-metamorphic iron ores. Lead deposits and copper prospects occur in the Canyoncito district, north of Socorro, near San Acacia, and indications of metals are found in the Lemitar and Ladrones ranges.

MOGOLLON DISTRICT.

LOCATION AND HISTORY.

The Mogollon Mountains (Pls. IX, A; X, A), constituting the most important group of southwestern New Mexico, are situated near the western boundary of the Territory, in the southwestern part of Socorro County. At their western base, close to the Arizona line, flows San Francisco River, and at some distance away, on the east and south, is Gila River. From the vicinity of Pinos Altos and Santa Rita northwestward the country is a much-dissected and partly gravel-covered plateau made up of volcanic rocks. Rising high above the level of this plateau, which has an elevation of 5,000 to 7,000 feet, the Mogollon Mountains attain a maximum altitude of about 10,500 feet. Although appearing at a distance as a moundlike mass, the mountains are found, on nearer approach, to have exceedingly rugged relief. Numberless sharply incised valleys of steep gradient on the high slopes unite and form, lower down, deep, steep-walled canyons, separated by sharp, craggy ridges. Although the topography is extremely youthful, the original constructional features have been much more effaced than in the volcanic mass of the Valles Mountains to the northeast. Owing to the roughness of the country and consequent difficulty of access, these mountains were for years the stronghold of the Apaches. Well up on the southwestern flank of the mountains are situated the camps of Mogollon and Cooney. They lie on opposite sides of a sharp ridge, Mogollon in the canyon of Silver Creek and Cooney, farther north, in Mineral Creek canyon. Transportation is effected by wagon to Silver City, which lies 85 miles to the southeast, at the terminus of a branch of the Atchison, Topeka and Santa Fe Railway. Water is scarce during part of the year and little pumping is required in any of the mines.

The first knowledge of the presence of ore deposits in the Mogollon Mountains is said to have been gained in 1870.[a] In 1875 James C. Cooney, a sergeant stationed at Fort Bayard, discovered rich copper-silver ore near the place that now bears his name. With associates from the mining camps of Central and Georgetown, in Grant County, Cooney began the location and development of claims. Good ore was encountered in several places, but the hostilities

---

[a] Jones, F. A., op. cit., p. 129.

of the Apaches hindered the development of the region for several years. During the late eighties and throughout much of the nineties mining was carried on actively. Up to the time of visit in 1905 the total production of the district was estimated at about $5,000,000 in silver, copper, and gold. At that time most of the mines were closed; only one, the Last Chance, was producing or being actively developed. In 1909 encouraging developments were reported from the district. The Last Chance mine of the Ernestine Mining Company was said to be developed to a depth of 1,200 feet and to be producing steadily. The Socorro Mines Company was producing from the Little Fanny and Champion mines. The depth attained was here 700 feet. The vein is reported to be 30 feet in width. A 30-stamp mill with cyanide plant is attached to the property.

GENERAL GEOLOGY.

The geology of the Mogollon Mountains is little known. The portion seen by the writer, embracing only the south end of the range, is composed wholly of effusive or flow rocks with an occasional dike or sill but is otherwise lacking in intrusive rocks. It has been stated[a] that the Mogollon Range is a mass of granite, but it is possible that the reddish color of the flow rocks, when seen from a distance, may have been the reason for this conclusion.

The rock most abundant in the south end of the range, at least, is commonly reddish and of distinctly rhyolitic aspect, though free quartz is lacking. It contains about equal amounts of orthoclase and of plagioclase that ranges from andesine to labradorite. Phenocrysts of these feldspars and of biotite, together with grains of magnetite, are held in a groundmass which in some places consists mainly of very minute laths of feldspar and probably interstitial grains of quartz; in others it is an indeterminable mass with spherulitic texture; and in still others it is practically an amorphous glass. Minute crystals of apatite and zircon are included in some of the feldspars.

Some idea of the general composition of the rock may be gained from the following partial analysis of a specimen from the canyon wall directly above Cooney, made by W. T. Schaller, of the United States Geological Survey. This analysis shows the rock to be a rhyolite rather richer in soda than the normal, and it is consequently designated soda rhyolite. A very similar rock occurs farther west in the Clifton district, Arizona.[b]

Partial analysis of soda rhyolite from the Mogollon district.

SiO_2	67.83
CaO	2.10
K_2O	5.46
Na_2O	3.30

Flow structure is well exhibited in many specimens by the arrangement of the feldspar laths; and vesicles, commonly filled with calcite and some chlorite, emphasize the surficial nature of the rock. The spherulitic texture of the rock at certain places has been well brought out by weathering. Practically none of the rock is entirely fresh. The feldspars are altered partly to kaolin and partly to calcite, and from the biotite chlorite has developed plentifully. Epidote has been produced sparingly, and the groundmass has been partly converted into a mosaic of quartz grains. The magnetite has been oxidized and the resulting product, spreading away from the original grains as a stain, is the cause of the red color of the rock. This alteration, which is essentially propylitic in type, is prevalent in a moderately advanced stage, and is especially marked in the vicinity of the veins; in such places pyrite is commonly present in addition to the other minerals.

Andesite or latite is less abundantly exposed in the Mogollon-Cooney region than soda rhyolite, but is to be seen at numerous places. It differs in appearance from the soda rhyolite in being darker, usually chocolate brown or brownish green, and showing more prominently against the darker groundmass the small phenocrysts of feldspar. Amygdaloidal structure is also more common, and in some places a considerable portion of the rock consists of blebs of calcite and chlorite. Plagioclase feldspar is the principal constituent. It is of the composition

---

[a] Keyes, C. R. [b] Lindgren, Waldemar, Prof. Paper U. S. Geol. Survey No. 43, 1905, p. 87.

of either andesine or, less commonly, oligoclase. Phenocrysts of pyroxene were present, but are now much decomposed, the usual product being quartz, chlorite, and iron ore (magnetite?). Some of this material may originally have been hornblende or even olivine. The groundmass consists principally of minute plagioclase laths with profuse grains of magnetite, now much altered to limonite. Glass is much less common than in the soda rhyolite. Alteration, while of the same character as that which has affected the soda rhyolite, has had a much more pronounced effect on the andesite. In many places the rock has been greatly bleached to a greenish gray color; quartz has formed plentifully in the groundmass and has replaced the calcite of many of the amygdules. Calcite is abundantly developed in places, however. Where the most alteration has taken place pyrite, chalcopyrite, and bornite are present in scattered grains and in tiny veinlets. The general composition of the rock is expressed by the following partial analysis of a specimen from the Cooney mine, made by W. T. Schaller:

Partial analysis of latite from the Mogollon district.

SiO_2	48.00
CaO	7.72
K_2O	3.28
Na_2O	1.95

It is difficult to account for the amount of potash shown by the analysis, as by far the greater part of the feldspar appears certainly to be plagioclase. The analysis was repeated and found correct. Probably most of the so-called andesites of this region are in reality latite, and the potassium is probably contained in the groundmass.

Basalt is not plentiful in the vicinity of the mines. It occurs mainly as thin sheets and perhaps in some places as intrusive sills between the other rocks. The rock is dense, fine-grained, and very dark gray. In texture it is finely porphyritic, with scattered phenocrysts of pyroxene and possibly olivine also. Plagioclase laths, pyroxene, and iron ore form a granular groundmass. The latest mineral to crystallize was an unstriated feldspar, probably plagioclase but possibly orthoclase; it contains many apatite needles. The rock has undergone but little alteration, probably owing to its comparative impermeability.

Alternating with quiescent flows of molten rock that spread away from the vents and produced the massive beds of these rocks were explosive outbursts of shattered and comminuted rock that fell back to the surface as breccia and tuff. A large part of the soda rhyolite is thus brecciated and occurs in fragments that range in size from those several centimeters in diameter down to the finest dust. Owing, however, to the uniform composition and character of the component particles, the fragmental nature of the rock is not obvious in most places and the tuff is easily mistaken for the corresponding massive rock. The andesite, as would be expected from its more basic character, was extruded more gently than the soda rhyolite and only here and there is it brecciated. The product is much coarser than most of the tuffaceous soda rhyolite and the fragmental character is, as a rule, readily discernible. It contains some pieces of soda rhyolite. The basalt, the most basic rock developed, appears to have remained sufficiently fluid during its eruption to flow out gently, and in consequence it is not brecciated.

All these rocks occur as a series of beds which in most places observed have not been greatly disturbed from the nearly or quite horizontal position which they initially assumed. The bedded character is generally best exhibited on the sides of the steep canyons, though even in such places it is not everywhere apparent, and on the flatter slopes and the craggy prominences the structure is not uncommonly difficult of detection. The thickness of the flows and tuffs undoubtedly aggregates some thousands of feet, but the individual sheets, to judge from the structure now exhibited by differential erosion and weathering, were comparatively thin. Except in the case of the basalt, the tuffaceous beds are more prominently exposed than the massive flows. The tuffs are in places "cross-bedded" at low angles, after the manner of shallow-water deposits. As shown especially by the uneven thickness of dark bands of basalt, there were brief erosion intervals during the accumulation of the series. The result is indicated by figure 12. Andesite is found most commonly on the lower slopes and is probably older than the soda rhyolite.

Although folding and tilting appear to have affected the rocks of the Mogollon-Cooney region but little, the results of faulting can be observed at many places, and it is probable that careful study of the district would show that faulting is very common. As will be shown, there may be a close relation between the veins and the faults. The throw of the faults, where determinable, is small, but unless some distinctive bed is cut by the fault plane the amount and direction of throw can not generally be determined. Figure 12 is a sketch of the northwest wall of Cooney or Mineral Creek canyon, where faulting is well brought out by the interruption of the basalt beds. There is no question that the three layers of basalt shown in each fault block are parts of three originally continuous beds. The amount of throw is thus well shown. Possible faults of greater magnitude will be mentioned in the description of the veins.

The rocks of the Mogollon Mountains are no doubt related in time and in origin to the similar rocks found in so many other places in New Mexico and elsewhere that were extruded during the Tertiary, probably early in that period.

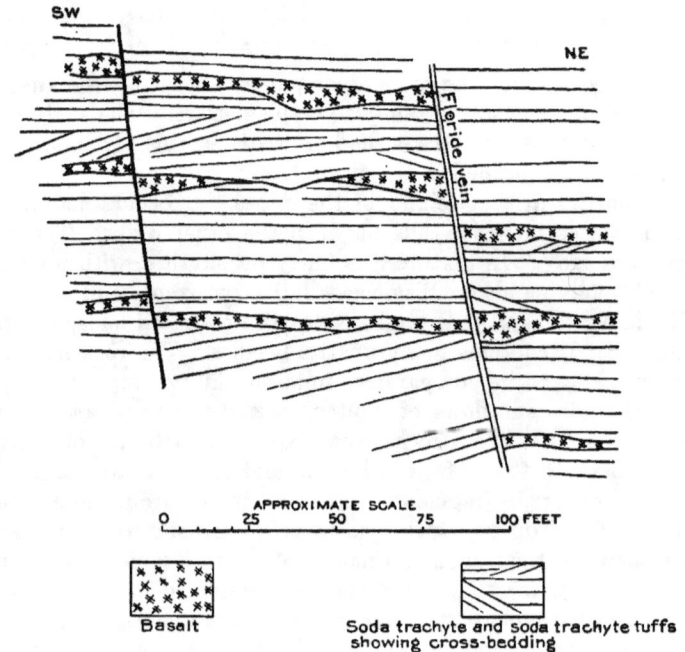

FIGURE 12.—Sketch of vertical section showing on northwest wall of Cooney Canyon just below Cooney, Mogollon district.

ORE DEPOSITS.

GENERAL STATEMENT.

The ore deposits of the Mogollon district consist of veins that cut these volcanic rocks. The veins are broadly similar to those found in so many other regions of Tertiary effusive rocks. Quartz is the predominant gangue mineral and calcite is also important. In places considerable fluorite is present. Adularia is widespread in small amount. Silver is the chief valuable metal in most veins and occurs in its original form probably as argentite. Gold is also present in important quantity in several mines. Bornite and chalcocite are present in only small quantities if at all in most of the veins, but in one or two veins these minerals are important and in these copper makes up a large part of the total value.

STRUCTURE AND ROCK ALTERATION.

The veins proper are for the most part fillings of cavities in the rocks, but replacement to a greater or less degree has gone on in the adjoining wall rock, and in places this altered rock carries values. The openings which the vein matter filled were irregular, but were more persistent and regular than those in many regions of effusive rocks. The veins are therefore distinctly tabular, of considerable extent and of fairly constant strike and dip. The width, however, is variable and may range in the same vein from a few inches to 20 feet or more. The cause of this irregularity of width lies apparently in the fact that the fractures along which the veins have formed, instead of being definite, individual cracks throughout their extent, are in many places very complex; the rock along certain tabular zones has been shattered and crushed, and in the numerous crevices so formed the vein materials have been deposited. It is commonly, though not always, in such places that the vein is widest; it includes innumerable little "horses," fragments of the shattered country rock that have been surrounded by the vein filling. At such places, where the ratio of exposed surface to mass is comparatively high, replacement of the country rock by the vein solutions has of course been particularly favored. A common product of this silicification is a greenish-gray or a jasper-colored hornstone. In some localities the replacement has been almost complete, the remnants of the rock fragments being so nearly converted into quartz as to be barely discernible, and in a few places the process has apparently gone to completion and no trace of the former presence of fragments of country rock remains except perhaps a suggestive arrangement of the deposited sulphides or the evidence of intense replacement at the sides of the vein.

The alteration of the country rock by the vein-forming solutions has already been outlined, for the complete replacement by quartz with some carbonates and sulphides, as described above, represents the final stage of alteration, and the development of carbonates, chlorite, quartz, epidote, and sparingly of sulphides, mentioned in the description of the rocks, represents the initial effect of the mineral-bearing solutions. Between these extremes are phases in which the rock is traversed by narrow streaks of quartz that has replaced the rock along incipient fractures. In these veinlets are found calcite and pyrite, locally also chalcopyrite and bornite. The gradual increase in number and width of these veinlets finally results in complete replacement of the rock.

The widespread presence of this alteration by the vein solutions and the analogy with other districts in which the source of these solutions can be traced back to the magma which had previously yielded the rocks that are thus altered suggest that here also the veins are genetically related to the parent magma from which the soda rhyolite, andesite, and basalt were derived. In this connection it is interesting to note that the chemical processes of weathering appear to have had little effect in the Mogollon district in changing the composition of the rocks. The strictly mechanical processes, however, have left their mark on the region and are mainly responsible for the intensely rugged topography. The highly silicified portions of the rock and the quartz-filled veins have of course been much less easily abraded than the type of altered rock containing chlorite, calcite, epidote, and less quartz. Being, as a rule, related to planes of fracture or permeability, these siliceous masses form bold, craggy lines of outcrop that receive the familiar names of "ledge" or "blow-out" from the miners.

The fractures along which the veins have formed appear to belong to two systems. The accompanying sketch, figure 13, shows the approximate position, location, and relations of the principal veins of the district as indicated by the position of claims on these veins. The King and Queen veins are grouped together as north-south veins; the other important veins are always spoken of as having a northwest strike, though in reality two or three of them trend more nearly east and west. There is a suggestion of radial arrangement of these northwest veins about a center some distance east of the Queen vein. Some fissures occur east of the Queen vein, but everything of value has been found on the west. The Queen vein (Pl. IX, A) itself carries values, but they are too low to permit working; better results have been obtained from the King vein, but the production of the district has come almost entirely from the northwest veins between the King and Queen. The King and Queen veins resemble the others of the district, except that on the whole they are much poorer, wider, and more persistent and are marked by more prominent outcrops. The King and Queen veins dip steeply to the east; the northwest veins dip about 70° to 75°, most of them to the northeast; the Little Fanny is one of the few that dips to the southwest.

One vein, the Floride, certainly occupies a fault fissure (see fig. 12), and many and perhaps all the vein-filled fractures may be of the same character. As has already been pointed out, positive evidence of faulting in these rocks is not easily gained except in favorable places. The prevalence of crushing and shattering along the fissures is indicative of important movement, and the manner in which so many veins seem to end against the King and Queen veins is also suggestive. The actual junction at such places was not seen by the writer. The cutting off of the Little Fanny vein by the Queen vein is said to be shown in the Champion workings, but these were not accessible at the time of visit.

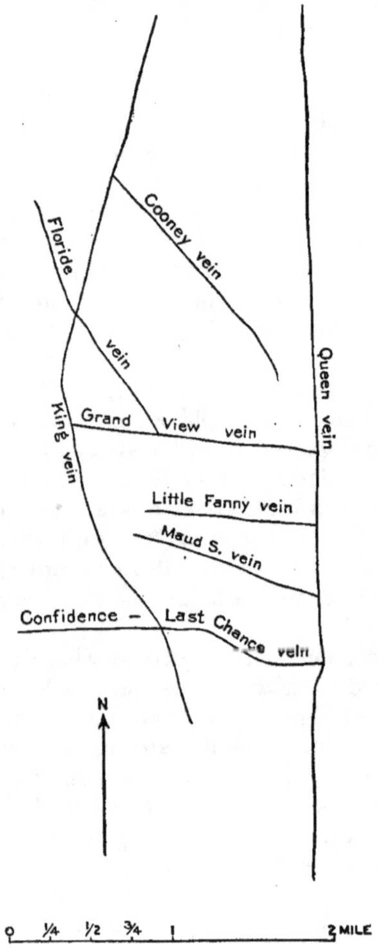

It is probable that the fault fractures were formed just before the vein solutions occupied them, and it may be that the King and Queen faults occurred after many of the northwest veins had been formed, and, though before vein formation had ceased, were so late that the solutions then carried very little of the valuable metals.

In internal structure the veins vary considerably. Probably the most common structure is a somewhat granular quartz with whatever calcite is present arranged in clusters of very thin hexagonal plates. This form of the calcite is best brought out by the cavities resulting from the solution of the mineral. The "hackly" quartz thus formed is common at and near the surface of a number of veins in the district. Irregular grains of a mineral corresponding to adularia (potash orthoclase) are scattered sparingly with the quartz. The sulphides are commonly not plentiful and are scattered irregularly through the gangue minerals in strings and bunches of small grains. A position where the deposition of sulphides seems more than ordinarily favored is at the boundaries of included wall-rock fragments that have been more or less replaced by quartz. In certain places where these fragments have thus been almost or quite replaced, their presence is revealed by the dark circling zone of minute sulphide grains. Such specimens resemble the typical ore of the Cochiti and Red River districts. Probably less commonly than in the arrangement that produces the "hackly" quartz, the calcite is present in narrow and interrupted bands parallel to the plane of the vein. In such places the sulphides are likely also to be arranged in streaks so that the vein has a distinctly banded or ribbon structure, but this is not persistent and may give way within a few feet to a systemless arrangement of the various minerals.

FIGURE 13.—Sketch showing approximate location of principal veins of Mogollon district.

Druses are rather common; they are generally small and as a rule are flat, but some are irregular in shape. They are most plentiful where a vein has banded structure, or where the lode consists of a network of veinlets inclosing fragments of wall rock, or between the flat plates of "hackly" quartz (distinct from the cavities formed by solution of the calcite tables). These druses are commonly lined with small quartz crystals; in some places thin hexagonal plates of calcite similar to that intergrown with the denser quartz are intimately associated and intergrown with these quartz crystals, and here and there groups of fluorite octahedrons are intergrown with the other two minerals. Quartz is not present in a few of the druses, which are lined with crystals either of calcite or, less commonly, of fluorite. In several places the three minerals are present in successive layers; the order of deposition seems invariably to have been first quartz, then calcite, and finally fluorite, although there was some overlap in the deposition of each mineral and consequently some intergrowth. In general the fluorite is confined to the druses, indicating that it was deposited late in the period of vein formation, but locally it occurs in the midst of solid vein matter, and in one specimen it was found replacing the wall rock an inch or more from the vein.

MINERALOGY.

On the basis of mineralogy the veins may be divided into two fairly distinct classes, though they all present so many features in common that there is no doubt that they had a single source. The group to which most of the veins belong may be designated as the silver veins. They carry sulphides in small amounts, as a rule, and in grains grouped in bunches and streaks. These clusters of finely divided sulphides are commonly of dark color. Because of the small size of the individual grains, positive identification of all the minerals present is not possible. Pyrite, chalcopyrite, and bornite can be distinguished. Among the black minerals are believed to be argentite, chalcocite, and possibly tetrahedrite and zinc blende. It is said that ruby silver, pyrargyrite, is sometimes found. Pyrite, chalcopyrite, and to some extent bornite, seem to occur preferably in the less-altered rock, whereas the darker minerals are apparently deposited chiefly in the very much silicified fragments of wall rock. The places that were originally open spaces do not appear to have been so favorable for precipitation. All the sulphides are contemporaneous with the gangue minerals, being inclosed within individuals of the quartz, calcite, or fluorite, probably in the adularia also, and occurring in semicrystalline grains on the exterior of well-formed crystals of the three first-named minerals. Native gold and silver are present in places. The elemental state of the silver appears certainly to be an original condition and not the result of oxidation or redeposition; the silver is found intergrown with unaltered sulphides below the zone where oxidation can be detected. It is present in the characteristic curved, crystal-like forms and in the wires and bundles of fibers so commonly seen. The free gold may be due to decomposition by oxidation of some combination with other substances, for the only specimens of free gold seen by the writer were oxidized.

In the veins of the other type copper is predominant. The same minerals as in the silver veins are probably present, but chalcopyrite and especially bornite and chalcocite, also fluorite, are much more abundant than in the silver veins; adularia appears more rare. Another point of difference is the larger proportion of sulphides to gangue. Bornite is probably the most abundant sulphide, with chalcocite next in importance, chalcopyrite less plentiful, and pyrite still less common. In places tetrahedrite is important and commonly is banded with chalcopyrite and bornite. In one specimen minute reddish specks on a scraped surface of tetrahedrite were regarded as probably included grains of pyrargyrite. Rosettes of specularite are found occasionally in veinlets of calcite and fluorite. An interesting occurrence that is seen in many specimens is intergrown bornite and glance completely surrounded by chalcopyrite. The gangue minerals are the same as in the silver veins, but calcite and especially fluorite are more important. The druses are larger and the crystals contained in them more perfect. In the largest druses much of the calcite is developed in an unusual habit—finely formed stout hexagonal prisms terminated at both ends by basal planes which differ from the bright prism faces by being dull like ground glass. These crystals reach several centimeters in diameter. The fluorite is commonly bright green and developed in small octahedrons, but in some individuals there is an outer layer of deep purple color; another manner of occurrence is in small, elongated grains grouped in radial aggregates whose outline approaches that of an octahedron. These are of pale green or faint pink color. Much of the quartz is slightly colored brownish red. The same order of deposition, quartz followed by calcite and then fluorite, is apparent, but the practical contemporaneity of the three is certain.

In one small cavity were found sharp little octahedrons of pyrite projecting from "dogtooth" crystals of calcite. In another flat druse imperfect sphenoids of chalcopyrite had grown upon and partly intergrown with octahedrons of fluorite. Bornite, chalcopyrite, and pyrite are sometimes found wholly inclosed in perfect prisms of quartz.

Oxidation has not greatly affected the Mogollon ore deposits, and only in exceptional places where fractures have permitted especially easy access have the veins been oxidized for any important distance from the surface. The general level of the ground water is below the depths of present operations, and the shallowness of oxidation must therefore be attributed to the geologic youth of the region and to rapid erosion. Limonite, malachite, cerargyrite or horn silver, and (probably) native gold, stated in order of decreasing abundance, are the chief oxidation products of the ore minerals. Malachite was never found in commercial quantities. Horn silver, to judge from some reports, was of considerable importance in the early days of mining in the district.

GENESIS.

The origin of the ore deposits can not be conclusively determined from the evidence obtained. As may have been inferred from the remarks concerning alteration, it is the writer's belief that the solutions which deposited the ores and caused the widespread alteration of the rocks were one of the final products of differentiation from the soda rhyolite, andesite, and basalt magma, and that these solutions appeared and the deposits were formed shortly after the consolidation and fracturing of the rocks. The evidence indicates that all the important minerals of the veins were practically contemporaneous. There is no reason whatever to ascribe the bornite or the chalcocite to secondary enrichment or to any result of oxidation. All the sulphide and gangue minerals are believed to be primary.

The abundance of druses and the shattered condition of the rocks confirm the conclusion to be drawn from the nature of the rocks themselves, that the portions of the veins now exposed were at no great depth below the surface at the time of their formation; or, in other words, that since the deposition of the veins the thickness of rock removed by erosion has not been great. On the whole, the mineralogy of these veins corresponds with these conditions of deposition at shallow depth. The quartz, calcite, adularia, pyrite, chalcopyrite, and argentite, and the nature of the rock alteration are characteristic of deposits comparatively near the surface. But the presence in the copper veins of bornite and specularite suggests that in spite of shallow depth the temperature may have been higher than is common in deposits of this class. If this view is correct, it may be assumed that the copper veins were formed nearer to the boundary between the solidified rock and that which was still molten than were the silver veins. A matter worthy of note, in view of the presence of potash in the vein solutions, as indicated by the adularia, is the absence of sericitic alteration of the wall rocks.

VALUE.

Ore of workable grade is commonly found in shoots in the vein, and over the rest of its extent the vein is of very low grade or even barren. The shoots may be large or small, but in the mines that have made important production they have been of considerable extent. No general rules as to location, pitch, etc., were observed. Intersecting veins or channels are generally not to be found at these places.

Little definite information could be gained as to the value of the ores worked. The writer was told that ore richer than $30 a ton was usually shipped to the smelters, while ore of lower grade was treated in local mills. It was reported that $8 ore was about the lowest that had been worked profitably. The average value of the ore produced during the last few years has been about $20 to $25 a ton, but this may not be fairly representative, as the output during that period has been derived chiefly from a single mine. It is stated that in the silver veins, on the average, silver constitutes about two-thirds of the value and gold about one-third. Ore is sometimes found rich in both gold and silver and of a value of several hundred dollars to the ton. The copper veins carry only from a few cents to a few dollars to the ton in gold, but important quantities of silver in places and copper up to 20 per cent. The relative proportion of the various metals is shown in the following published statement of returns from shipments in 1903-4 by the Mogollon Gold and Copper Company, the only operator of a copper vein in the district: Gold, $3,648; silver, at 55 cents, $42,926.95; copper, at 13 cents, $138,438.80.

The greater part of the ore that has been produced in the district has been milled, the high transportation charges precluding the shipment of any but high-grade ore. In the early days much of the silver was obtained by amalgamation, mostly in pans. Cyaniding is now chiefly employed. At the only mine in operation at the time of visit, the Last Chance, the ore was stamped in cyanide solution and then concentrated, the concentrates being shipped and the coarser tailings leached, while the slimes were treated by decantation. Loss was said to occur in the treatment of the slimes, and a filter process was about to be installed to remedy this defect. The copper ore of too low grade to be shipped was concentrated without other treatment.

MINING DEVELOPMENTS.

The Last Chance mine (Pl. IX, *B*) of the Ernestine Mining Company is situated on a branch of Silver Creek, a short distance southwest of Mogollon camp. Although actively developed only in recent years, the mine has been an important producer, with a record up to 1909 probably as large as any other in the camp, and for the last five years has been almost the only important producer of the district. In 1905 the workings consisted of a crosscut tunnel to the vein. From this level a 300-foot inclined shaft was sunk on the vein and continued upward to the surface, where the hoist is located. There were 11 levels, each about 200 to 300 feet long, the levels being run close together to permit rapid development during the sinking of the shaft, which was then in progress. The mill had 20 stamps and standard tables, Dorr classifiers, and cyanide vats. The daily capacity, 50 tons, has since been increased.

The mine is located on the Confidence-Last Chance vein, near the Queen vein. The vein, which strikes nearly east and west and dips about 71° N., is a quartz-calcite vein in a shattered zone in andesite and probably also in soda rhyolite. In the present workings, which are below the tunnel level, the vein averages about 7 feet in width. One stope in the old workings, near the surface and above the tunnel, is said to have been 15 to 20 feet wide.

The ore is typical of that of the silver veins of the camp. The values, concerning which little information was given for publication, are said to average about two-thirds in silver and one-third in gold. The concentrates, which represent about 1 per cent of the ore, are said to run $800 to $1,100 a ton; how much is obtained from the cyanide solutions can not be stated. In some places the vein is banded and in others large surfaces show the intergrowth of quartz and calcite that, on solution of the calcite, produces "hackly" quartz. Pyrite and a black sulphide carrying silver, probably argentite, are the principal metalliferous minerals.

Three ore shoots are known in the mine. One, but little developed, occurs where the crosscut tunnel strikes the vein. The main shoot is opened by the inclined shaft and is the same as the one worked in the old upper levels. This shoot is at least 200 feet long. The workings are said not to have gone out of the shoot, although the values diminish toward the ends of the drifts. Values in the bottom of the mine were said to continue good, and the shaft has since been considerably deepened. The third shoot was somewhat developed in the old workings west of the main shoot. All the shoots appear to pitch steeply in the vein, possibly a little to the east.

The Confidence and Black Bird mines of the Helen Mining Company are situated on the same vein as the Last Chance and lie, respectively, west and east of the King vein. They have been considerably developed and have the deepest workings in the district, 1,030 feet. A 30-stamp pan amalgamation and concentration mill located at Graham, at the foot of the mountains on the road to Silver City, was in 1905 being remodeled for the cyanide process. The production is variously reported at $1,000,000 to $1,300,000. The mine had not been in operation since 1902 and was not visited, but ore said to be typical of the mine resembled that from the Last Chance.

The Maud S. and Deep Down mines are situated on Silver Creek just below Mogollon camp. They both lie on the Maud S. vein, which strikes northwestward and dips about 70° NE. As shown in a stope reaching to the surface on the Deep Down claim, the vein is about 4 feet wide and has fairly regular walls. The ore is typical of the silver veins. The Deep Down mine was developed only near the surface, but was said to have produced over $75,000. The three claims of the Maud S. Company, adjoining the Deep Down on the northwest, were said to be developed to the 500-foot level and to have produced about $750,000. Some of the ore was of high grade.

The Little Fanny mine, comprising the Little Fanny and Confidence claims, is situated well up on the south side of the divide between the Mineral Creek and Silver Creek canyons. The Little Fanny vein strikes about N. 75° W. and dips about 68° SW. In places it is 15 to 20 feet wide, but averages about 8 feet. The ore is said to have carried more gold than most of the silver veins and to have averaged higher in grade, running in excess of $20 a ton, but in mineralogical character it is similar to the Last Chance ore. The workings are said to have extended to a depth of 480 feet, still in good ore, and an ore shoot, whose upper surface pitches flatly to the west, has been developed for a length of 800 feet, with stoping ground almost continuous. The output is reported as having been $1,250,000 up to 1895, when production ceased, owing to litigation, and it is said that as much more is in sight. Large ore bodies are said to be blocked out in the Champion claim close to the intersection with the Queen vein, but wholly on the Little Fanny vein.

The Grey Hawk mine, situated a little northwest of the Little Fanny mine and on a parallel vein, is credited with a production of $75,000.

The Cooney mine, sometimes called the Silver Bar, is owned by the Mogollon Gold and Copper Company. It is situated in the canyon of Mineral Creek just above Cooney camp. The claim was among the earliest locations, having been staked by the discoverer of the district. In the early days the precious metals only were sought, the copper not being recovered by the extraction processes then employed. In recent years copper has been the chief product of value. The total output up to 1905 was stated to have been over $1,000,000. At that time the mine was idle, but production has since been resumed. The workings consist of an adit at the level of Mineral Creek which follows the vein to connect at a depth of 100 feet with an inclined shaft on the vein. Below the adit level there were in 1905 six levels, the lowest being 600 feet below the surface at the shaft. The shaft has since been extended to a depth of 760 feet. The mill equipment consisted of rolls, Huntington mills, Wilfley tables, and Frue vanners, with a capacity of 100 tons daily.

The predominating rock is andesite, much of it brecciated; soda rhyolite is present also, though the two rocks can not be distinguished underground. Both rocks show the alteration common in the district. The mine is located on the Cooney vein, which strikes about northwest and has an average dip of about 72° NE. It is a quartz-calcite-fluorite vein and has practically been described in the statement regarding the copper veins. The proportions of copper, silver, and gold are given on page 199. The principal ore body extends downward from the surface a little southeast of the shaft. It has been stoped through most of its extent. On the first level, 115 feet below the adit, the stope was 100 feet long and 12 to 15 feet wide; and on the second, 50 feet lower, the stope was of similar dimensions. On the third level, 50 feet below the second, the stope was 60 feet long and 25 to 30 feet wide, but not all the ore was mined out. It was at this point, especially in a stringer 3 to 8 inches wide lying in the hanging wall a foot or two from the main vein, that were found the finely crystalline specimens described on page 197, and consisting of calcite, fluorite, and quartz with bornite, chalcocite, and chalcopyrite. This streak was very rich in copper. The stope on the third level was said to be at the junction of the Twig vein, which lies just northeast of the Cooney vein, but the writer was unable to observe any junction. A crosscut to the northeast revealed, about 35 feet from the Cooney vein, a narrow streak carrying calcite and bornite, but this appears to be simply a stringer running east and west. Little pay ore had been encountered in a long drift on the fourth level. The two lowest levels were not visited. It is reported that in the new workings at the bottom of the mine bornite and chalcocite have practically disappeared and pyrite, carrying some gold, is more abundant than it is above. Such a statement at once suggests secondary enrichment, but the writer believes that if this condition is general through the mine it is due to difference in original deposition resulting from unlike temperatures and pressures at different depths from the surface.

Among the properties which have been somewhat developed and are reported as showing good values are the Little Charlie, on the continuation of the Little Fanny vein; the Enterprise and the Great Western on the King vein, south of the Floride vein, which show decomposed pyrite carrying gold; the Floride, on the Floride vein just west of the King vein, which occupies a definite fault fracture and is reported as carrying very good values in silver and gold. The northwest end of the Cooney vein, near the King vein, appears to split up into several branches. Some of these branches carry principally copper values and others gold; they have been but little developed.

The advent of a railroad, which has long been discussed, would be of great value to the district and would undoubtedly make it an important factor in the total production of the Territory.

ESTEY DISTRICT.

The Estey district is situated on the lower eastern slope of the Sierra Oscura, in eastern Socorro County. It is about 15 miles slightly north of west from Oscura station on the El Paso and Southwestern System. The presence of copper in the region had been known for years, but it was not till about 1900 that efforts at mining were made. At that time the Estey Mining and Milling Company was formed to work the deposits. Elaborate surface improvements were made, including a large reduction plant. The mill consists of a crushing section with a capacity of 200 tons daily, a leaching section for treating 125 tons, and an electrolytic-precipitation section. There is also a water-jacketed smelting furnace. After heavy expenditures and practically no production by this company, it was reorganized into the Dividend Mining and Milling Company, the present holder. Very little ore has been milled and only a few carloads have been shipped. In 1905 no operations were being carried on by this company, but a little work was being done on other properties.

The Sierra Oscura is one of the several monoclinal blocks that constitute the central axis of the Territory, from the Sandia Mountains on the north to the Franklin Mountains on the south. The strata of the Oscura block dip eastward and the east slope of the mountains practically coincides with the dip of the bedding. The eastern or lower half of the east slope comprises three distinct ridges roughly parallel with the general axis of the mountains—that is, a little west of north. Estey lies at the eastern base of the highest or westernmost ridge. From this ridge westward to the crest of the main range the rocks are well exposed and, though the section was not traversed by the writer, appear to consist wholly of limestone, like the upper eastern slope of this west ridge. In any event, the crest of the Sierra Oscura is made up of stratified rocks; pre-Cambrian or intrusive rocks are reported to be exposed on the western or scarp side. Fossils collected from probably the upper portion of this limestone by H. W. Turner[a] were identified as upper Carboniferous. Overlying the limestone and outcropping in a band just west of the Estey town site are sandstones and arkose grits that pass upward into a series of red beds, at least 500 feet thick. The lower portion of the red beds consists of rather fine-grained and almost quartzitic sandstone; the upper portion of red or pinkish shale. The rocks are covered between the western and the middle ridges, but the steep west slope of the middle ridge again shows red beds, and a similar exposure appears in the corresponding west face of the easternmost ridge. Standing at Estey, the observer sees that these two red-faced ridges gradually merge a few miles to the north into a mesa whose steep south side is marked by a single band of red strata lying almost horizontal. There can be little doubt that the three bands of red beds shown by a cross section passing through Estey are a repetition by faulting of a single series of red beds. The western faces of the eastern and middle ridges are thus modified fault scarps. The westernmost ridge suggests by its form that it may be a similar uplift along a fault, but if such is the case this western fault scarp has been more eroded and either the throw was greater than in the two faults farther east or else the valleys between the first and second and

---

[a] Trans. Am. Inst. Min. Eng., vol. 33, 1903, p. 678.

the second and third faults are deeply filled with detritus. At any rate there has been, along the eastern foot of the Sierra Oscura, repetition, on a smaller scale, of the monoclinal structure exhibited by the main range.

The copper-bearing beds are identical, so far as lithologic correlation is concerned, with those at the Blake mine in the Tecolote district, namely, the arkose formation lying above the Carboniferous limestones and below the sandstones of the "Red Beds," probably equivalent to the Abo sandstone of Lee,[a] which is regarded as Upper Pennsylvanian. That this formation is actually equivalent, stratigraphically, with the one in San Miguel County is, however, uncertain. The arkose is mostly dark colored, with numerous grains of red orthoclase showing prominently. Grains of a nonmagnetic iron ore, probably ilmenite, are abundant. Between the larger grains of the sandstone are many smaller ones, and all are held together by a scanty cement which in most places appears to be quartz, but locally is calcite. Some of the material is so fine grained as to be almost a shale. In places a fine-grained limestone breccia is found, and layers of massive limestone are interbedded with the arkose. The principal ore mineral is malachite, present along seams and in the pores of the rock, in part replacing the calcite cement. A little azurite is also present. As in so many similar deposits, however, the copper carbonates are due to oxidation, and chalcocite, the cuprous sulphide, was originally the chief copper-bearing mineral. Bornite and chalcopyrite are also present sparingly, and associated with them is a little pyrite. Chalcocite is found especially in rounded nodules ranging up to over an inch in diameter in the limestone breccia, for example, on the Lucky Jack and Baby claims. Thin sections of these nodules show that they are not solid but spongy, evidently owing to the fact that the replacement by chalcocite was selective and not complete. A little bornite is intergrown with the chalcocite in these nodules. Another occurrence of chalcocite is associated with charcoal found in the arkose and probably representing remains of woody fiber. Where not copper bearing, this material is traversed by a honeycomb network of calcite veinlets; in the cupriferous portions both the charcoal and the calcite have been partly replaced by copper sulphide, now mostly converted into malachite. It is interesting to note that this malachite is almost invariably accompanied by a considerable amount of somewhat earthy hematite that has the appearance of having been formed by oxidation of an iron-bearing sulphide. Though the eastern and middle ridges are said to contain similar deposits beneath the "Red Beds," only the western ridge, that nearest Estey, was visited by the writer. He was obliged to go over most of the ground without local guide, and as the developments were meager, observations as to the distribution of the ore were not wholly satisfactory. In the vicinity of Estey the copper-bearing beds strike about N. 20° W., and the dip, though variable, probably averages about 25° E. According to Turner,[b] the rocks of the western ridge contain at least three cupriferous beds, but this relation was not made evident to the writer in the few hours spent in the district. The copper-bearing layers are thin, for the most part from a few inches to 2 or 3 feet thick. Turner states than on the Just Before claim, in the middle ridge, there is a cupriferous band 7 feet thick. Ore also occurs in cross fractures. On the Copper Queen and Lucky claims, about 4 miles north-northwest of Estey, carbonates and glance are found in prominent fractures of a jointing system that strikes about north and dips steeply. The Lucky Jack veinlet of this kind strikes N. 81° W. and dips 85° E. It is not improbable that these fractures are related to the fault system of the range. According to Turner there are also fissures that contain chalcopyrite. The same authority quotes assays of two shipments of picked ore in which the copper amounted to 9.9 and 10.5 per cent and the silver to 1.2 ounces and 0.9 ounce, respectively, with a trace of gold in each. He also states that the average copper content of the reefs as determined from a series of assays is about 4 per cent. It seems extremely doubtful, however, if any important extent of ground averaging as rich as that will be found.

Among the properties that have made any production may be mentioned the Colorado, near Estey, which extracted, from a short incline, arkose ore with carbonaceous fragments

---

[a] Lee, W. T., and Girty, G. H., The Manzano group of the Rio Grande valley, N. Mex.: Bull. U. S. Geol. Survey No. 389, 1909.
[b] Op. cit., p. 679.

partly replaced by copper; the Copper Boy, farther north, from which a carload of selected malachite ore said to run 12 per cent was taken from a 15-inch bed of sandstone, and the Copper Queen, already mentioned. The two last-named properties are owned by the Chicago Copper Company; the Colorado by the Dividend Company. The Sierra Oscura Company is said to have shipped a little ore from the deposits at the west base of the middle ridge.

On the grounds of association with organic material and of distribution in layers corresponding to stratified beds, Turner concluded that the copper was deposited from the waters which laid down the sediments, and that subsequent faulting has cut the cupriferous formation into blocks whose edges are exposed along the three ridges mentioned. The writer is of the belief, on the other hand, that the main faults antedated the deposition of the copper, which was brought in sulphide form by solutions ascending along the fault channels and, spreading out in the adjoining rocks, was deposited mainly as chalcocite, but in part also as bornite and chalcopyrite, at such places and in such beds as favored precipitation.

IRON DEPOSITS IN THE NORTHERN SIERRA OSCURA.[a]

To the observer looking southward from Bursum's 27-mile ranch, which has an elevation of 5,300 feet and is situated about 27 miles due east of San Antonio, the Sierra Oscura presents a steep westward-facing fault scarp front about 2,000 feet high, composed mostly of gently dipping lead-gray or blue Carboniferous (?) limestone with interbedded shale, resting on brownish granitic rock, probably of pre-Cambrian age. To the north the limestones dip below the surface, and the "Red Beds" and younger formations, lying in a broad north-northwestward-plunging syncline, appear as low mountains or hills. About 4 or 5 miles east of these hills the Carboniferous is again brought up, exposing a thickness of nearly a thousand feet, and forms an eastward-facing fault scarp along the edge of the east limb of the syncline, and thence continues with more or less prominence northward throughout the higher part of the Sierra, so far as the writer's observations extend. On the east side of the crest line, northeast and east of 27-mile ranch, the trend of the topography and structure veers from north and south to N. 70° W. The general elevation of the summit of the Sierra is about 7,000 feet.

The main feature of interest, from an economic standpoint, is the deposit of magnetic iron ore described below.

Along the west front of the Sierra, especially in the region of 27-mile ranch, the Carboniferous (?) limestone contains lead ores at many places, and is dotted with prospect dumps; some prospectors were at work in this vicinity.

Copper in small amounts is also present in these limestones. More important copper deposits occur in an outlying ridge a couple of miles west of the front of the Sierra, too remote and too detached to be called foothills. This ridge is composed of red sandstones or shales, probably of upper Carboniferous or Triassic age, which at Hansonburg contain notable amounts of copper ores.

The Jones iron deposits have been known for about eight years. They are located in eastern Socorro County nearly 47 miles about due east from San Antonio, a station on the Santa Fe Railway that is 12 miles south of Socorro and probably about 30 miles northwest of Carrizozo, on the El Paso and Southwestern System. They lie on the eastern slope of the northern part of the Sierra Oscura, at an elevation of about 6,700 feet. The nearest settlement, a few miles to the north, is Bursum's 43-mile ranch, where the elevation is about 6,500 feet.

The prevailing rock is a limestone, probably Carboniferous, with anticlinal structure trending nearly east and west. Along the crest of the small anticline is intruded a dike of monzonite, which has the same strike; it has a width of a quarter to half a mile and has been traced for several miles. The deposits were followed for a linear extent of about 3 miles, throughout which outcrops, cuts, shafts, and tunnels show or indicate magnetite to be present in commercial amount; they are reported to extend throughout the length of the district, which is about 9 miles.

---

[a] By F. C. Schrader.

The magnetite occurs in irregular bodies along the contact of limestone and gypsum with the monzonite. The limestone is lead gray or blue and is probably of Carboniferous age; the portion that forms the crest of the ridge is heavy bedded or massive. It contains fossil remains, which, however, could not be identified. Locally the limestone contains irregular cherty or ferruginous nodules or masses. Below the limestone the section exposes about 30 feet of indurated fine-grained sandstone or quartzite. Below the quartzite the rock is generally shaly and contains considerable soft, disintegrated gypsiferous limestone, buff or brownish in color, and near the base of the same slope on the south side of the ravine there occur in this material two beds of white gypsum, aggregating 5 or 6 feet in thickness. The limestone occurring with the iron or close to it has a steeper dip; it extends to and seems to rest upon the lower slopes of the monzonite dike on either side. In part it is a dense dark-gray limestone, in places light colored and probably gypsiferous, but appears to show little or no evidence of contact metamorphism.

As stated above, the structure is anticlinal; the limestone on the north side of the monzonite dike dipping 50° to 60° NNE.; on the south side the dip at the dike is nearly vertical, but it rapidly flattens to 20° SSW. on the low ridge south of the principal workings.

The monzonite forms the entire crest and most of the slopes of the axial ridge in which it occurs. The ridge is about one-fifth of a mile in width at its base and rises about 200 feet above the gulches on either side. It is flanked on both sides by the upturned sedimentary rocks covering its lower parts and in some places extending nearly to the crest. The monzonite is probably intrusive into these sedimentary rocks, but no sharp contacts of the fresh rocks were found. The monzonite is a granular rock of medium grain, gray or reddish-gray in color, and has the general appearance of a diorite. It is cut by jointing—more so than the limestone—and a set of these joints trend about east and west parallel with the dike, and dip steeply north or south. Its principal microscopic features are an abundance of lathlike crystals of andesine or labradorite with a smaller amount of interstitial orthoclase. Between the laths lie prisms or anhedrons of light-green augite, partly altered to chlorite and light-green hornblende. The accessories are magnetite and apatite, the former present in no unusual amount.

The iron ore on the south side of the dike, near Jones camp, is a partly oxidized magnetite and is comprised in an east-west belt about one-fourth of a mile long and about 150 feet wide. In this belt are two principal ledges. They are 10 to 40 feet thick and consist of massive beds of the iron ore standing at steep dips, with some interbedded impure altered limestone or gypsum. They probably extend downward below the present shafts, which are 30 to 40 feet in depth. Between the iron ore and the fresh monzonite lies a zone about 40 feet wide, in which the igneous rock is softened, altered, and bleached, probably by circulating surface waters.

By croppings, shafts, and prospect holes the deposits are seen to extend about a mile east of Jones camp and about 2 miles west of it; the monzonite seems to extend about 4 miles east of the camp and possibly much farther.

An additional structural feature in the iron deposits is a sheeting which dips steeply north or is almost vertical, and which is prominent in the Jones mine, on the south side of the dike, and in the deposits about one-half to three-fourths of a mile east of Jones camp, on the north side of the dike.

The question of the extent of the ore must of course be settled by actual exploration. But it is proper to mention that if, as seems probable, the ore is the result of contact metamorphism of the limestone by the monzonite dike, the mineralization has probably taken place only in certain strata of the limestone. It is not likely that the ore will accompany the contact indefinitely in depth and it is not probable that the ore will be found in quartzitic or shaly strata, which may readily be conceived to underlie the limestone.

Associated with the iron-ore deposits, only locally, however, occurs a small amount of copper in the form of malachite or chalcocite.

These deposits of magnetite have been described by F. A. Jones in "Mines and minerals of New Mexico," a book frequently referred to above, and also by N. W. Emmens.[a] According

---

[a] Min. Mag., vol. 13, 1906, pp. 109-116.

A. VIEW ACROSS COONEY CANYON, MOGOLLON DISTRICT, SOCORRO COUNTY.

Showing volcanic flows, faulted on left side. Looking west.

B. GRANITE SPIRES OF LOS ORGANOS, DONA ANA COUNTY.

West slope of Organ Mountains, between Stephenson-Bennett and Modoc mines.

to the last-named author the average ore contains 60.59 per cent of metallic iron, 2.53 per cent of silica, 0.152 per cent of phosphoric acid, and 0.203 per cent of sulphur.

C. R. Keyes[a] describes deposits of magnetite and hematite which occur on the Chupadera Mesa along dikes of "trachyte" (probably monzonite) under conditions very similar to those of the Jones camp, described above. The iron ores appear on both sides of the dikes, which are from 100 to 250 feet wide, at places where the dikes intersect Carboniferous (?) limestone. The dikes do not cut adjoining Cretaceous sandstones, and from this fact the conclusion is drawn that the intrusions preceded the deposition of these sandstones. No determinations of horizon are given. The ore appears to be 50 feet wide on one side of the dike and a shaft 100 feet deep has been sunk in it.

DONA ANA COUNTY.

By Waldemar Lindgren.

GENERAL FEATURES.

Dona Ana County, in south-central New Mexico, occupies an area of 3,818 square miles. It is bordered by Mexico and Texas on the south, Luna and Sierra counties on the west, Otero County on the east, and Sierra and Socorro counties on the north. The Rio Grande flows through it diagonally from northwest to southeast. The western and central portions of the county lie in the basin of that river valley and are characterized by gently sloping high plains, dotted with smaller ridges and hills, mostly of volcanic origin. In the southern part, along the Mexican line, are Quaternary basalt flows and craters.[b]

The south end of the Caballos Range is in the northwestern part of the county, north of Rincon. East of the Rio Grande extend the broad plains of the formerly dreaded Jornada del Muerto, and beyond, along the western margin of the county, the narrow San Andreas Range, continued to the south by the Organ Mountains, forms an almost continuous wall, at whose eastern foot the Sacramento Valley spreads out with its salt marshes, white gypsum hills, and black basalt flows. The San Andreas Range presents a gentle slope to the west and a steep scarp to the east and is built up mainly of westward-dipping Carboniferous limestones, underlain along the eastern slope of the range by pre-Cambrian granites. The Organ Range is less plainly monoclinal in structure and consists of the same Paleozoic limestones intruded by a large faulted batholith of quartz monzonite.

The greatest elevation in the county is attained in Organ Peak, due east of Las Cruces, which reaches 9,108 feet. The elevations of the valleys and high plains range from 3,500 to 4,500 feet.

The principal mining district is that of the Organ Mountains, described in considerable detail in the following paragraphs. The ores carry mainly lead and copper, with less gold and silver, and are contained in fissure veins, replacement veins, and contact-metamorphic deposits. Of late years the annual production of copper has varied between 100,000 and 1,000,000 pounds; about the same amount of lead, and a few thousand ounces of silver.

The Hembrillo district, which was not visited, is situated about 30 miles north of Organ, in the San Andreas Range, and is at present somewhat difficult of access. The ores are said to contain copper; the developments are slight.

ORGAN MOUNTAINS.

GENERAL FEATURES.

The Organ Mountains, named by the Mexicans Sierra de los Órganos on account of the conspicuous granite spires along their western front (Pl. X, B), form a link in the long chain of north-south ranges extending from the Sandia Range, near Albuquerque, on the north, to the Franklin Range, terminating at El Paso, in the extreme west corner of Texas, on the south. The Organ Range is separated from the Franklin Range by a gap southeast of Las

---

[a] Iron deposits of the Chupadera Mesa, New Mexico: Eng. and Min. Jour., vol. 78, 1904, p. 632.
[b] Lee, W. T., Afton craters of southern New Mexico: Bull. Geol. Soc. America, vol. 18, 1907, pp. 211-220.

Cruces, and continues northward for about 30 miles, its width varying from 5 to 15 miles. In its southern part the western slope is abrupt, in places scarplike. The eastern slope is, as a rule, more gradual. The range culminates in Organ Peak with an elevation of 9,108 feet (Wheeler). Fifteen miles northeast of Las Cruces, at the eastern base, is Organ, a small settlement and the center of the principal mining operations. Near this place the range ends at San Augustin Pass (elevation, 5,654 feet), north of which rises the prominent peak of the same name (elevation, 6,850 feet). The main highway between the Rio Grande valley and eastern New Mexico leads through this pass. North of the pass begins the San Andreas Range, marked by a relatively gentle western slope and a more abrupt eastern scarp. (See fig. 14.)

The foothills of the Organ Mountains are bare, while the higher ridges support a scant growth of pines and juniper. The principal part of the range is mapped on the Las Cruces sheet of the United States Geological Survey.

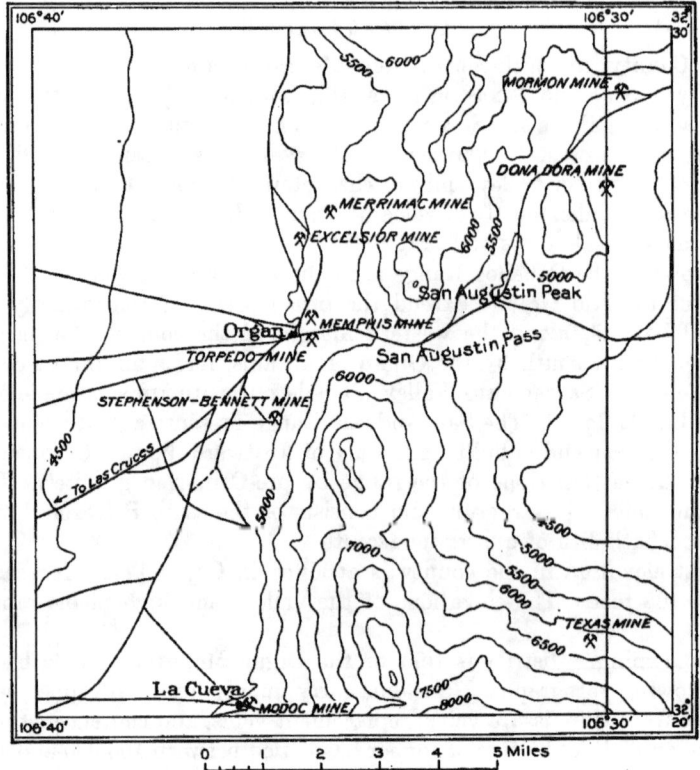

FIGURE 14.—Sketch map of the northern part of the Organ Mountains, near San Augustin Pass.

GEOLOGY.

The range has been visited by several geologists, though no detailed work has been done. The first description is given by Antisell, who traveled through this part of New Mexico in 1854.[a] He mentions the Carboniferous limestone flanking the range near San Augustin and the granitic rock which makes up the mass of the southern part of the range. This granite he considers older than the Carboniferous. On the east side of the range the granitic rocks prevail, but Antisell mentions Carboniferous limestone in the foothills, capped by 300 feet of a gypsiferous sandstone, which he regards as Triassic. He also mentions the Stephenson mine, which had just been opened at that time. The argentiferous galena obtained in this mine was smelted in an adobe furnace at Las Cruces. Antisell also states that the Barilla mine, at the mouth of Soldado Canyon (probably the Modoc), had been worked some years previously, and that the

---
[a] Pacific R. R. Surveys, vol. 7, 1857, pp. 161 et seq.

nonargentiferous galena produced was smelted in a furnace erected at the mine He was unable to visit this mine, as several lodges of Apaches were camped close to it.

C. R. Keyes [a] briefly describes the region in a recent paper, and like his predecessors considers the granite of pre-Cambrian age.

The mass of the Organ Mountains from a point about 4 miles north of San Augustin Pass to a point at least 10 miles south of it is made up of a coarse granitic rock, which has a somewhat porphyritic structure by the development of large feldspar crystals, especially emphasized near some of the contacts; for instance, at Organ where it also tends to assume a finer grain. The outcrops at many places form bold cliffs and spires of brownish-gray color. Surface decomposition makes it difficult to obtain fresh specimens of the rock. Heavy bowlder fields and broad débris fans mark the base of the granitic mountains.

A typical specimen obtained near the main contact, about 500 feet south of the Merrimac shaft, 4 miles north-northeast of Organ, is a reddish-gray granular rock of medium grain, containing some larger ill-developed orthoclase crystals up to 15 millimeters in length. The mass of the rock has an average grain of 3 or 4 millimeters and is composed of anhedrons of orthoclase, rarely with microperthite, laths of oligoclase, some andesine and small quartz grains. A tendency to micropegmatitic intergrowth is noted here and there. There is a small percentage of chestnut-brown biotite and an equal quantity of pale-green augite, in places roughly prismatic in form and partly converted to greenish hornblende. Accessory minerals are magnetite, apatite, and titanite; the titanite is visible by the naked eye in all specimens. The structure is typically hypidiomorphic.

Analysis of quartz syenite near Merrimac shaft, 3 miles north of Organ.

[George Steiger, analyst.]

SiO_2	61.12	ZrO_2	0.04
Al_2O_3	15.78	CO_2	.22
Fe_2O_3	2.69	P_2O_5	.45
FeO	3.15	SO_3	None.
MgO	1.90	S	.05
CaO	3.95	MnO	.09
Na_2O	4.14	BaO	.07
K_2O	4.48	SrO	.04
H_2O-	.32		
H_2O+	.56		100.35
TiO_2	1.30		

The rock is evidently a quartz monzonite.

On the whole there is little variation in this type. In places the roughly porphyritic orthoclase crystals are more prominent. At the Modoc mine the rock contains somewhat more quartz and less oligoclase than the specimen just described. The augite is almost wholly converted to hornblende. The biotite is fresh. At the Galloway prospect, a mile east of San Augustin Peak, the rock is similar to that of the Modoc mine. At the Mormon mine, on the east side of the range, about 8 miles east-northeast of Organ, the rock is composed chiefly of oligoclase with some orthoclase and quartz; the ferromagnesian silicates are entirely decomposed.

The monzonite of the Organ Mountains contains many dikes, among which, however, no pegmatites were found. They are usually fine-grained, light-colored rocks, either of typical aplite composed of quartz and orthoclase with a few grains of albite, or else of a syenite porphyry or monzonite porphyry, with decomposed slender prisms of hornblende or augite in a groundmass of interlocking grains of orthoclase. At the Mormon mine a dike of granular, almost pure hornblende rock was found.

The Paleozoic sedimentary rocks form a narrow, interrupted belt at the western base of the range. North of Organ they are much more prominent as heavy-bedded limestones, dipping 20° to 30° W. or NW., and 3 or 4 miles north of the pass they overlap the range as a domelike anticline. For a considerable distance north of the pass the range is built up chiefly

---

[a] Geology and underground water conditions of the Jornada del Muerto, New Mexico: Water-Supply Paper U. S. Geol. Survey No. 123, 1905, p. 28.

of limestone. Lack of time prevented detailed study of the Paleozoic rocks. Their aggregate thickness is very considerable. Limestones predominate, but 4 miles north of the pass they contain a bed of black shale of moderate thickness, which may represent the Devonian.

The Paleozoic rocks of the Franklin Mountains have been studied in detail by G. B. Richardson,[a] and consist of the following formations:

Paleozoic rocks of Franklin Mountains.

System.	Formation.	Thickness (feet).
Carboniferous:		
Pennsylvanian	Hueco limestone	3,000
Mississippian	Absent.	
Devonian	Absent.	
Silurian	Fusselman limestone	1,000
Ordovician:		
Upper and Middle	Montoya limestone	250
Lower	El Paso limestone	1,000
Cambrian	Bliss sandstone	300

The Bliss sandstone is said to rest unconformably on the pre-Cambrian and the whole series is cut by post-Carboniferous granitic rocks.

No mineral deposits are known from the Franklin Mountains, except an occurrence of tin ore in granite, noted in the first paper cited above.

It remains to describe the relations of the granitic rock and the sedimentary series in the Organ Mountains. The literature shows that up to the present time the former has been considered as a pre-Cambrian basement on which the limestones were deposited. The facts do not sustain this view. On the contrary, the granitic rock forms a stock or batholith, in places with features of a laccolith, which is distinctly intruded into the limestones, and which, beyond all doubt, is of post-Carboniferous age. It is possible that pre-Cambrian rocks may be present in the Organ Mountains, as indeed they are in the Franklin Range, but they probably do not occur in the vicinity of the mining districts here described. The intrusive origin of the granitic rocks is proved by the structural relations at the contacts and by the contact metamorphism of the limestone. It is also supported by the fact that dikes of aplite, syenite porphyry, and dark diorite cut the sedimentary rocks as well as the main intrusive mass, and that a dike of granite or quartz syenite was observed in the limestone close to the shaft at the Stephenson-Bennett mine.

The course of the contact is best seen near the Merrimac mine, 2½ miles north of Organ, where it cuts the sediments very irregularly, breaking across both shales and limestones. Considered as a whole, the limestones north of Organ appears to overlie the granitic rock. They lap over it as a flat arch, dipping away from the central axis of the mountains on both sides and covering its summits entirely a few miles north of Organ. The igneous rock has been intruded underneath, the sediments bulging them in the manner of a laccolith. The phenomena of contact metamorphism are shown most strikingly all the way from Organ to the vicinity of the Merrimac mine. Near the Torpedo mine at Organ the limestones have been silicified at the contact, and this zone of silicification continues southward to the Stephenson-Bennett mine. About a thousand feet south of the Torpedo mine this silicified limestone adjoins the granitic rock and contains, a few hundred feet from the contact, a small area of garnetized limestone inclosing a dike of granite porphyry. At the Torpedo mine, in the main gulch, about 500 feet from the contact, the sediments strike N. 47° E. and dip 40° NW. They consist apparently of benches of limestone with some strata of quartzite, but closer examination shows that the limestone is wholly converted to a fine-grained mass of garnet and colorless pyroxene. Nearer to the contact, which is poorly exposed, the rock is a typical fine-grained hornfels, consisting of garnet, colorless pyroxene, and a feldspar closely corresponding to anorthite. There is also some residuary calcite.

---

[a] Contributions to economic geology, 1905: Bull. U. S. Geol. Survey No. 285, 1906, p. 146; Paleozoic formations in trans-Pecos Texas: Am. Jour. Sci., 4th ser., vol. 25, 1908, pp. 474–484.

Across the gulch northward from the Torpedo is the Memphis mine, where the crumbling granitic rock is in immediate contact with a coarse-grained metamorphosed limestone now consisting of garnet, quartz, specularite, oxidized copper minerals, and residuary calcite. On the hill above the Memphis the contact zone is narrow; the granitic rock is adjoined by a bench of coarsely crystalline limestone. Next to this lies a cupriferous garnet bed, which is in turn adjoined on the west by almost normal hard dense blue limestone. Ordinarily, however, the altered zone ranges up to several hundred feet in width. From the Memphis mine the contact makes a wide detour around the broad flat, underlain by granitic rock, which extends up to the granite bluffs of San Augustin Peak. One mile north of the Memphis property the contact bends eastward again and is splendidly exposed at the Excelsior and other prospects. The contact zone here is 200 or 300 feet wide and consists of heavy benches of garnet rock alternating with hornfels (metamorphosed shale) and coarsely crystalline limestone. The garnet commonly shows copper stains.

The general structural features of the range are simple. On the north, as stated above, the limestones lap over the intrusive rock as a broad, low arch extending northward as far as can be seen. At Black Mountain, on the east side, near the Mormon mine, the contact phenomena will probably be found as well exposed as on the west side. South of Organ the limestone area runs out to a point along the foothills and there is evidence of a pronounced faulting movement, which probably has produced the western scarp of the range between Organ and the Modoc mine. The great mass of the mountains consists of the intrusive granitic rock.

MINERAL DEPOSITS.

GENERAL FEATURES.

The mineral deposits of the Organ Mountains are of several well-defined types, described in the following paragraphs.

(1) Fissure veins in intrusive rocks. These are usually narrow, sharply defined, quartz-filled veins with east-west strike and steep dip. Most of them are on the east side of the range, and many occur along dikes, either of syenite porphyry or of dark basic hornblende rock. Most of them contain silver with galena, zinc blende, and tetrahedrite; others, like those at Gold Camp or on Texas Creek, contain chiefly gold associated with pyrite. Probably very little of the gold is free. Molybdenite is reported from at least two places.

(2) Replacement veins in limestone. This type is best represented by the Stephenson-Bennett mine, but there are also less well developed examples half a mile west of the contact at Organ and on the main ridge north of San Augustin Peak. The deposits are probably connected with fissures but generally follow the stratification of the limestone; at the Stephenson-Bennett mine they outcrop a few hundred feet west of a fault contact between syenite and unaltered limestone. There is no alteration of the limestone, and the scant gangue consists of quartz, coarse-grained calcite, barite, and fluorite. Abundant wulfenite indicates molybdenite as an original mineral, besides the prevailing galena and some pyrite and zinc blende. The oxidation extends to a depth of several hundred feet. The galena contains some silver.

(3) Contact-metamorphic deposits. These deposits carry chiefly copper and occur for several miles along the main contact of quartz monzonite and limestone from the Torpedo mine at Organ northward. They form irregular masses, roughly following the stratification, and the primary chalcopyrite is intimately associated with a gangue of yellowish-green garnet, some epidote, quartz, calcite, and specularite. Zinc blende is present in places and molybdenite is reported. The oxidation is irregular, but in places extends to depths of several hundred feet; elsewhere sulphides may be found near the surface. The Modoc deposit is of a peculiar type. It occurs at the contact of limestone and andesite and contains nonargentiferous galena in epidote gangue. The deposits of this third type appear to contain very little gold and silver.

All these deposits, except that of the Modoc mine, are probably closely related and were formed in one epoch of mineralization, during or shortly after the intrusion of the great batholith. In general the copper and zinc, with some gold and silver, appear to be deposited at the contacts. The lead deposits seem to have preferred a certain distance from the intrusive contact. The gold and silver were deposited chiefly in the quartz veins in the later part of the epoch of mineralization, shortly after the consolidation of the intrusive magma.

Reliable estimates of total production are difficult to obtain. Josiah Bond [a] estimates a total yield of $2,500,000 prior to 1900, of which $1,250,000 came from lead ores, $1,000,000 from copper ores, and $250,000 from silver ores.

GOLD AND SILVER VEINS IN GRANITIC ROCKS.

A small but characteristic silver vein occurs in the Ruby property in the granitic rock, 2 miles north of Organ and only a few hundred feet from the contact with limestone. It is a narrow quartz vein with east-west strike and nearly vertical dip. The quartz contains disseminated tetrahedrite, galena, and zinc blende. The developments ore slight. Some ore has been concentrated on hand jigs and shipped.

A number of similar silver veins having about the same strike are found on the east side of San Augustin Peak; some of them occur along dikes of porphyry or aplite.

In the so-called Gold Camp, situated about 8 miles east-northeast of Organ, between Mineral Hill and Black Mountain, the values are principally in gold. The same east to west strike of the veins prevails. The Dona Dora, said to be a promising property, lies on the north side of Mineral Hill, and development work was in progress here in 1905. At the foot of Black Mountain is the Mormon mine, which is opened on a vein that follows a persistent dike of hornblende rock in a granitic light-colored rock, composed chiefly of oligoclase and quartz. The strike is N. 80° W. and the dip 70° S. The vein is opened by several shafts along a distance of about 500 feet, the principal shaft being 100 feet deep. The quartz vein lies in part between the dike and the syenite, but stringers also occur in the dark-green hornblende rock. The quartz is accompanied by some calcite and contains pyrite and chalcopyrite. The gold is free only near the surface in the comparatively shallow zone of oxidation. Water was encountered in the shaft. The equipment consists of a gasoline hoist and a small stamp mill with concentrating table. The mine has a small production to its credit, but was idle when visited in 1905.

The Texas prospect, which was not visited, is situated in Texas Canyon, on the east side of the Organ Mountains, about 10 miles southeast of Organ. It is said to be a vein in granitic rock up to 8 feet wide, and the specimens inspected contain pyrite in quartz gangue. Molybdenite is reported from this deposit. The gold values are stated to range from $8 to $15 a ton.

LEAD-SILVER VEINS IN LIMESTONE.

The most prominent among the mines of the Organ Mountains is the Stephenson-Bennett (Pl. XI, B), situated in the western foothills 1½ miles south of Organ. Jones [b] states that it was discovered in 1849 by the Mexicans and, as cited above, Antisell notes that in 1854 the ores were smelted in an adobe furnace at Las Cruces. Since then it has been worked fairly continuously. The production between 1854 and 1857 is estimated to be $80,000 or $90,000. The adobe furnace was discontinued with the advent of more modern methods of smelting, and the ore has been shipped to various smelters. At present it is smelted in El Paso. The value of the total production to 1906, inclusive, is estimated to be $600,000. Jones estimates that from 1890 to 1900 the production was $200,000. The mine is owned by the Stephenson-Bennett Mining Company, which was reorganized in 1909. A small concentrating plant has been erected at the mine.

The developments consist of a tunnel in the first foothills, which opens the two principal veins; also a winze sunk 150 feet below this level. This tunnel has an elevation of about 4,900 feet. The croppings are 360 feet above it, and the upper old workings were opened from an upper tunnel about 100 feet below the croppings. In 1905 a vertical shaft was sunk in solid limestone 50 feet above the main tunnel with the intention of developing the mine to a depth of 500 feet. The aggregate length of development is considerable, but the ore shoot is not very

---

[a] Mining World, March 17, 1906.   [b] Jones, F. A., New Mexico mines and minerals, 1904, p. 73.

A. MODOC MINE AND MILL, ORGAN MOUNTAINS, DONA ANA COUNTY.

Dump of mine in background.

B. STEPHENSON-BENNETT MINE, ORGAN MOUNTAINS, DONA ANA COUNTY.

long. Much water was encountered below the tunnel level; the pumps installed were expected to handle 500 gallons a minute.

The foothills in which the deposit is contained consist of a hard, dense blue limestone dipping 70° W. The prominent mass of silicified limestone showing at the Torpedo mine continues southward to this property, but here lies 500 feet west of the vein croppings and apparently has no connection with the mineralization.

About 100 feet east of the croppings the limestone borders against the main granitic mass with sharp contact, the plane of which, so far as can be seen, is vertical and surely represents a fault, which probably continues for a long distance southward along the western base of the range. No metamorphism is noted at the main contact, and it is clear that there is a downthrow of the west side amounting to not less than 500 feet, and probably much more. A dike of granitic rock was noted at the shaft, and another is reported to occur in the southern part of the mine, but was not seen; the main contact is not exposed by the workings.

The deposits consist of two bedded veins following the stratification, at least approximately. The eastern and principal vein outcrops at several places on the steep limestone hill back of the mine, but the croppings are barren and the deposit is indicated only by a number of seams without ore or gangue. Immediately below the croppings rich masses of cerusite were opened, the old stopes being in places 15 feet wide. According to G. P. Merrill[a] this part of the mine contained caves of later origin than the deposit, from the floor of which masses of oxidized lead ores were shoveled out. This vein has also been opened on the main tunnel level, but was not cut in 1905 on the winze level. The west vein was opened on both levels, showing wide stopes of mixed lead carbonate and galena.

The primary ore consists of galena with some yellow zinc blende, pyrite, and probably molybdenite in a scant gangue of massive or comb quartz with a little coarse-grained calcite and some barite and fluorite. Secondary minerals are cerusite and abundant wulfenite in fine crystals. Some of the carbonate ores mined were extremely rich. The mixed ore mined in 1905 contained about 5 per cent of lead and 20 to 30 ounces of silver to the ton, but no gold. The prospects for finding ore on lower levels appeared bright in 1905 and have been substantiated by rather heavy shipments up to 1908.

There are several places in the limestone near Organ where prospects containing galena have been opened. They are situated from a quarter to half a mile from the contact, and no contact-metamorphic minerals appear in them. The Jim Fisk mine, which was not visited, is located on the main ridge about 3 miles north of Organ, perhaps half a mile from the main granite contact. Some galena has been shipped from it at intervals.

CONTACT-METAMORPHIC DEPOSITS.

The most prominent of the belt of copper deposits which follow the contact for a distance of 2 miles north of Organ is found at the Torpedo mine. It is situated practically on the contact 800 feet east of Organ, the shaft being located on a little hill south of the principal gulch. It is owned by the Torpedo Mining Company. A considerable quantity of copper ore of good grade has been shipped. In 1905 the mine was worked by lessees, who shipped some ore with about 15 or 16 per cent of copper. In 1906 and 1907 shipments continued rather heavy, averaging over 5 per cent in copper.

The mine is developed by a tunnel at creek level and a shaft 150 feet deep situated close by; a second shaft on the hill 30 feet above the tunnel which was 200 feet deep in 1905; and a third shaft close by, in limestone, 300 feet deep in 1905. The water is very heavy; at a depth of 150 feet 700 gallons a minute was encountered—in fact, more than the pump could handle. Presumably a pumping plant of greater capacity has been installed. Only the tunnel level could be inspected in 1905.

The sedimentary rocks cropping on the hill and along the gulch consist of garnetized beds of fine-grained limestone with well-preserved planes of stratification dipping 40° NW. At the shaft a thick mass of entirely silicified limestone adjoins the contact and lies between the granitic

---

[a] Oral communication.

rock and the unaltered blue limestone. The actual contact is poorly exposed. The ore deposit appears to be contained in this silicified limestone. The second shaft, 200 feet deep, was, according to accounts, sunk entirely in ore.

The ore body consists of a large mass of chrysocolla ore, containing, according to reliable statements, only 2 ounces of silver to the ton. It lies between the limestone and the granite, but the exact relations could not be ascertained. The gangue is a soft kaolinized and brecciated material, partly replaced by chrysocolla.

The Memphis mine, owned by the Stephenson-Bennett Mining Company, lies on the north side of the creek and is developed by a shaft 175 feet in depth. Some good ore has been shipped. About 1884 a small water-jacket smelter was operated for some time on Memphis ore. In 1905 the mine was worked on a small scale by lessee. The contact is exposed all along the claim, the sedimentary series consisting of alternating beds of coarsely crystalline limestone and heavy cupriferous garnet rock, all of which dip westward at moderate angles. The shaft is sunk in limestone and intercepts the ore-bearing stratum in depth. It contained about 50 feet of water. On the deepest level the ore-bearing stratum is said to have been 10 feet thick and to dip 15° W. The ore is partly oxidized and contains malachite, chalcocite, and chalcopyrite. Its average composition is said to be 40 per cent of silica, 15 to 20 per cent of iron, 9 to 27 per cent of copper, 6 to 15 ounces of silver to the ton, and a trace of gold. Molybdenite is reported. The gangue consists of garnet and quartz with some specularite, the copper minerals being intimately intergrown with these gangue minerals. Galena and zinc blende are rarely found.

The Excelsior property, idle in 1905, lies 1¼ miles south of Memphis, on the main contact. It is developed by a 60° incline 175 feet deep, partly filled with water. The owner is said to be George E. Wood, of Chicago. The ore evidently occurs in a garnetized stratum alternating with coarsely crystalline limestone. The body is reported to be up to 20 feet wide. The minerals noted were malachite, chalcocite, and chalcopyrite, the last intergrown with garnet. At this mine the contact swings eastward, and the next property of note is the Ruby, one-fourth mile from the Excelsior. There is here evidently a tongue of limestone projecting eastward into the granitic rock. The developments are slight; the ore is oxidized and similar to that from the Excelsior.

The Merrimac property is situated in Merrimac Gulch, three-fourths of a mile east of Excelsior, and is said to be owned by Godfrey & Felton, of El Paso. It is developed by a vertical shaft 200 feet deep and was idle in 1905. On following up the gulch the principal rock seen is unaltered limestone, dipping north or northwest. The shaft is situated on a small ridge about 200 feet north of the contact. At the shaft the limestone assumes a coarse texture and contains interbedded benches of garnet rock. The croppings show only copper-stained rock rich in iron, but the vertical shaft encountered the same stratum, which has the prevailing dip of 30° NW. and consists of a heavy garnet rock containing residual calcite as well as fresh, brown zinc blende with some chalcopyrite. It is said to contain a little gold and silver.

From the Merrimac the main contact turns southeastward and crosses the main ridge. No copper prospects of importance are reported beyond this point.

The copper-bearing stratum of the Merrimac is underlain by a bed of black shale, which in turn rests on coarsely crystalline limestone. The Little Buck claim is located here close by the Merrimac, and its ore is of a rather unusual character. Rich silver ore with some gold occurs in irregular bunches in this coarsely crystalline limestone and is accompanied by a gangue of a little quartz and fluorite. The principal ore minerals are probably horn silver and argentite. Some rich pockets have been extracted, principally from surface cuts, and the total value of the production is given as $42,000.

The ore-bearing stratum continues eastward across Merrimac Gulch into the Jim Fisk and Black Prince claims, but is here farther from the granite contact and the ore changes to galena.

The steep escarpment of the western slope of the Organ Mountains extends southward for 12 miles from San Augustin Pass. The granitic bluffs as a rule rise directly from the débris fans at the base of the mountains, but in many places a narrow belt of limestone lies at their foot.

The Modoc property (Pl. XI, A) is situated 6 miles almost due south of Organ. The mill at the creek level has an approximate elevation of 5,500 feet. The mine is higher up on the slope, about 700 feet above the camp, almost at the foot of the prominent granitic bluffs. The mine is mentioned by Antisell in his report on New Mexico, as stated above, and has been producing lead ores at intervals. In 1905 some ore was shipped to the smelter at Deming and production was continued in 1906. Very little water is available, and two unsuccessful attempts have been made to install dry concentrators. The developments consist of an incline 185 feet long on the dip of the deposit, and a tunnel cutting the shaft at a depth of 100 feet. A rope tramway connects with the mill.

Above the masses of heavy granite bowlders at the mill the spur on which the mine is situated is made up of heavy benches of blue unaltered limestone, probably Carboniferous, striking N. 70° W. and dipping 45° SW. At the mine this series is cut by a broad belt of porphyry, above which the granitic bluffs rise. This granitic rock is quartz monzonite, almost identical with that near Organ. The porphyry is a fine-grained brownish-gray rock with abundant small crystals of orthoclase and minute needles of hornblende. The microscope shows predominant orthoclase phenocrysts in a partly glassy trachytic groundmass with small needles of a soda-lime feldspar. Calcite and epidote are the main secondary minerals. The rock is unquestionably of an effusive type—that is, it formed part of a lava flow consolidated near the surface, and probably was erupted from this vent between the syenite and the limestone. It should be classed as a hornblende trachyte or as a latite. Dikes of this or a similar rock occur in the limestone, close by the mine. This magma has caused some metamorphism of the limestone at the immediate contact, and in this narrow zone the ore is found. It forms a bed up to 15 feet thick, following the dip of the limestone—almost 45° SW.—at the contact. The limestone has here been converted to a crystalline rock of typical contact-metamorphic structure consisting of quartz and residuary calcite, filled with small prisms of yellow epidote. The galena is intimately intergrown with these minerals, showing contemporaneous deposition, and presents the peculiar feature of having parallel orientation over large areas. The structure is similar to that called poikilitic by petrographers. A little pyrite and zinc blende accompanies the galena, which is poor in silver.

It is very rare to find contact metamorphism exerted by an effusive magma, and the writer does not know of any other ore similar to that just described.

SIERRA AND CENTRAL SOCORRO COUNTIES.

By C. H. Gordon.

GENERAL RELATIONS.

LOCATION AND EXTENT.

The region covered by this section is bounded approximately on the east by the Rio Grande and on the west by the first guide meridan west, and extends from Rio Salado, about 18 miles north of Socorro, southward to Deming. It includes the middle portion of Socorro County and the larger part of Sierra County. It is approximately 160 miles long north and south, 65 miles wide at the north end, and 35 miles wide at the south, and comprises an area of about 8,000 square miles.

A general review of Socorro County has been given on pages 190–191 of this volume. Sierra County has an area of 3,081 square miles and is bounded on the north and east by Socorro County, on the west by Socorro and Grant counties, and on the south by Luna and Dona Ana counties.

Its eastern part is traversed from north to south by the Rio Grande. To the east of that stream lie the narrow longitudinal Cristobal and Caballos ranges. In the western part of the county rises the Black or Mimbres Range, from which a number of smaller streams join the main river.

There is a small coal field in Sierra County, near Engle.[a] Some copper and gold veins occur in the Caballos Range, and gold placers are found at the foot of its steep western scarp.

---

[a] Lee, Willis T., The Engle coal field, New Mexico: Bull. U. S. Geol. Survey No. 285, 1906, p. 240.

The principal metal mining districts are, however, contained in the Black Range; from north to south they are as follows: Black Range, Apache, Hermosa, Kingston, Hillsboro, Carpenter, Tierra Blanca, and Lake Valley. Most of the deposits are silver-lead replacement deposits in limestone, but silver-copper veins in andesite are mined at Chloride and gold-bearing veins in porphyries and monzonites at Hillsboro. Of late years the production of this county has been small, but in former days some of the districts, particularly Lake Valley, have yielded large quantities of rich silver ore.

The remaining 5,000 square miles of the area here discussed lie in southern and central Socorro County west of the Rio Grande and embrace the Magadalena and San Mateo ranges. In these ranges the principal mining camps are the important zinc-lead district of Magdalena, where the ores occur as replacement of limestone, and the Rosedale district, which contains gold-quartz veins in rhyolite. In former years the Magdalena camp yielded a considerable production of lead, but in the last few years its great resources in zinc ores are beginning to be utilized.

The short time available for the study of this large and complex area precluded more than preliminary or reconnaissance work. It was the aim to visit all mines and the most important prospects, but obviously some of them could receive but scant attention, and some prospects of value may have been overlooked altogether. The fact that a particular deposit is not mentioned is not to be taken as an indication that it lacks value.

Much of the information contained in this report was obtained from miners and prospectors familiar with the region, and to these men the thanks of the author are due. He is under special obligation to Mr. C. T. Brown, of Socorro, mining engineer and operator, whose intimate knowledge of the region has been of great assistance and who has rendered valuable service in other ways; Messrs. F. H. Gregg and L. S. Ferry, managers of the Graphic and Kelly mines, respectively; Hon. Quinby Vance; Mr. E. James; and Mr. S. J. Macy.

PREVIOUS INVESTIGATIONS.

Almost no detailed geologic work has been done in this region. Such data as are available are comprised in the reconnaissance memoirs of the earlier surveys and in separate papers by Cope, Silliman, Endlich, Springer, and Clark, together with the more recent publications of Herrick, Hill, Jones, Keyes, and others. A list of publications relating to the area follows:

BRADY, F. W. A valuable bat cave in New Mexico. Mines and Minerals, vol. 26, 1905, pp. 97-98.
    Describes occurrence of bat guano in caves in basaltic lava in vicinity of Lava and Engle.

BRINSMADE, R. B. Kelly, New Mexico—A zinc camp whose ores have been made available by modern metallurgical methods. Mines and Minerals, September, 1906, pp. 51-53.
    Gives a brief description of the geology of the district and discusses the nature and occurrence of the ores and the mine workings.

CLARK, ELLIS. The silver mines of Lake Valley, New Mexico. Trans. Am. Inst. Min. Eng., vol. 24, 1894, pp. 138-169, Pls. I-IV, 14 figs.
    Describes the geology of the district and the occurrence of the ores.

COPE, E. D. Geology of the Lake Valley mining district, New Mexico. Am. Naturalist, vol. 15, 1881, pp. 831-832.
    Gives a brief description of the geology of the district.

—— Geological age of the Lake Valley mines of New Mexico. Eng. and Min. Jour., vol. 34, 1882, p. 214.
    Discusses age of the ore-bearing limestones and uses for them the term "Lake Valley limestones."

—— Additions to the knowledge of the Puerco epoch (Eocene). Proc. Am. Philos. Sóc., vol. 20, 1883, p. 545-565; vol. 21, 1883, pp. 309-324.
    Describes in all 74 species of vertebrates occurring in the Puerco formation.

—— On the distribution of the Loup Fork formation in New Mexico. Proc. Am. Philos. Soc., vol. 21, 1883, pp. 308-309.
    Describes beds along the Rio Grande as far south as Socorro and in the vicinity of Hatch station, which he identifies with the Santa Fe marl and refers to the "Loup Fork" formation.

DUTTON, C. E. The volcanoes and lava fields of New Mexico. Bull. Philos. Soc., Washington, vol. 7, 1885, pp. 76-79.
    Describes the great flows of lavas, both old and young, occurring in the Great Plateau region, especial consideration being given to the vicinity of Mount Taylor.

ENDLICH, F. M. The mining region of southern New Mexico. Am. Naturalist, vol. 17, 1883, pp. 149-157.
    Describes the geology of the Lake Valley region and the occurrence of the silver ores. Gives profile section (Pl. III) showing stratigraphic relations

FAIRBANKS, H. W. The physiography of southern Arizona and New Mexico. Jour. Geology, vol. 11, 1903, pp. 97–99.

Describes the great débris-covered plains of the Southwest and discusses their probable origin, concluding that the materials were accumulated under water, which extended far inland from the sea.

GORDON, C. H. Notes on the Pennsylvanian formations in the Rio Grande valley, New Mexico. Jour. Geology, vol. 15, 1907, pp. 805–816.

Discusses the character and classification of the Pennsylvanian formations in the Rio Grande valley with especial reference to the lower division, for which the name Magdalena group is proposed.

—— Mississippian ("Lower Carboniferous") formations in the Rio Grande valley, New Mexico. Am. Jour. Sci., 4th ser., vol. 24, 1907, pp. 58–64.

Describes the occurrence of Mississippian rocks at Lake Valley, Hillsboro, Kingston, and Cooks Peak and gives lists of the fossils found in them. The name Graphic-Kelly applied by Herrick to these rocks in the Magdalena district is changed to Kelly limestone.

GORDON, C. H., and GRATON, L. C. Lower Paleozoic formations in New Mexico. Am. Jour. Sci., 4th ser., vol. 21, 1906, pp. 390–395; abstract in Jour. Geology, vol. 15, 1907, pp. 91–92.

Describe the occurrence of formations belonging to the Cambrian, Ordovician, Silurian, and Devonian systems in southwestern New Mexico. Discuss character of the formations. Names for formations proposed: Cambrian, Shandon quartzite; Siluro-Ordovician, Mimbres limestone; Devonian, Percha shale; Quaternary (Pleistocene), Palomas.

HERRICK, C. L. The so-called Socorro tripoli. Am. Geologist, vol. 18, 1896, pp. 135–140, Pls. IV, V.

Describes the occurrence on the east side of the Rio Grande, nearly opposite San Antonio, of a "tripoli" which has been worked to some extent as a polishing powder. Gives analyses and discusses origin.

—— Applications of geology to economic problems in New Mexico. Proc. Intern. Min. Cong., 4th sess., 1901, pp. 61–64.

Describes some of the geologic features and the occurrence of economic minerals.

—— The geology of a typical mining camp in New Mexico. Am. Geologist, vol. 19, 1897, pp. 256–262. Pls. XIII–XX.

Describes the general geologic features of the Magdalena Mountains and the fault phenomena of the region.

—— Papers on the geology of New Mexico. Bull. Denison Univ. Sci. Lab., vol. 11, 1898, pp. 75–92, Pls. IX–XII; reprint, Bull. Univ. New Mexico, vol. 1, 1899, pp. 75–92.

Describes the geology of Socorro Mountain, Lemitar, and Mount Magdalena; includes an account of the basic eruptions of the Magdalena district.

—— The occurrence of copper and lead in the San Andreas and Caballos mountains. Am. Geologist, vol. 22, 1898, pp. 285–291, 1 fig.; reprint, Bull. Univ. New Mexico, vol. 1, 1899, pp. 285–291, 1 fig.

Describes the occurrence of the ore deposits in Carboniferous rocks and discusses their origin. Describes briefly the phenomena of faulting and displacement in the Caballos Mountains.

—— Report of a geological reconnaissance in western Socorro and Valencia counties. Am. Geologist, vol. 25, 1900, pp. 331–346, Pls. VIII–IX; reprint, Bull. Univ. New Mexico, vol. 2, pt. I, 1900, pp. 82–99, Pls. I–II.

Describes the general geologic features of the region and the occurrence and character of the Carboniferous and Cretaceous rocks.

—— and BENDRAT, T. A. Identification of an Ohio Coal Measures horizon in New Mexico. Am. Geologist, vol. 25, 1900, pp. 234–242; reprint, Bull. Univ. New Mexico, vol. 2, 1900.

Describes the occurrence, character, and fauna of the formations in the Sancha Mountains east of Albuquerque.

—— Secondary enrichment of mineral veins in regions of small erosion. Min. and Sci. Press, vol. 87, 1903, p. 97.

—— A Coal-Measure forest near Socorro, New Mexico. Jour. Geology, vol. 12, 1904, pp. 237–251, 10 figs.

Describes the general geologic structure of the Rio Grande valley and the occurrence, character, and fauna of coal-measure strata in vicinity of Socorro.

—— Laws of formation of New Mexico mountain ranges. Am. Geologist, vol. 33, 1904, pp. 301–312, 393, 2 pls.

Describes the geologic structure and physiographic features of several mountain ranges of New Mexico.

—— The clinoplains of the Rio Grande. Am. Geologist, vol. 33, 1904, pp. 376–381, 1 fig.

Describes the character, occurrence, and origin of clinoplains in the vicinity of Socorro.

JEWETT, J. J. Notes on the topography and geology of New Mexico. Trans. Kansas Acad. Sci., vol. 19, 1905, pp 141–149.

Describes in a general way the topography and geology of the Territory. The Rio Grande valley considered to have been occupied by a lake or series of lakes in middle and late Tertiary time. During the glacial period cataclysmic rains frequent and canyons and slopes deeply eroded. Probably in this age the Rio Grande freed its channel from the natural bars of the Pliocene eruptions.

JOHNSON, D. W. Block mountains in New Mexico. Am. Geologist, vol. 31, 1903, pp. 135–139.

Discusses the phenomena of block faulting and describes the occurrence of examples of this type of structure in the central New Mexico region.

JONES, F. A. New Mexico mines and minerals (World's Fair edition), Santa Fe, N. Mex., 1904, 349 pp., 50 figs.

Includes a brief account of the general geology and observations on the occurrence, geologic relations, and character of ore deposits, mining and production of minerals, etc. Gives a list of minerals occurring in New Mexico.

KEYES, CHARLES ROLLINS. Geological structure of New Mexican bolson plains. Am. Jour. Sci., 4th ser., vol. 15, 1902, pp. 207-112.

    Describes the character of bolson plains, with special application to the Jornada del Muerto, and discusses the geologic history of the region.

—— Ephemeral lakes in arid regions. Am. Jour. Sci., 4th ser., vol. 16, 1903, pp. 377-378.

    Notes the occurence of ephemeral ponds and lakes in Mexico and New Mexico and describes the structure and origin of the bolson plains where they are found in connection with these lakes.

—— A remarkable silver pipe. Eng. and Min. Jour., vol. 76, 1903, p. 805.

    Discusses the occurrence and origin of "pipe veins" and describes an occurrence in central New Mexico.

—— Note on block mountains in New Mexico. Am. Geologist, vol. 33, 1904, pp. 19-23.

—— Bolson plains and the conditions of their existence. Am. Geologist, vol. 34, 1904, pp. 160-168.

    Describes the characters of bolson plains and discusses their origin.

—— Remarkable occurrence of aurichalcite. Proc. Iowa Acad. Sci., vol. 11, 1904, p. 253.

    Describes an occurrence of aurichalcite in the Magdalena Mountains.

—— Certain basin features of the high plateau region of southwestern United States. Proc. Iowa Acad. Sci., vol. 11, 1904, pp. 254-257.

    Describes features of bolson plains of New Mexico and discusses their origin.

—— Note on the Carboniferous faunas of Mississippi Valley in the Rocky Mountain region. Proc. Iowa Acad. Sci., vol. 11, 1904, pp. 258-259.

    Notes the identity of many of the fossils from the two regions, although they have been described under different names.

—— Unconformity of the Cretaceous on older rocks in central New Mexico. Am. Jour. Sci., 4th ser., vol. 18, 1904, pp. 360-362, 2 figs.

    Describes the relations of the Cretaceous rocks to the underlying formations. Includes a table giving a general geologic section for New Mexico, showing the sequence, thickness, and lithologic character of the formations.

—— Structures of basin ranges. Jour. Geology, vol. 13, 1905, pp. 63-70, 5 figs.

    Discusses systems of faulting and the general geologic structure of the basin ranges of New Mexico, and the physiographic development of the New Mexican region.

—— The fundamental complex beyond the southern end of the Rocky Mountains. Am. Geologist, vol. 36, 1905, pp. 112-122.

    Discusses age, relations, and character of igneous and altered clastic rocks occurring in the New Mexican portion of the Rocky Mountains.

—— Ore deposits of the Sierra de Los Caballos. Eng. and Min. Jour., vol. 80, 1905, pp. 149-151, 3 figs.

    Describes the general geology of the region and the occurrence and character of the lead deposits.

—— Zinc carbonate ores of the Magdalena Mountains. Min. Mag., vol. 12, 1905, pp. 109-114, 5 figs.

    Describes the geology of the range and the occurrence and relations of the zinc-ore deposits.

—— Geology and underground water conditions of the Jornada del Muerto, New Mexico. Water-Supply Paper U. S. Geol. Survey No. 123, 1905, 42 pp., 9 pls., 11 figs.

    Describes the physiographic character of the region, the geologic structure, the occurrence and relations of Archean, Algonkian, Carboniferous, Jurassic, Triassic, Cretaceous, and Quaternary deposits and of the eruptive rocks, and the underground-water resources.

—— Triassic system in New Mexico. Am. Jour. Sci., 4th ser., 1905, vol. 20, pp. 423-429, 1 fig.

    Discusses the geologic position of the "Red Beds" of the Great Plains and the Southwest, and the occurrence and relationships of the Carboniferous and Triassic "Red Beds" in New Mexico.

—— The Jurassic horizon around the southern end of the Rocky Mountains. Am. Geologist, vol. 36, 1905, pp. 289-292, 1 fig.

    Discusses the stratigraphic and time relations of some Mesozoic formations in New Mexico.

—— Northward extension of the Lake Valley limestone. Proc. Iowa Acad. Sci., vol. 12, 1905, pp. 169-171.

    Describes the occurrence of lower Carboniferous rocks in New Mexico.

—— Genesis of the Lake Valley silver deposits. Trans. Am. Inst. Min. Eng., 39, 1908, pp. 139-169.

LEE, W. T. Water resources of the Rio Grande valley in New Mexico. Water-Supply Paper U. S. Geol. Survey No. 188, 1907.

    Describes briefly the geography and geology of the Rio Grande valley and discusses the conditions relating to underground-water supplies.

—— Notes on the red beds of the Rio Grande region in central New Mexico. Jour. Geology, vol. 15, 1907, pp. 52-58.

    Describes the red sandstones and shales including the gypsiferous beds usually classed as a part of the complex called the "Red Beds" of the Rocky Mountain region. Discusses their relations and shows that there are two red formations, one of which is Pennsylvanian and the other Upper Cretaceous in age.

—— The Manzano group of the Rio Grande valley. U. S. Geol. Survey, Bull. 389, 1909.

MILLER, S. A. Sub-Carboniferous fossils from the Lake Valley mining district of New Mexico, with description of new species. Jour. Cincinnati Soc. Nat. Hist., vol. 4, 1881, pp. 306-315.

    Gives determination of fossils collected by E. D. Cope and describes several new species.

SHUMARD, G. G. The geological structure of the Jornada del Muerto, New Mexico. Trans. St. Louis Acad. Sci., vol. 1, 1860, pp. 341–353.

An abstract of the geologic report of the expedition under Capt. John Pope, U. S. Topographic Engineers, for boring artesian wells along the thirty-second parallel. Gives the itinerary of a trip across the Jornada and discusses the geologic structure of the region.

SILLIMAN, BENJ., JR. Geological age of the Lake Valley mines of New Mexico. Eng. and Min. Jour., vol. 34, 1882, p. 214.

Letter in reply to E. D. Cope concerning age of the ore-bearing limestones at Lake Valley.

——— Mineral regions of southern New Mexico. Trans. Am. Inst. Min. Eng., vol. 10, 1882, pp. 424–444; Eng. and Min. Jour., vol. 34, 1882, pp. 199–200, 212–219.

Describes the geology of the region west of the Rio Grande from Socorro southward and comments on the ores and minerals found.

SPRINGER, F. M. On the occurrence of the lower Burlington limestone in New Mexico. Am. Jour. Sci., 3d ser., vol. 27, 1884, pp. 97–103.

Gives section of the rocks occurring at Lake Valley, together with a list of lower Carboniferous fossils collected from the upper limestone beds. Faunal relations discussed and regarded as establishing conclusively the lower Burlington age of the limestones.

STEVENSON, J. J. Note on the Laramie group of southern New Mexico. Am. Jour. Sci., 3d ser., vol. 22, 1881, pp. 370–372.

Describes the occurrence of coal in the San Pedro locality, 23 miles south-southeast of Socorro (Carthage coal field). Discusses the stratigraphic relations and refers the beds containing the coal to the Laramie.

STORRS, L. S. Rocky Mountain coal fields. Twenty-second Ann. Rept. U. S. Geol. Survey, pt. 3, 1902, pp. 415–471.

Describes the coal fields of the Rocky Mountain region, including (pp. 449–453) those of New Mexico. Those of the Carthage field, like the others, are considered to be of Laramie age.

WHITE, C. A. Note on fossils from Lake Valley, New Mexico. Am. Naturalist, vol. 15, 1881, p. 671.

CLIMATE AND VEGETATION.

The belt here discussed lies within the arid region and owing to the diversity of altitude shows considerable variety in climate and vegetation. In the Rio Grande valley the climate is warm and dry, the average temperatures ranging from 77.8° in July to 36.3° in January, with an annual average of about 57.7°. The normal precipitation is under 10 inches in the valley and ranges from 10 to 20 inches along the slopes of the Black Range. At Socorro, as shown by the reports of the United States Weather Bureau, the average precipitation for the six years preceding August, 1902, was 8.03 inches, the greatest being 10.61 inches in 1896–97 and the least 4.75 inches in 1899 and 1900. For the year ending with September, 1905, the precipitation reached the unprecedented figure of 21.05 inches. Except for one or two months in midsummer and during the early spring, when strong southwesterly winds prevail, the climate is very pleasant.

In the elevated regions to the west of the Rio Grande valley the fall of rain and snow is much more frequent and the average temperature lower, giving a summer climate that for salubrity and comfort is equaled in few other regions. The snowfall ranges from a few inches to several feet, and the higher parts of the mountains are usually covered with snow from two to three months each year.

The differences in meteorologic conditions are naturally attended by corresponding differences in vegetation. In the valley near the rivers the mesas are but scantily clothed with vegetation, but there is a notable increase in the growth of range grass and other plants toward the foothills. On the mesas and bolson plains the larger forms of vegetation consist chiefly of cacti, greasewood, yucca (soapweed), etc. In the river bottoms are cottonwoods, with a varying undergrowth of smaller shrubs. With increase of elevation there appears a scattered scrubby growth of oak, cedar, and juniper, which are "at first gnarled, stunted, and timid, at length bold, exuberant, and well favored."[a] The higher elevations are covered with a varying growth of timber, consisting chiefly of yellow pine, interspersed with cedar, juniper, etc. The best timber is found on the upper slopes of the Black Range. Of less importance is that on the Magdalena Range and the San Mateo Mountains, though in some parts of these mountains the stand is good.

---

[a] Dutton, C. E., Sixth Ann. Rept. U. S. Geol. Survey, 1885, p. 125.

CULTURE.

There are few towns within the district. The most important of them are Socorro, the county seat of Socorro County, and San Marcial, both of which are located on the Albuquerque and El Paso line of the Atchison, Topeka and Santa Fe Railway. Magdalena is the terminus of a small branch which connects with the Santa Fe at Socorro, and Lake Valley is the terminus of another small branch which connects with the Silver City branch at Nutt. There are a number of small interior towns, the most important of which are Hillsboro, the county seat of Sierra County, Fairview, and Kelly. Kingston and Hermosa were at one time thriving mining towns, but with the closing of the mines these places have declined until little remains but decayed remnants of their former activity.

The Albuquerque and El Paso line of the Atchison, Topeka and Santa Fe Railway parallels the eastern border of the district, sending off a branch at Socorro to Magdalena, 28 miles west, and another at Rincon to Silver City and Lake Valley. The lack of adequate transportation facilities in the west-central part of the Territory constitutes a serious obstacle to the development of its natural resources. Undoubtedly mine production would be notably increased in the Black Range and Mogollon districts were it not for the great cost of marketing the ores.

MINING DISTRICTS.

The chief mining districts in this belt, with their principal metals, are given in the following list. The most important metals are indicated by italics.

Socorro	*Silver*, gold.
Canyoncito	Lead, copper.
Magdalena	*Zinc, lead.*
Rosedale	*Gold.*
Black Range	Gold, silver, copper.
Apache	*Silver, copper*, gold.
Hermosa (Palomas)	*Silver.*
Hillsboro (Las Animas)	*Gold, silver, copper*, lead, manganese, iron.
Kingston	*Silver*, lead, copper.
Carpenter	Lead, zinc.
Tierra Blanca	*Silver, gold* (worked intermittently).
Lake Valley	*Silver*, lead, manganese.
Pittsburg	Gold placers.
Caballos	Copper, lead.

PHYSIOGRAPHIC FEATURES.

DRAINAGE.

With the exception of the Rio Grande, there are no permanent streams in the region. A number, however, are supplied by springs which furnish a flow of water that may extend for some distance but that eventually sinks into the sand, to reappear, possibly, at intervals farther down the valley.

The Rio Grande approximately follows the eastern boundary of the belt as far south as San Antonio, a distance of 30 miles. Here the stream bends to the west and then swings eastward around the south end of the Caballos Mountains in the vicinity of Rincon. In this westward swing the stream, which toward the north has followed the eastern foot of the Lemitar Range and Socorro Peak, passes to the west of the Cristobal and Caballos ranges.

To the east of the two last-named mountain ranges lies a great flat-bottomed basin or bolson plain tilted to the south, called the Jornada del Muerto, or "journey of death," from the dangers experienced by travelers across these waterless plains before the advent of the railroad. At the north and south ends of the Cristobal Range, and farther south at Palomas Canyon and south of the Caballos Range, the western rim of the basin is being dissected by the lateral tributaries of the Rio Grande.

In the vicinity of San Marcial an extensive bed of basalt covers the mesas on the east side of the river, its western edge forming a bold escarpment opposite the town. This flow, of comparatively recent date, evidently extended across the valley to the west and dammed the stream for a time. Farther north, at San Acacia, another flow of basalt temporarily turned the stream from its course. Later the stream cut its way through the barrier, forming a gorge with a small mass of basalt on the northwest side.

The present valley varies in width from 2 to 4 miles. In the spring the river carries an abundant supply of water from the melting of snow in the mountains of Colorado. Freshets may also swell its volume to dangerous proportions at other times. During the dry season, however, the flow dwindles and almost disappears on the surface.

The drainage of the district is effected by a network of arroyos connecting with the Rio Grande. The most important of these tributaries within the limits of the district are Alamosa River, Cuchillo Creek, Palomas Creek, Arroyo Seco, Rio Las Animas, Rio Percha, and Mimbres River. Alamosa River has its origin in the union of two creeks from warm springs in the basin to the west of the San Mateo Mountains, flows southeastward, and passes out into the Rio Grande valley through a canyon (Cañada Alamosa) in the gap between the San Mateo and Cuchillo mountains. Just above the entrance of the stream into the canyon is located the old Fort Ojo Caliente, where fine springs of warm water occur. Mimbres River rises west of the Black Range, in Grant County, and flows to the south, but is soon lost in the sand of the plains that border the mountains on the south in Luna County. Most of the remaining streams have their source along the eastern slopes of the Black Range; permanent supplies of good water occur in springs along their upper courses and can usually be obtained at a moderate depth below the surface in the lower valleys. Owing to the large run-off during times of heavy precipitation, the volume of water carried by the streams is great, resulting in vigorous sheet and stream erosion.

The area here described lies at the southeast border of the great plateau country, where it swings northward along the Rio Grande. Writing in 1885 Dutton [a] says:

The plateau country attains its greatest southing in the southwestern part of New Mexico. Little is known of its geology there. Its boundary then swings around to the northeast, and finally to the north, where it touches the Rio Grande. The river now becomes for a time the definite and unmistakable boundary line. At length, some 20 miles north of Albuquerque, that remarkable sharpness of limit which thus far has characterized its edges is no longer seen, and its separation from the region which embraces the Rocky ranges of Colorado and of northern New Mexico is more or less arbitrary.

In this part of its course the Rio Grande appears to follow a great fault zone along the foot of scarps which from Socorro northward face the east, while southward from that point they face the west. The relations of the larger structural valley are too complex and the data in hand are insufficient to permit an adequate description. The average width of this great structural trough is nearly 20 miles. In places over it appear small eminences representing remnants left by the erosion of small blocks of tilted strata in some places and of flows of basalt or other eruptive rock in others. The most pronounced displacements along this main zone of faulting are represented on the north by the Lemitar and Socorro mountains, where the blocks are tilted to the west, and farther south by the Cristobal and Caballos ranges, where the blocks are tilted to the east.

East of the river in the northern part of the area, rising above the general level of the plain formed of fluviatile wash deposits, presumably of Pleistocene age, occur blocks of older formations tilted in various directions and worn by erosion. In some places knobs of granite appear in relations which suggest the presence of that rock generally at no great depth below the valley plain.

About 20 to 25 miles west of the displacement marked by the Caballos and Cristobal ranges, is a lesser one, which toward the north is represented by the Sierra Cuchillo. Here also the fault scarp faces the west. The range ends not far south of Fairview, but the extension of the

---

[a] Dutton, C. E., Sixth Ann. Rept. U. S. Geol. Survey, 1885, p. 116.

fracture zone appears in minor displacements and faults as far south as Hillsboro. Although the chief displacement evidently took place in Tertiary time, evidences of later movement appear in places in the faulting of the Palomas gravel. Flows of basalt occur along this belt, also along the larger fault to the east.

MOUNTAINS.

The general trend of the mountain ranges is north and south. Socorro, Lemitar, and Polvadera are names applied to different parts of the range which extends for a distance of 25 miles along the river northward from Socorro. Socorro Peak is the highest point in this range, with an elevation of 7,280 feet.

To the west and south is the Magdalena Range, about 20 miles long, the highest point, known as Old Baldy, rising to a height of 10,798 feet.

The San Mateo Mountains lie to the southwest of the Magdalena Range, and cover a roughly elliptical area of about 800 square miles, sharply trenched on all sides by lateral drainage.

Rising out of the San Augustin Plains, to the west of the San Mateo Mountains, and running southward for a distance of nearly a hundred miles is a system of elevations belonging to the great Continental Divide, called the Black Range. Some of the highest peaks are Victorio's Outlook and Hagen's Peak, near Chloride; Mimbres Peak, southwest of Hermosa; Hillsboro and Sawyer's peaks, near Kingston. The south end of the range proper is sometimes referred to as the Mimbres Range. The range flattens out considerably to the south of Lake Valley, but renews itself again in Cooks Range, an outlying series of elevations, the highest of which is Cooks Peak. The following are some of the elevations as recorded by the Wheeler Survey:

Elevations in Black Range.

	Feet.		Feet.
San Mateo Peak	10,209	Pass at head of Rio Percha	8,264
Mimbres Head	9,832	Cooks Peak	8,330
Hillsboro Peak	10,061		

East of the river in Sierra County rise the Cristobal and Caballos ranges, 20 and 30 miles long, respectively. They are separated west of Engle station by a broad gap through which passes the stage road to Fairview and other points on the west side of the river.

The types of mountain structure represented in the district may be classed as (1) the tilted mountain, whose primary feature is due to the displacement of a crustal block; (2) that due to the upthrust of a granitic core; and (3) that resulting from the accumulation of volcanic material. The first two are usually modified by igneous activity resulting in extensive accumulations of volcanic products and by intrusions. The modifications effected by erosion in all these types have been extensive, resulting in the removal of vast deposits of the lavas from large areas. In general the volcanic activity seems to have been subsequent to the disturbances which primarily outlined the mountain, but in some places it evidently accompanied these disturbances.

To the first or faulted-block type belong several ranges, among them the Lemitar, Socorro, Magdalena, Cristobal, Caballos, and Cuchillo, though in all of these the present relief has been accentuated by vast outflows of lava which erosion has carved into a system of rugged peaks and deep canyons.

Of the second type are the Black and Mimbres ranges, in which the uplifted strata slope away on both sides from a central mass of granite. At present, however, the most prominent features of the ranges are due to the piling up of vast deposits of volcanic material. In this type the removal of either limb by erosion would produce a mountain apparently indistinguishable from a monocline, but the competency of erosion to bring about this result is questioned.

In the third type the mountain is constructed wholly of volcanic material. Of this type the most prominent examples are the San Mateo Mountains, which occupy a position intermediate between the Magdalena and the Black ranges. These mountains consist almost entirely, so far as the exposed portions are concerned, of rhyolites and their associated volcanic products.

Rising out of both mesas and bolson plains in a number of places occur isolated hills, usually of conical shape, though locally elongated and irregular in outline, which represent the remnants left by the erosion of extensive flows of lava. Examples of such hills occur west of Magdalena, where Cat Mountain and the Tres Montosos are prominent, and in places along the west side of the Rio Grande valley southward from Socorro. Along with these occur the lenticular elevations due to the tilting of small blocks of strata, which have not yet suffered complete destruction by erosion.

BOLSON PLAINS.

As defined by Hill[a] and further elaborated by Tight,[b] the bolson is a plain formed by the more or less complete filling of an intermontane basin with detritus derived from adjacent eminences. "Along the margins of these plains are talus hills and fans of bowlders and other wash deposits brought down by freshets, while scattered over the interior may be deposits of lacustral origin." The bolson is therefore a constructional plain and is genetically a type of the gradation plain. As shown by Tight, the essential feature of the bolson is that the plain is bordered by mountain forms or plateau escarpments that have supplied the detritus with which it is floored. The definition therefore does not concern itself so much with the formation of the original trough or basin as with the conditions which produce a greater contribution of material than preexisting drainage can remove. A more or less completely closed basin is, therefore, an essential feature of a bolson plain.

One of the most prominent and characteristic plains of this type is the San Augustin Plain west of Magdalena. This plain is about 60 miles long and 12 to 20 miles wide, the longer axis extending in a northeast-southwest direction. Along the borders rounded hills of volcanic rock rise out of the débris filling the basin. The plain is inclosed on the west and northwest by the Datil Range, on the south by the Mogollon and San Mateo mountains, on the east by the Magdalena Mountains, and on the north and northeast by the Gallina and Bear Spring mountains. The mean elevation of the plain, according to the Wheeler Survey, is 6,900 feet, but the mountains forming the rim on the north and west are from 1,500 to 2,000 feet higher. The basin receives all the drainage of the slopes bordering on the north and west sides, but toward the east and south the plain connects over low cols with adjoining smaller bolsons, some of which merge with those of the Rio Grande valley.

The Rio Grande valley has been described as a series of bolson plains more or less dissected by the present stream and its tributaries. R. T. Hill[c] states that the river "from its entrance to the San Luis Valley in southern Colorado to where it cuts the Sierra del Carmen in longitude 103°, just east of the apex of the great bend, flows almost continuously through a chain of ancient bolson plains connected by canyons which progressively increase in length and depth toward the southeast," but this view of the valley is disputed. Of these so-called basins the Jornada del Muerto is perhaps one of the most noted.

As shown by W. T. Lee,[d] however, the Jornada del Muerto is rather a nearly level detrital valley plain 10 to 20 miles or more in width extending from San Marcial southward to Las Cruces and bounded on the east by the San Andreas Range and on the west by the Caballos and Cristobal ranges. The altitude of the plain at the north end near San Marcial, according to Lee, is about 4,700 feet and at its south end 4,250 feet. The southward slope of the plain is therefore 450 feet in 100 miles of length, or an average gradient of 4.5 feet to the mile.

The sedimentary rocks exposed in the mountain slopes on either side dip toward the axis of the valley, apparently forming a syncline, and the relations have been described as such in various papers.[e] It is doubtful if the structure of the region is as simple as this. In the interior the plain is covered with detritus to a depth of at least several hundred feet, no wells

---

[a] Hill, R. T., Topographic Atlas U. S., folio 3, U. S. Geol. Survey, 1900, p. 8.
[b] Tight, W. G., Am. Geologist, vol. 36, 1905, pp. 271-284.
[c] Op. cit., p. 9.
[d] Water-Supply Paper U. S. Geol. Survey No. 188, 1907, p. 10.
[e] Keyes, C. R., Water-Supply Paper U. S. Geol. Survey No. 123, 1905, p. 26.

having penetrated its full thickness. Various facts point to the inference that this was formerly the course of the Rio Grande and that, as in other parts of the main valley, the underlying strata have been subjected to block faulting, though no indications of this appear through the heavy covering of detritus.

Another sloping plain occupies the side of the valley north of San Marcial and a more extensive one occurs south of that point, spreading out in a broad, steeply sloping plain from the south end of the San Mateo Mountains. When viewed narrowly these plains correspond essentially with Johnson's[a] "gradation plains," which are by Herrick[b] termed "clinoplains." The more appropriate term alluvial plain has been adopted for them by the United States Geological Survey. Two types of alluvial plains are represented. Along the side of the valley resting against the flanks of the mountains is the talus plain, or, as Johnson calls it, the "débris apron." This plain is made up by the confluence of the alluvial fans, together with deposition from sheet-flood erosion along the foot of the bordering mountain slopes. The material composing this plain has been derived from the adjoining heights and its surface has a relatively high grade, the result primarily of the ground absorption of the run-off on meeting the lessened slope of the valley plain.

As stated by Herrick, "in some places the talus plain reaches the bottom, but generally it passes abruptly into the plain of the second type or clinoplain proper." The surface of this plain is but slightly inclined and is broken by arroyos with steep walls and nearly parallel courses, the interstream areas being intersected by an intricate network of small, shallow tributaries. In places this plain also reaches to the bottoms, the bordering bluffs varying in height with the fluctuations in the course of the stream. Along the border of this plain occur parts of a degradational plain formed by the river at some former stage of its existence. A prominent remnant of this plain occurs on the west side of the Rio Grande in the vicinity of Cuchillo, where it has a width of several miles, and in it head many small ravines and arroyos that lead to the river. The inclination of this plain is less than that of the upper plain. Its level varies according to location, but in general it appears to be about 100 to 150 feet below that of the outer plain. Its height above the river bottoms is apparently a little more. Toward the south the alluvial plain laps about the south end of the Mimbres Range and merges with the great plain which extends westward to the Arizona line.

Within and behind the lower elevations along the west side of the valley occurs a series of small more or less closed alluvial plains, most of which are deeply dissected by the headwaters of tributaries of the Rio Grande that reach the main valley through narrow gaps in the foothills. Two of them, one lying to the west of the San Mateo Mountains and the other to the east, connect with the San Augustin Plain, as previously described. The first of these extends to the south of Fairview. Its northern half is dissected by the headwaters of Alamosa River, which passes out through a gap in the rim at Ojo Caliente, the southern half is drained by Cuchillo Creek and its branches. Another basin of the same type lies between the foothills and the main range at Hillsboro.

The original surface of these bordering plains, which were covered in many places by flows of basalt, reaching over into adjoining portions of the main valley, has been much dissected by the drainage of the streams that have cut deeply into the gravel deposits, leaving buttes capped here and there by remnants of the basalt flow.

The surface of these smaller plains is intermediate in elevation between that of the great San Augustin and Rio Grande plains. The elevation is approximately 6,000 feet, with a general slope toward the south.

GEOLOGY.

GENERAL RELATIONS.

The area here considered lies at the southeast border of the Great Plateau country,[c] and like the great physical unit of which it forms a part, is characterized geologically by the

---

[a] Johnson, W. D., The high plains and their utilization: Twenty-first Ann. Rept. U. S. Geol. Survey, pt. 4, 1901, pp. 609–741.
[b] Herrick, C. L., Clinoplains of the Rio Grande: Am. Geologist, vol. 33, 1904, pp. 376–381.
[c] Dutton, C. E., Sixth Ann. Rept. U. S. Geol. Survey, 1885, p. 116.

horizontal position of its beds; displacement or dislocation of strata takes place with little or no folding and without well-marked anticlines and synclines. Where deformation has been intense, especially near the margin of the province, the displaced portions tend to assume the form of tilted blocks.

System.	Series.	Thickness.	Groups and formations.		Character of rocks.	
Quaternary.	Recent.	Feet.			Alluvial sands, clays, and gravels.	
	Pleistocene.	900	Palomas gravel		Conglomerates and gravels.	
Tertiary.					Eruptives, andesites, rhyolites, etc.	
Cretaceous.	Upper Cretaceous.				Yellow sandstones and shales with deposits of coal.	
Carboniferous.	Pennsylvanian.	500–600	Manzano group.	San Andreas limestone.	Limestones.	
		500–1000		Yeso formation.	Vermilion, pink, and yellow sandstones, some shales and limestone, and deposits of gypsum.	
		400–800		Abo sandstone.	Red sandstones and conglomerates, with some shales and limestones.	
		1000–1200	Magdalena group.		Sierra County.	Socorro County.
					Limestones and some shales.	Madera limestone. Limestones with some shales and sandstones.
						Sandia formation. Shales, limestones, and quartzitic sandstones.
	Mississippian.	300	Lake Valley limestone.		Crinoidal, blue, and nodular limestones.	Kelly limestone. Granular limestones, 125 feet.
Devonian.		200	Percha shale.		Gray fossiliferous shales above; black fissile shales below.	
Silurian-Ordovician-Cambrian (?).		900	Mimbres limestone.		Limestones, mostly Ordovician. Silurian fossils in places at top. Lower part may be Cambrian.	
Cambrian.		200	Shandon quartzite.		Quartzites and siliceous shales.	
Pre-Cambrian.					Granites, gneisses, and schists.	

FIGURE 15.—General section of geologic formations in Sierra County and central Socorro County.

The district abounds in volcanic rock. It constitutes a part of the great volcanic area which, as shown by Dutton, characterizes the marginal portion of the plateau country. Most of the eruptions appear to belong to middle or late Tertiary time, but no attempt was made to determine their age definitely. More recent (early Quaternary) eruptive rocks are represented in the basalt flows which cap the mesa gravel beds (Palomas gravel) in many places along the Rio Grande valley.

The sedimentary formations represented in the region constitute a nearly complete section of Paleozoic rocks. Beginning with the Cambrian, which rests unconformably upon older rocks, the Paleozoic is represented also by the Ordovician, Devonian, and Carboniferous (both Pennsylvanian and Mississippian), and probably by the Silurian. Jurassic and Triassic rocks are not known to be present in this district; some formations heretofore assigned to these divisions are now proved to be upper Carboniferous. It is possible, however, that some of these rocks may be represented here. Cretaceous strata occur east of the Rio Grande, but have little or no development on the west within the boundaries of the belt described. They are well represented, however, in the western part of Socorro County. Tertiary sedimentary rocks are of small extent and importance, the next succeeding deposits being Quaternary. Over considerable areas the sedimentary deposits are covered by volcanic products, from beneath which, however, they crop out here and there most unexpectedly. The most complete sections were observed in the southern part of the Black Range about Hillsboro, Kingston, Lake Valley, and Cooks Peak, and in the Caballos Mountains near Shandon. A general columnar section of the rocks in this region is given in figure 15. The profile section (fig. 16) will give an approximate idea of the formations as they occur in this part of the district.

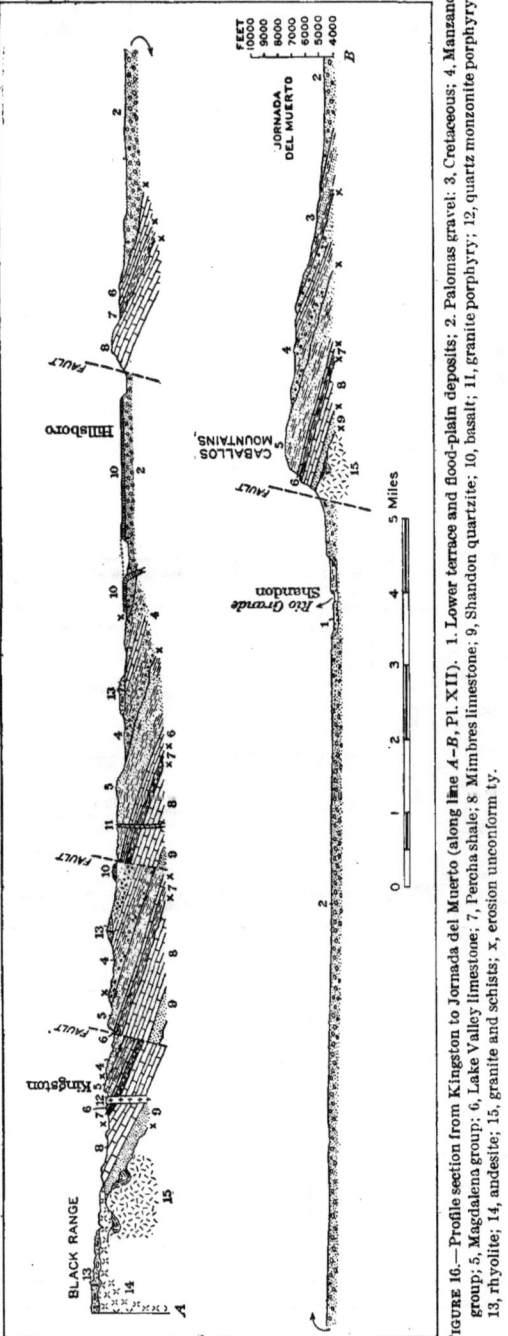

FIGURE 16.—Profile section from Kingston to Jornada del Muerto (along line A-B, Pl. XII). 1, Lower terrace and flood-plain deposits; 2, Palomas gravel; 3, Cretaceous; 4, Manzano group; 5, Magdalena group; 6, Lake Valley limestone; 7, Percha shale; 8, Mimbres limestone; 9, Shandon quartzite; 10, basalt; 11, granite porphyry; 12, quartz monzonite porphyry; 13, rhyolite; 14, andesite; 15, granite and schists; x, erosion unconformity.

Exposures of the lower Paleozoic formations are known to occur only in the southern portion of the district. The Paleozoic rocks appear at the surface at Lake Valley, and northward along the Black Range to Chloride. Northward of this point the sedimentary formations occur only in more or less detached areas, being for the most part hidden by the covering of volcanic rock and recent deposits. In the Magdalena, Socorro, and Lemitar ranges the upper Paleozoic rocks appear in more or less restricted areas. In the Caballos and Cristobal ranges the lower Paleozoic strata are brought to view along their western face by faulting. The upper Carboniferous "Red Beds" occur in many places along the Black Range in narrow bands adjoining the limestone on the east. They occur also in several localities in the northern part of the district and along the eastern slopes of the Caballos and Cristobal ranges.

The Pleistocene is represented by an extensive development of consolidated gravel and bowlder beds made up chiefly of volcanic débris, to which the name Palomas gravel has been given.

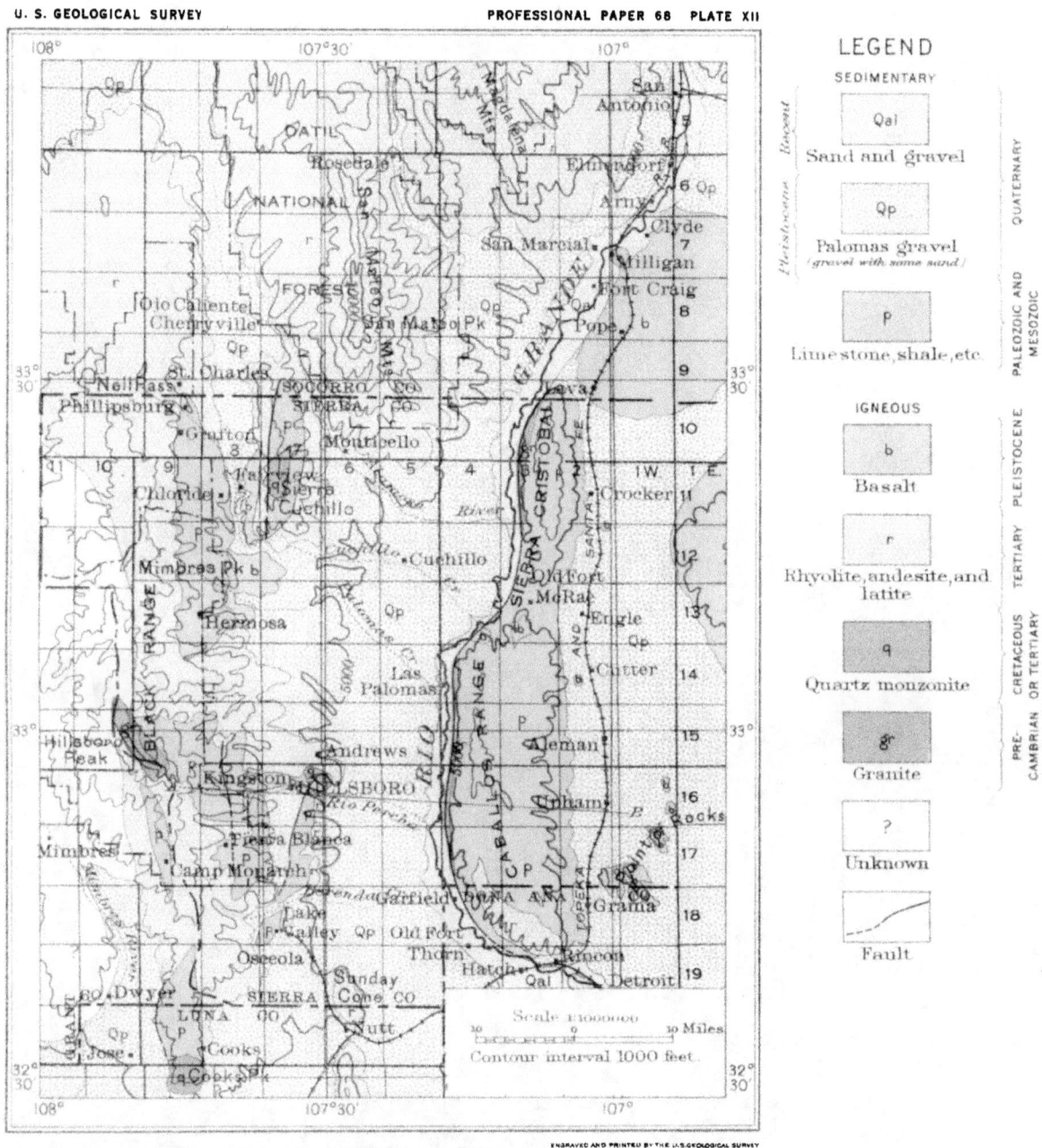

GEOLOGIC SKETCH MAP OF PARTS OF SAN MATEO MOUNTAINS AND THE BLACK, CABALLOS, AND SIERRA CRISTOBAL RANGES, NEW MEXICO

By C. H. Gordon
1909

These beds have a thickness of 700 to 900 feet along the western border of the Rio Grande valley. (See Pl. XII.)

PRE-CAMBRIAN ROCKS.

Wherever uplift and erosion have brought to view the lower Paleozoic rocks they are shown to rest upon a foundation of granite and crystalline schists. The prevailing rock is a coarse red granite cut by numerous dikes of fine-grained granite, aplite, pegmatite, and others of a more basic type. Associated with the granite in places, and inclosed in it, are hornblende schists and granite gneiss. In some localities the granite is undergoing rapid disintegration, and satisfactory hand specimens are difficult to obtain. Good exposures of these pre-Cambrian rocks were observed along the western base of the Caballos Range and the Black Range, near Kingston. A part of the central mass of the Magdalena Mountains is also of pre-Cambrian age, though a considerable portion is a later granite porphyry of intrusive origin. Near Shandon, in the Caballos Range, the granite is associated with hornblende schists and granitic gneiss, and besides being cut by numerous acidic and basic dikes, is intersected at short intervals by quartz veins having for the most part a general northwest-southeast course.

In the Florida Mountains a good exposure of these rocks occurs at the foot of Capitol Dome, 10 miles southeast of Deming. Here the rock is a very coarse granite, in places much decomposed and crumbling rapidly under the attack of the weather. A few miles northwest of Kingston, and at other places along the axis of the range, the granite outcrops over considerable areas. It is here but little decomposed and is cut in different directions by dikes of greenstone, aplite, etc. Upon the irregular surface of the granite rests unconformably the overlying Cambrian quartzite, to which the name Shandon quartzite has been given, from its typical exposures at Shandon.

CAMBRIAN SYSTEM.

SHANDON QUARTZITE.

Until recently little evidence had been offered showing the occurrence of the lower Paleozoic beds within the limits of New Mexico. The investigations of 1905,[a] however, have revealed the presence of a nearly complete sequence of these rocks in this part of the Territory.

Overlying the pre-Cambrian rocks, wherever they are exposed to view, is a considerable thickness of quartzite, followed upward by limestone, the lower beds of which have a tendency to crystalline granular structure and constitute a fair quality of marble. The transition zone between the quartzites and the limestones, having a thickness of 25 to 50 feet, is marked by alternation of thin bands of quartzite and limestone. Farther to the northwest, as shown by L. C. Graton, these alternating beds extend through a thickness of about 1,000 feet. Three distinct phases of the quartzite were noted in the district. Resting upon the granite usually is a coarse, dark, iron-stained quartzite, the lowermost beds of which are more or less conglomeratic and contain much feldspar in large irregular grains, as well as rounded pebbles of white quartz. The white pebbles appear also at intervals through the formation. These lower beds are generally massive and their maximum thickness as observed in the Florida Mountains is about 60 feet. At this place they are dark brown and are overlain by about 75 feet of a coarse white quartzite which, on exposed surfaces, has a cellular structure due to the weathering out of some of its constituent minerals.

At the base of the Caballos Range near Shandon the Cambrian has a thickness of more than 50 feet. The section on page 226 was taken there.

In this section beds 2 to 5, inclusive, have been referred to the Cambrian from the fact that certain layers of No. 5 contain linguloid shells which have been identified by Charles D. Walcott as *Obolus* (*Westonia*) *stoneanus* Whitfield, an upper Cambrian form found at Newton, N. J., and in Wisconsin. In the lower part of No. 5 there is a layer of green glauconitic shale, with small lenses of limestone containing fragments of trilobites.

---

[a] Gordon, C. H., and Graton, L. C., Lower Paleozoic formations in New Mexico: Science, new ser., vol. 23, 1906, pp. 590–591; Am. Jour. Sci., 4th ser., vol. 21, 1906, pp. 390–395.

The total thickness of the quartzite formation is estimated to be 150 to 200 feet, though the greatest thickness observed in any one locality was about 135 feet.

In the El Paso region to the south, Cambrian rocks in similar position are known as the Bliss sandstone.

Section near Shandon.

Character.	Thickness (feet).	Age.
15. Limestones (not examined); probably upper Carboniferous.		Pennsylvania.
14. Cherty siliceous limestones.	50	Mississippian (Lake Valley limestone ?).
13. Subcrystalline limestones.	15	
12. Shales, fissile below, appearing sandy above.	50	Devonian (Percha shale).
11. Siliceous limestone, in places a flint breccia, or a drusy quartz rock. Thickness variable; apparently belongs to the formation below. Surface uneven.		Mainly Ordovician — Mimbres limestone.
10. Limestone; some beds massive, others thin and seamed with chert. In places the beds are coarsely brecciated, blocks of a calcareous gritstone appearing along with those of the limestone.	450	
9. White or pinkish quartzite.	15	
8. Crystalline granular limestone.	300	
7. Limestone filled with thin seams of chert, giving on weathered surfaces a banded or platy appearance.	150	Ordovician (?) or Cambrian (?).
6. Calcareous shales and shaly limestones.	15	
5. Dark ferruginous shales, with flaggy beds of quartzite above, containing in certain layers *Obolus* (*Westonia*) *stoneanus* Whitfield.	40	Cambrian (Shandon quartzite).
4. Massive dark, ferruginous quartzite.	5	
3. White quartzite.	3–5	
2. Dark ferruginous quartzite, absent in places.	5	
1. Coarse red granite, in places associated with hornblende schists, and cut by dikes of pegmatite, aplite, and diabase (?).		Pre-Cambrian.

MIMBRES LIMESTONE.

Overlying the Shandon quartzite conformably is a series of limestones approximately 1,000 feet in thickness. In the Shandon section these beds are represented by Nos. 6 to 11, inclusive. Sections corresponding to this were observed in the Mimbres Range in the vicinity of Kingston. (See Pl. XIII.) Lithologically the series presents a twofold aspect, the lower half consisting in large part of crystalline granular limestones and the upper of massive limestone, in many places coarsely brecciated.

At the base are about 15 feet of calcareous shales and shaly limestones, above which in places occurs limestone in thin beds interlaminated with thin seams of chert, presenting on weathered surfaces a peculiar banded appearance. The latter beds have in some localities a thickness of more than 150 feet. Similar beds have been described by F. L. Ransome[a] as a characteristic feature of the Cambrian Abrigo limestone in the Bisbee quadrangle, Arizona. No fossils were obtained from these beds by the writer, but in beds apparently similar in the Silver City district L. C. Graton obtained Ordovician fossils. Beds 6 and 7 of the Shandon section may therefore belong in the Cambrian.

These shaly and cherty members (Nos. 6 and 7 of section) have been observed in only a few places, and their persistence is a matter of doubt. The larger part of the lower division consists of a granular limestone (No. 8 of section) in beds of variable thickness, the total being more than 300 feet. These limestones (No. 8 of section) were seen in connection with the laminated cherty beds (No. 10 of section) only at Shandon, and there they appear to be separated from those beds by a pinkish-white quartzite or calcareous grit, the relations of which are, however, not absolutely certain.

Where regularly bedded, the upper division consists of blue and white subcrystalline limestones (No. 10 of section), usually in rather massive beds. In most of the exposures

---

[a] Prof. Paper U. S. Geol. Survey No. 21, 1904, p. 31, Pl. VI, *B*.

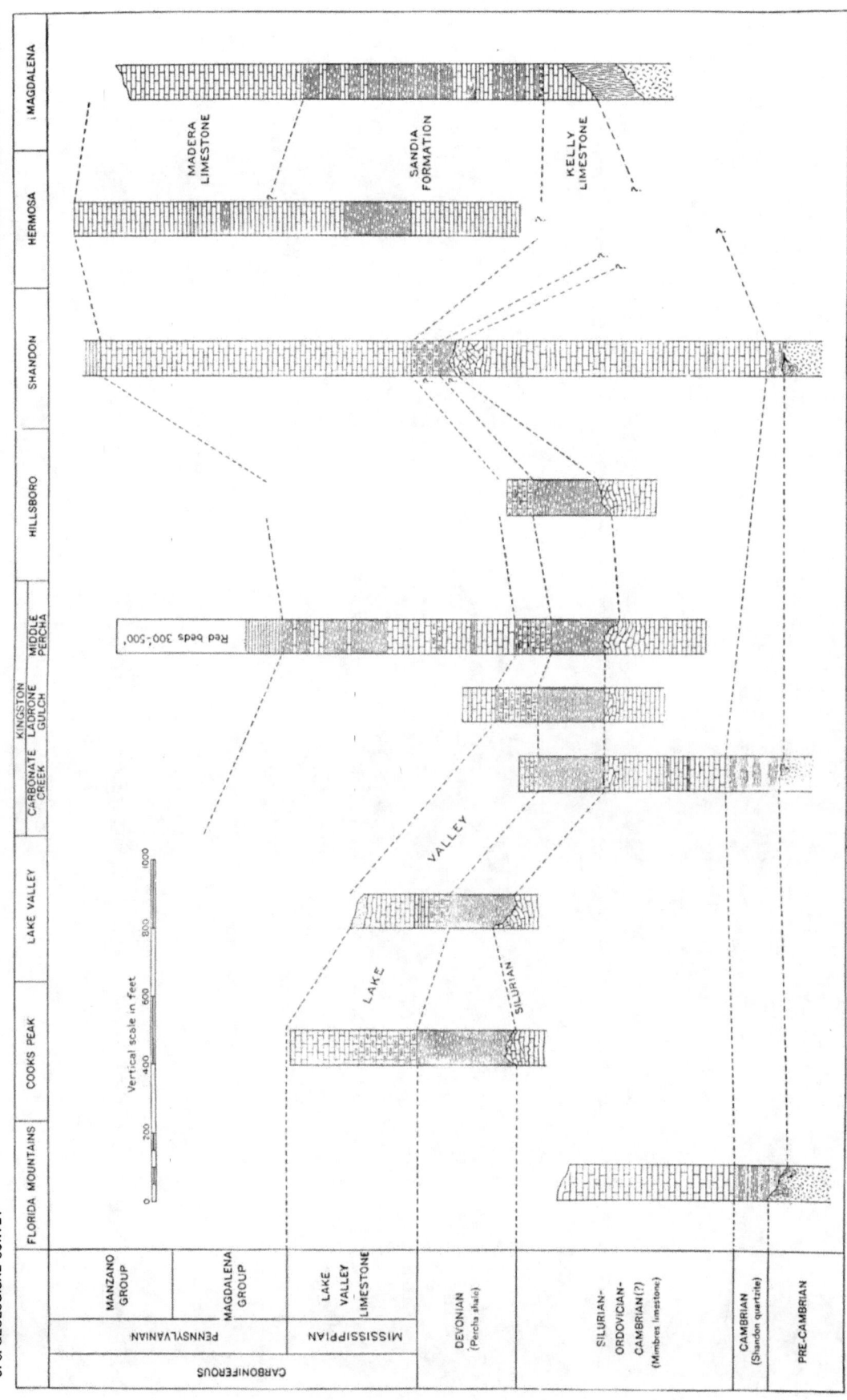

CORRELATION TABLE: CENTRAL SOCORRO, SIERRA, AND LUNA COUNTIES.

observed, however, these beds have been broken and the fragments, mostly from 1 to 5 or 6 feet in diameter, have been cemented together by calcareous material derived from the wear of the rock itself. In some places the breccia includes blocks of white grit or sandstone. Only at Shandon was the contact of these beds (No. 10) with those below observed, and there it has the appearance of a slight unconformity.

The thickness of the upper limestone (No. 10) varies greatly, but probably averages 350 or 400 feet. Its upper surface is uneven and in places there is a well-marked erosional unconformity at this horizon. This is seen in the walls of the canyon of Rio Percha about a mile east of Hillsboro, where the limestones (No. 10 of section) are overlain unconformably by a bed of red and white conglomeratic quartzite 6 to 10 feet thick, containing small nodules resembling a phosphatic substance which possibly indicates, as suggested by E. O. Ulrich, that this bed belongs at the base of the Devonian. The beds overlying the quartzite consist of a reddish-brown, compact siliceous rock which, so far as can be told from its structure, may have been derived from the Devonian black shales by the action of heated waters.

The upper portion of the limestone bed (No. 10) has also suffered more or less silicification, which has produced a highly siliceous, drusy quartzose rock, often called a "quartzite" or a "limestone-quartzite." These beds constitute the "foot-wall lime" for the ores at Cooks, Kingston, Tierra Blanca, and elsewhere.

In places the beds carry fossils, chiefly corals, in considerable numbers. A collection from the upper limestone (No. 10) near Hillsboro was examined by E. O. Ulrich, who reports as follows concerning them:

The fossils consist entirely of corals, of three genera and species—*Columnaria alveolata*, *Favosites asper* (late Ordovician variety), *Stromatocerium* cf. *pustulosum*. The horizon indicated by these species is the coral bed of Richmond (= late Ordovician) age recognized in the last two years in Missouri, Iowa, Wyoming, Utah, and in the Robledo and Caballos mountains of New Mexico. The same association of coral species occurs in the latest Ordovician (Borkholm limestone) in the Russian Baltic provinces, where, as in this country, it is accompanied by early representatives of the Silurian fauna.

Outcrops of the Ordovician limestones occur in many places along the east face of the Black Range from Cooks to Hermosa, and along the west side of the Caballos Range they constitute a narrow band along the face of the escarpment. From Hillsboro southward the upper siliceous beds, including the silicified portions of the overlying shales, appear at many places in irregular reefs projecting above the surface of the ground or as ledges capping the escarpments, from which circumstance they are sometimes called the "rim rock."

In the Lake Valley district there is at the top of the Ordovician limestone a bed of quartzite 5 to 10 feet thick overlain by 12 feet of reddish or pink limestone, from the outcrop of which on the west side of the Berenda Valley were obtained some imperfect fossils that, according to E. O. Ulrich, indicate the Silurian age of the rocks.

The limestone has not been recognized elsewhere in this region, but a quartzite equivalent to the one below the reddish limestone was observed in several localities. That outcropping in the canyon below Hillsboro is probably the same, though, as already stated, Ulrich suggests that the bed may belong to the Devonian.

The silicification which characterizes the contact with the overlying shale affected these beds to a marked extent, giving them a highly siliceous appearance on fracture. They much resemble quartzite and are usually filled with drusy cavities.

In the El Paso quadrangle, to the southeast of this area, in the interval between the Cambrian and the Devonian three distinct formations are recognized [a]—the El Paso limestone (Lower Ordovician), Montoya limestone (Middle and Upper Ordovician), and Fusselman limestone (Silurian). It is possible that when detailed work is done in New Mexico these subdivisions will be recognized in the area treated in this report. In the absence of such detailed studies, however, the general term Mimbres limestone, which has been applied to these limestones in Sierra and Socorro counties,[b] will be used here.

---

[a] El Paso folio (No. 166), Geol. Atlas U. S., U. S. Geol. Survey, 1909.
[b] Gordon, C. H., Jour. Geology, vol. 15, 1907, pp. 91-92, 809; Am. Jour. Sci., 4th ser., vol. 24, 1907, pp. 58-64.

DEVONIAN SYSTEM.

PERCHA SHALE.

Resting upon the eroded surface of the Mimbres limestone is a shale formation, varying in thickness from 50 to 250 feet, to which the name Percha shale[a] has been applied. In general the formation is about 200 feet thick. The lower half is a black fissile shale; the upper half is a blue clay shale, weathering to buff or brown, with little or no fissility. The black shales are well exposed at Lake Valley, but the upper division is mostly covered by talus. This formation is well exposed at Cooks, Kingston, Hillsboro, and elsewhere. The black shales contain no fossils and observers at Lake Valley have heretofore generally agreed in including them with the overlying limestones in the lower Carboniferous.

Two miles east of Hillsboro and at Kingston the blue shales were found to contain near the top well-preserved fossils in abundance, chiefly brachiopods. George H. Girty and E. M. Kindle, of the United States Geological Survey, to whom the collections were referred, state that the fauna is characteristically Upper Devonian. Doctor Girty adds that "it is one of peculiar interest, inasmuch as it is the same which was discovered years ago in the Ouray limestone in southwestern Colorado by Endlich. It is characterized by the large and striking species *Camarotœchia (Plethorhyncha) endlichi* Meek, and heretofore has not been recognized outside of the San Juan Mountains."

Mr. Kindle reports as follows concerning the fossils collected at Hillsboro and Kingston:

The fauna of the two localities is essentially the same, a large proportion of the species being common to both localities. The fauna includes the following species:

Zaphrentis sp.	Camarotœchia contracta (Hall)?
Spirorbis sp.	Athyris coloradensis Girty.
Leptæna rhomboidalis Wilckens.	Pugnax pugnus Martin.
Productella coloradensis var. plicatus Kindle.[b]	Spirifer whitneyi Hall.
Productella spinigera Kindle.	Syringospira prima Kindle.[b]
Productella sp.	Spirifer whitneyi var. animasensis (Girty).
Schizophoria striatula var. australis Kindle.[b]	Reticularia spinosa Kindle.[b]
Camarotœchia (Plethorhyncha) endlichi (Meek).	Meristella barisi Hall?

The species listed above represent the Upper Devonian fauna of the Ouray limestone of Colorado. The composition of the New Mexico fauna is strikingly similar to that of the Colorado fauna, and leaves no doubt as to the identity of the two. In Southern Colorado this fauna is followed by a fauna of Carboniferous age, the two faunas occupying the same limestone formation.

Portions of the shales at the bottom are in many places altered, presumably by heated siliceous waters, to a fine-grained, usually red to black siliceous rock resembling flint or jasper, corresponding evidently to the rock known as lydian stone or lydite. The rock usually presents to a greater or less degree the character of a breccia in which locally the fragments may be white, red, gray, or black in a matrix of like substance but different color. It is generally more or less seamed with quartz and filled with drusy cavities.

Owing to its susceptibility to erosion the formation is generally found in basins or troughs skirting the base of lower Carboniferous escarpments, and is usually more or less hidden by talus from the adjoining slopes.

The name Percha shale, which has been applied to this formation, is taken from Rio Percha, along which the best exposures of the fossiliferous beds occur.

MISSISSIPPIAN SERIES ("LOWER CARBONIFEROUS").

GENERAL FEATURES.

Exposures of rocks belonging to the Mississippian series ("Lower Carboniferous") occur at a number of places in this region. They have long been known to occur at Lake Valley, from which circumstance they early received the name Lake Valley limestone.[c] The observations of the writer show that exposures of the Lake Valley limestone occur in many places

---

[a] Gordon, C. H., Jour. Geology, vol. 15, 1907, pp. 91–92; Am. Jour. Sci., 4th ser., vol. 24, 1907, pp. 58–64.
[b] See Bull. U. S. Geol. Survey No. 391, 1909.
[c] Cope, E. D., Eng. and Min. Jour., vol. 34, 1882, p. 214.

about the southern extension of the Black and Mimbres ranges, and that rocks lithologically similar occur in the Caballos Mountains, though fossil evidence of their age is not at hand. In Socorro County there are but two areas in which small outcrops of supposed Mississippian rocks occur—one in the Magdalena Mountains, where they constitute the principal ore-bearing formation of the lead and zinc mines at Kelly, and another on Arroyo Salado at the base of the Sierra Ladrones, discovered in 1905 by W. T. Lee, of the United States Geological Survey. The limestones at Kelly are seemingly unfossiliferous, though Mississippian crinoids are reported[a] to have been found in them. In the absence of satisfactory data, however, these beds can not well be correlated with the Lake Valley limestone, which is known to be of Mississippian age. Herrick[b] gave to them the name "Graphic-Kelly" limestone, but a hyphenated name of this kind is objectionable, and Kelly limestone from the town in the vicinity of which they occur has been adopted instead.[c]

North of Socorro County the Mississippian rocks disappear by overlap, the Pennsylvanian series resting directly upon the pre-Cambrian.

LAKE VALLEY LIMESTONE.

Above the Percha shale in western Sierra County is a series of limestones with some shales, the upper beds of which at Lake Valley and Hillsboro are filled with fossils. The Carboniferous age of these beds was first recognized by E. D. Cope,[d] who, on the authority of C. A. White, referred them to the "Middle Carboniferous" (August, 1881). A little later (January, 1882) Benjamin Silliman, jr.,[e] on the authority of Arnold Hague and C. D. Walcott, assigned these beds to the "Lower Carboniferous." This conclusion was strengthened in 1883 by the investigations of F. M. Endlich,[f] whose excellent section of the formations at Lake Valley is reproduced on page 278. In a paper published in 1884, F. M. Springer[g] describes the Lake Valley beds and gives a list of the fossils obtained from them.

Ten years later Ellis Clark, at that time manager of the Lake Valley mines, published a paper[h] illustrated by map and sections, in which the stratigraphy and ore deposits are described with considerable detail.

The following is a section of these beds as exposed in the escarpment northwest of Lake Valley:

Section northwest of Lake Valley.

	Feet.
9. Capping of andesite.	
Lake Valley limestone:	
8. Coarse, subcrystalline, yellowish-white limestones in moderately thick beds, more shaly below. Abound in crinoids and other fossil forms. Some beds cherty. Full thickness not shown	60
7. Blue shale including thin beds of bluish limestone containing the same fossils as No. 9, but crinoids not so abundant	75
6. Grayish-blue hard, compact limestone, more or less siliceous at top. This is called the "Blue" limestone, and locally is known as the "foot-wall lime" from the fact that it underlies the ore deposits. It is overlain in places by a bed of flint breccia. The flint fragments, some of which carry silver, are of gray, brown, chocolate, pearl, and green colors, the green flints yielding ore of a higher grade than those of the other colors. At the base is a bed of coarse, crystallized yellowish-white limestone 5 feet thick	25
5. Compact grayish limestone filled with nodular chert. Shale partings usually rather thick	50
Percha shale:	
4. Grayish-yellow and blue shales	60
3. Black fissile shale	100
Mimbres limestone:	
2. Pink limestone, upper beds siliceous with drusy cavities; contains Silurian fossils	12
1. Basal quartzite and limestone (thickness not shown in section).	

---

[a] Herrick, C. L., Am. Geologist, vol. 33, 1904, p. 310; Jour. Geology, vol. 12, 1904, p. 138; Keyes, C. R., Proc. Iowa Acad. Sci., vol. 12, 1904, p. 169.
[b] Loc. cit.
[c] Gordon, C. H., Am. Jour. Sci., 4th ser., vol. 24, 1907, pp. 62–63, 807.
[d] Am. Naturalist, vol. 15, 1881, pp. 831, 832.
[e] Trans. Am. Inst. Min. Eng., vol. 10, 1882, pp. 424–444.
[f] Am. Naturalist, vol. 17, 1883, pp. 149–157.
[g] Am. Jour. Sci., 3d ser., vol. 27, 1884, pp. 97–103.
[h] Trans. Am. Inst. Min. Eng., vol. 24, 1894, pp. 138–169.

The recognition of the Mississippian ("Lower Carboniferous") facies of the fossils obtained from beds 7 and 8 is confirmed by G. H. Girty, of the Geological Survey, to whom the fossils collected by the writer were referred. In the collection from the blue shaly beds (No. 7) Doctor Girty identified the following species (723 a and b):

Zaphrentis sp.
Favosites sp.
Platycrinus pileiformis Hall.
Megistocrinus evansii? Owen and Shumard.
Physetocrinus planus Winchell.
Trematopora americana Miller.
Fenestella sp.
Pinnatopora sp.
Crania sp.
Leptæna rhomboidalis Wilckens.
Rhipidomella? sp.
Productus semireticulatus Martin.
Productus burlingtonensis? Hall.
Productus aff. scabriculus Martin.
Productus aff. arcuatus Hall.
Productus sp. a.
Productus sp. b.

Camarotœchia metallica? White.
Rhynchopora aff. pustulosa White.
Spirifer imbrex? Hall.
Spirifer aff. grimesi Hall.
Spirifer aff. peculiaris Shumard.
Delthyris nova-mexicana Miller.
Syringothyris sp.
Reticularia temeraria Miller.
Athyris lamellosa L'Eveillé.
Athyris aff. incrassata Hall.
Cleiothyridina roissyi L'Eveillé.
Retzia sp.
Platyceras sp.
Orthoceras sp.
Goniatites? sp.
Phillipsia peroccidens? Hall and Whitfield.

The following were obtained from the overlying beds (No. 8):

Zaphrentis sp.
Crinoid stems.
Megistocrinus evansii? Owen & Shumard.
Platycrinus sp.
Platycrinus peculiaris.
Platycrinus parvinodus Hall.
Dorycrinus unicornis Owen and Shumard.
Stegonocrinus sculptus Hall.
Physetocrinus lobatus Wachsmuth and Springer.
Periechocrinus whitei Hall.
Fenestella sp.

Trematopora americana Miller.
Rhombopora sp.
Leptæna rhomboidalis Wilckens.
Schizophoria swallowi? Hall.
Productus aff. arcuatus Hall.
Spirifer aff. grimesi Hall.
Spirifer imbrex? Hall.
Reticularia cooperensis Swallow.
Athyris lamellosa L'Eveillé.
Athyris? sp.

Crinoids are more abundant in this bed, both in species and individuals, than in the shaly beds below. Concerning the faunal relations of this fauna Doctor Girty says: "The crinoid-bearing beds at Lake Valley have long been known to paleontologists, and they are generally regarded as of lower Burlington age."

Beds 5 and 6 contain corals, crinoid plates, and stems, but good specimens are difficult to obtain.

Two miles east of Hillsboro a good exposure of these beds shows the lower beds (5 and 6) to be lacking, and the upper crinoidal beds (7 and 8) rest upon the eroded surface of bluish-gray calcareous shales that carry an abundant Devonian fauna. The following fossils were obtained from the crinoidal beds at this locality:

Michelinia? sp.
Zaphrentis sp.
Amplexus sp.
Periechocrinus whitei Hall.
Rhodocrinus wortheni var. urceolatus Wachsmuth and Springer.
Cactocrinus multibrachiatus Hall.
Cactocrinus proboscidialis Hall.
Steganocrinus pentagonus Hall.
Platycrinus sp.
Platycrinus subspinosa Hall.
Physetocrinus lobatus Wachsmuth and Springer.
Physetocrinus copei Miller.
Trematopora americana Miller.
Leptæna rhomboidalis Wilckens.

Rhipidomella dalyana Miller.
Productus semireticulatus Martin.
Productus aff. scabriculus Martin.
Spirifer imbrex? Hall.
Spirifer aff. peculiaris Shumard.
Spirifer aff. forbesi Norwood and Pratten.
Spirifer aff. grimesi Hall.
Delthyris nova-mexicana Miller.
Spiriferina sp.
Athyris lamellosa L'Eveillé.
Cleiothyridina sp.
Platyceras, 3 sp.
Phillipsia aff. peroccidens Hall and Whitfield.
Phillipsia aff. loganensis.

At Kingston a series of limestones resting upon black fossiliferous shales (Devonian) yielded the following species:

Zaphrentis sp.
Crinoid indet.
Fenestella sp.
Rhombopora? sp.
Leptæna rhomboidalis Wilckens.
Schizophoria swallowi? Hall.
Productus aff. scabriculus Martin.

Spirifer imbrex? Hall.
Spirifer aff. peculiaris Shumard
Syringothyris sp.
Athyris lamellosa L'Eveilllé.
Athyris aff. incrassata Hall.
Cleiothyris roissyi L'Eveillé.
Orthoceras sp.

Similar beds at Cook yielded the following:

Zaphrentis sp.
Crinoid stems.
Schizophoria swallowi? Hall.
Productus semireticulatus Martin.

Productus sp.
Spirifer centronatus Winchell.
Athyris lamellosa L'Eveillé.
Cleiothyridina sp.

Exposures of the Lake Valley limestone evidently occur at Shandon and Hermosa, but no fossils were obtained from these localities.

KELLY LIMESTONE.

The Kelly limestone consists of about 125 feet of white crystalline limestone in thick beds practically devoid of shale partings. Near the middle of the formation is a dark-bluish, weathering to yellowish-drab, noncrystalline layer 5 feet thick, which in the miners' parlance is known as the "Silver Pipe" limestone. This bed furnishes a convenient guide in following the ore bodies. The areal extent of these beds appears to be small, as they are known to occur only in detached areas due to faulting about the north end of the Magdalena Range. The limestone is believed to be of Mississippian ("Lower Carboniferous") age, but conclusive evidence on this point is lacking.

PENNSYLVANIAN SERIES ("UPPER CARBONIFEROUS").

GENERAL FEATURES.

In this part of New Mexico the Pennsylvanian ("Upper Carboniferous") is represented by a series of limestones, shales, and sandstones having an aggregate thickness of 3,000 to 5,000 feet. These rocks may be separated into two main divisions or groups, to the lower of which the name Magdalena group has been applied,[a] from the Magdalena Mountains, where they are well exposed; the upper division comprises the formations, chiefly sandstone, to which Herrick[b] applied the term Manzano, from the mountains of that name, and the overlying limestones which Lee[c] has shown to belong in the group.

In the southern part of the district, where the Pennsylvanian rocks and the Lake Valley limestone (Mississippian) occur together, the relations are those of apparent conformity. The Mississippian rocks thin out toward the north in Socorro County and are not known to occur north of the Ladron Mountains. In the Magdalena Mountains the basal bed of the Pennsylvanian is a conglomerate, which rests with apparent conformity upon the Kelly limestone (Mississippian?). To the north in Bernalillo County the Pennsylvanian rests unconformably upon granites and other rocks of supposed pre-Cambrian age. A well-marked unconformity occurs also at the base of the Manzano group east of Socorro, as shown by Herrick,[d] who first recognized it and who states that it is evidently an unconformity by overlap. Lee[e] has shown that the Manzano group is separated from overlying formations by a great erosional unconformity. The twofold division of the Pennsylvanian is sustained, therefore, both by

---

[a] Gordon, C. H., Jour. Geology, vol. 15, 1907, pp. 805-816.
[b] Herrick, C. L., Jour. Geology, vol. 8, 1900, p. 115; Am. Geologist, vol. 25, 1900, p. 337; Bull. Univ. New Mexico, vol. 2, fascicle 3, 1900, p. 4; Jour. Geology, vol. 12, 1904, p. 244.
[c] Lee, W. T., Jour. Geology, vol. 15, 1907, pp. 53-54.
[d] Herrick, C. L., Jour. Geology, vol. 12, 1904, p. 244.
[e] Lee, W. T., Jour. Geology, vol. 15, 1907, p. 54.

faunal distinctions, according to George H. Girty,[a] and by relations of uncomformity as well. The faunal studies thus far made do not show any marked change in the life represented within each group and the subdivision of these groups is based entirely on lithological distinctions.

MAGDALENA GROUP.

General outline.—In general the Magdalena group may be said to be characterized by the predominance of limestone, while on the other hand sandstones constitute the most prominent feature of the Manzano group. In Sierra County the Magdalena group consists for the most part of massively bedded blue and gray limestone, interstratified with which are thin-bedded limestone and dark-blue shales. Here and there a thin bed of sandstone may be seen. At Kingston, on the east slope of the Black Range, the basal strata consist of about 300 feet of dark-blue and gray limestone in thick beds with thin shale partings. The upper portion has about the same thickness and consists chiefly of blue and drab shales interstratified with several limestone formations varying from 15 to 20 feet in thickness. Resting unconformably upon the shale are the red sandstones and shales of the Manzano group.

At Palomas camp, 20 miles north of Kingston and 2 miles east of Hermosa, Palomas Creek has cut a gorge 1,000 feet deep through the sedimentary formations. The walls of the canyon are nearly vertical and consist almost wholly of blue and gray limestone of the Magdalena group. The lower half of the escarpment consists of limestone and shale in about equal amounts; the upper portion is made up of hard, massively bedded gray limestone. About halfway up the cliff a few thin beds of quartzite are interstratified with the limestone. The overlying red beds of the Manzano group, together with some of the upper beds of limestone have been removed by erosion at this point, but the sandstones occur in considerable development a short distance to the northwest.

In the Caballos Mountains, 35 or 40 miles east of Kingston, the group is represented chiefly by limestones with some shale beds in the basal part, as at Hermosa, but the writer is unable to give details of the group at this locality. The beds were observed to rest with apparent conformity upon other limestones considered to be of Mississippian age, and below these, separated from them by a thin formation of shale (Percha?), is 900 feet of limestone belonging to the so-called Mimbres formation.

The data at hand are insufficient to warrant an attempt to subdivide the group in this region. The character of the beds at Hermosa suggests the twofold division observed farther north and, in connection with the character of the formations in adjoining districts, indicates a gradual transition in sedimentation from shallow-water conditions at the north to deeper waters toward the south. With the progress of time the entire region assumed conditions apparently of deep and clear waters wherein were deposited the extensive limestone formation which constitutes the upper part of the group and to which the name Madera limestone has been applied.

In Socorro and Bernalillo the Magdalena group comprises from 1,000 to 1,300 feet of sediments, the character of which in their typical locality will be seen from the section on page 224.

This group is characterized throughout by a typical Pennsylvanian fauna, which shows no essential variation from the base to the top of the group, but on lithologic grounds the group is readily separable into two formations. To the lower of these Herrick[b] gave the name Sandia, from the Sandia Mountains, where it was first studied; the upper is known as the Madera limestone.[c]

Sandia formation.—In Socorro County the Sandia formation consists of alternating beds of blue and black clay shales, compact earthy limestones and conglomerates, and vitreous sandstones or quartzites, the shales and limestones predominating. In places the shales are highly carbonaceous and some beds show traces of coal, but thus far no coal beds of importance have been discovered in this formation. The sandstones are usually hard, with a vitreous

---

[a] Idem.
[b] Herrick, C. L., Jour. Geology, vol. 8, 1900, p. 115; Am. Geologist, vol. 25, 1900, p. 235; vol. 33, 1904, p. 310.
[c] Keyes, C. R., Water-Supply Paper U. S. Geol. Survey No. 123, 1904.

fracture, and present the characteristic appearance of quartzites. The beds are in many places conglomeratic, the included pebbles consisting for the most part of pure white quartz. The basal beds of the formation in the Magdalena Mountains, having a thickness of 10 to 15 feet, consist of a moderately coarse conglomerate interbedded with dark shale. Although these beds rest with apparent conformity upon the limestones below, the relations are undoubtedly those of unconformity. About 125 feet above the base of the formation is a coarse white quartzite or conglomerate in massive ledges separated by thin beds of shale. Some of the quartzite beds are filled with pebbles. Overlying this is a limestone, 80 to 90 feet thick, in which appear some thin beds of shale and quartzite.

At the base of this limestone is 6 feet of thick-bedded dark-blue subcrystalline limestone overlain by about 25 feet of shales and thin limestones, above which is 2 feet of quartzite, succeeded by 50 feet of compact bluish earthy limestones. It is along the contact of the limestone with the conglomerate below that the upper or surface ore deposits occur.

The Sandia formation is well exposed on the east slope of Socorro Peak [a] and to the north in the Lemitar Mountains. It also appears on the east side of the river opposite Socorro and elsewhere. To the south there is a marked decrease in the proportion of sand and clay beds accompanied by an increase in calcareous sediments. The shaly beds which constitute the lower part of the escarpment near Hermosa are evidently the equivalent of the Sandia formation as represented in Socorro County, but farther south the distinction between the upper (Madera) and lower (Sandia) divisions of the Magdalena group appears to be lacking.

The thickness of the sediments referred to this formation varies from 500 to 700 feet.

Madera limestone.—Overlying the Sandia formation in Socorro and Bernalillo counties is a dark-blue limestone, for the most part in thick beds alternating with thin shaly beds and blue shales. The limestones contain many fossils of the same type as those found in the underlying Sandia formation, but owing to the extreme hardness of the rock specimens are difficult to obtain.

These limestones constitute the top of the ridge above Kelly, where they are about 300 to 500 feet thick, having been partly removed by erosion. On Socorro Peak they reach a thickness of over 600 feet, and good exposures of the beds appear also in the Lemitar Mountains. The Madera formation constitutes the great limestone plateau along the back slope of the Sandia Mountains, and upon this limestone plateau is located the little Mexican town of La Madera, from which the formation is named.[b]

Herrick [a] has described the lower part of the formation in the Sandia Mountains as consisting of dark limestone, and the upper of beds of massive gray siliceous lime with an intervening sandstone or conglomerate of inconstant thickness (40 feet at the maximum) which he calls the Coyote sandstone from the canyon of that name in the south end of the range.

As already shown, the distinctions on which the subdivision of the Magdalena group in Socorro and Bernalillo counties is based fail toward the south, though at Hermosa the Sandia formation is probably represented in the lower shaly beds and the Madera constitutes the upper part of the great limestone escarpment.

MANZANO GROUP.

General outline.—Above the Madera limestone and resting unconformably upon it in the Rio Grande region is a series of red and pink sandstones and shales with deposits of gypsum, capped by a prominent formation of limestone. These red sandstones and shales constitute in part the beds usually regarded as the equivalent of the Kansas "Red Beds," called Permian, while some of the upper beds were thought by Herrick to be "Jura-Trias" in age.

The writer's examination of these formations, though slight, led him to regard them as belonging to the Pennsylvanian series, a conclusion supported by the more detailed studies

---

[a] Herrick, C. L., Jour. Geology, vol. 8, 1900, pp. 114-115.
[b] Herrick, C. L., Bull. Univ. New Mexico, vol. 1, 1899, p. 104.

of W. T. Lee.[a] The series of variegated sandstones, shales, and gypsiferous beds was described by Herrick[b] from the Manzano Mountains, northeast of Socorro, in 1900, and by him named the Manzano "series." In the southern part of the region a limestone formation several hundred feet thick overlies the red beds, and from it Lee obtained a large collection of fossils, which are distinctly allied to those obtained from the red series, the whole, as stated by Girty,[c] being distinctly Pennsylvanian, though differing markedly from the fauna of the underlying Magdalena group. Farther north in Bernalillo County the upper limestone is absent, and variegated sandstones and shales, probably of Morrison age, rest upon the pink sandstones with an intervening erosion unconformity.

Lee[d] subdivides the Manzano group into three formations, the lowest of which he names the Abo sandstone, the middle the Yeso formation, and the upper the San Andreas limestone.

Several miles northeast of Socorro, in plain sight from the town, are a number of minor elevations called the Cerrillos del Coyote. At this locality the Abo sandstone is well exposed in the west face of the hills, the strata dipping sharply to the northeast. (See Pl. XIV, A.) The red sandstones of the Abo occur on the west side of the Magdalena Range south of Kelly and along the east side of the Black Range from Fairview south to Kingston. No rocks that might be considered the equivalent of the Yeso and San Andreas formations were observed in this region.

These red sandstones have been at times confounded with the "Red Beds," supposedly of Permian and Triassic age, but according to G. H. Girty,[e] who has made a study of the fossils collected from them by W. T. Lee, they are Pennsylvanian and correspond to the upper part of Richardson's Hueco limestone in Texas and the Aubrey group in the Grand Canyon region.

In the Mount Taylor region, 60 miles west of the Rio Grande, there are 1,200 feet of Pennsylvanian red and yellow sandstones, according to Dutton, who identified them with the Aubrey and used that name for them.

Abo sandstone.—The Abo sandstone consists principally of dark-red sandstones interstratified with red sandy shales and some thin beds of bluish-drab earthy limestones. At the base is a limestone conglomerate, the pebbles of which were derived from the underlying Madera limestone. Above this, as seen in the hills east of Socorro, is a coarse red granitic quartzite conglomerate. The thickness of the formation varies from 400 to 800 feet.

Yeso formation.—The Yeso formation consists of alternating strata of yellow, pink, and white sandstones and shales with lenses of gypsum and a subordinate amount of limestone. In places the gypsum is massive and reaches a thickness of 140 feet. This formation ranges from 500 to 1,000 feet in thickness.

San Andreas limestone.—The San Andreas consists for the most part of gray limestone in which an abundant fauna occurs. It is well developed in the mountains east of Socorro and in the San Andreas Mountains east of Engle, but was not observed on the west side of the Rio Grande. The thickness of the formation is from 300 to 500 feet.

NOMENCLATURE.

Although a great many papers relating to the region under consideration have been published, very little detailed work has been done. Some of the names applied to different formations have not been at all defined, and for others the descriptions are so ambiguous as to be interpreted with difficulty. A few names, however, are well established. Herrick[f] applied the name Sandia to the alternation of shales, limestones, and sandstones which constitute the lower

---

[a] Jour. Geology, vol. 15, 1907, pp. 52–58.
[b] Jour. Geology, vol. 8, 1900, pp. 115–116; Bull. Univ. New Mexico, vol. 2, pt. 1, 1900, p. 4.
[c] Quoted by Lee, Jour. Geology, vol. 15, 1907, p. 54.
[d] Bull. U. S. Geol. Survey No. 389, 1909, p. 9.
[e] Bull. U. S. Geol. Survey No. 389, 1909, pp. 43, 44. In this bulletin, which has appeared since the foregoing was written, Girty suggests the possible correlation of the Manzano with the Whitehorse sandstone member of the Woodward formation and the Quartermaster formation of Oklahoma and Texas, saying: "On the whole, however, it seems quite probable that the Manzano and Whitehorse-Quartermaster horizons will prove to be in a general way equivalent." As the Permian age of the Woodward-Quartermaster formations is generally accepted, and recent studies by Beede, David White, and others seem to establish the Permian age of the underlying Albany and Wichita formations, it follows that if the correlation suggested by Girty can be established, the Manzano should be referred to the Permian.
[f] Herrick, C. L., Jour. Geology, vol. 8, 1900, p. 115; Am. Geologist, vol. 25, 1900, p. 235; Jour. Geology, vol. 12, 1904, pp. 237–251.

A. COYOTE BUTTES, SOCORRO COUNTY.

Eight miles northeast of Socorro. Beds of Manzano group dipping to the northeast.

B. SOCORRO PEAK.

Main mass andesite; sedimentary rocks (Madera limestone) at "*a*." Torrance mine in foreground. Looking northwest.

half of the Magdalena group in Socorro and Bernalillo counties in 1900, and his descriptions are so clear that no difficulty is encountered in the application of this term. At the base of the formations east of Socorro is a bed of clay containing Carboniferous plants to which Herrick gave the name Incarnacion fire clay, but he expressly states that "there would seem to be no reason for separating the fire clay from the Sandia formation, it being but a local variation."[a]

The limestone overlying the Sandia formation, however, he appears to have left unnamed, but to the bed of sandstone which occurs near the middle of the formation in the vicinity of Coyote Spring, Bernalillo County, he gave the name Coyote sandstone.[b] The same author applied the name Manzano[b] to the series of red sandstones and other beds which overlie the rocks of the Magdalena group in the Manzano Mountains and adjacent regions. His description of these beds does not make it altogether clear whether he meant to apply the name to the lower red sandstones alone or to the whole series, including the gypsiferous beds, the chocolate-colored sandstones, and their accompanying shales and earthy limestones. It would seem, however, that the latter was his intention. He does not seem to have included under this name the limestone overlying the pink and yellow sandstones, which, as shown by Lee,[c] belongs in the group.

In a paper published in 1906 Keyes[d] presents a classification of the Carboniferous rocks of New Mexico in which several names appear that are thus far without adequate definition. The shale bed at the base of the Sandia formation he separates, giving it the name Alamito, and makes it the equivalent of a series to which the name Ladronesian is applied. No evidence in support of this separation is given, and in view of Herrick's specific statement that none exists these terms can hardly be accepted. For the remaining part of the Magdalena group Keyes uses Herrick's term Manzano; or, as he gives it, Manzanan. Herrick's description is sufficient to show that these are not the beds for which the name was originally proposed. Keyes subdivides the "Manzanan," using in addition to Herrick's Sandia beds and Coyote sandstone the name Montosa for the limestone below the Coyote sandstone, and Mosca for that above. These formations have not yet received sufficient study in the field to warrant this refinement of classification. The writer's observations lead him to conclude that the Coyote sandstone is merely local and the subdivision to be unsustained by field evidence.

Overlying his so-called Manzanan, Keyes[e] places a limestone formation which he calls Maderan. Evidently he regards this as the same formation which in an earlier paper he says is, in the Sandia Mountains, called the Madera limestone, and adds that it forms by far the most important portion of the Carboniferous in all the mountain ranges mentioned. From Herrick's description of the geology of the Sandia Mountains, which is corroborated by the studies of W. T. Lee,[f] of the United States Geological Survey, it is clear that the great limestone formation of the Sandia Mountains is the limestone which constitutes the upper half of the Magdalena group and comprises the formations to which in the same table Keyes gives the names Montosa and Mosca, with the included Coyote sandstone. Inasmuch as the name Madera very appropriately applies to the limestone overlying the Sandia beds in the Sandia Mountains, it may well be retained for the upper formation of the Magdalena group. The names applied to the subdivisions of the Manzano group are those adopted by Lee.[f]

---

[a] Jour. Geology, vol. 12, 1904, p. 242.
[b] Idem, vol. 8, 1900, p. 115.
[c] Lee, W. T., Jour. Geology, vol. 15, 1907, pp. 53–54.
[d] Keyes, C. R., Jour. Geology, vol. 14, 1906, pp. 147–154.
[e] Water-Supply Paper U. S. Geol. Survey No. 123, 1905, p. 22.
[f] Bull. U. S. Geol. Survey No. 389, 1909, p. 14.

In the following table are presented in convenient form for comparison the classifications of the Pennsylvanian rocks in the Rio Grande region made by the different authors mentioned:

Classifications of Pennsylvanian rocks in Rio Grande region.

Lee and Gordon, 1907.			Herrick, 1900.		Keyes, 1904.a		Keyes, 1905.b		Keyes, 1906.c	
Pennsylvanian.	Manzano.e	San Andreasd—limestone, 300 to 500 feet.	Jura-Trias.	Limestone.	Triassic.	Cimarronian.	Permo-Carboniferous.	Red Beds.	Permian.	Cimarronian.	Moencopie shales. Sandstone. Shales.
		Yesod—pink or vermilion and yellow sandstones, shales, and earthy limestones, and gypsum beds, 500 to 1,000 feet.		Manzano.							
		Abod—red sandstones and conglomerates, with some earthy limestones, 400 to 800 feet.	Permian.			Bernalillo shales.				Guadalupian.	Capitan limestone. Eddy sandstone.
		Unconformity.	Coal measures—Permo-Carboniferous.	Limestone.	Carboniferous.	Madera limestone.	Pennsylvanian.	Madera limestone.		Maderan ("Hueco in part").	Maderan limestone.
		Maderab—limestones with some shales and sandstones, 600 to 1,000 feet.		Coyote sandstone.							
				Limestone.							Mosca limestone.
	Magdalena.f	Sandiad—alternating shales, limestones, and sandstones: toward south nearly all limestones; 500 to 600 feet.		Sandia beds or series.		Sandia limestone.		Sandia quartzite.		Manzanan.	Coyote sandstone.
											Montosa limestone.
											Sandia shales.
										Ladronesian.	Alamito shales.

a Keyes, C. R., Am. Jour. Sci., 4th ser., vol. 18, 1904, p. 360.
b Keyes, C. R., Water-Supply Paper U. S. Geol. Survey No. 123, 1905, p. 22.
c Keyes, C. R., Jour. Geology, vol. 14, 1906, p. 154.
d Lee, W. T., Bull. U. S. Geol. Survey No. 389, 1909, p. 12.
e Herrick, C. L., Bull. Univ. New Mexico, vol. 2, 1900, fasc. 3, p. 4; also Jour. Geology, vol. 8, 1900, pp. 112–128.
f Gordon, C. H., Jour. Geology, vol. 15, 1907, pp. 805–816.

TRIASSIC (?) SYSTEM.

In some places in New Mexcio red sandstones and shales considered to be of Triassic age overlie the Carboniferous, but as yet little is known of them. No exposures of these beds are known to occur in this part of the Rio Grande valley. It is possible that some of the red sandstones observed along the front of the Black Range may belong here, but there is at present no satisfactory evidence that rocks higher than the Carboniferous occur in that part of the region.

CRETACEOUS SYSTEM.

Extending southward from Colorado as far as Engle is a series of sandstones and shales in which Benton fossils were collected by W. T. Lee.a These beds are found exposed near Engle and farther north, but were not recognized in the Black Range, though further observations may disclose their presence there. In the vicinity of Elephant Butte, according to the same writer, these strata are overlain by an extensive series of Cretaceous sandstones containing beds of coal at the base and great quantities of fossil wood at higher horizons. The upper member of this series of Cretaceous sandstones consists of red sandstone and shale beds containing vertebrate fossils belonging to the genus *Triceratops*, thus indicating their late Cretaceous age.

a Water-Supply Paper U. S. Geol. Survey No. 188, 1907, p. 26; Jour. Geology, vol. 15, 1907, pp. 56–57.

TERTIARY SYSTEM.

With the close of the Cretaceous period in western New Mexico came an uplift giving rise to land conditions. A period of great erosion followed, resulting in the planing down of the land surface and the distribution of the material over the bottoms of adjoining lakes and seas. To this epoch belong the basal Eocene deposits which occur in the northwestern part of New Mexico. Within Sierra County and the central part of Socorro County the Tertiary beds, if occurring at all, are probably confined to the vicinity of the Rio Grande. To the Tertiary should doubtless be referred the highly tilted sandstones and conglomerates observed by W. T. Lee on the west side of the Rio Grande along the Arroyo Salado.[a]

This epoch was followed by the great crustal displacements which outlined in their larger features the Great Plateau and the Rio Grande valley. During this period the region was the scene of great volcanic activity which resulted in the accumulation of vast deposits of volcanic ejecta, the earlier of which were andesite and the later rhyolite.

The Santa Fe marl, named and described by Hayden[b] in 1869, and said to have an extensive development in the Rio Grande valley, was by him referred to the upper Tertiary.[c] Cope, writing in 1883,[d] referred to this formation marl and sandstone beds observed in the Rio Grande valley as far south as Socorro and similar beds cut by the Atchinson, Topeka and Santa Fe Railway west of Rincon. As yet, however, knowledge of these beds is very incomplete. No exposures of Tertiary sediments are known to occur south of Arroyo Salado, the beds near Socorro and those near Hatch mentioned by Cope being probably of later age.

QUATERNARY DEPOSITS.

PALOMAS GRAVEL (PLEISTOCENE).

Filling the Rio Grande valley to a depth in places of more than 2,000 feet, as shown by well records, is a deposit of gravels and sands derived from the wearing down of the adjacent slopes. The local source of the material is well shown by the variation in the composition of the deposit with the change in character of the rocks in the adjacent valley sides. These "wash" deposits, which cover the greater part of the Rio Grande valley and its tributary valleys, are of Quaternary age and represent two distinguishable formations, to the older of which, usually called the "mesa gravels," the name Palomas has been applied,[e] from Palomas Creek, where it is well exposed. It is coarsest along the valley sides, grading rapidly to finer materials toward the middle of the basins. On the side of the valley between Fairview and Hillsboro the materials, which are coarse and only partly worn, are consolidated into a firm hard conglomerate. In composition and geologic horizon this formation is apparently identical with that in Arizona to which Gilbert has given the name Gila conglomerate.[f] This conglomerate is extensively developed along the east foot of the Black Range, extending up the tributary valleys to a height of 1,400 feet above the level of the river. Palomas Creek and other streams have sunk their channels to a depth of 800 feet or more into the formation, but nowhere have they exposed its full thickness. (See Pl. XVI, B.) Patches of this formation occur on some of the interior slopes of Socorro Peak as remnants left by the erosion of filled basins at an elevation of a thousand feet or more above the valley plain.

East of the Rio Grande in Socorro County, where these deposits derive their material largely from sedimentary formations, they consist of incoherent sands, sandstone, gravel, and marly beds which have a tendency to weather into a characteristic badlands topography. These beds were correlated by Cope with the beds above Albuquerque, with which they appear to be identical and to which Hayden applied the term Santa Fe marl.

---

[a] Water-Supply Paper U. S. Geol. Survey No. 188, 1907, p. 14.
[b] Prel. Rept. U. S. Geol. Survey Colorado and New Mexico, 1869, pp. 66–69.
[c] Idem, p. 90.
[d] Proc. Am. Philos. Soc., vol. 21, 1884, pp. 308–309.
[e] Gordon, C. H., and Graton, L. C., Jour. Geology, vol. 15, 1907, p. 92.
[f] Gilbert, G. K., U. S. Geog. Surveys W. 100th Mer., vol. 3, 1875, p. 540.

RECENT DEPOSITS.

Of later age are the sand and gravel deposits which form some of the intermediate terraces representing a filling of channels excavated in the older (Palomas) deposits, and youngest of all are the flood-plain deposits along the present course of the river.

IGNEOUS ROCKS.

ANDESITE AND LATITE.

The earliest eruptions in this region are represented by extensive flows, tuffs, and breccias of andesite and intermediate rocks like latite. The Black Range consists in large part of latite over which there are usually more or less extensive flows of rhyolite. In some places there are indications of two or more eruptions of andesite and latite, but the evidence at hand is not conclusive.

There are no known exposures of andesite in the San Mateo Mountains, but these rocks are well represented in the Magdalena and Socorro mountains and to the north in the Polvadera mountains. They are also reported from the Cristobal and Caballos ranges, but no examination was made of these localities.

RHYOLITE.

In many places the andesite is overlain by rhyolite, evidently representing more or less extensive flows. Associated with the flows are considerable deposits of ash, tuffs, and breccia, and locally of volcanic glass, belonging to the same period of eruption. The rhyolites have a great development in the San Mateo Mountains, where they constitute apparently the whole of the range. They are generally present unless removed by erosion, overlying the andesite in the Black Range, the Magdalena Mountains, the Socorro Mountains, and farther north. Large flows occur in the south end of the Caballos Range and evidently also in the Cristobal Range.

BASALT.

Large flows of normal basalt occur at many points along the west side of the Rio Grande valley from Socorro and southward and on the east side of the river near San Acacia and San Marcial and in the mountains to the south. So far as observed, all these flows are of relatively recent date and were poured out subsequent to the deposition of the Palomas gravel.

At least two distinct epochs or periods of eruption may be recognized, separated by a long period of erosion. The first epoch evidently followed the filling of the valley and spread vast sheets of basalt over the surface of the Palomas gravel. Succeeding this, as shown by W. T. Lee,[a] came a long period of erosion, followed by another eruption, which is represented by the San Marcial flow and that farther south on La Mesa, west of Mesilla Valley. The lava beds have, for the most part, suffered greatly from erosion, which has cut deeply into the underlying beds. They have been removed from considerable areas, leaving scattered remnants capping buttes and benches.

Two miles west of Hillsboro, Rio Percha has cut a deep gorge through a mass of basalt, which appears to overlap the sandstones on the west, while on the east side the Palomas gravel abuts against it. The contact dips about 45° E. and the planes of stratification in the gravel beds bend rather abruptly upward along the contact. Whether this represents a vent from which the lava covering the Palomas gravel of adjoining areas has come or a flow antedating the deposition of the gravel beds was not determined, though the former hypothesis is regarded as the more probable.

INTRUSIVE ROCKS.

Igneous intrusions are numerous and extensive throughout this district and occur in the form of dikes and sills, together with larger bodies which correspond to the type of the chonolith as defined by Daly.[b] Most of these intrusive masses are of intermediate char-

---

[a] Water-Supply Paper U. S. Geol. Survey No. 188, 1907, p. 22.
[b] Daly, R. A., The classification of igneous intrusive bodies: Jour. Geology, vol. 13, 1905, p. 485.

acter, standing between the granites and the diorites. According to recent usage they are designated as monzonite porphyry.

As examples of such intrusions may be cited the granite and monzonite porphyry masses near Magdalena and north of Hillsboro, the similar igneous masses occurring in the Cuchillo Range east of Fairview, the granites near Kingston, and the monzonite porphyry composing Cooks Peak. Both monzonite porphyry and granite porphyry occur very commonly as dikes, and with one or the other are associated most of the important ore deposits of the region.

Basic and acidic dikes are present very generally in the mining districts, but so far as observed have exerted little influence in the formation of ore deposits.

MINERAL DEPOSITS.

INTRODUCTION.

The mineral deposits occurring in this region show considerable diversity both in character and occurrence. They contain lead, zinc, copper, gold, silver, iron, and manganese. The gold is found in placers and along with silver in fissure veins. Silver is the principal metal and most of it is derived from replacement deposits in the limestone. In these deposits it occurs as native silver, as chloride, or as a sulphide (argentite), also as a constituent of sulphidic copper or lead ores. Lead and zinc occur in considerable quantities as carbonates in the upper parts of replacement deposits in limestone; in the deeper workings are the sulphides of the same metals, in many places accompanied by those of copper. Lead sulphide occurs also in veins, though as yet these are of comparatively little importance.

Copper sulphides are associated with those of lead and zinc in the Magdalena district and occur as replacements of the shattered country rock in shear zones at other points. They are common in many of the silver-gold veins, and in these are usually argentiferous, especially the bornite. Deposits of iron and manganese are present toward the south end of the Black Range, at Hillsboro and elsewhere, but they have not been developed and little is known of them. A deposit of iron is reported to occur in the northern part of the Cuchillo Range, but this was not examined. Near Hillsboro some deposits of wulfenite (molybdenate of lead) and vanadinite (vanadinate of lead) are found.

Up to the present time most of the production has been derived from replacement deposits in limestone. Among these deposits there is considerable variety as to the nature of the ores and the horizon where they occur. In the Magdalena, Cooks Peak, and Carpenter districts the ores contain chiefly lead or zinc, or both, but in the belt extending from Hermosa to Lake Valley, including Kingston and Tierra Blanca, silver ores predominate. Some of these deposits occur in caves, but so far as observed the cave deposits are always in the oxidized zone and are apparently of secondary origin.

SOCORRO PEAK DISTRICT.

HISTORICAL NOTE.

At one time Socorro Peak was the scene of considerable activity in mining operations, the results of which are now seen in abandoned prospect holes and tunnels all along the east face of the mountain.

It is reported that where first prospected in 1867 evidences were found of ancient workings, presumably by the Spaniards. The Torrance mine, the principal producer of the district, is said to have been discovered by Hanson, one of the former owners of the Kelly mine at Magdalena. Next in importance is the Merritt mine, which adjoins the Torrance on the south. These two mines have furnished most of the ores mined in the district. Operations other than annual assessment work have ceased for many years. According to the best available sources the production of the district has been $760,000.[a]

GEOLOGY.

Socorro Peak (Pl. XIV, B) is composed chiefly of eruptive material and evidently represents an old volcano now eroded into irregular jagged outlines. It is elongated in a north-south

---

[a] Jones, F. A., New Mexico mines and minerals, 1904, p. 111.

direction, and toward the north connects with similar mountain masses called the Lemitar and Polvadera mountains. The eruptions evidently occurred at different places along the range. The cone within Blue Canyon (Pl. XV, B), which Herrick calls Pyramid Mountain,[a] and Strawberry Peak, at the north end of the mountain, may represent two such vents.

The oldest and apparently the greatest eruptions were of andesite, approaching a latite. These occur at the base and along the east side of the mountain and appear in the interior in a number of places where erosion has cut through the later flows. A period of erosion evidently followed the eruption of the andesite. Then came an eruption of an intermediate rock (trachyte or latite), represented in Pyramid Mountain. Under the microscope sections of this rock show a sparing amount of phenocrysts of orthoclase, plagioclase, and hornblende in a groundmass composed of glass and microscopic tabular feldspars. A partial analysis of the rock in the laboratory of the United States Geological Survey by George Steiger gives the following result: SiO_2, 65.56; CaO, 3.65; Na_2O, 3.83; K_2O, 3.39. This was succeeded by a flow of rhyolite and glassy lavas, remnants of which appear in many places over the mountain. South of Blue Canyon is a fine example of flow structure which has been contorted by movement of the mass just prior to final consolidation. (See Pl. XV, A.)

At a much later date a flow of basalt from the southwest side was spread over the inclined surface of the Palomas gravel and extended for some distance out into the Rio Grande valley through the gap on the south side of the mountain.

In the east face of Socorro Peak is an exposure of 800 to 1,000 feet or more of shales and limestones representing the lower part of the Pennsylvanian series. Toward the north these beds are overlain by the red sandstones and shales that represent the higher beds of the same series. The Madera limestone constitutes over 600 feet of the exposed section and is cut off abruptly on the west and south by andesite.

South of the principal mines are extensive deposits of blue and white clays, which are evidently due to the kaolinization of tuffs and breccias.

About half a mile north of the Torrance mine, at the bottom of a small canyon cut in the limestone, is exposed a mass of coarse granular rock composed chiefly of orthoclase, plagioclase, and hornblende and corresponding in its mineral composition to monzonite.

ORE DEPOSITS.

The mineralized zone occurs near the contact of the andesite tuffs with the overlying flows, extending parallel with the base of the mountain and dipping westward with it at an angle of 40° to 50°. The principal mine is the Torrance, which is adjoined on the south by the Merritt.

The vein, which started in andesite or latite, terminated abruptly on reaching the clayey tuffs below. It is developed by an incline with five levels, also two short tunnels. The disappearance of the ore was supposed to be due to faulting or slipping along the contact between the tuff and the andesite, but efforts made to locate the extension of the vein were unsuccessful.

The chief values were in silver, with little or no gold in a gangue of quartz and barite. The ores are said to be highly siliceous. The shoot occurred mainly on the south side of the incline. A production of $760,000 is claimed for the district, which has been idle for many years.

CANYONCITO DISTRICT.

GEOLOGY.

The Canyoncito district is situated approximately 2 miles east of the Rio Grande, about midway between San Acacia and La Joya. At this point an area of red granite gneiss is exposed, bordered on the north and west by sedimentary beds. The gneiss is coarse and incloses bands of dark hornblende schist. Overlying it is a considerable thickness of quartzite, which is succeeded by limestones overlain by red sandstone and shales of the upper Carboniferous. At this point the sedimentary beds dip about 40° NW. Toward the east the underlying rocks are covered by a deposit of rhyolite and on the south by basalt. The gneiss is intersected by numerous dikes of fine-grained granite and aplite.

---

[a] Herrick, C. L., Bull. Univ. New Mexico, vol. 1, 1899, p. 77.

A. SOUTH SIDE OF BLUE CANYON, SOCORRO MOUNTAINS.
Showing flow structure in rhyolite. Looking south.

B. BLUE CANYON, SOCORRO MOUNTAINS.
Socorro and Rio Grande 4 miles in distance. Looking east.

ORE DEPOSITS.

A vein which is 3 to 5 feet wide cuts the gneiss nearly parallel with the strike of the limestones and associated beds. The ore (galena) occurs in bunches distributed in a gangue of quartz, barite, and fluorite, which occupies the full width of the vein. The walls are usually well defined and are covered to a thickness of one-fourth to one-half inch with clay gouge. At the Dewey mine a shaft has been sunk on the vein to a depth of 300 feet. No water was encountered. In the upper 230 feet the dip is about 65° SE.; below that it is 75° to 80°. Above the change of dip the ore mostly follows the hanging wall, but below that point it occurs principally along the foot-wall side. No shipments have been made from the mine, but about 100 tons of ore now lie on the dump. It is a low-grade galena ore carrying very little gold or silver. The well-defined character of the vein and the fact that it maintains this character to the bottom of the shaft indicate its continuation to a point considerably below the present workings.

SAN LORENZO DISTRICT.

Reports were current in the press during 1909 of the discovery of oxidized copper ores in fissures in an andesitic rock in the San Lorenzo district, 20 miles northwest of Socorro and 6 miles west of Alamilla station.

MAGDALENA DISTRICT.

GEOGRAPHIC RELATIONS.

The Magdalena mining district is situated in Socorro County and comprises a portion of the north end of the Magdalena Range, from which it takes its name. Included within its limits are the two towns of Magdalena and Kelly. The former is located in the plain near the north side of the district, about 26 miles a little north of west from Socorro, with which it is connected by the Magdalena branch of the Atchison, Topeka and Santa Fe Railway. Kelly lies at the west foot of the main range, about 3 miles southeast of Magdalena, and at an altitude of 1,000 feet above that town. (See Pl. XVI, A.) It is near the head of a narrow plain with steep but uniform slope that opens northward into the plain upon which the town of Magdalena is situated.

About midway between Magdalena and Kelly are the offices and ore bins of the Graphic mine, and a spur of the railroad for handling ore extends to this point (Graphic station). The Graphic mine is farther east and is connected with this station by a tramway. The Kelly, Juanita, and other mines are located just above Kelly to the east, and their ore is hauled to the cars by wagon. The recently opened Key mine lies about three-fourths of a mile east of the Graphic, on the east side of the range. To facilitate the marketing of the ore from this mine a 2,000-foot tunnel is being driven through the mountain to connect with the Tip Top tunnel in the gulch above Kelly. With the exception of the Key mine, practically all the mines and prospects of any importance in this district occur in a strip about 1½ miles wide by 4 miles long, extending nearly due north and south along the west slope of the main range. Kelly is near the west center of this area, and within a radius of a mile from this point are located all the principal mines of the district. Plate XVII is a sketch map of the district showing the location of the principal claims.

HISTORICAL SKETCH.

The discovery of ore in the Magdalena Mountains is said to have been made in 1866 by Col. J. S. Hutchason, who is still living in Magdalena. The first claim to be staked out was the Juanita, and three weeks later the Graphic was located. The first ores to be mined were the "sand carbonates," or lead ores, which were smelted in an adobe furnace, the product being hauled to Kansas City by ox teams. In 1881 Gustav Billings, then in control of the Kelly mine, erected a smelting plant at Socorro, where the product of the mine, as well as custom ore, was treated. This plant remained in operation until 1893. The Graphic smelter at Magdalena was built in 1896 and ran intermittently until the exhaustion of the red-lead ores in 1902.

Between 1894 and 1902 the Kelly and Graphic mines were worked in desultory fashion by lessees or owners. In 1903 C. T. Brown and J. B. Fitch, lessees of the Graphic mine, sent some smithsonite to Missouri, and the ore proved so satisfactory that more was asked for. Shortly afterwards the Sherwin Williams Paint Company purchased the Graphic mine, and in 1904 the Kelly mine was sold by Mrs. Billings to the Tri-Bullion Mining and Smelting Company.[a] The production of the Kelly and Graphic mines up to that time is said to have been approximately $5,800,000, and for the whole district the production is said to have been $8,700,000.[b] Brinsmade gives the figures of total production as $6,600,000, of which the Kelly and Graphic mines yielded 90 per cent. The following figures indicate the production of the district since 1904; they are taken from "Mineral resources of the United States," published by the United States Geological Survey:

Metallic production of Magdalena district, Socorro County, N. Mex., 1904–1908.

Year.	Gold.		Silver.		Copper.		Lead.		Zinc.		Total value.
	Quantity.	Value.	Quantity.	Value.	Quantity.	Value.	Quantity.	Value.	Quantity.	Value.	
	Fine oz.		*Fine oz.*		*Pounds.*		*Pounds.*		*Pounds.*		
1904	7.44	$154	5,024	$2,906	3,200	$416	588,209	$29,593	13,493,835	$674,692	$707,761
1905			4,800	2,899	320,000	49,920	390,000	18,330	14,630,051	863,173	934,322
1906					64,560	12,460	238,088	13,571	17,044,163	1,039,694	
1907							707,981	37,523	676,356	39,905	
1908	3.48	72	460	244	2,605	344	16,073	675	3,325,446	156,296	157,631

The deeper workings have revealed the presence of large bodies of zinc blende along with galena and chalcopyrite. With the beginning of the utilization of zinc ores in the Cordilleran States mining in the district revived, and it is now the chief zinc-producing district in New Mexico. A large part of the ore is utilized for the manufacture of zinc-lead paint. Until recently it was shipped crude, but since 1907 attention has been devoted to devising suitable processes of concentration. In 1909 three mills were ready for concentration work. About 1908 the Mistletoe and Magdalena Tunnel Company erected a mill for the dry pneumatic process. In 1909 the Tri-Bullion Company continued shipments of sulphide ores from the Kelly mine, but the upper levels, containing oxidized ores, were operated by lessees. Auxiliary to the wet concentration plant of this company is a plant for roasting and magnetic treatment of concentrates. The Ozark Smelting and Mining Company, operating the Graphic mine, has also erected a concentrating plant, which is situated near the old smelter. Mining developments on this property in the last three years have been slight, pending the erection of the plant. It is reported that a large body of milling ore, said to have been opened on the Waldo tunnel level, contains about 20 per cent of zinc, 4 to 5 per cent of lead, and 30 to 40 per cent of iron, much of the iron being in the form of magnetite.

At one time a considerable amount of money and labor was expended in the attempt to establish paying mines on the east side of the mountain near Water Canyon, but the efforts did not meet with success.

TOPOGRAPHIC RELATIONS.

The Magdalena Mountains constitute a range about 25 miles in length, extending in a north-south direction parallel with the Rio Grande and 15 to 20 miles west of it. About midway between this range and the river lies Socorro Peak and the smaller elevations which constitute its extension toward the south.

The highest point in the Magdalena Range, Old Baldy, is some distance south of the Magdalena mining district. It has an elevation of about 10,800 feet. At the south end of the Magdalena district the crest of the range attains an elevation of 9,600 feet. East of the Graphic mine it reaches 9,130 feet, and a mile farther north 8,400 feet. North of this point the range descends rather abruptly to the level of the plain in which Magdalena is situated. Lower out-

---

[a] Brinsmade, R. B., Mines and Minerals, vol. 27, 1906, p. 49. [b] Jones, F. A., New Mexico mines and minerals, 1904, p. 119.

A. MAGDALENA MOUNTAINS, SOCORRO COUNTY.

Kelly mining camp at extreme right margin. Looking south from Magdalena.

B. VIEW ON ARROYO DE LA PARIDA, SOCORRO COUNTY.

Seven miles northeast of Socorro; characteristic erosion forms in Palomas gravel (Pleistocene).

lying hills mark the northward extension of the range for several miles. The height above sea level at Magdalena is about 6,500 feet, or 1,900 feet higher than the railroad station at Socorro.

The town of Magdalena is situated on the south side of the valley of a small stream which drains a subordinate basin lying between the low elevations constituting the eastern boundary of the San Augustin Plain and the north end of the Magdalena Range. This stream leads out to the northeast into Arroyo Salado, a distance of 20 or 25 miles.

South of the town, and bordering the west side of the narrow valley in which Kelly is situated, are erosion remnants of volcanic flows known as Magdalena Mountain and Elephant Butte. In the vicinity of Graphic station this valley has a width of about a mile. A small channel carries the ordinary run-off, but in times of heavy rainfall this channel is wholly inadequate and the water spreads in a wide sheet over the plain, sweeping down through the town below, sometimes flooding it to a depth of several feet. The altitude of Kelly is about 7,400 feet.

Magdalena Mountain consists of rhyolite and, with Elephant Butte to the south, belongs to the sheet that covers the west side of the range from Kelly southward. On the east side of the small valley here described, at the foot of the range, is a line of low hills which represent remnants probably of the same flow. The face of the main range is gashed by numerous ravines and gulches. One of the largest is Kelly Gulch, which drains a small basin in the mountain above Kelly. At several places along the foot of the range springs of good water occur.

GENERAL GEOLOGIC RELATIONS.

Within the limits of the district the Magdalena Range consists essentially of a block of sedimentary formations tilted to the west at an angle of 30° to 55° and resting on a basement of crystalline schists and granite. The beds have not suffered folding to any notable extent. Faulting is a characteristic feature of the district. The basal stratified rocks consist of about 125 feet of the Kelly limestone in which, it is reported, Mississippian ("Lower Carboniferous") crinoids have been found. Overlying the limestone conformably is a series of shales, quartzites, and limestones about 600 feet in thickness, which, as shown by Herrick,[a] correspond to the beds in Bernalillo County to which the name Sandia formation has been applied. Above these beds, capping the ridges above Kelly, is a heavy calcareous formation, considered the equivalent of the limestone (Madera) overlying the Sandia formation in Socorro Peak. The thickness of this formation in the Magdalena district is from 300 to 500 feet. Some distance south of Kelly exposures of the overlying Carboniferous red beds occur, but these were not examined. A columnar section at Kelly is shown in figure 17.

The area in which the sedimentary formations appear was evidently at one time covered more or less completely by flows of lava (andesite and rhyolite), most of which have been removed from this part of the range by erosion.

Intrusions of both acidic and basic type occur, the former being predominant, and are considered to have special significance in connection with the formation of the ore bodies.

STRATIGRAPHY.

PRE-CAMBRIAN ROCKS.

Greenstone schists, presumably of pre-Cambrian age, outcrop in several places within the district. Associated with them are granites which are evidently intrusive in the schists. The sedimentary formations rest unconformably upon these rocks.

In most of the places where the base of the limestones is exposed, there intervenes between the granite and the basal limestone a grayish-green to dark-green schist, known locally as "greenstone." The schist is as a rule a compact fine-grained rock showing on weathered surfaces a distinct banding running approximately parallel with the limestone contact. Jointing is well developed and under the influence of the weather the rock breaks up into rather small flat-sided polyhedrons to such an extent that good specimens are difficult to obtain. The rock is very hard and tough and is penetrated with difficulty in mining. The surfaces of the joint blocks are usually more or less covered with a thin coating of epidote which has been deposited

---

[a] Herrick, C. L., Am. Geologist, vol. 33, 1904, p. 310.

in the fractures. In places the schist has a coarser grain and carries crystals of hornblende visible to the unaided eye.

Under the microscope sections of the banded schists show the presence of numerous small but well-defined grains of quartz and a few plates of muscovite in a microcrystalline groundmass, composed largely of quartz, in which appear abundant scales of chlorite, some muscovite, and a few minute grains of twinned feldspars. The larger quartz grains are, for the most part, elongated and angular in outline with parallel orientation and are more thickly crowded along certain planes, which accounts for the banded structure seen in weathered surfaces. In the coarser phases the microscope shows much hornblende in relatively large crystals, and patches more or less altered to chloritic substances. There appear also rather large grains of black iron ore (ilmenite), in large part altered to leucoxene. The remaining portions are made up of decomposition products, some areas indicating obscure traces of crystal outline and suggesting derivation from feldspathic minerals.

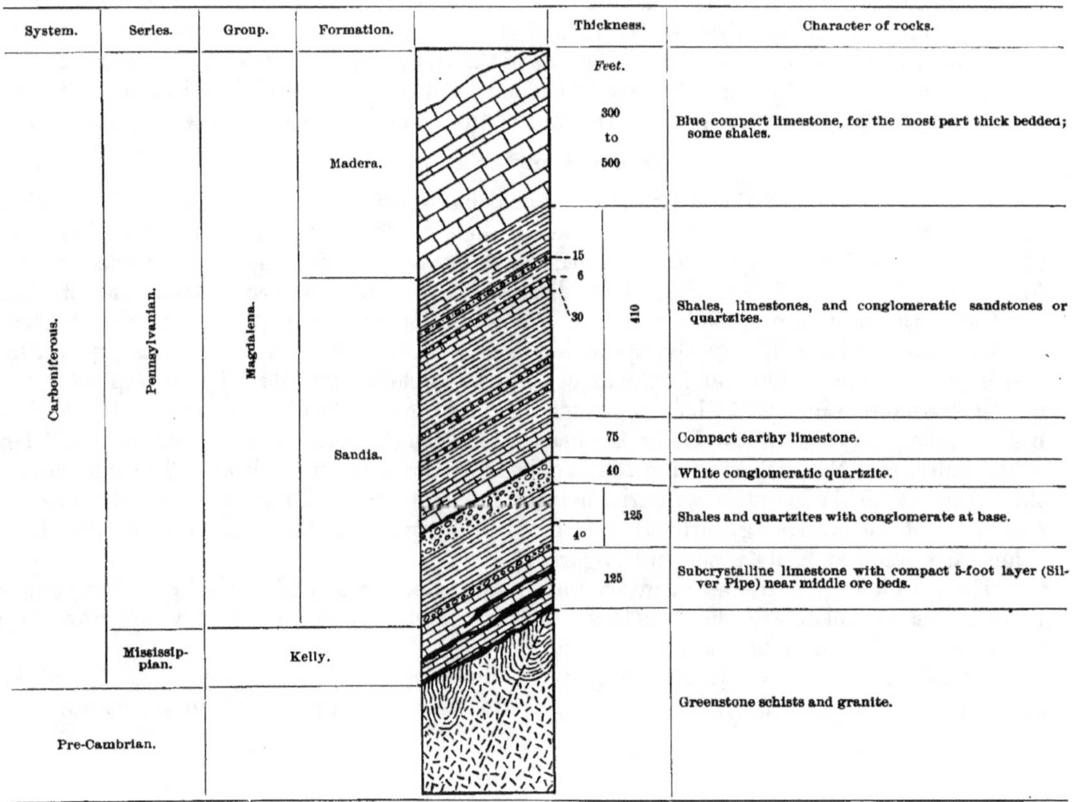

System.	Series.	Group.	Formation.		Thickness.	Character of rocks.
					Feet.	
Carboniferous.	Pennsylvanian.	Magdalena.	Madera.		300 to 500	Blue compact limestone, for the most part thick bedded; some shales.
				15 6 30	410	Shales, limestones, and conglomeratic sandstones or quartzites.
			Sandia.		75	Compact earthy limestone.
					40	White conglomeratic quartzite.
				4°	125	Shales and quartzites with conglomerate at base.
					125	Subcrystalline limestone with compact 5-foot layer (Silver Pipe) near middle ore beds.
	Mississippian.	Kelly.				
Pre-Cambrian.						Greenstone schists and granite.

FIGURE 17.—Geologic column at Kelly, Magdalena district.

Exposures of the greenstone schists occur in a number of localities on both sides of the range. One of these just east of Kelly covers an area of perhaps 20 acres and is deeply trenched at the middle by Kelly Gulch. On the west, north, and northeast the greenstone schist disappears under the limestone and overlying formations, which dip away along the contact border at angles varying from 35° or 40° on the west to 10° on the northeast. In the vicinity of the Kelly mine the contact is marked by more or less faulting and slipping, and the reversal of dip on the northeast side is evidently due to this cause.

About three-fourths of a mile north of the Kelly area, adjoining the Graphic mine on the east, is another area of the greenstone schists. As shown by the mine workings these belong to the same rock mass, which is hidden from view in the intervening area by the covering of limestone, shales, and quartzites.

A third area lies on the east side of the range, about 800 feet below the crest, in the vicinity of the Key mine. At the entrance to the tunnel the greenstone schists are in contact with the limestone on the west and with granite on the east. In the interior workings, however, the schists are cut off and the granite occurs in nearly vertical contact with the limestone.

Another exposure of the schists occurs on the east side of the range north of Water Canyon. In all these areas the surface exposures are roughly elliptical in outline, with the longer axes parallel with the axis of the range.

At the surface more or less oxidation has taken place along the contact of the schists with the overlying limestone, and, as previously stated, in some places there has been more or less slipping. In the Key and Greyhound tunnels the contact is not apparent except on close observation, but there, as elsewhere, unconformity is shown by the discordance between the planes of stratification in the limestone and the surface of the greenstone schists.

The character and relations of the schists indicate that they have been derived from ancient sediments, apparently tuffaceous, which have been intruded by basic dikes and later highly metamorphosed through the agency of heat and pressure.

CARBONIFEROUS ROCKS.

Kelly limestone.—Coarse white or gray subcrystalline limestones, considered to be of Mississippian ("Lower Carboniferous") age, constitute the lowermost division of the unaltered sedimentary formations. These limestones are for the most part heavily bedded and approximate 125 feet in thickness. Near the middle of the formation is a bed 4 to 5 feet thick, of compact, fine-grained, bluish-black siliceous limestone, which weathers to a grayish-white or buff color. This bed, called the "Silver Pipe" limestone by the miners, is a characteristic feature of the ore-bearing limestone formation and is relied upon as a guide in the search for ore. Below it is the principal ore body, known as the lower or Silver Pipe lode. The upper limestone member does not differ essentially from the lower either in character or thickness. Near the middle of the upper beds occurs the upper or west lode which, next to the Silver Pipe lode is the most important ore body in the district.

The Kelly limestone rarely contains fossils, and its assignment to the Mississippian is tentative, being based on stratigraphic relations and the reported discovery of crinoids,[a] on the strength of which it is probably to be correlated with the Lake Valley limestone farther south. No satisfactory description of the occurrence and identification of these fossils has been found, however, and in the absence of convincing evidence of this character, the correlation of these beds with the Lake Valley limestone can not be considered as conclusively established. In this connection it is to be noted that in the Black Range at Kingston and vicinity the lowermost 300 feet of the Sandia formation consists of massively bedded limestones which contain fossils of Pennsylvanian age, and these limestones rest conformably upon other limestones containing the Lake Valley fauna. As stated on page 229, the discovery of the Lake Valley fauna in the Ladrones Mountains, about 20 miles north of Magdalena, strengthens the suggestion that the lower limestones at Kelly are the equivalent of the Lake Valley limestone.

Along the contact the strata are locally broken and brecciated, and alterations effected by percolating waters are plainly apparent. At one place on the Kelly property a layer of the limestone much altered through silicification was observed to be caught in a nearly vertical position between the schists and limestone. At another point a coarse white quartzite which occurs above the Kelly limestone rests on the "schists" with the ledge of altered limestone between. These occurrences, apparently due to a slipping along the contact plane, have doubtless caused the erroneous representation on the Kelly mine map of a "quartzite dike" extending down into the schists. On the south side of Kelly Gulch a tunnel has been opened on the contact between the schists and the Kelly limestone. Here the surface of the schists dips nearly west (S. 83° W.) at about 65°, while the limestone dips 15°. Adjoining the schists is a zone of breccia 3 to 4 feet

---

[a] Keyes, C. R., Proc. Iowa Acad. Sci., vol. 12, 1905, p. 170. "Of special interest is the fact of the recent discovery of the lower Burlington fossils, such typical forms as *Batocrinus subæqualis* (Hall), in the Magdalena Range more than 100 miles north of Lake Valley."
Herrick, C. L., Am. Geologist, vol. 33, 1904, p. 310. "Sub-Carboniferous crinoids, however, in two cases, are said to have been found in the ore-bearing horizons and more light may be eventually expected on this point."

wide consisting of limestone, in part altered to limonite and ferruginous clay, and a thin clay gouge lies along the surface of the schists. For a distance of 8 or 10 feet from the contact the limestone is decidedly siliceous and filled with cavities lined with drusy quartz. In places it is altered to a siliceous iron ore. Along with the quartz in the brecciated zone occur barite, some manganese, and scattered stains of copper.

Where the Kelly limestone rests upon the schists unconformity is shown in many places, accentuated apparently by slipping and alteration by surface waters, but in other places the plane of separation between the two rocks is not clearly marked. In the Key mine a thin bed of conglomerate intervenes between the schists and the limestone. This conglomerate is composed of rounded grains and pebbles of quartz and feldspar (microcline) in a matrix of crystallized lime (calcite). Likewise in the Greyhound tunnel the limestone at the contact is seen to include rounded and angular grains of feldspar (microcline and microperthite) but no quartz. Instead there appear garnets, showing double refraction, both in idiomorphic form and in granular areas. Some epidote is present and there are some irregular microcrystalline areas, the particular nature of which could not be determined but which may represent fragments of the schists.

Magdalena group.—Under the name Sandia formation is included a series of shales, limestones, and quartzites overlying with apparent conformity the Kelly limestone, and having an approximate thickness of 600 feet. At the base of the formation is a quartzite conglomerate which in the mines has sometimes been mistaken for a "quartzite dike." The character of the formations may be seen in the detailed section of the rocks of the Magdalena district on page 244.

Overlying the Sandia formation is the Madera limestone, consisting of blue compact limestones, for the most part massively bedded and approximating 300 to 500 feet in thickness. These limestones appear at the surface in the hills above Kelly and farther north at the Waldo tunnel, where they have a thickness of perhaps 100 feet, the upper beds having been removed by erosion. Near the foot of the slope they are terminated abruptly on the west by an intrusion of quartz monzonite porphyry.

South of Kelly the red sandstones of the Manzano group occur above the Madera limestone, but these beds have been entirely removed from the area in which the ore deposits occur.

Distribution.—The surface exposure of the Carboniferous formations within the limits of this district constitutes an elongated area extending from a point near the middle of sec. 30, T. 2 S., R. 3 W., diagonally across the range toward the southeast, past Water Canyon. It is approximately 7 or 8 miles in length by 1½ miles in width. Outside the limits of the main area on the north and east are several smaller ones entirely surrounded by granite or porphyry.

On the west the area of sedimentary rocks is bounded by quartz monzonite porphyry, as previously noted; on the southwest, south, and southeast by eruptive rocks; on the east by granite and greenstone schists, and on the north by quartz monzonite and quartz monzonite porphyry. Within the area erosion has uncovered several small areas of crystalline rocks.

The rocks in general dip 30° to 45° S. 78° W. Considerable variation occurs locally, especially along the boundary of the schists, where more or less disturbance is manifest. The strike of the beds is S. 12° E., a trend which carries the section across the range to the southeast, as noted above. On the east side of the range opposite Chihuahua Gulch, one-half mile south of Kelly, a small area of sedimentary rocks has been faulted to the east and is now separated from the main area on the west by an interval of granite which here constitutes the summit of the range. In this area the beds have the same dip as those on the west side.

Erosion has removed the sedimentary formations from the north end of the range with the exception of a small area of the Kelly limestone at the point where the Hardscrabble mine is located. This small mass apparently rests upon and is surrounded on all sides by quartz monzonite porphyry and granite porphyry.

POST-CARBONIFEROUS SEDIMENTS.

No indurated formations of later age than the Carboniferous were observed in the district, though Cretaceous beds are reported to occur along the west side of the Magdalena Range, and rocks of this age are known to have a large development in the western part of Socorro County.

At the north end of the range the rocks are covered not at all or to a very slight depth only with soil. The valleys, especially in their lower portions, are filled to a moderate depth with rock waste, derived from the adjoining slopes. Where stripped of this covering, the rock floor underlying the valleys and portions of the plains adjoining is seen to be composed usually of andesite or rhyolite.

IGNEOUS ROCKS.

The intrusive rocks of the region consist of granite, quartz monzonite, and quartz monzonite porphyry in the form of stocks and dikes, and the sedimentary formations are cut at numerous places by dikes of basic rock (greenstone) allied to diabase.

The east face of the range northward from Water Canyon is composed chiefly of reddish granite of medium-grained texture closely resembling the granites found elsewhere in the Territory and known to be of pre-Cambrian age. Under the microscope the rock is seen to be composed principally of an alkali feldspar and quartz, with a considerable amount of chlorite resulting from the alteration of the ferromagnesian constituents. Especially notable is the prevailing development of micropegmatitic intergrowth between the feldspar and the quartz.

The exposures of granite occur chiefly along the east side of the range. Southeast of Kelly, above Chihuahua Gulch, it appears at the crest of the range. It is not known to occur on the west side of the range except on the south side of Kelly Gulch, where it constitutes a lens-shaped area bounded by "greenstone" on the west and limestone on the east. That the granite is intrusive in the "greenstone" is clearly evident from an exposure of their contact in Kelly Gulch, where portions of the "greenstone" are inclosed in the granite and small stringers of the granite extend into the "greenstone." The relation of the granite to the limestone, however, is not so clear. At some places near the contact with granite the limestones are somewhat altered, as indicated by a high degree of silicification, together with the development of such minerals as pyroxene, amphibole, and garnets. Whether these occurrences are the result of intrusions or of heated solutions following the plane of contact was not determined. It will require a detailed study of the region with the aid of a good map to determine the exact relations of the granites to the Carboniferous rocks.

North of the Graphic mine the sedimentary formations are terminated abruptly by an area of monzonite, irregularly circular or elliptical in outline, with its largest diameter measuring half a mile or more in length. In the hand specimen the rock appears holocrystalline, gray, and granular, and composed chiefly of feldspar and some ferromagnesian constituent, with little or no indication of quartz. In thin sections the constituents named in the order of abundance are found to be orthoclase and plagioclase in about equal proportions, though in some sections plagioclase appears to predominate, a colorless pyroxene, green hornblende, biotite, and a little interstitial quartz. Magnetite and apatite are the principal accessory constituents. A partial analysis of this rock in the laboratory of the United States Geological Survey by George Steiger gave the following result: SiO_2, 57.67; CaO, 5.89; Na_2O, 3.47; K_2O, 3.54.

Adjoining this monzonite area on the north, in the vicinity of the Hardscrabble mine, is an area of quartz monzonite porphyry whose relations to the quartz monzonite and to granite porphyry which lies to the east were not ascertained. A dike 50 feet wide of rock of essentially the same type is intersected by the stream channel three-fourths of a mile west of Water Canyon. A partial analysis of this rock made by Mr. Steiger gave the following results: SiO_2, 67.54; CaO, 2.65; Na_2O, 3.63; K_2O, 3.82.

These porphyries are characterized by large phenocrysts of feldspar and a few of quartz in a holocrystalline groundmass. Under the microscope the phenocrysts are seen to consist of orthoclase, plagioclase, and quartz, in the order named, and the groundmass to consist chiefly of orthoclase and plagioclase with some hornblende and muscovite, with magnetite, titanite, and apatite as accessories. Pegmatitic intergrowths of quartz and feldspar appear in places along the boundaries of the large feldspar phenocrysts. The presence of three generations of orthoclase is witnessed by the occurrence within the orthoclase phenocrysts of perfectly contoured individuals of the same mineral with unlike orientation. Considerable epidote and chlorite, with some quartz, appear as secondary products.

Closely allied to the quartz monzonite porphyry of the Hardscrabble mine is a mass of granite porphyry which cuts the limestones on the west approximately parallel with the trail along the foot of the slope from Kelly to the Graphic mine. The extent of this intrusive mass is not known, as within a short distance it disappears from view under the remaining basal portion of the great andesite flow which once covered this region. The rock is holocrystalline, porphyritic, with phenocrysts of varying size in a microcrystalline groundmass. The phenocrysts consist of orthoclase, quartz, albite (?), hornblende, and ilmenite (leucoxene) in about the order named. The groundmass is composed of minute laths of feldspar in a matrix whose nature was indeterminable with the material at hand. The rock is much altered, rendering diagnosis difficult. A partial analysis of this rock was made in the laboratory of the Survey by George Steiger, with the following result: SiO_2, 69.32; CaO, 0.53; Na_2O, 2.45; K_2O, 5.54.

A small area of granite porphyry lies along the east side of the greenstone schists in Kelly Gulch. It is included between the greenstone schists and the limestones to the east and appears to connect with the granite area on the south. Where this porphyry lies in contact with the schists portions of the schist are included in it. Granite porphyry also constitutes the top of the range above the Hardscrabble mine.

A number of dikes of a dark-green rock have been encountered in the mines, all apparently of similar composition, but owing to alteration the mineral constituents are not clearly recognizable in hand specimens and the term "greenstone" appears applicable to the rock. The dikes are usually narrow, the largest in the Graphic mine being 12 feet wide. Under the microscope the rock is found to consist largely of plagioclase and hornblende, with an abundance of decomposition products, including chlorite, epidote, etc. Traces of the original structure indicate that the rock was holocrystalline and coarsely ophitic and probably a rock allied to diabase.

The extrusive rocks consist of andesites and rhyolites. Two centers of eruption have been noted, one on the northwest side of little Baldy Peak and the other probably represented in Big Baldy. The highest point of Little Baldy is composed of tilted strata, but Big Baldy is made up wholly, so far as its exterior shows, of volcanic rock. The flows from these centers evidently covered to a considerable depth the adjoining slopes, leaving the top of the range from Little Baldy northward bare or covered to a slight depth only. The first eruptions consist of andesite or latite tuffs, breccias, and flows, over which were spread the products of the succeeding eruption, consisting of rhyolite. Trachytes were not observed, though it is not improbable that the order of succession prevailing here is similar to that in Socorro Peak, and further observation may reveal the presence of trachytic rocks in the Magdalena Mountains.

The andesite is dark colored, fine grained or aphanitic, and usually appears brownish gray from weathering. Much of it shows tabular crystals of plagioclase in the dense groundmass. In Mill Canyon, where it incloses deposits of copper ores, the phenocrysts of plagioclase are usually large, and to this circumstance is due the name "bird's-eye porphyry" applied to it. In thin section the large plagioclase phenocrysts are inclosed in a groundmass of smaller crystals of plagioclase.

Rhyolite caps many of the lateral ridges and occurs along the higher slopes. Magdalena and Elephant buttes are composed mostly of rhyolite tuffs and glasses, and the top of the ridge between Hop and Mill canyons is capped by a sheet of whitish-gray rhyolite marked by prominent phenocrysts of quartz and sanadine, the latter fresh and vitreous in appearance.

Basalt is not known to occur within the range, but covers considerable portions of the plains on the west.

STRUCTURAL FEATURES.

That portion of the Magdalena Range included within the Magdalena district consists essentially of a block of Carboniferous strata dipping westward at a high angle (30° to 50°) and resting upon a core of granite. (See fig. 18.) To the south the sedimentary formations are entirely obscured by lava flows, which form the most prominent parts of the range. The eastern face of the mountains marks the location of a fault zone, rather than one great fault. In describing the phenomena to the north, similar except that the fault scarps face west. Dutton[a] says that "it seems better here to lay aside the conception of fault and conceive of

---

[a] Dutton, C. E., Sixth Ann. Rept. U. S. Geol. Survey, 1885, p. 194.

the strata as having been rent asunder along a general north-south line, and one portion pushed over and backward by the uprising boss of granite from beneath." Although in its general features this portion of the Magdalena Range corresponds to the block-mountain type of structure, in the crustal readjustment the main block has been broken into a number of small blocks more or less faulted among themselves, and further essential modifications appear as a result of erosion. Faults are numerous and constitute an especially important feature in the district. The faults are almost wholly of the so-called normal type and include both strike and transverse faults. An example of the transverse fault is the "big fault" in the Key mine, which extends N. 65° W. It dips about 45° N. and has produced in the westward-tilted limestones a horizontal displacement of 129 feet. In a letter dated January 29, 1907, the secretary of the company owning the Key mine states that in the development of this mine three series of faults have been encountered, in two of which the throw appears to be horizontal, one to the east and the other to the west, while in the third the throw is vertical. Indications of other cross faults occur in the gulches between the Kelly and Graphic mines and elsewhere.

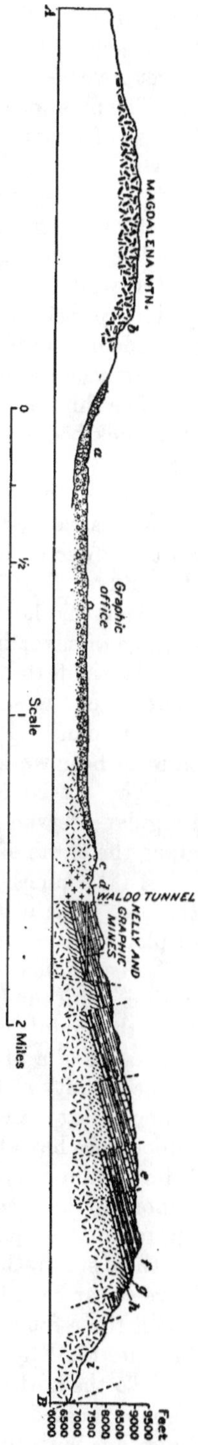

In general, however, the north-south or "strike" fault is the most prominent and characteristic structural feature. Well-marked indications of strike faults can be seen between the "greenstone" areas and the top of the range, especially along the top of the spur running westward midway between the Kelly and Graphic mines. Along this ridge occur a series of low eastward-facing escarpments of the Madera limestone. In these places the fault plane runs apparently parallel with the trend of the range, giving rise to so-called "distributive faults."

Keyes[a] has contended that these faults are due to differential weathering, but Johnson[b] and Herrick[c] have maintained the view held by most local observers that the repetition of the limestones is due chiefly to a succession of faults and only in part to differential weathering, and the writer's observations lead him to concur in this conclusion. Further study in the vicinity of the greenstone schist areas shows clear evidence of a local drop of the strata to the east, also local variations and reversals of dip which support still further the conclusions here maintained.

The great displacement along the east side of the range is probably not the result of one great fault but of a series of faults, undoubtedly of greater size than those of the west side.

Herrick estimates that the limestone beds already mentioned, occurring to the east of the axial fault and separated from the main body by an interval of granite, have been displaced over 300 feet. He assumes that the major outlines of the range, as they now appear, have been determined by the dropping of the strata on the west, due to vast outflow of basic lava in that direction. Whatever of truth there may be in these views, the writer is disposed to think that the strata composing the back slope of the Magdalena Range owe their present attitude chiefly to the dislocations that attended the orogenic movements resulting in the uplift of the Great Plateau, with which is to be associated the relative sinking of the valley of the Rio Grande.

FIGURE 18.—Generalized profile section across the Magdalena Range at Kelly (along line A-B, Pl. XVII). a, Alluvium; b, rhyolite; c, andesite; d, granite porphyry; e, Madera limestone; f, Sandia formation; g, Kelly limestone; h, schists; i, granite.

---

[a] Keyes, C. R., Notes on block mountains in New Mexico: Am. Geologist, vol. 33, 1904, p. 22.
[b] Johnson, D. W., Block mountains in New Mexico: Am. Geologist, vol. 31, 1903, p. 137.
[c] Herrick, C. L., Laws of the formation of New Mexico mountain ranges: Am. Geologist, vol. 33, 1904, p. 301.

ORE DEPOSITS.

General features.—The lead and zinc ores of the Magdalena district do not occur in veins in the proper significance of that term, though in local parlance the ore bodies are usually referred to as such. Ore is found at five separate horizons, the first, second, and third of which have as yet revealed no important deposits.

The first and lowest horizon is along the contact of the schists or the granites with the limestones. In some places also small amounts of copper sulphides fill cracks in the greenstone schists. Only in the Hardscrabble mine has there been any notable quantity of ore derived from the limestone and schist contact. The recently reported strike of ore at the Key is on the limestone and granite contact, as were also the small deposits of the Ambrosia.

The second horizon is at the contact of the limestones with the quartz monzonite porphyry on the west. The only ore body yet opened at this horizon was cut recently in driving the Waldo tunnel to reach the lower workings of the Graphic mine. A considerable body of sulphide ore containing lead, zinc, and copper was encountered, but as yet nothing is known of its extent or value.

The third horizon is in the Sandia formation, about 160 feet above the base, at the contact of a thick-bedded quartzite with a bed of limestone, 6 feet thick, overlain by shales and limestones.

Most of the ore is found at the fourth and fifth horizons, where it occurs in deposits of irregular shape, distributed more or less capriciously along the bedding planes of the Kelly limestone. At the lower or fourth horizon occurs the Silver Pipe lode in the lower member just below the "Silver Pipe" limestone stratum. The most important ore bodies of the district are found at this horizon. The lode takes its name from the occurrence of a considerable percentage of silver in one large ore body.

At the fifth horizon, about 30 feet above the "Silver Pipe" limestone stratum, occurs the "outer" or "west" vein. Some important ore bodies have been opened at this horizon in the Graphic mine. It does not appear in the Kelly, but it is the only ore-bearing bed that has thus far been worked in the Juanita.

The ore bodies are roughly lenticular in shape and locally of large size. They occur at irregular intervals along the bedding planes of the limestones, the principal bodies lying apparently along the crests of low arches running transverse to the strike of the beds.

In the mines small faults are of common occurrence and cause more or less trouble in following the ore. One of the largest is that of the Key mine, already described, where the horizontal displacement measures 129 feet.

Distribution.—The principal mineral-bearing area may be described as roughly triangular in outline, with the longer side approximately coinciding with the axis of the range and extending from the middle of the north line of sec. 30, T. 3 S., R. 3 W., to the southeast corner of sec. 7, a distance of a little more than 4 miles. The west angle of the triangular area may be placed in the vicinity of North camp, where the Waldo tunnel is located. The triangular district thus defined has an area of about $2\frac{3}{4}$ square miles. Over the larger part of this area no ore bodies of importance have been found. Thus far the ores mined have come chiefly from the two narrow belts within which appear the outcropping edges of the ore-bearing limestones. The more important of these belts lies along the west or longer side of the triangle, covering an area approximately half a mile wide by $1\frac{1}{2}$ miles long. Within this belt are located, in order from north to south, the Graphic, Kelly, Juanita, and South Juanita mines. The Graphic and Kelly have so far been the largest producers of the district. The other ore belt lies parallel with and about three-fourths of a mile northeast of this one, along the east side of the mineral area, and in it are the Key, Ambrosia, and other claims.

To the north of the Ambrosia is a small isolated area of limestone where the Hardscrabble mine is located. In this mine the ore deposits, now exhausted, occurred at the contact of the limestone with the schists. This deposit, the only important one thus far discovered at this horizon, yielded about $250,000 in lead and silver. The Key, which lies south of the Ambrosia, was worked on a moderate scale years ago, but the ore bodies being small work was abandoned. With a reorganization of the company new workings have been projected and energetically prosecuted. The entrance to the tunnel lies on the east side of the range, about 400 feet below the crest, near the contact of the ore-bearing limestone with the greenstone schists. To provide facilities for working the ore the company was, in January, 1907, driving a tunnel about 1,200 feet long through the mountain, opening toward the southwest on the Tip Top property.

About 1½ miles south of Kelly, on the southward extension of the Graphic-Kelly belt, are located the Cavern, Young America, and Imperial properties. At one time considerable work was done in that vicinity, but the ore bodies were small and work has been long discontinued. Farther south, on the line of the eastern or Key ore belt, disconnected outcroppings of the ore-bearing limestones occur, but as yet these have not been proved to contain workable deposits.

Structural relations.—In the Graphic-Kelly belt erosion has brought to view the underlying schist in two areas, between which it is covered by the Sandia and Madera formations. So far as can be determined, the width of this belt of schist at the middle is about 600 to 800 feet, but this does not represent its full thickness. On the west the limestones dip away from the contact at angles of 30° to 45°. The occurrence of the greenstone schist in this position is evidently the result of the faulting which they have suffered along with the strata overlying it. The beds have been faulted along the east side of the schist area, the sedimentary block to the east having dropped to the extent of 100 to 200 feet or more. In a few places in the basin above Kelly, erosion has brought to view the ore-bearing limestones on the east side of the schist area, but no ore bodies of importance have been opened in that locality.

In the Graphic-Kelly area the principal bodies of ore occur mainly in parallel shoots, marking apparently the upper part of broad, shallow arches that pitch with the dip of the beds, toward the west. The axes of these undulations in the beds are about 1,500 feet apart. In order from north to south these shoots may be known as the Graphic, Kelly, and Juanita shoots from the mines located on them. Ore bodies occur also in the intervening areas, but they are usually smaller and more sporadic in occurrence. Disturbance is plainly evident, both on the surface and in the underground workings, and consists in the fracturing and crushing of the rocks, accompanied by much faulting on a minor scale. In this region some good bodies of oxidized ore were taken out of the Kelly mine. They occupied well-marked watercourses along two prominent fractures, of which one runs east and west, the other northwest and southeast. The work of development has not yet progressed to the point where the exact form and nature of the ore shoots and bodies can be clearly deciphered, and the absence of good maps of the present workings increases the difficulty.

Besides the faults, of which there are many in the mines, dikes of basic rock cut the beds in different directions. Some of them extend roughly parallel with the strike of the beds, thus forming troughs that have assisted in the concentration of the oxidized ores, which the miners assert to be better developed locally on the upper side of the dike than below it.

In the Graphic mine a dike of basic rock 4 to 5 feet thick appears on the third level, cutting the limestones at right angles to the strike (S. 80° W.) and dipping about 60° N. On the fifth and sixth levels this dike and the inclosing rocks have been cut by two faults. The plane of one runs N. 42° E., with a horizontal displacement of 15 feet to the south; the other extends S. 62° W. There has been considerable crumpling along this line, and the complication thus resulting has led to the supposition that there were two dikes, whereas but one seems to be present. Ore occurs on both sides of the dike. According to Mr. Gregg, manager of the Graphic mine, the ore on the upper side is chiefly zinc carbonate, but native and oxidized ores of copper appear with the carbonates along the lower side.

Toward the west the stratified beds are cut off by the granite porphyry mass already described, the contact following the foot of the limestone slopes from a point about 100 yards northwest of the Graphic tunnel through North camp and the upper part of Kelly. Along this line the Madera limestone constitutes the surface rock except where erosion has cut through it to the beds below. The limestones and the porphyry are separated by a seam of talcose clay varying from almost nothing to 10 inches in thickness and containing rock fragments, among which rounded blocks of porphyry are numerous. Just at the contact the dip of the limestone in places appears nearly vertical, but within a short distance to the east it gradually resumes its ordinary angle of about 35° to 40°. The position of the plane of contact could not be clearly determined, but it appeared to dip about 80° E. The thickness of the middle of the block thus included between the schists and the granite porphyry, measured horizontally, is approximately 1,800 feet, while the distance measured along the strata, if the dip is assumed to be 40°, would be about 2,400 feet. The opinion seems to prevail among the miners that slipping has occurred along some of the bedding planes of the limestones. It is not improbable that some movement may have occurred along the mineralized zone, but there is no conclusive evidence of this.

In the eastern or Key belt the ores are situated on or near the contact of the limestones with the granite. In the Key mine the second and third horizons have been extensively prospected, but thus far without results. On crosscutting to the contact of the limestone with the granite on the east indications of ore were found, and reports indicate that a considerable body of ore was discovered at this horizon. Recent information is to the effect that after exposing a very substantial body of cerusite and a corresponding body of smithsonite, work in the Key mine was halted pending the completion of the tunnel to give an outlet for the ore by way of the wagon road leading down between the Graphic and Kelly mines to the railroad at Graphic station.

In the Ambrosia mines (now idle) similar conditions prevail. At the Hardscrabble the ore was found at the limestone and schist contact. As yet it does not appear that any important deposits have been found at the Silver Pipe or west lode horizons east of the east limits of the Graphic-Kelly belt.

Ores.—As a whole the ore deposits contain lead, zinc, iron, copper, gold, and silver. The chief ores of this district are the carbonates and sulphides of lead and zinc. Oxides of copper occur in small quantities in the upper workings. Deeper workings show considerable amounts of sulphide of zinc associated with the sulphides of lead and copper. Gold and silver are usually present in very small amounts only. Until within two or three years efforts were confined to the extraction of the lead carbonate ores ("sand carbonates") occurring near the surface. The yellowish or brownish gray mineral (zinc carbonate) associated with the lead was not recognized as having commercial value and was either left untouched in the mines or thrown out with the waste. The discovery of the value of these ore bodies revived the industry, which had begun to wane with the exhaustion of the more available "sand carbonate" ore bodies.

In the Graphic mine the lead and zinc carbonates have been largely exhausted and the chief ore in sight at present is zinc sulphide. In the deepest workings (sixth level) both the oxides and the sulphide of copper appear in connection with the sulphides of lead and zinc.

The ore body recently struck in the Waldo tunnel consists almost wholly of the sulphides of copper, lead, and zinc, the proportion of copper being somewhat less than that of the others.

In the Kelly mine operations are still confined chiefly to the extraction of the carbonate ores, though large bodies of the sulphides have been reached in the deeper workings.

The Juanita is at present working "sand carbonate" ores, while the zinc carbonate and the sulphide remain untouched.

Relation of the ores.—The ores of the Magdalena district replace the limestone beds in which they occur. In the oxidized zone a concentric arrangement characterizes many of the ore bodies, sand carbonate occurring within and zinc carbonate constituting a more or less irregular shell on the outside. Between the two carbonates there is usually a zone of iron or copper oxides and in places remaining portions of sulphides. The outer portion of zinc carbonate is not of uniform thickness, but is usually much thicker on the under side of the deposit than elsewhere. In its outer periphery there is a gradual transition from ore to unaltered limestone. In this zone the original structural features of the limestone are well preserved. In one place in the Kelly mine a mass of limestone is inclosed by zinc carbonate. The same relations may be

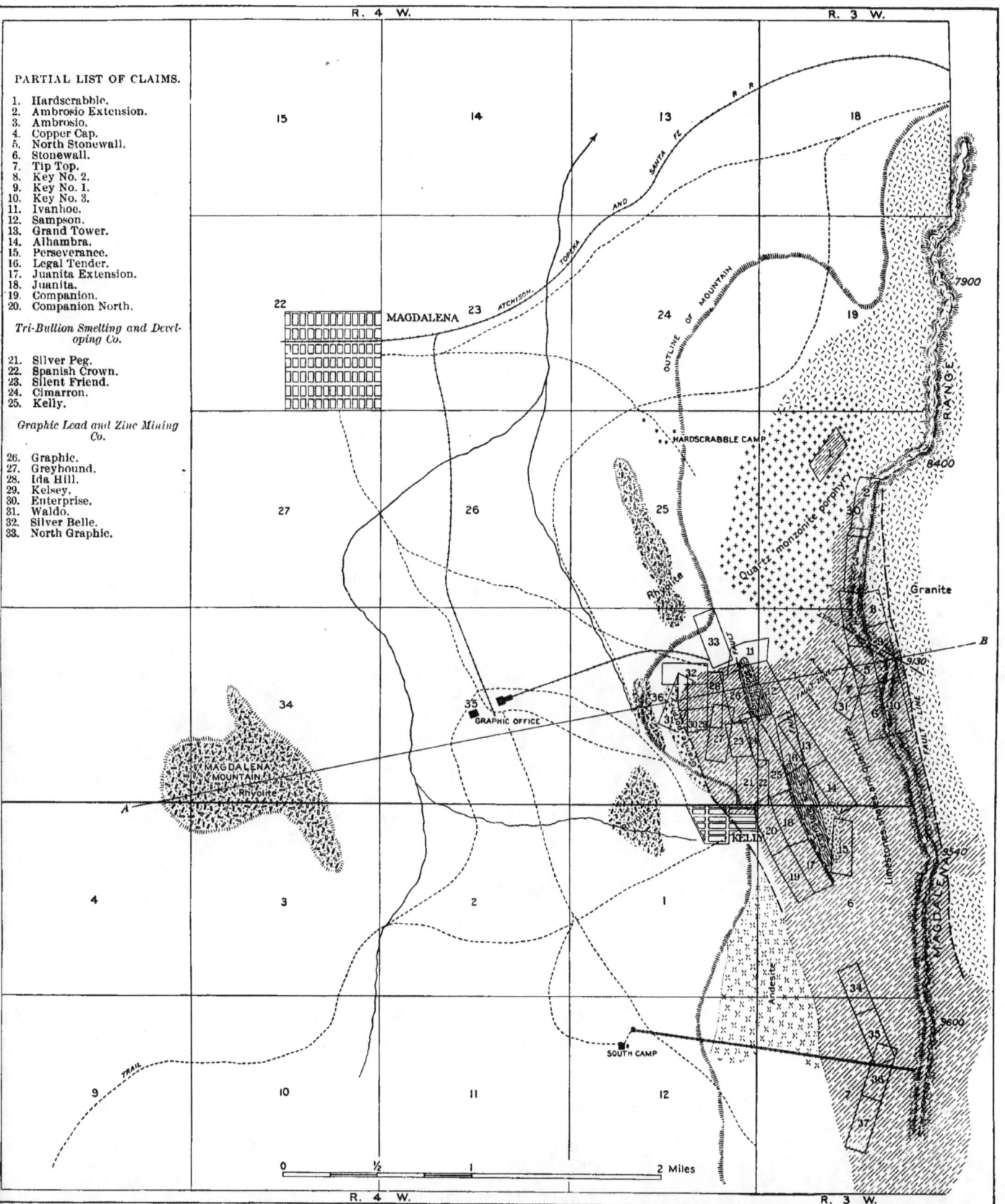

SKETCH MAP OF MAGDALENA MINING DISTRICT, SOCORRO COUNTY.

THIN SECTIONS OF ORE FROM GRAPHIC MINE, KELLY, SOCORRO COUNTY.

A. *q*, Quartz; *c*, calcite; *m*, magnetite; *h*, hematite; *sp*, specularite; *x*, quartz and calcite intergrown. B. *c*, Calcite; *m*, magnetite; *s*, sphalerite; *x*, pyroxene and calcite intergrown; *p*, pyroxene.

seen in some specimens a few inches in diameter. In many of the upper beds galena occurs in association with the carbonate ores.

In the lower (sulphide) ore bodies no definite arrangement of the different constituents was observed, but the well-preserved structural features of the limestone furnish convincing proof of the origin of these ore bodies through metasomatic replacement of the limestone. Except in the less proportion of silver the occurrence of the ores in the Magdalena district has a close analogy in that of the Elkhorn mining district in Montana, described by W. H. Weed.[a]

Values of ores.—No average values can be given for the different ores for any of the mines. The average of several assays of sulphide ores from the Graphic mine, including about 280,000 pounds, gave 23.4 per cent of zinc, 12.8 per cent of lead, 1.6 per cent of iron, 2 per cent of copper, 16.7 per cent of silica, 2 ounces of silver to the ton, and a trace of gold. For profitable shipping the zinc carbonate ores can not have less than 35 per cent zinc. For ores carrying 35 per cent or more of zinc the price paid f. o. b. at Magdalena in September, 1905, was $12.75 a ton.

Minerals of the ore bodies.—The Magdalena district is noted for the beautiful specimens of minerals which occur in the mines. The dumps of the Kelly mine alone would supply a sufficient number of fine specimens of copper and zinc carbonates to stock a dozen museums. The principal zinc minerals are zinc carbonate or smithsonite (accompanied by hydrozincite and monheimite), which occur in brownish to green botryoidal or stalactitic masses showing a radiate crystalline structure. It is found within cavities and in yellowish-brown or buff masses, due to metasomatic replacement of the limestone. The primary sulphide from which the oxidized zinc ores are derived is sphalerite. Associated with it are pyrite, galena, and chalcopyrite.

The lead minerals are chiefly the carbonate (cerusite), the sulphide (galena), and the sulphate (anglesite).

The oxidized copper minerals are represented by the red oxide, cuprite, with here and there a little native copper, and beautiful specimens of the blue and green carbonates, azurite and malachite. The sulphides are represented chiefly by chalcopyrite. Of special interest is the local occurrence in pockets of the rather rare mineral aurichalcite, $2(Zn, Cu)CO_3 + 3(ZnCu)(OH)_2$, a basic carbonate of zinc and copper that occurs in clusters of pearly green acicular crystals constituting scaly incrustations in drusy cavities, and in some places completely preserved by being enveloped in crystals of calcite which thus take on a beautiful green color. The occurrence of chalcophanite $(Mn, Zn)O + 2MnO_2 + 2H_2O$, which appears in steel-blue aggregates associated with smithsonite, is also of interest.

The gangue minerals, occurring in cavities, are especially quartz, calcite, aragonite, and some iron ores, as hematite (specularite) and locally magnetite and pyrite. The association of the sulphide ores with quartz, specularite, epidote, and pyroxene (Pl. XVIII) is noteworthy. Under the microscope the zinc sulphide appears in granular areas alone or associated with specularite inclosed in another mineral, partly green and partly colorless, marked throughout by a pronounced fibrous structure. Between crossed nicols this fibrous mineral resolves itself into an irregular intergrowth of green pyroxene with the characteristic extinction of augite and cloudy gray to colorless amphibole (tremolite). The specularite occurs in grains and masses of considerable size, usually associated with pyroxene and tremolite and with quartz. In thin section a mass of black iron will appear, made up of a skeleton framework of specularite with quartz occupying the interspaces. Along the borders the specularite crystals project into a crystalline groundmass composed of calcite, pyroxene, tremolite, and small amounts of quartz, with possibly some magnetite and scattered masses of zinc sulphide (Pl. XVIII).

Chalcanthite occurs along with pyrite as a greenish-white efflorescence covering the walls of the tunnels in many places. Garnets have not been noted among the gangue minerals, but they occur in the limestone adjoining the contact with the greenstone schists in the Greyhound tunnel as described on page 246.

Alteration of the country rock.—Aside from the development of magnetite, specularite, and silicates which have attended the deposition of the ores, the limestones have suffered little

---

[a] Twenty-second Ann. Rept. U. S. Geol. Survey, pt. 2, 1901, p. 399.

alteration. At some places where they come into contact with the granite they appear highly silicified; at others the contact is marked by decomposition products (clay, etc.), though these may have resulted from faulting. The presence of such minerals as pyroxene with the sulphide ores is indicative of hot waters under great pressure, and the silicification of the limestone along the granite contact may be due to the same agency.

Mode of formation of the ores.—That the occurrence of these ores is the result of metasomatic replacement of the limestone through the agency of heated solutions appears to be conclusively proved by the structure, mineralogy, and general relations of the deposits.

Sometimes, on breaking a piece of zinc carbonate ore, unaltered limestone will be found at the center, and what appeared to be a large mass of zinc carbonate ore inclosed in "sand carbonate" will prove to be mostly limestone with an outer shell of zinc carbonate. The large sulphide bodies show plainly in places the structural features of the limestone.

The formation of the zinc carbonate and the sulphide ore bodies is evidently due to a metasomatic interchange between the limestone and the ore-bearing solutions, though in the former case it is evidently attributable to descending meteoric waters, while in the latter it is believed to be due to ascending heated solutions which, prevented from rising by the overlying impervious strata, have sought out passages along the bedding planes of the limestones.

The evidence that the primary sulphide ores were formed by hot ascending waters is found (1) in their maximum occurrence in lenses below impervious strata along the crest of low arches; (2) in the evidence presented by the gangue minerals; and (3) in their stratigraphic relations.

The compact "Silver Pipe" limestone stratum appears to have served as an impervious cover for the solutions, except where fracturing and faulting permitted these to escape into the limestone above. In this way is to be explained the sporadic occurrence of the deposits of the "outer" or "west" vein in the Graphic mine, which appear about midway between the "Silver Pipe" stratum and below the impervious shale covering the limestone. These merge with those of the "Silver Pipe" limestone below the sixth level. No deposits occur at this horizon in the Kelly mine.

Evidence of the heated condition of the solutions is found in the character of the associated gangue minerals, and the geologic relations of the masses of quartz monzonite porphyry and granite porphyry favor the conclusion that the solutions were derived from these bodies. As yet the workings are confined chiefly to the oxidized zone, where the change from sulphides to oxides is a secondary process unquestionably due to leaching and concentration by waters. The evidence on which is based the prevailing opinion that the Magdalena ores are due to descending waters is apparently all derived from the oxidized zone. The occurrence of considerable bodies of ore along the troughs on the upper side of some of the dikes is held by some as proof of this origin. As this applies only to the oxidized ores which have been concentrated undoubtedly by descending waters, it furnishes no evidence bearing on the original derivation of the ores. Moreover, the fact that some of the dikes cut the sulphide ore bodies indicates that the deposition of the ores took place before the dikes were formed.

Age of the ore deposits.—If the theory of the igneous origin of the ores is accepted, from the structural relations of the ore deposits, it is evident that the solutions from which they were derived must have accompanied or followed the intrusion of the granite porphyry bounding the ore belt on the west. The discovery of a large body of sulphide ore along this contact, together with the fact that the principal deposits are confined to the belt adjoining this intrusive mass, is of special significance. Except that it is post-Carboniferous, the data at hand relating to the age of this intrusion are not conclusive. The fact that the principal intrusions of acidic lavas in this and adjoining districts represent the closing stages of the great volcanic outburst of Tertiary time leads the writer to assign the intrusion of the granite porphyry to that epoch. It is believed that the movements resulting in the monoclinal uplift began long before. Whether the major faulting took place prior to the deposition of the ore is not altogether clear. The occurrence of the ore along the fault planes in the Cavern mine and at the junction of the faults in the Juanita (p. 257), as also possibly the merging of the two veins in the Graphic, would seem to support that view. The fact, however, that the ore found at the fault planes in the Juanita and

A. KELLY MINING CAMP, SOCORRO COUNTY.

Kelly mine and dump in foreground. Elephant Butte beyond, with Cat Mountain showing over the north end of the ridge. San Mateo Mountains in the background. Looking southwest from point above Kelly mine.

B. NORTH ENTRANCE TO GRAPHIC MINE, SOCORRO COUNTY.

Kelly limestone overlain by Sandia formation; rocks inclined S. 65° W.

Cavern mines is "sand carbonate," and undoubtedly secondary, makes this evidence of doubtful value. On the other hand is the evidence afforded by the dikes, which cut the ore beds and are themselves faulted. In this connection it is of interest to note that in the Hop Canyon district, a few miles south of Kelly, similar dikes of basic rock cut the rhyolite.

That there has been displacement at different times is probable, but the major part of the faulting doubtless occurred as a result of the stratigraphic readjustments which attended or followed the uplift, and the introduction of the ores may have marked the closing events of the principal igneous eruptions of Tertiary time.

Mine workings.—The only mines in the district that are prepared to ship ore are the Kelly, Graphic, and Juanita. The South Juanita was operated to some extent in 1905, and it is reported that the Key will begin shipment as soon as facilities are provided for getting the ore to the cars. The main workings of the mines consist of tunnels driven parallel with the bedding planes of the limestones, and the ore is removed by overhead stoping. The lack of order or arrangement of the ore bodies, together with the faulting of the beds, has made the exploration work more or less irregular, and as the limestone may show no indication of the proximity of ore, it has sometimes happened that tunnels were driven close to an ore body whose presence remained unrecognized. The roof is usually solid and in only a few places has timbering been found necessary. Levels are driven about 60 feet apart, measured on the incline of the bedding. Owing to the irregular shape of the ore bodies, the stopes are very irregular and range in size from openings a few feet across to great caverns 50 to 80 feet high and several hundred feet long.

In the Graphic mine (Pl. XIX, *B*) the main (third level) tunnel is driven into the hill southward along the strike of the Kelly limestone, close to the contact of the limestone with the greenstone schists. About 200 feet from the entrance a crosscut extends each way, that to the east connecting with a tunnel that extends along the lower side of the "Silver Pipe" limestone (lower main lode) and the other extending to the top of the incline, where the hoist is located. The incline descends 175 feet parallel with the strike of the beds, cutting across them and into the greenstone schists, the contact here having a westward inclination of about 45°. The bottom of the incline is on the sixth level about 75 feet east of the contact. The elevation at the entrance to the third level is about 7,600 feet. To facilitate the extraction of ore a new tunnel, the Waldo, completed in June, 1906,[a] crosscuts the beds and reaches the "Silver Pipe" limestone about 200 feet vertically below the sixth level. The entrance to this tunnel, which is 1,700 feet long, is at North camp, at the west foot of the slope. The workings of the mine are so connected with this tunnel that all the ore can be extracted by overhand stoping and removed on a single track by horses. Near the south end of the Graphic property the Greyhound tunnel crosscuts the beds at the third level, but no ore bodies were encountered in driving this tunnel. It connects with the southward extension of the third level. It is said that there are in all about 7 miles of tunnels in the Graphic mine. The mine is owned and operated by the Graphic Lead and Zinc Mining Company, with offices located in Cleveland, Ohio. About 100 men are employed, the capacity of the mine being about 65 to 70 tons of ore a day, about 50 tons of which are carbonates. Until recently the ores were shipped to the smelter at Coffeyville, Kans., but the plant has been moved to Joplin, Mo., and the ores are now shipped to that point.

The Kelly mine (Pl. XIX, *A;* figs. 19, 20) adjoins the Graphic on the south, the fourth and lowest level being about at the same elevation and connected with the third or main level of the Graphic mine. The entrance to the mine workings is on the north side of Kelly Gulch, half a mile above the town of Kelly. The main tunnel is driven northward along the strike of the beds, between the greenstone schists and a basic dike which runs nearly parallel with the strike of the beds but dips about 85° W. At some distance in from the entrance a roll occurs in the surface of greenstone and at this point the dike apparently cuts off the intervening limestones and meets or intersects the greenstone schists. A shaft (the Paschal) for hoisting ore has just been completed northwest of the old entrance of the Silver Peg lode. This shaft starts on the shales of the Sandia formation and from the bottom a crosscut reaches the Silver Pipe ore bodies

---

[a] Brinsmade, R. B., Mines and Minerals, September, 1906, p. 52.

at levels considerably below the present workings of the mine. Large bodies of sulphide ores have been uncovered in the lower workings of the mine, but in 1905 only 2 carloads had been shipped and these for testing purposes. The Kelly mine is owned and operated by the Tri-Bullion Smelting and Development Company, with offices located in Chicago. The capacity of this mine is about 100 to 125 tons a day. About 100 men are employed.

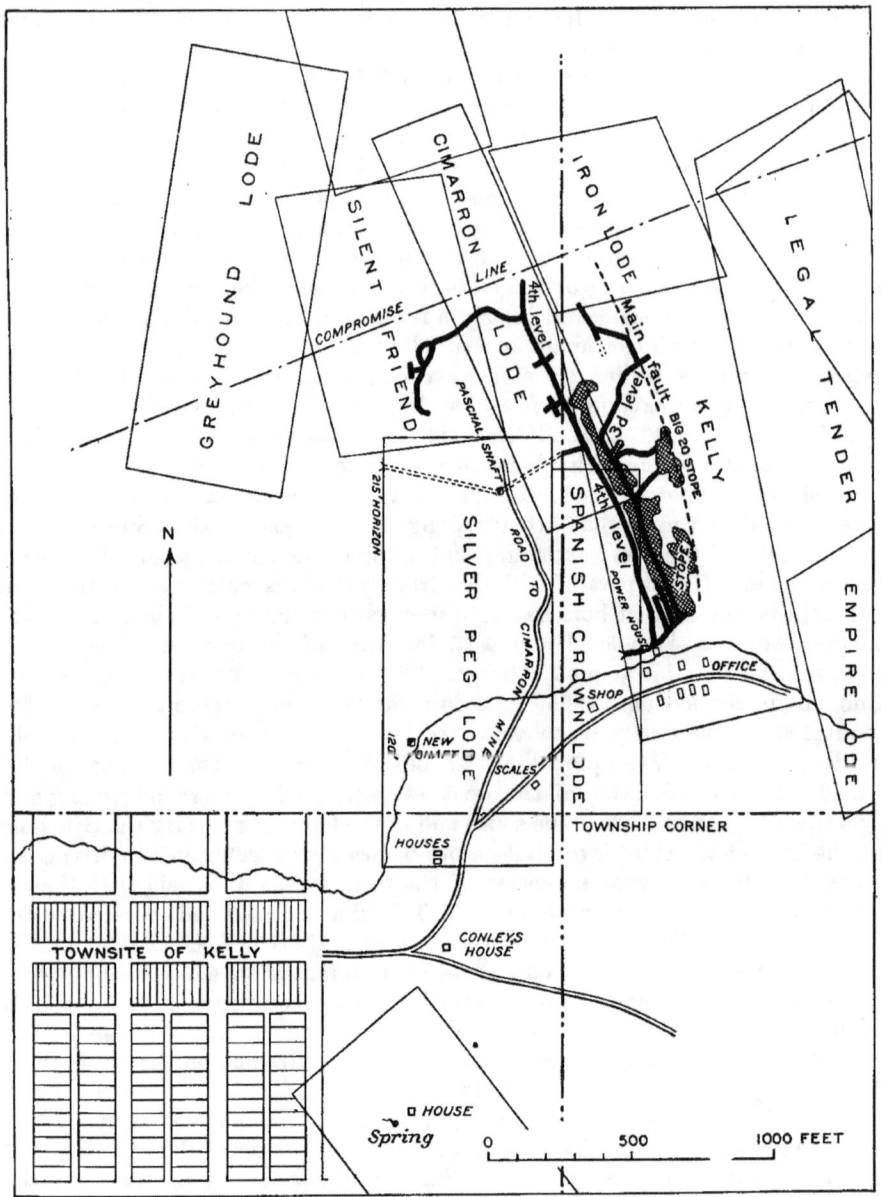

FIGURE 19.—Sketch showing part of workings of Kelly mine, Magdalena district.

The Juanita mine lies to the south of the Kelly property, the surface workings lying on opposite sides of Kelly Gulch. In this mine the ore horizon is reached by a tunnel 380 feet long, crosscutting the Sandia formation. The tunnel enters the hill above the town about one-fourth of a mile south of the Kelly. Near the end of this tunnel an incline at the horizon of the west lode follows the bedding planes, at the top of which is located the hoist for extracting the ore.

An extension of the entrance tunnel for 115 feet to the "greenstone" shows the limestone to have a thickness of about 100 feet. This tunnel did not encounter ore on the Silver Pipe lode. Measured along the incline the first level is about 180 feet below the entrance and the second 40 feet below the first. The second and third levels appear to locate the position of two north-south fault planes, which, about 60 feet north of this entrance, are cut at right angles by another. Near the intersection of these faults are located the deposits of "sand carbonates" which were mined in 1905. Other faults occur in the mine, one a north-south fault, appearing in the main tunnel just as it enters the ore-bearing limestones. No systematic effort has yet been made to prospect the Silver Pipe horizon below the entrance tunnel. The workings in 1905 extended to a depth of 260 feet below the entrance, measured on the incline. This mine is owned

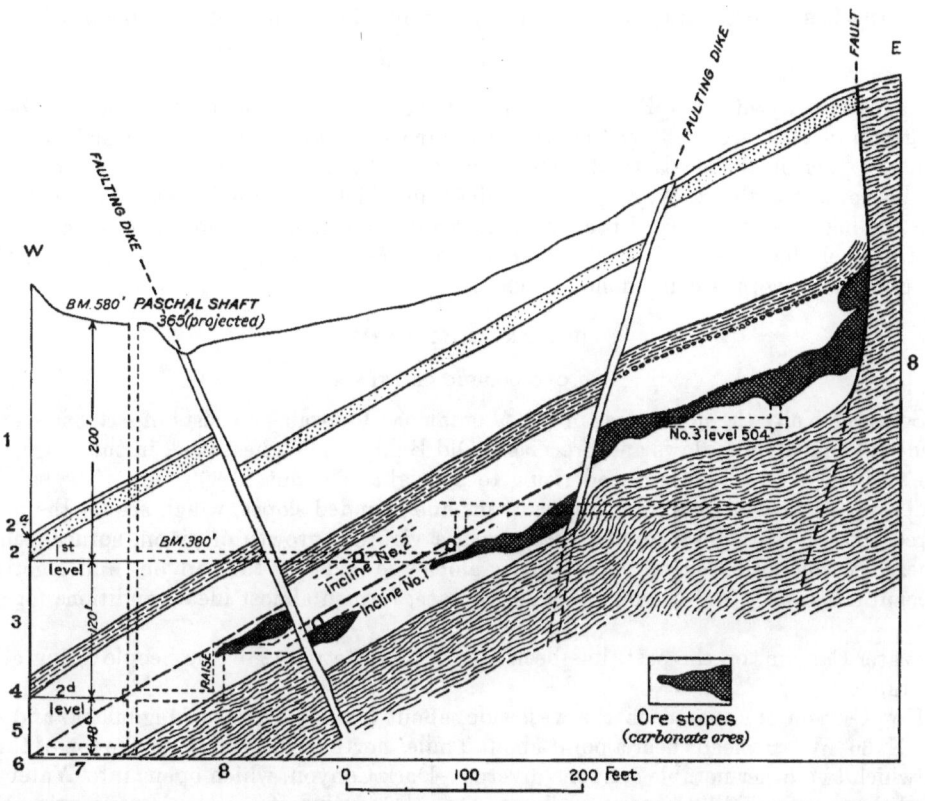

FIGURE 20.—Cross section of Kelly mine, Magdalena district, showing portion of ore bodies. (After Philip Argall.) 1-4, Sandia formation (Pennsylvanian): 1, Shales, limestones, and quartzites; 2, quartzite (contact vein just above); 2ᵃ, upper ore horizon; 3, shale and limestone; 4, shale. 5-7, Kelly limestone (Mississippian): 5, Upper limestone; 6, "Silver Pipe" limestone; 7, lower limestone. 8, Greenstone schists.

by T. B Catron, of Santa Fe, N. Mex., and is operated under lease by F. J. Thomas. At present about 10 men are employed, who take out from 10 to 15 tons of ore a day.

The Juanita Extension mine joins the Juanita on the south. The presence of a number of basic dikes has given considerable trouble in the working of this mine. In other respects conditions are much the same as in the Juanita. No work is being done on this property at present.

In the Key mine work, in 1906, had not progressed beyond the prospecting stage. A small amount of zinc carbonate had been marketed. The entrance to the workings is at the contact of the limestone with the greenstone schist on the east side of the range, and the tunnel has been driven south within the schist for about 100 feet. From this point it is driven west for 75 or 80 feet, to a point where the "Silver Pipe" limestone is encountered, thence S. 12°–15° E. along the lower side of this bed for 500 feet to the big fault trending N. 52° W., which is followed for a distance of 129 feet before the "Silver Pipe" appears again on the south. The greenstone schist evidently gives out somewhere between the first and second crosscuts and southward from this point the limestones abut directly against the granite. The developments measure over 1,200 feet in length. Thus far the principal ore is zinc carbonate. The Key mine is controlled by the Mine Development Company.

SILVER MOUNTAIN (WATER CANYON) DISTRICT.

GEOGRAPHIC RELATIONS.

The Silver Mountain district lies on the east side of the Magdalena Range, adjoining the Magdalena district on the south, and includes within its boundaries the southward extension of the sedimentary belt. According to F. A. Jones,[a] the first claims were located here in 1868. The district has unfortunately suffered much from ill-advised promotion enterprises.

MINERAL DEPOSITS.

In the vicinity of Little Baldy a considerable amount of development work has been done on a group of claims owned by the Abbey Mining Company. The values are said to be in lead and copper on the contact between limestone and porphyry. Situated in a lower zone on a similar contact is the Buckeye group of mines, in which the valuable metal is chiefly copper. Along the higher slopes toward the south the values are mainly in gold and silver. The principal work here has been done by the Timber Peak Mining Company. At present little more than assessment work is being done in this district.

HOP AND MILL CANYONS.

GEOGRAPHIC RELATIONS.

South of Kelly the main mass of the Magdalena Mountains consists of extensive eruptions of andesite and rhyolite flows and breccias. Old Baldy, the highest peak in the range, is 6 or 7 miles south of Kelly, its bare head rising to a height of about 10,800 feet. The sides of the mountain have been carved into deep canyons and rounded slopes, which are for the most part covered, except the highest and steepest, with a vigorous growth of piñon, spruce, cedar, and juniper. Springs of excellent water appear along the sides of the canyons and together with the beautiful scenery and the invigorating climate, furnish almost ideal conditions for summer resorts.

Water Canyon, on the east, has been long a favorite resort of the people living along the Rio Grande.

Hop Canyon is situated on the west side, about 6 miles south of Magdalena, and extends about N. 60° W. It heads near a point about 2 miles north of Old Baldy, at the top of the range, from which two other notable canyons diverge—Dark Canyon, which opens into Water Canyon toward the east, and Mill Canyon, leading toward the southwest. In its lower portion the sides of Hop Canyon are steep and rocky, but toward the head a strong growth of timber has developed.

Mill Canyon lies south of Hop Canyon and is separated from it by a prominent divide capped with rhyolite porphyry.

GEOLOGY.

In its lower portion the slopes of Hop Canyon are occupied mostly by rhyolite and rhyolite tuffs and breccias. Andesite tuffs and breccias appear wherever the overlying rhyolite has been cut through by erosion. In its upper half on the east side along the axis of the ridge the eruptive rocks have been removed, exposing granite. On the southwest the high escarpment constituting the divide between Hop and Mill canyons is a porphyritic rhyolite composed of prominent quartz phenocrysts in a grayish-white felsitic groundmass. Dikes of basic rock cut the rhyolite in different directions. One of considerable thickness was encountered in the principal tunnel.

---

[a] New Mexico mines and minerals, 1904, p. 126.

A. CLIFFS OF RHYOLITE TUFF IN CHLORIDE CANYON, BLACK RANGE.

B. NORTH END OF SAN MATEO MOUNTAINS, SOCORRO COUNTY.

Showing characteristic tree-covered slopes. Mountains composed of rhyolite flows and tuffs. View taken after slight snowfall.

MINERAL DEPOSITS.

A considerable amount of prospect work has been done on both sides of the canyon, following small veins, but as yet without material success. The Hop Canyon Mining Company was vigorously conducting prospecting operations in 1905 on that part of its property on the south side of the canyon, consisting of the Marguerite, Oshkosh, Lucy, and Lookout claims. The ore here lies in a shear zone cutting the rhyolite in a direction about N. 20° to 30° W. As nearly as can be made out the dip is about 70° NE. In all about 2,400 feet of work has been done on this property, of which about 1,100 feet is in one tunnel driven in a direction S. 20° E. to crosscut the main vein at about 500 feet below the outcrop, and the rest is in open pits, shafts, and tunnels in several places.

The surface workings show thin seams of more or less oxidized ore but no workable deposits. At the point where the main tunnel should intersect the vein the fracture was tight and no ore was found though assays are said to have given some results in gold. Conditions here do not appear to warrant the expectation of finding any important bodies of ore.

On the east side of Mill Canyon a considerable body of cuprite has been opened close to the surface in the Wheel of Fortune mine. It occurs in a fissured zone about 24 feet wide running N. 20°–25° W. The country rock is called "birdseye porphyry" by the miners. This rock is an andesite with large phenocrysts of plagioclase in a groundmass composed chiefly of small laths and grains of feldspar and a sparing amount of ferromagnesian constituents. Little or no evidence of flow structure can be seen. The copper appears as a replacement of the shattered porphyry, which has suffered a great deal of alteration along the vein. Its present concentrated form is apparently due to oxidation and secondary enrichment. Very little work had been done on the property at the time of visit (1905) and no conclusion as to the extent of the ore body could be drawn.

ROSEDALE DISTRICT.

GEOGRAPHIC RELATIONS.

Rosedale is situated on the northeastern slope of the San Mateo Range, about 25 miles southwest of Magdalena and about 30 miles west by a little north of San Marcial, on the Rio Grande. The little mining town of Rosedale lies well within the recesses of the hills at an elevation of about 7,100 feet. The surface is one of strong relief, with scrub oak covering the lower slopes and a fairly good growth of piñon, cedar, and juniper on the higher elevations and in the intermontane valleys. (See Pl. XX, B.) A mile or two east of Rosedale the rugged, hilly country gives place to the gravel-filled valley plain which lies between the San Mateo Mountains and the Magdalena Range on the northeast. This valley leads outward to the southeast from the east end of the San Augustin Plain to the Rio Grande valley. The valley is floored by the Palomas gravel and later wash deposits from the adjoining slopes. The beds are composed largely of subangular pieces of rock, chiefly rhyolite and andesite, as a rule loosely consolidated. The rock fragments are large and more angular along the border of the valley and become gradually finer away from it.

HISTORICAL NOTE.

The discovery of gold in the San Mateo Mountains dates from 1882. A rush to the district soon followed, but it was checked by the frequent incursions of the Apache Indians under Geronimo, and after several massacres most of the prospectors left the country.

The only mine is the Rosedale, which is said to have the distinction of paying dividends simply from the ore encountered in sinking and drifting on the vein.

GEOLOGY.

So far as present observations extend no important bodies of sedimentary rocks occur within the San Mateo Mountains. The range apparently consists wholly of rhyolite with its associated breccias, tuffs, and devitrified glass. Underlying all, probably, is a basal deposit of andesite. At the north end of the range a small exposure of stratified tuffs was observed at the foot of the slope, and near Estalina Spring, on the ranch of A. Clemens, a small mass of limestone is inclosed on all sides by rhyolite. It is not improbable that other and larger areas of sedimentary rock may appear on the west side through the stripping away of the rhyolite by erosion.

ORE DEPOSITS.

General character.—The deposits at Rosedale contain principally free-milling ore in a gangue of bluish-white quartz, carrying in places small amounts of oxide of iron and manganese dioxide. The latter appears to be associated with the richer grades of ore. The ore in the deepest workings is said to be base. Repeated assays of the mill run are said to give an average value of $10 a ton in gold. Very little silver is present.

Geologic relations.—The ores occur in a well-marked fissure vein in the rhyolite having a strike of about N. 20° W. and dipping at a high angle (80°) toward the east. The vein is from 3 to 5 feet wide at the surface, but is said to reach in places a width of 12 feet. The rhyolite shows in hand specimens easily recognized crystals of quartz in a well-crystallized groundmass, giving the rock a granitoid appearance.

Two ore shoots have been located, one immediately below the hoist house, the other some distance to the south. They approach each other with increasing depth and by some are supposed to join below.[a]

The ground-water level occurs at a depth of 726 feet, above which the ores are all oxidized; free-milling sulphides appear below this level.

Workings.—Twelve men were employed when the mine was last in operation and an average of about 450 tons of ore was produced monthly. "This was treated on the ground by crushing with stamps, amalgamating on copper plates, and treating by cyanide. The cyanide slimes were sent to El Paso, after drying, for treatment." The workings are shown in the accompanying diagram. They consist of a shaft 732 feet deep and levels approximating 4,644 feet. The mine is owned by the W. H. Martin Company.

BLACK RANGE AND APACHE DISTRICTS.

GEOGRAPHIC RELATIONS.

The Black Range and Apache mining districts (fig. 21) constitute an area about 24 miles long by 14 miles wide lying along the east slope of the Black Range, about 30 miles southwest of Rosedale and 85 miles southwest of Magdalena. The crest of the range, which here represents the Continental Divide, with an altitude ranging from 7,500 to 8,500 feet, constitutes the west boundary of the district.

The Black Range district lies along the north boundary of Sierra County, extending a short distance into Socorro County, and includes as its chief mineralized areas the Phillipsburg and Grafton lodes. The Apache district adjoins that of the Black Range on the south and in it the chief centers of mineralization lie within or adjoining the drainage basin of Chloride Creek. Fairview, a town of about 100 inhabitants, is situated near the east side of the area, and 3 miles to the southwest is Chloride, formerly a mining town of considerable importance but now reduced to a small hamlet. A daily stage connects Fairview with Engle, a station on the Albuquerque and El Paso line of the Atchison, Topeka and Santa Fe Railway, 50 miles to the southeast, and a good wagon road extends to Magdalena, 90 miles to the north. The drainage is carried by eastward-flowing tributaries of Cuchillo Creek, which empties into the Rio Grande in the vicinity of Elephant Butte. The west half of the area is included in the Gila River National Forest.

HISTORICAL NOTE.

According to F. A. Jones,[b] the discovery of ore in this region is attributed to Harry Pye, a driver in the United States Army, who in 1879, while transporting military supplies, found a

---

[a] The writer was unable to enter the mine, operations having ceased a short time before, preliminary to the transfer of the property to a new company, and is indebted to Mr. F. A. Jones for most of the facts touching the underground conditions.

[b] New Mexico mines and minerals, 1904, p. 99.

piece of float in the canyon near the present site of the Chloride post-office which when assayed was found to carry high values in silver. Later, having fulfilled his freight contract with the Government, Pye with some friends located what is known as the Pye lode. From the character

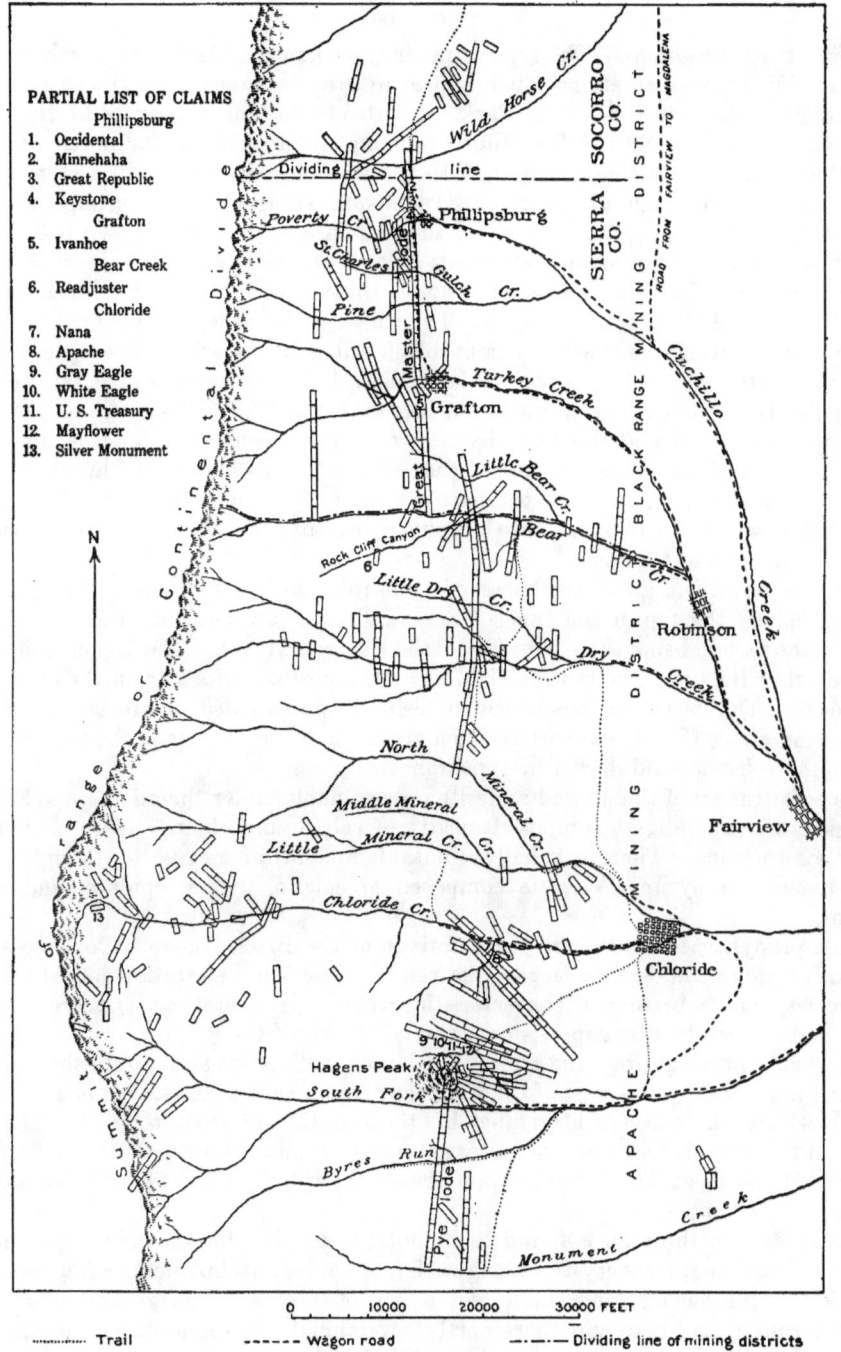

FIGURE 21.—Sketch map of Black Range and Apache districts, reduced from old map of mining districts.

of the silver ore the new camp was named Chloride, and shortly afterward, in 1880 and 1881, with the incursion of the main body of prospectors, Fairview was started a few miles to the northeast.

During the early eighties the Apache Indians were committing depredations throughout this western New Mexico country, and in 1881 several prospectors were killed in the Apache district, among them Pye, the discoverer of the district.

GEOLOGY.

Sedimentary formations.—The region under consideration marks the northern limit of the exposures of Paleozoic formations which appear in irregular areas along the eastern slope of the Black Range as far south as Cooks Peak. North of Chloride exposures of these formations are limited to small patches of limestone occurring at intervals as far north as Phillipsburg. Near Chloride and in the Cuchillo Range, to the east of Fairview, the limestones of the Magdalena group occur, overlain by the lower red sandstones of the Manzano group (Abo sandstone). At Chloride the Manzano group, which in this region contains one or two beds of limestone, dips to the east and abuts against a narrow belt of andesite tuff that intervenes between the sandstones and the Palomas gravel underlying the valley between the Black Range on the west and the Cuchillo Range on the east. The width of the valley here is about 5 or 6 miles.

The Cuchillo Range represents a fault block tilted to the east. The dislocation resulting in this uplift is thought to have occurred later than that of the main range, but no direct proof on this point is at hand. The throw is not less than 1,000 feet. This site was evidently once covered with sheets of andesite and rhyolite, reaching downward from the main range, but erosion has stripped away the covering of the sedimentary rock from the higher points, leaving portions as lesser eminences along the east side and over adjoining parts of the plain. Just east of Fairview a mass of monzonite porphyry has invaded the sedimentary beds and now appears at the crest of the range.

Eruptive rocks.—A great thickness of andesite, consisting of flows, tuffs, and breccias, extends all along the top of the Black Range and, except where removed by erosion, reaches down into the valley, being there covered by the Palomas gravel. The higher ridges are capped usually by rhyolite and rhyolite tuffs (Pl. XX, *A*), deposited evidently on the eroded surface of the andesite. The lower portions of the andesite where revealed by erosion along the bottom of the canyon above Chloride are altered to a grayish-green rock, but at the top of the range the rock is notably fresher and darker in appearance.

The constituents of the altered andesite determinable under the microscope are plagioclase and augite in a groundmass composed mostly of calcite and chlorite, with some epidote and very little serpentine. There is usually a notable amount of magnetite, though it is in many places changed to pyrite. Veinlets composed of calcite, pyrite, epidote, and adularia are abundant.

These propylitic andesites are characteristic of the Black Range region and have a large development along the eastern face of the range. The same alteration has affected the tuffs and breccias, which present a characteristic grayish-green mottled appearance on exposed surfaces and constitute vast deposits over the north end of the range.

The rhyolite resting upon the andesite is largely tuffaceous and rarely shows indication of flow structure. Some miles west of Chloride the canyon cuts across an old channel in the andesite tuffs which was subsequently filled by the eruption of rhyolite tuffs. The deposits of rhyolite on this side of the range are far less extensive than those of andesite, and stream erosion has dissected the sheet, leaving only cappings on the ridges or the filling of old depressions in the surface of the andesite.

In thin section the rock is found to be notably porphyritic, showing large phenocrysts of corroded and broken quartz crystals along with more numerous but smaller fragments of crystals of orthoclase, plagioclase (andesine), and a few of biotite. The groundmass is glassy, in places spherulitic, and here and there partly crystalline. In some places the phenocysts predominate over the groundmass, giving the rock in hand specimen a pseudogranular structure.

MINERAL DEPOSITS.

General character.—The ores occur as vein deposits, chiefly in the andesite but partly in the rhyolite. Most of the veins may be included in two systems—a north-south system and an east-west system—which are distinguished from each other in a general way not only by the direction of their strike but by the character of their ores. Those of the north-south system contain chiefly gold, silver, and locally copper; the other system is characterized by silver-copper ores. This distinction applies only in a broad way, however, for exceptions are not uncommon.

A belt about 2 to 3 miles wide, extending from Phillipsburg on the north to Hagens Peak on the south, following approximately the contact of the limestone and eruptive formations, includes the major part of the gold-bearing veins. In this belt some of the strongest veins strike nearly due north and south. Others, such as the Apache and U. S. Treasury veins, near Chloride, strike northwest, and there are a few veins which extend northeast. These veins consist of quartz, locally with barite, and are as a rule fairly persistent. Some of them may be traced for several miles. A number of these veins consist of a single fissure which may have a width of 2 to 5 feet; others occupy broken or sheared zones in andesite and andesite tuffs. The ores are free milling, usually with some sulphides.

The silver-copper ores occur to the west of the gold belt, nearer the crest of the range. These veins conform even less to any order or system, but in general they extend more or less transverse to the axis of the range. They are less persistent than the gold veins and carry but a small amount of quartz. They constitute shear zones along which silver-bearing copper sulphides, chiefly bornite, have been deposited, partly as a filling of the fractures but more largely as an impregnation or replacement of the more or less crushed rock along the sides of the fissures. There is usually but one well-defined wall—the hanging wall. On the other side the ore fades out gradually into the country rock.

Phillipsburg.—The principal workings here are on a north-south quartz vein called the Republic, from 2 to 5 feet in width. In general the walls of the vein are composed of andesite breccia, but about midway between the Keystone and Republic mines it bends sharply to the east, apparently along the contact between a small mass of limestone on the west and andesite tuff on the east. Along Poverty Creek, west of this point, occur two small exposures of limestone inclosed on all sides by andesite breccia. In one of these, half a mile above Kingsbury's cabin, the limestone strata have a sharp anticlinal flexure, and limestone fragments occur in the andesite breccia above and below the limestone mass.

Considerable work has been done on the Republic vein, but no great depth has been attained. Some of the principal properties on this vein are the Minnehaha, Republic, and Keystone groups, named in order from north to south. A few years ago a stamp mill was erected and rather extensive plans were inaugurated for developing the Great Republic mine. A small amount of ore was extracted, but work was then abandoned. The vein has a width of about 3 feet filled with white amethystine quartz, much of which shows a ribbon structure. The quartz occurs also as a lining to cavities or as lenslike masses presumably as a filling of former spaces. The vein matter is frozen to the wall throughout most of its course. The ore occurs usually in small lenses along one or the other wall. The shoots for the most part are of moderate size and apparently pitch to the south. The dip of the vein is 55° E. On the Keystone property the vein has a fine exposure along the west side of a gulch, but no work of importance has been done here.

A zone of mineralization extends southward from this point, but the relation of the veins to each other was not clearly determined. About a mile south of the Great Republic occur a number of prospects known as the Gold Bug group. In one of these—the Chicago—the sulphide ores occur in a gangue of quartz, calcite, and barite along the contact between limestone and andesite tuff. Other small exposures of the limestone occur in the adjoining slope. The vein dips about 75° W. One thousand pounds of ore shipped from this mine is said by the owners to have yielded 13 per cent in copper, 2.1 ounces of gold, and some silver.

Grafton.—Grafton is situated on Turkey Creek about 3 miles due south of Phillipsburg and 10 miles northwest of Fairview. At one time it was the center of a considerable mining industry, but it is now practically deserted. The only property on which work is now in progress is the Ivanhoe, of which Robert G. Ingersoll was at one time a part owner.

The Ivanhoe vein strikes about N. 23° E., apparently crossing, some distance down the hill, another—the Emporia—which comes from the northwest. It is the belief of the miners that the two veins come together some distance north of the camp, but this could not be verified. The vein varies in width from 4 to 12 feet and dips 70° E. The ore is partly free milling and also contains sulphides. The gangue consists of predominant quartz, with calcite and barite. The country rock is andesite and andesite breccia. Along the vein the rock is much altered. Silliman,[a] who visited this mine in 1882, speaks of the Great Mother Lode as composed of a "vitreous and variegated copper, blue and green malachite, calcite, cerusite, free silver, silver chloride and gold in a quartzose gangue." Streaks of black ore brilliant with free gold are found.

The vein is usually frozen to both walls and possesses in places a well-marked ribbon structure. The ore shoot is about 50 to 60 feet wide along the vein, with a well-marked pitch to the south. At the 100-foot level occurs a clearly defined watercourse lined with calcite. The ore occurs in small, irregular lens-shaped masses, and according to the miners is generally associated with the barite.

The workings consist of a 270-foot tunnel and a winze, 285 feet deep, which starts about 165 feet from the portal. Levels extend out at intervals of 50 feet. A very little ore is being taken out. It is said to assay 40 ounces in silver and $1 in gold to the ton and $1\frac{1}{2}$ to 2 per cent in copper.

Mahoning group.—South of Grafton a good deal of superficial prospecting has been done on all the creeks crossing the mineralized belt. The most important prospect in this vicinity apparently is on the property known as the Mahoning group. The claims of this group are on the south side of Rock Cliff Creek, a tributary of Bear Creek. They are located on a well-defined vein 3 to 5 feet wide, cutting rhyolite and rhyolite tuff in a direction about N. 10° W. The rhyolite has a marked granitoid texture, in hand specimens resembling nevadite. It constitutes the capping to the hills in this vicinity, and rests upon the irregular surface of andesite tuffs. The upper portion of these tuffs is noticeably stratified.

The vein, which dips 75° E., has been opened by a shaft to a depth of 208 feet. Up to 1905 no large bodies of ore had been found. The walls of the vein are well defined, and clay occurs in considerable amount, both as a selvage along the walls and in places nearly filling the vein. Some of the clay pockets extend out into the wall rock. Masses of ore and some fragments of the wall rock occur embedded in the clay.

The ore contains sulphides of copper, carrying both gold and silver, in a gangue of quartz, barite, and clay. An average of the dump is reported by the owner, Quinby Vance, to assay 28 ounces of silver and $1.96 in gold to the ton and 2.5 per cent of copper. Selected ore runs as high as 200 ounces in silver, $2 to $4 in gold, and 3 per cent of copper. About 5,000 tons of ore has been shipped from this locality.

Chloride Creek.—The mineralized belt crosses Chloride Creek a short distance west of the town of Chloride. The town is located at the west border of the Palomas gravel, where it overlaps the andesite breccia and tuff. A short distance to the west the creek cuts through a narrow belt of red sandstones which are bordered on their east, north, and west sides by andesite tuffs and breccias. The west contact of the sandstones with the tuffs appears to be faulted. A lode following up the stream from this point for 3 miles above Chloride crosses the valley in a direction S. 40° E. This lode, which is called the Apache, is really an aggregation of small quartz veins occupying a shear zone 3 to 5 feet wide. There is a fairly well-developed hanging wall, which dips about 75° to 80° SW. In places a clay selvage occurs along the hanging wall. Quartz constitutes the chief gangue mineral, filling the main fissure and extending out in stringers into the adjoining rock. There was at one time much mining activity in this vicinity, but very little is being done at the present time.

In the Nana mine, which is opened on the Apache vein, the ore shoot has a length of about 40 feet along the vein, and in places shows a thickness of 2 to 4 feet of low-grade ore. Selected

---

[a] Silliman, Benj., jr., Trans. Am. Inst. Min. Eng., vol. 10, 1882, p. 441.

dark ore runs 144 ounces in silver and $14 in gold to the ton and 9 per cent in copper, according to the statement of E. James, one of the owners.

About 2 miles south of the Nana is a group of claims, including the Gray Eagle, White Eagle, U. S. Treasury, and others, which are located on a well-defined quartz vein running parallel with the Apache vein (N. 42° W.) and dipping about 75° to 80° SW. The extensive deposits of andesite tuffs lying to the west are here bordered on the east by a band of limestone, and along the east side of this is a similar band of red sandstone. The contact between these two bands of sedimentary rocks is covered by a deposit of tuff and breccia. The vein runs transverse to the boundary between the andesite tuff and limestone. In the andesite the vein is generally from 4 to 6 feet wide, but in places it is wider. In the U. S. Treasury it has a thickness locally of 24 feet.

The gangue is a white granular quartz, carrying more or less free gold. Mr. James reports that assays show values of about $10 to $11 in gold and $5 in silver to the ton. In the limestone the vein is said to carry chiefly lead in a gangue of quartz, barite, and calcite.

To the east of the U. S. Treasury group is the St. Cloud group, and south of this are the Colossal and Midnight properties. The Colossal is said to have produced over $70,000 worth of ore. The U. S. Treasury vein evidently carries a large amount of low-grade milling ore and under proper management might develop into a paying property.

About a mile south of the U. S. Treasury location is Hagens Peak, on the north, east, and south sides of which are a number of veins that appear to radiate from the mountain. One of the most prominent is the Pye lode, which may be traced for a long distance toward the south. Owing to their isolation very little has been done toward the development of the mines in this vicinity, and with the exception of assessment work no active mining operations are at present in progress. In his description of the mineral regions of southern New Mexico in 1882, Silliman[a] refers to Hagans lode, probably meaning thereby the Pye lode, and says that it is "chiefly valuable for silver with a little gold and some yellow copper in a cross-grained crystalline and cellular quartz." He reports ores from different claims as yielding from 118 to 168 ounces of silver, chiefly as chloride, from a trace to 1 ounce of gold, and from 1 to 7 ounces of copper.

Silver Monument.—About 10 miles west of Chloride, at the head of Chloride Creek, is the Silver Monument mine, the only one operating in the district in 1905. The mine is located on a shear zone in andesite running N. 73° W. The elevation at the top of the shaft, which is 325 feet below the crest of the range, is about 7,625 feet. The vein is from 2 to 6 feet in width, and dips about 70° N. Along the walls of the vein the dark-colored andesite is changed to a greenish-gray rock, showing the propylitic alteration characteristic of the andesites of the region.

The ore occurs in a fault breccia of altered andesite, mingled with which is more or less quartz. The hanging wall is well defined, but there is no apparent foot wall. A selvage of clay, which in places reaches a thickness of 2 feet, occurs along the hanging-wall side. The ore shoot has a width along the vein of about 70 feet, and pitches toward the east. The ore minerals are chalcopyrite and bornite. The silver occurs chiefly in the bornite. The workings consist of a 260-foot shaft with several levels. Another shaft, now abandoned, is said to have reached water level at 390 feet.

Under economical management the ore bodies would probably permit profitable working, but the distance from the railroad makes it possible to handle only the richest ores. Heretofore the ore has been hauled by wagon 64 miles to Engle and shipped to the El Paso smelter. A mill has recently been located near the mine for treating ores, but at the time of visit it had not begun operation. It is reported that prior to closing in 1893 the Silver Monument shipped $100,000 worth of ore. Work on this property was renewed recently and is steadily progressing. A tunnel being driven from Chloride Canyon just above the mill will cut the vein 200 feet below the deepest workings, thus furnishing a convenient exit for the ores.

New Era.—The New Era mine is located on the south side of Chloride Creek about 1½ miles from the Silver Monument. The geologic conditions here correspond to those at that mine.

---

[a] Silliman, Benj., jr., Trans. Am. Inst. Min. Eng., vol. 10, 1882, p. 441; Eng. and Min. Jour., vol. 34, 1882, p. 212.

The ore occurs in a shear zone approximately 5 feet wide running almost parallel with the Monument vein (N. 72° W.). The shaft has been put down 350 feet, with levels at 90, 140, 200, and 350 feet. The water level occurs about 30 feet below the top of the shaft, and the difficulty of handling the water caused a suspension of operations. C. T. Brown, who had charge of the mine, states that the sulphide ore gave returns of about 17 per cent of copper and 161 to 256 ounces of silver to the ton.

Late developments.—In the Apache and Black Range districts there was much activity in 1909 and several promising developments are reported. On the U. S. Treasury mine, which has now attained a depth of 200 feet, developments were in progress and ore was shipped to the El Paso smelter. Accounts current in the press state that the higher-grade ore, besides containing much silver, has proved to be rich in gold. Other properties worked are the Silver Monument, where an adit tunnel has now attained 1,700 feet in length; the Polar Star, on Wild Horse Creek; the Eureka, on Poverty Creek; and the Keystone, in the northern part of the district. A new town, called Fluorine, is located at the north end of the district; the Great Republic mine is in operation in that vicinity.

HERMOSA (PALOMAS) DISTRICT.

GEOGRAPHIC RELATIONS.

Hermosa is situated on Palomas Creek about 15 miles south of Chloride and about 35 miles nearly due west from Engle, on the Atchison, Topeka and Santa Fe Railway. The stream and its tributaries have here cut deep channels in the sedimentary formations, which rise almost vertically to a height of 1,000 feet above the stream bed. Two centers of mineralization are situated near Hermosa, both at one time the scene of active mining operations. One is just north of the site of the old town, and the other (Palomas camp) on the left bank of the stream about a mile below.

GEOLOGY.

Hermosa lies on the western boundary of a belt of sedimentary rocks which have been exposed by the removal of extensive deposits of lavas that once covered the region. On the west are volcanic rocks, chiefly andesite, with scattered patches and embayments of limestone. Igneous rocks, both intrusive and eruptive, occur also within the sedimentary belt. At Hermosa the area of sedimentary rocks is 3 or 4 miles wide. About 2 miles below Palomas camp these beds are terminated abruptly by the Palomas gravel, which abuts directly against the cliff formed by the erosion of the limestone. Along the sides of the valley here the Palomas gravel has been deeply trenched, exposing a thickness of approximately 800 to 900 feet. North of Palomas Creek are several buttes capped with basalt.

At Palomas camp the walls of the canyon are about a thousand feet in height and are composed in large part of massively bedded limestone. In the lower half some shale and quartzite beds alternate with the limestones. At the base of the escarpment is a thick-bedded fine-grained grayish-blue limestone overlain by dark carbonaceous shales. The ores at this camp occur in pockets and shoots in the limestone below the shale. Fossils obtained from the limestone were identified by George H. Girty as forms occurring near the base of the Pennsylvanian series. From this it would appear that the base of the section at Hermosa is to be correlated with the Sandia formation at Magdalena and farther north, indicating a change in the character of the sedimentation toward the south whereby the shales and quartzites give place to limestones.

The strata dip from 20° to 30° N. 62° E. At Palomas camp the general slope of the beds is interrupted by a low arch which pitches to the north. At this point two systems of faults occur, the major faults running N. 38° W., and a series of lesser breaks trending approximately with the dip. One of these, the "Kendall break," contains one of the principal ore deposits of the district.

About 2 miles above Palomas camp, toward the west, the limestone is cut off by andesite tuff. Along the boundary occurs a remnant of an earlier flow of andesite about 600 feet wide, which runs in a nearly north-south direction. On the west side of the mass the surface rock is andesite tuff, showing well-marked stratification planes which bend up abruptly on meeting the old ridge. The south fork of the river on which the town is situated cuts through the mass just north of the hotel. North of the stream the greater hardness of the andesite causes it to stand out as a ridge, which decreases in height toward the north and finally disappears under the covering of andesite tuff. Between this ridge of andesite and the limestone to the east is a depression underlain by the Palomas gravel, which here evidently constitutes a filling in a previously eroded valley. The mines of the Antelope group are located in the limestone near the contact with the andesite.

THE ORES.

General character.—The ores of this district are principally silver-bearing sulphides of lead, copper, and iron. Native silver is common, but the chief silver values, which run from 300 to 700 ounces to the ton, occur in galena and bornite.

Geologic relations.—The principal mines at Palomas camp are comprised in what is known as the Pelican group. In this group are included two patented claims, the Palomas Chief and the Ready Cash. Others not patented are the Happy Annie, Columbia, and Mountain Chief. All these claims are located along the north side of the creek in the limestone at the base of the low arch already described (p. 226). The ore deposits consist of pockets and pipes filled with ore in a talcose clay gangue in the beds of limestone below the shale. Some of the ore bodies occur in flat-lying lenses along the contact with the overlying shale; others of irregular shape extend downward to varying depths into the limestone. They vary in cross section from a few inches to 10 or 20 feet. They are closely connected with the faults and fractures already described and the largest bodies of ore usually occur at the intersections of these breaks. Some of the most important deposits were found along what is called the "Kendall break." Thin seams of ore occupy the fissures that connect the larger ore bodies and the miner is often able to locate the larger bodies by following these seams.

The fractures with which the ore is connected are not apparent in the overlying shale, and no ore is found in the shale or in the overlying beds. As stated above, the low arch or fold under which the ores have been deposited pitches at a low angle to the north. Owing to this pitch the ore-bearing limestones occur at a higher level in the walls of the canyon on the south side than on the north. Some deposits of ore have been opened on the south side, but the chief production has been from a few mines on the north side of the creek. No opportunity was afforded to observe the occurrence of the ores in the Antelope group, but a surface examination shows that they are in the limestone at or near the contact with the andesite. In the Antelope, which was at one time a large producer, the ores were reached by a shaft and tunnels, the condition of which at the time of visit did not permit inspection underground.

Mine workings and production.—In the early days this camp was a center of great mining activity, but the mines have been idle for many years and the workings are mostly dismantled and caved in. Work was resumed on the Palomas Chief mine in 1905 by a newly organized company. From a letter[a] it is learned that in a new tunnel the Kendall break was struck at 925 feet from the portal and with it a considerable body of ore which assays up to 1,500 ounces of silver to the ton. The break at this point is reported to be about 8 feet wide, only a part of which is ore.

The district is said to have produced about $1,250,000 up to January 1, 1904, principally in silver. Of this amount nearly $500,000 is credited to the Palomas Chief. In 1909 work was in progress on the Ocean Wave claim, where the ore is treated in a small concentrating plant. The Palomas Chief mine was also in operation.

In the gulch which joins Palomas Creek near the hotel, about half a mile south of that point, were located the American Flag and Flagstaff properties, which, several years ago, were producers of lead ore (sulphide). Reports place the output of these mines at $25,000 to $50,000, but no work is being done on this property at present.[a]

---

[a] Letter from C. T. Brown, dated March 29, 1907.

Mode of formation of the ores.—That the ore deposits at Palomas camp have been formed by ascending waters seems to be well attested. Briefly stated, the evidence supporting this conclusion is as follows:

1. The occurrence of the ores immediately below, nowhere above, an impervious stratum of shale.

2. Their direct connection with systems of fractures in the limestone which furnished channels for the ascending solutions. As the shales would not permit the formation of open fissures, there could be no passage of the ore-bearing solutions through them.

3. The occurrence of the principal bodies of ore along the arch of a low-pitching fold. The presence of the ores in this position underneath an impervious stratum of shale precludes the assumption that they have been derived from descending waters. The conclusion seems warranted, therefore, that the solutions came from below. There is strong probability that the source of the ore-bearing solutions was in a mass of intrusive rock lying at no great depth. The data at hand with reference to the ores of the Antelope group do not warrant a definite statement, but it is thought that the conditions there are the same and that further investigation will disclose the presence of intrusive masses from which the solutions may have come.

KINGSTON AND ADJOINING DISTRICTS.

GEOGRAPHIC RELATIONS.

The town of Kingston, which is near the center of the mineralized zone, is situated near the head of Middle Percha Creek about 20 miles due south of Hermosa. A daily stage connects the camp with Hillsboro, 9 miles to the east. From Kingston a trail to points on the west side of the range ascends the steep slope to the summit, which is here carved into strong and rugged relief. (See Pl. XXI.)

HISTORICAL NOTE.

According to F. A. Jones [a] the discovery of silver at Kingston was made in 1880 by a party of prospectors. The two first locations were the Iron King and the Empire. In 1881 the area along the east slope of the mountains from Kingston to Grafton, embracing a belt 50 miles long from north to south and 20 miles wide from east to west, was organized into one mining district, called the Black Range district. In this area were a number of mining camps and out of it several districts have been formed.

GEOLOGY.

The axis or core of the Black Range consists of granites, gneisses, and schists flanked by Paleozoic strata which dip away on both sides. Extensive masses of andesites and rhyolites, as tuffs, breccias, and flows, lie upon the eroded surface of these formations. The volcanic rocks extend far eastward and connect on the west with the great lava fields of this part of the Territory. The Paleozoic strata exposed by the denudation of the overlying eruptive rocks comprise formations ranging from the Cambrian to the upper Carboniferous, and it is not improbable that Triassic and Cretaceous sediments may occur in the region. A few miles west of Hillsboro appear red shales and sandstone overlain by soft yellowish-white conglomeratic sandstones which may represent this horizon. In the absence of conclusive evidence as to their relations it is thought best to consider these beds as belonging to the Manzano group of the upper Carboniferous. The relation of the sedimentary formations is shown in the profile section on page 224.

On the higher slopes, where pre-Cambrian granites and schists have been exposed in places, the Cambrian quartzites rest upon the eroded surface of the crystalline rocks. Along the axis of the range the sedimentary beds have a dip of 30° to 40°, but east of Kingston the dip flattens out to about 20°. The eastward dip brings the red sandstone of the Manzano group to the level of the stream in the eastern part of the town, but the lower beds are again brought up by a north-south fault about half a mile east of the post-office. The fault scarp is about 500 feet high and indicates a displacement of not less than 1,000 feet.

---

[a] New Mexico Mines and Minerals, 1904, p. 94.

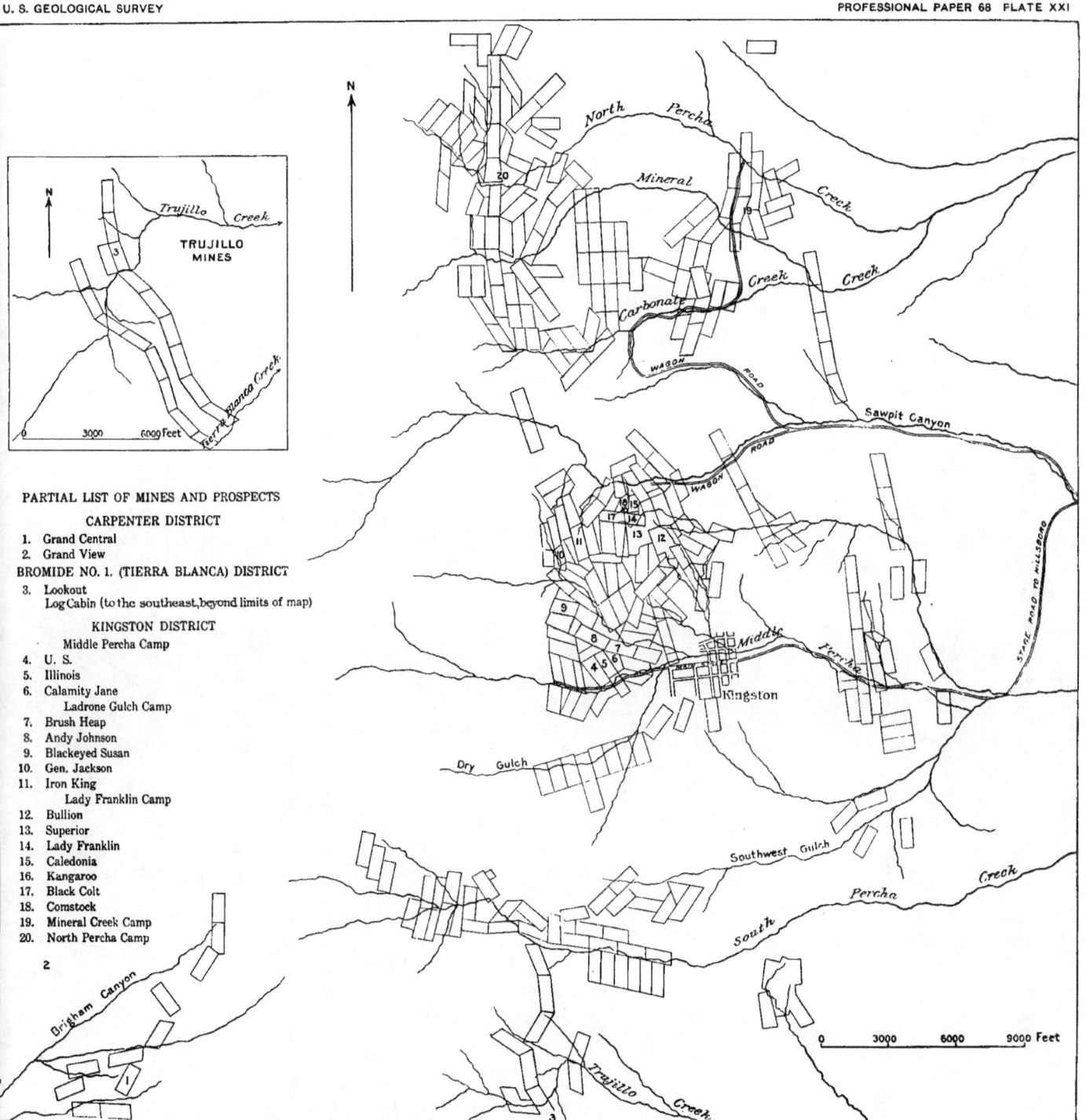

SKETCH MAP OF KINGSTON, CARPENTER, AND TIERRA BLANCA DISTRICTS, SOCORRO COUNTY.

West of Kingston the tributaries of Rio Percha have cut through the belt of limestone, giving good exposures of the beds. In this locality the beds are affected by folding to a minor extent, producing low arches which appear to have controlled in part the deposition of the ores. A short distance west of the town the limestones and associated beds have been cut by a dike of monzonite porphyry. This dike is about 400 feet wide on the north side of the canyon. On the south side it expands into an irregularly shaped area of considerable extent. It can be traced northward for a distance of 1 or 2 miles. Some faulting seems to have accompanied or followed the intrusion. Excellent exposures of the stratified rocks occur along the walls of the canyon of the middle fork just above Kingston and in Ladrone Gulch, adjoining it on the north.

On Carbonate Creek, about 3 miles to the northwest, the limestones below the Percha shale are underlain by quartzites which rest upon the irregular surface of red granite. By combining the sections obtained in these three localities we have the following:

Section in vicinity of Kingston.

Character.	Thickness (feet).	Age.
14. Red sandstones and shales..	600–800	Manzano group (Pennsylvanian).
13. Blue and drab shales with some beds of limestone..	325	Magdalena group (Pennsylvanian).
12. Compact blue limestone, thick bedded...	75	
11. Blue clay shales..	15	
10. Thick-bedded blue limestones, some thin chert and shale beds..............................	400	
9. Shales and thin-bedded limestones, lower Carboniferous fossils............................	25	Lake Valley limestone (Mississippian).
8. Subcrystalline limestones with nodular chert, lower members thick bedded.................	80	
7. Black fossiliferous shales, calcareous above, containing *Camarotœchia endlichi*, etc.......	200	Percha shale (Devonian).
6. Compact blue limestone...	15	Mimbres limestone (Silurian-Ordovician-Cambrian?).
5. Thin-bedded cherty limestones...	30	
4. Gray crystalline granular limestones...	300	
3. Calcareous shales with some beds of limestone...	100	
2. Dark-red quartzite, with 3-foot bed of shale near middle.................................	75	Shandon quartzite (Cambrian).
1. Granites and schists...	....	Pre-Cambrian.

According to George H. Girty the fossils collected from the beds numbered 10 in the above section belong in the Pennsylvanian, probably in the lower part. Two small lots (No. 10) apparently represent, according to the same authority, the horizon from which C. A. White noted some fossils in 1881, and a small collection obtained from a 15-foot bed of limestone near the base of No. 13 is said to be suggestive of the fauna of the Hermosa formation in southwestern Colorado.

ORE DEPOSITS.

In their general character these ores correspond with those of the Palomas district. Lead and copper sulphides carrying silver and gold along with free silver in a clay gangue represent the principal ores. Zinc sulphide is present generally. In addition to these, chloride of silver appears in places. The gangue is mostly talc with some quartz.

Geologic relation.—The ores occur in pockets and pipes in the limestone just below the Percha shale. The principal deposits are located in a relatively narrow belt bordering the porphyry dike on the west. No deposits are known to occur along the east side of the dike. The deposits are not uniformly distributed throughout the mineralized belt, but occur in groups separated by ground that is either barren or nearly so. One such group occurs on the north side of Middle Percha Creek about half a mile above the town. The principal mines in this group are the United States, Illinois, Calamity Jane, Brush Heap, and Andy Johnson.

About three-fourths of a mile farther north is a mineralized area in which is located another important group of mines, consisting of the Lady Franklin, Kangaroo, Caledonia, Black Colt, Comstock, and others. Both camps are located on the apexes of upward swells and the principal deposits of ore occur under the arches. The ore bodies are said to be distributed at somewhat regular intervals (about 50 feet apart) along fractures running about N. 40° W. Another series of minor fractures or joints intersect these, but usually carry no ore. The largest ore bodies occur at the intersections of the two series of fractures. Narrow seams or stringers of sulphide ore connect the ore bodies, and by following these the larger bodies are located. Most of the deposits extend downward for some distance into the "under lime," their size and shape being rather irregular. Some of them connect laterally and are known as "pipes." One such deposit was followed for a distance of 400 feet or more. Only here and there is the ore in contact with the overlying shale, being separated from it nearly everywhere by a thin layer of the limestone. A thin seam of sulphide ore usually leads from the shale contact to the ore body. In general the deposits do not extend to a distance of more than 100 feet from the shale. With increasing depth the ores diminish in value.

In some places are caves lined with ore and crystals. Along the contact between the limestones and shales the beds have suffered a marked silicification. In places it is the limestone, elsewhere the shale, which appears to have been more affected by the alteration. The silicified stratum, which is very irregular in thickness, consists locally of a breccia of white or pink flint. The shales yield a dark-gray or black flint breccia. The siliceous rock is in places more or less quartzose and drusy, with numerous cavities lined with quartz crystals. Locally it is often called "quartzite." It is observed to mark this horizon over a large part of the region. At Lake Valley a like alteration has taken place along the top of the ore-bearing limestones (Lake Valley).

Here and there the silicification extends downward along the sides of fractures into the limestones. It is reported that this contact rock usually carries low values in gold.

Mode of formation of the ore.—The ores here, like those at Hermosa, are probably to be attributed to ascending hot solutions, and it is likely that the solutions were derived from the intrusion of porphyry with which they are associated. The warrant for this conclusion lies in the following conditions:

1. Their occurrence under low arches of small folds or domes in the limestone overlain by a stratum of impervious shale.

2. The presence of the principal deposits in the limestone along the west side of the porphyry dike. Hot solutions proceeding from the porphyry would find their ascent hindered by the shale and would naturally follow the arches, which rise with the strata toward the west.

3. Their relations to the systems of fractures which furnished passageways for the ascending solutions.

Although it is believed that the original deposition of the ores occurred as a replacement of the limestone through the agency of heated solutions, it is evident that oxidation has played an important part in bringing the deposits into their present condition. Along with other features the relatively large proportion of free silver present in the talcose clay would seem to indicate this agency.[a]

Mine workings and production.—In some places erosion has removed the overlying beds and the ores are mined in open cuts or pits. Generally they are opened by tunnels in the sides of the hills or by shafts of no great depth.

As the ore values diminish with increasing depth, mining operations did not extend far below the surface, all the workings thus far being above water level. No work is being done here at the present time. Most of the ores taken out came from the two localities above mentioned, both of which were at one time large producers.

Kingston is credited with a larger production of silver ore than any other camp in New Mexico. Up to January, 1904, the estimated production, as given by F. A. Jones,[b] was $6,250,000, nearly all of silver.

---

[a] As the mine workings have long been abandoned there was no opportunity to inspect the ore occurrences underground, and for information concerning them the writer is under obligations to Mr. Ray, an experienced miner whose habits of careful observation and long experience as an operator in the district give great weight to his statements.

[b] Op. cit., p. 98.

BROMIDE NO. 1 (TIERRA BLANCA) DISTRICT.

GEOGRAPHIC RELATIONS.

The Bromide No. 1, or, as it is better known, Tierra Blanca district, is situated about 6 or 7 miles southeast of Kingston, near the head of White Water Creek. (See Pl. XXI.) It is about 15 miles northwest of Lake Valley and 10 miles southwest of Hillsboro. It received the name Tierra Blanca from the prominent white capping of rhyolite in the adjacent hills.

GEOLOGY.

The geologic relations at Tierra Blanca are essentially the same as at Kingston. The belt of Paleozoic limestones extends southward along the range, though covered in many places by volcanic tuffs and flows. The formations have the same general dip to the east and in the vicinity of the mines are cut off in that direction by a prominent intrusion of monzonite porphyry. Toward the west the sedimentary formations are overlain by extensive deposits of andesites, with white tuffaceous rhyolite capping the hills. In the vicinity buttes and mesas composed of tuffs with intercalated flows constitute prominent features of the landscape. Several miles to the east occur remnants of basalt flows overlying the Palomas gravel.

Near the mines the limestones containing the ores are divided by fractures into many small blocks which have suffered some relative displacement. The prominent fractures are for the most part inclined to the porphyry contact. There are indications of low arches and rolls in the limestone, though they were not definitely determined.

ORE DEPOSITS.

Character of the ores.—Both silver and gold ores occur in this locality. The silver ores occur in the form of native silver and in combination with the sulphides of lead and copper. Silver is found also in the form of a sulphide, and here and there bromides and chlorides are seen.

Geologic relations.—As at Kingston, the silver ores occur chiefly in pockets and pipes in limestone along a narrow band bordering the porphyry. In general they are just beneath or not far below a bed of shale which overlies the limestone. This limestone is the Lake Valley, the horizon of the silver ore here corresponding to that of the deposits at Lake Valley. Though no conclusive evidence was obtained pointing to the actual distribution of the ore bodies with reference to folds, the general conditions seem to indicate that they are related.

Gold is found in a quartz reef at the top of the Mimbres limestone and, in the Log Cabin mine, in a vein of quartz connected with the silver deposits. This quartz reef outcrops in the valley below the Log Cabin mine and represents the upper silicified portion of the Mimbres limestone. The ledge is composed of pink and white quartz with more or less flint breccia and contains many cavities lined with drusy quartz. It is said to run $8 to $9 a ton in gold, but this was not verified. In connection with this statement it may be noted that the siliceous formation occurring along the contact of the Mimbres limestone with the overlying Percha shale is reported to carry generally more or less gold.

Mode of formation of the ores.—Both in type and mode of formation the silver ores of Tierra Blanca resemble those of Lake Valley, Kingston, and Hermosa. The source of the solutions carrying the ores may have been in the monzonite porphyry mass with which the deposits are associated.

Workings.—Very little mining is now (1905) being done in this district. The Log Cabin mine is operated in a small way intermittently. Just prior to the visit the mine had closed after a shipment of one carload of high-grade ore. No attention is being given to the gold-quartz ores at present.

One of the most important mines of the district is the Lookout, situated near the head of Trujillo Creek. It is reported that in both the Lookout and Log Cabin mines deposits rich in gold and silver were found at or near the surface.

CARPENTER DISTRICT.

GEOGRAPHIC RELATIONS.

The Carpenter district lies on the west slope of the Mimbres Mountains, as this portion of the range is called, about southwest of Kingston. (See Pl. XXI.) The district is difficult of access and there are as yet no producing mines in it. In two localities, however, some important prospects are being developed, which if provided with suitable transportation facilities may prove to be paying properties. One of these is the Grand View group, located about 7 miles southwest of Kingston, and the other the Grand Central group, 4 miles south by southeast from the Grand View camp.

GEOLOGY.

The range is flanked on the west at this point by lower Paleozoic (Ordovician) rocks covered in places by flows of andesites and rhyolites and cut by large dikes of granite porphyry and lesser ones of diabase. In the vicinity of the mines the sedimentary formations consist of blue and white crystalline limestones alternating with beds of quartzite evidently belonging near the base of the Mimbres limestone. Both limestones and quartzites show marked indications of metamorphism, the limestone in places resembling marble. About half a mile west of the Grand View mine these formations are cut on the west by a mass of granite porphyry, which constitutes a prominent ridge running north and south for 2 or 3 miles. A dike of a similar rock 20 feet wide was observed on the trail a short distance north of the mine, cutting across the limestones toward the southeast.

About 2 or 3 miles north of this locality occurs a cliff of limestone seamed with chert closely corresponding in appearance to the Abrigo limestone (Cambrian) described by Ransome [a] in his report on the Bisbee quadrangle, Arizona.

At the Grand View mine the sedimentary formations dip 30° to 35° W. They appear to flatten out somewhat toward the southeast and at the Grand Central locality the dip is not more than 15°.

ORE DEPOSITS.

Character of the ores.—At the Grand View camp the ores consist chiefly of galena, with some zinc sulphide and a large amount of iron pyrites in quartz gangue. The ore is of low grade and carries little or no gold or silver. The Grand Central properties show a higher content of zinc, with considerably less of lead and iron sulphides.

Geologic relations.—In both places the ores occur along shear zones cutting the crystalline limestones. At the Grand View the shear zone is about 30 feet wide and extends N. 53° E. for a distance of 1,000 feet or more. At the Grand Central mines the shear zone is about 25 feet wide and extends nearly due north. The ore is distributed somewhat irregularly along the zone, which appears to dip at a high angle to the east. The openings, which are superficial, are on the east side of a small valley; on the west side the westward-dipping limestones are covered by a flow of rhyolite. The limestone along the vein on the foot-wall side is altered to a hard siliceous rock containing bunches and stringers of ore. No intrusive rocks were observed, but the general relations strongly suggest their presence near at hand.

Workings.—In 1905 the operations in this district did not extend beyond the prospect stage. At the Grand View a tunnel 165 feet long meets the vein at a distance of 100 feet from the portal, follows the west side for 35 feet, and then cuts directly across it. Farther to the north is a shaft 100 feet deep. The Grand Central workings consist chiefly of superficial excavations which have not disclosed the extent of the ore bodies.

LAS ANIMAS (HILLSBORO) DISTRICT.

GEOGRAPHIC RELATIONS.

The Hillsboro district includes the mineralized belt that extends along the foot of the range and is bisected near the south end by Rio Percha. Within it, on the banks of this stream, is situated Hillsboro, the county seat of Sierra County, 18 miles west of the Rio Grande and a like distance by wagon road north of Lake Valley, the nearest railroad station.

---

[a] Ransome, F. L., Bisbee folio (No. 112), Geol. Atlas U. S., U. S. Geol. Survey, 1904.

HISTORICAL STATEMENT.[a]

The first discovery of gold in the Hillsboro district is said to have been made by two prospectors in 1877. In crossing the site of what is now called the Opportunity mine they picked up some float which, on being assayed at an old quartz mill on Mimbres River, ran $160 to the ton. A little later the Rattlesnake mine was discovered. In August of the same year the first house in Hillsboro was built.

Placer gold was found in Snake and Wicks gulches in November, 1877, and soon afterward adjoining areas were found to contain gold in paying quantities. It is said that during the winter of 1877-78 a miner named George Wells came to Hillsboro with $90,000 in gold dust and nuggets which he had taken from Wicks Gulch. With the exception of 5 tons hauled

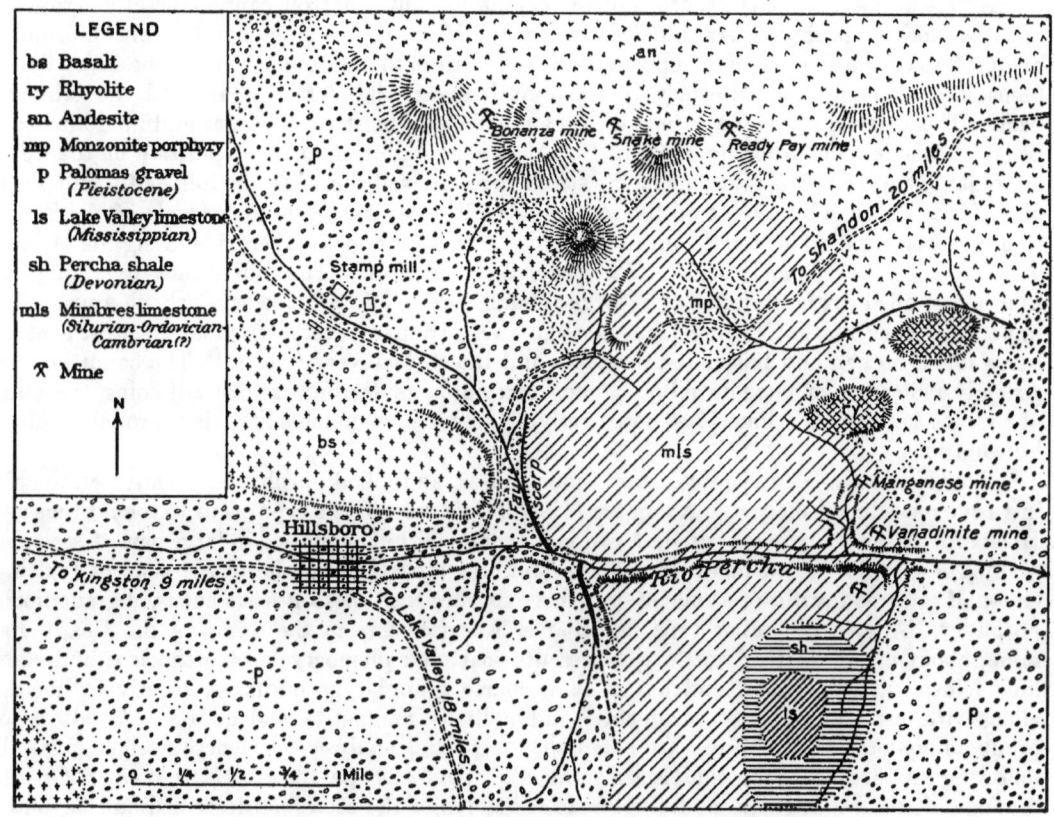

FIGURE 22.—Sketch map showing geologic features of Hillsboro district.

to the mill on Mimbres River, the first ores mined in the camp were worked in arrastres erected in Hillsboro in 1877. In 1878 there was erected on the old arrastre site a 10-stamp mill, which still stands, though long dismantled.

The work has been intermittent. In 1906 a revival of mining in the district was taking place and many old properties were being reopened. F. A. Jones estimates the total production of the camp to be $6,750,000.

GEOLOGY.

Stratigraphic relations.—Hillsboro is situated on a belt of the Palomas gravel intervening between a faulted block of the lower Paleozoic rocks on the east and the Manzano group of the upper Carboniferous, which makes its appearance about 2½ miles west of town. (See fig. 22.) After the Palomas gravel had been deposited it was covered more or less completely

---

[a] The historical notes are mainly taken from Jones, F. A., Mines and minerals of New Mexico, 1904, p. 81.

by a flow of basalt. Erosion has greatly dissected the plain surface, leaving masses of basalt here and there capping buttes, the lower parts of which are composed of the consolidated Palomas gravel.

About a mile east of town the beds have been faulted with upthrow on the east side, bringing to the surface the Mimbres limestone. The amount of displacement can not be less than 2,000 feet and is probably considerably more. The major part of the displacement here may have taken place prior to the deposition of the Palomas gravel, but the fact that this formation has shared in some of the movement indicates that the disturbance has extended down into comparatively recent time and in its later stages was probably connected with the extravasation of the basalt flows in the region.

About a mile east of Hillsboro the Rio Percha has cut a narrow canyon directly across the uptilted block, exposing several hundred feet of the Mimbres limestone overlain unconformably by 200 feet of the Percha shale, upon which rest in turn the crinoidal beds of the Lake Valley limestone. A slight unconformity is apparent between the Percha shale and the crinoidal beds. The absence of the lower part of the Lake Valley limestone (nodular and blue limestone beds) at this point is probably explained by their removal prior to the deposition of the upper crinoidal members. The relations of the strata here are well shown in the hill half a mile south of S. J. Macy's vanadium mine on the south side of Rio Percha. The upper beds of the Percha shale are well exposed here and contain Devonian fossils in large numbers, including the striking winged brachiopod *Camarotœchia (Plethorhynchus) endlichi*. The limestones overlying the shales are filled with crinoids and forms characteristic of the lower Carboniferous. The uppermost beds of the Mimbres limestone in this region are very irregular in stratification, being much broken and brecciated in places and the surface very uneven. The conditions are strongly indicative of unconformity. Along certain belts the formations adjoining the plane of contact between the limestone and overlying shales show much silification, probably due to hot solutions.

Intrusive rocks.—Monzonites, quartz monzonites, and allied porphyries constitute the chief intrusive rocks in the district. A number of bodies of rocks of this type occur to the northeast of Hillsboro. About a mile east of the town is a prominent hill composed in large part of a coarse-grained monzonite which has intruded the Mimbres limestone adjoining the zone of faulting. One-half to three-fourths of a mile east of this exposure the limestones are cut by a granodiorite porphyry. A rock of similar type but coarser was taken from the dump of the Ready Pay mine, about a mile farther north. It presents scattered phenocrysts of orthoclase and soda-lime feldspars, hornblende, and augite in a coarse granular groundmass, mostly orthoclase, along with lesser amounts of plagioclase and a little magnetite and quartz. As alteration products epidote, chlorite, etc., are common. A partial analysis by George Steiger gives the following proportions of the constituents named: SiO_2, 52.39; CaO, 6.40; Na_2O, 3.45; K_2O, 4.92. A range of hills capped with andesite lies to the north of the area just described and beyond it is another large area of coarse monzonite porphyry carrying large crystals of feldspar, many of which are twinned according to the Carlsbad law.

Effusive rocks.—The higher elevations lying to the northeast of Hillsboro represent remnants of extensive flows and tuffs, to the removal of which by erosion is due the exposure of the areas of limestone and porphyries. In thin section these rocks show a porphyritic texture, the phenocrysts consisting chiefly of plagioclase and augite; the latter shows much alteration to chlorite products. The groundmass is either microcrystalline trachytic or more or less glassy. Fluidal structure is in many places pronounced. This rock was first thought to be an andesite, but the analysis below, by George Steiger, shows that it is in reality a latite, standing between the andesites and the trachytes.

Analysis of effusive rock north of Hillsboro.

SiO$_2$	54.54	TiO$_2$	0.86
Al$_2$O$_3$	14.66	ZrO$_2$	None.
Fe$_2$O$_3$	4.20	CO$_2$	2.19
FeO	2.74	P$_2$O$_5$	.49
MgO	3.21	SO$_3$	None.
CaO	5.64	S	.01
Na$_2$O	3.47	MnO	.29
K$_2$O	5.28	BaO	.07
H$_2$O(−)	1.10	SrO	.05
H$_2$O(+)	1.87		
			100.67

A rock having the appearance of a dacite was observed in one locality about three-fourths of a mile east of the monzonite area.

ORE DEPOSITS.

Character of the ores.—Three kinds of deposits occur in the district—(1) those containing lead, iron, and manganese at the top of the Mimbres limestone; (2) those containing gold, silver, and copper in single veins or shear zones in the andesite; and (3) placers.

The placer deposits are practically exhausted, very little work being done on them at the present time. The lead deposits are largely limited to the claims adjoining Rio Percha, owned by S. J. Macy. The ores, which consist of the minerals wulfenite and vanadinite (endlichite), together with cerusite, lie in lenses distributed along the contact of the Mimbres limestone with the overlying Percha shale. Associated with the ores is a considerable amount of manganese, more or less quartz, and locally melanotekite, a silicate of lead and manganese. These mines have furnished most of the fine crystals of vanadium minerals for which Hillsboro is noted. The deposits are of moderate extent and little is being done on them at present.

Farther north manganese was at one time obtained at the same horizon, and at other points a siliceous iron ore (limonite) occurs, both of which were used in smelter operations then carried on at Hillsboro. Nothing is now being done on these deposits and their extent is unknown. The chief interest in the district attaches to the veins of gold and silver, which furnish the only producing mines within the district. The ores are base, chiefly sulphides of copper carrying gold and silver with some free gold.

Geologic relations.—The gold and silver ores occur in shear zones cutting the andesites which constitute the hills 2 miles north of the town. These hills, which are much dissected by erosion, rise in places to a height of over 1,200 feet above the level of the town. They are irregular in outline and together constitute a long, broken ridge, extending approximately in a northwest-southeast direction. Toward the northeast is a prominent elevation called Las Animas Peak.

The walls of the veins usually consist of a porphyritic andesite with large phenocrysts of feldspar. The rock is commonly mistaken for diorite porphyry. Under the microscope, however, it shows the characteristic texture of andesite. Along the shear zone it has suffered much alteration, being changed to a greenish-gray rock, sometimes erroneously referred to as "trachyte." The alteration is characterized by pronounced chloritization, together with the presence of considerable calcite. Little or no serpentinization appears. The veins cut the ridge of andesite and appear to converge toward a point in the vicinity of Las Animas Peak. In order from west to east the principal veins noted, named from their mines, are the Empire, Richmond (N. 47° E.), Bonanza (N. 37° E.), Snake (N. 22° E.), Opportunity, and Ready Pay (N. 20° E.).

The Empire Gold Mining and Milling Company controls the Bonanza and Empire properties and during 1905 was taking out ore on the Bonanza claim. The Snake mine, once actively worked but idle for many years, became the property of the Sierra Consolidated Gold Mining Company, which began operations on it in 1906, but soon closed down. The property of the

Consolidated Gold Mining and Milling Company is 6 miles nearly due north of Hillsboro, at Andrews post-office.

In 1909 some production continued in the Hillsboro district. The Bonanza mine is operated by the Ameranza Mining Company. The Wicks mine, operated by the Sigma Consolidated Company, has attained a depth of 335 feet and is producing milling and shipping ore. Some of the higher-grade shipping ore is stated to contain 12 ounces of silver and 2.75 ounces of gold to the ton, as well as 6 per cent of copper. Some vanadium ore is said to have been produced in the Hall mine. The press also contains reports of dikes of copper-bearing rock which are being worked on Percha Flat, 5 miles east of Hillsboro.

Bonanza mine.—The Bonanza is the only mine which was producing ore in 1905. The ores are free milling, with sulphides of copper carrying gold and silver values. Smelter returns of gold, silver, and copper are made. The lower-grade ores are treated in the company's mill, located in the valley three-fourths of a mile north of Hillsboro, and the concentrates, together with the shipping ores, are sent to the smelter at El Paso. The smelter returns on three car lots of rich shipping ore, as taken from the company's records, are as follows:

Smelter returns on ore from Bonanza mine, Hillsboro.

	Silver (ounces per ton).	Gold (ounces per ton).	Copper (per cent).	Iron (per cent).	Insoluble silica (per cent).	Sulphur (per cent).
1	27.0	8.20	10.8	13.4	55.80	17.2
2	32.8	8.35	14.7	16.6	45.01	17.1
3	37.1	9.40	16.5	17.1	41.60	16.2
Average	32.3	8.65	14.0	15.7	47.47	16.83

Quartz is the chief gangue mineral. In places the lode consists of a single vein from 2 to 8 feet in width; in others, of a branching system of thin seams occupying a shear zone from 2 to 7 feet wide. These lead out into the country rock on the foot-wall side. The dip is to the west at a high angle—nearly vertical. The ore bodies occur along the walls, here on one side and there on the other. Some distance in from the portal of the tunnel the vein branches, but the branches come together again farther on. At 2,000 feet from the portal of the lower tunnel the vein splits up into several small veins, branching out into the foot walls. At this point the main vein becomes barren, and at the time of visit (1905) crosscutting was in operation in the hope of locating the lode.

Mining operations here consist chiefly in drifting on the lode. Some of the upper tunnels extend through the hill and the lowest is about 2,000 feet in length. In 1905 forty men were employed, of whom twenty-five were at work underground. The mill, which had 10 stamps, was found insufficient and has been recently increased to 20 stamps. It began operations December 1, 1904, and up to September, 1905, the net returns on ores from the Bonanza mine were as follows: Bullion, $23,801.51; concentrates, $4,821.02; shipping ore, $7,099.29; total, $35,721.82.

LAKE VALLEY DISTRICT.

SITUATION AND HISTORY.

The principal mines in the Lake Valley district are located at Lake Valley, the terminus of a small branch of the Atcheson, Topeka and Santa Fe Railway, which connects with the Silver City branch at Nutt, 12 miles to the southeast. The district lies near the south end of the Mimbres Range, just within the line of rounded hills bordering the Rio Grande valley. (See fig. 23.)

The discovery of ore at Lake Valley was made in August, 1878. The fame of the discovery spread and resulted in a rush for the new camp. The best properties were soon absorbed by three companies, named from the groups of mines, the Sierra Grande, the Sierra Bella, and the Sierra Apache. For several years operations were conducted under the management of the Sierra Grande Company. The discovery in the early eighties of the Bridal Chamber, one of the richest single bodies of silver ore ever found, took place on the very day the general manager of the mines, George Daly, was killed by Apache Indians 6 miles from camp. Since 1894 there has been but little activity in the camp and very little production.

GEOLOGY.

General relations.—Extensive flows of andesite covered by rhyolite constitute the surface rock of the region. In many places these rocks have been stripped away by erosion, exposing considerable areas of the underlying Paleozoic formations. One such area begins at Lake Valley and extends northward in a narrow irregular band as far as Hillsboro. This exposure is evidently the result of the southward extension of the fault noted at Hillsboro.

Formations.—A short distance to the northwest of the town Mimbres limestone outcrops in a long, narrow ridge bearing to the northeast. Toward the east is a parallel ridge composed of Lake Valley (Mississippian) limestone, and the intervening valley is floored by the more easily eroded Percha (Devonian) shale. At the top of the Mimbres limestone here corals were obtained, which E. O. Ulrich identified as belonging to the Silurian. (See p. 227.) As Ordovician fossils were obtained at or near the top of the Mimbres limestone at Hillsboro, it is evident that the Silurian can have but a scanty development in the region. According to Ellis Clark,[a] a shaft sunk in the valley 1,200 feet northwest of the Grande workings passed through 130 feet of shale and 12 feet of red limestone, under which was quartzite. As a bed of quartzite 5 to 10 feet thick was observed at Hillsboro and elsewhere near the top of the Mimbres limestone, it is possible that this bed may represent the plane of separation between the Ordovician and Silurian rocks. The surface of the Mimbres limestone is characterized here, as at Hillsboro, by its uneven and siliceous character. Along the valley knobs of these siliceous beds appear at a number of places, protruding through the shales. A section across the Lake Valley district is given in figure 24.

The dip of the strata is about 20° SE., which carries the beds below the surface just east of the village. Half a mile west of town a small valley which is tributary to Berenda Creek follows the strike of the beds toward the northeast. By removing the easily eroded Percha shale the stream has been able to undermine the overlying hard rocks, thus shifting the escarpment gradually eastward. A similar process has attended the wearing down of the shales above the blue limestone, leaving hills of the upper or crinoidal beds of the Lake Valley limestone north of the town. The removal of the beds overlying the blue limestone facilitated the discovery of the ore deposits which occur in it.

Rhyolite constitutes a large elevation known as Porphyry Hill to the southwest of the town and others toward the axis of the range.

No intrusive bodies were observed. The "porphyrite" to which Clark ascribes the genesis of the ores is clearly a part of the andesite flow poured out upon the uneven eroded surface of the sedimentary formations.

The occurrence of the Mimbres limestone at the surface here is evidently due to a fault somewhat to the west of the outcropping ridge. Smaller faults occur in the vicinity of the mines; one of these, called the Columbia fault, is on the Columbia and Last Chance claims. It terminates the ore shoots where it crosses them, but the ores are said to appear again at the proper horizon on the other side, indicating that the ore deposition took place prior to the faulting. The displacement is not more than 50 to 100 feet. A similar slip appears also near the Bridal Chamber. The faults extend in a southeasterly direction.

ORE DEPOSITS.

Character.—A good description of the nature and occurrence of the silver ores of Lake Valley is given in the paper by Ellis Clark, already cited. He classes them as (1) siliceous, (2) neutral, and (3) more or less basic. The siliceous ores were obtained from the Bridal Chamber,

---

[a] Trans. Am. Inst. Min. Eng., vol. 24, 1894, p. 140.

278　ORE DEPOSITS OF NEW MEXICO.

Twenty-five Cut, and Thirty Stope; the Emporia furnished the neutral ores; and ores more or less basic were obtained from the Bunkhouse, the Columbia, and the Apache. Clark's descrip-

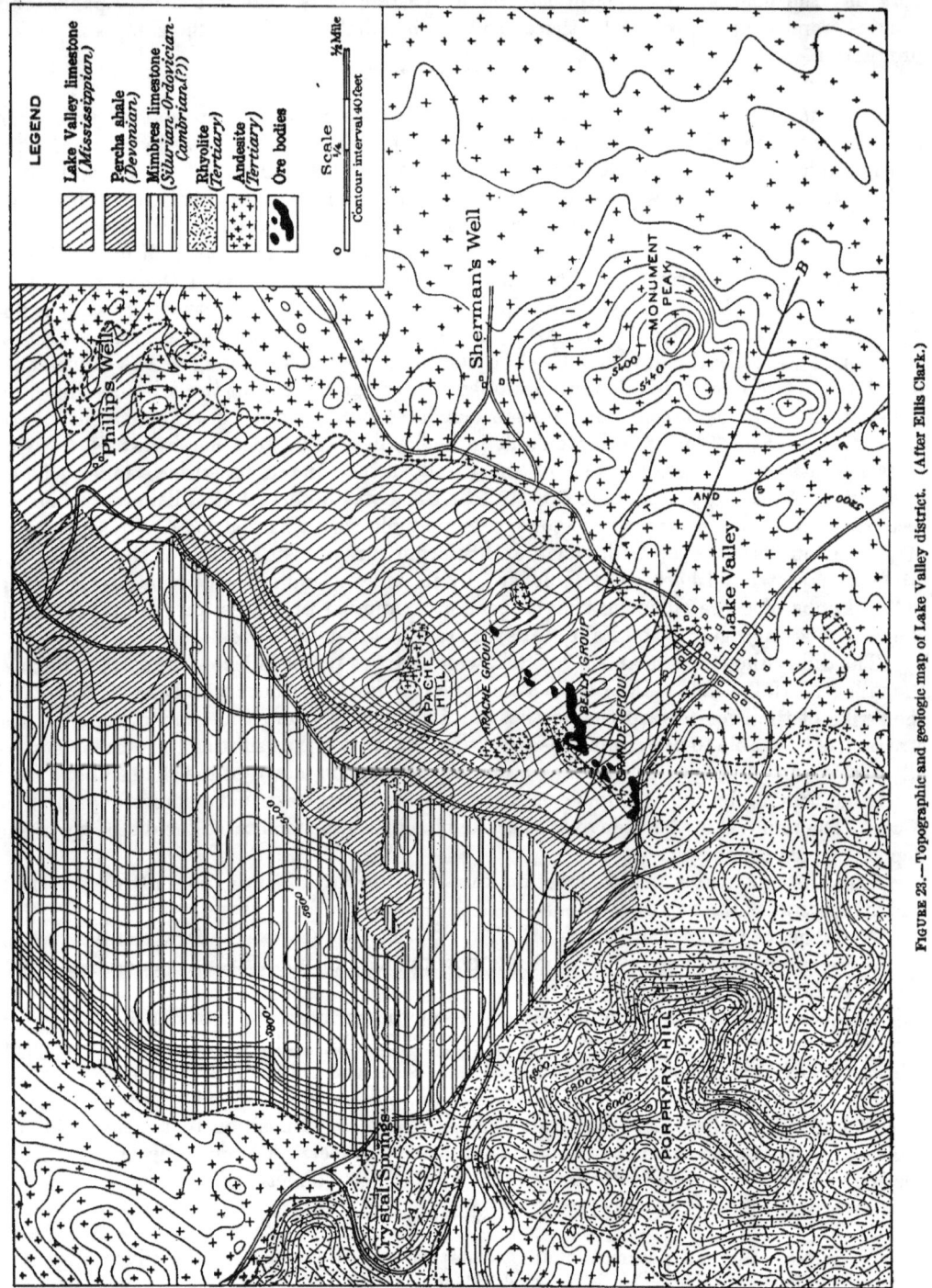

FIGURE 23.—Topographic and geologic map of Lake Valley district. (After Ellis Clark.)

tion[a] of the ores, based on a practical knowledge of the workings, is so complete that it merits reproduction.

---

[a] Clark, Ellis, Tran. Am. Inst. Min. Eng., vol. 24, 1894, pp. 148–149.

The ores from Thirty Stope consist principally of gray, brown, chocolate, pearl, and green flint. The green flint has generally yielded a grade of ore higher in silver than the other colors. The flint has been much broken and crushed, the fragments subsequently having been cemented by veins of transparent quartz. The manganese and iron contents are small, the silica contents large. An average of the ore from this working for the past few years would be 65 per cent silica, 6 per cent iron, and 12 per cent manganese, with 20 ounces of silver per ton. Fine specimens of pyrolusite showing crystalline structure are found in these workings.

Much of the Bridal Chamber ore consisted of cerargyrite; and the little ore that has been mined from the chamber in recent years has been similar to the Thirty Stope ore, but slightly less siliceous.

The Emporia Incline ore, when considered in carload lots, is a neutral ore, consisting of 30 per cent silica, 12 per cent iron, and 18 per cent manganese, the remainder being limestone. It is generally brownish black, with a tendency to brown, and frequently carries from 1 to 5 per cent of lead in the form of galena. Its contents in silver vary from 30 to 50 ounces to the ton.

Galena is one of the accessory minerals in this working, sometimes occurring massive and compact, with a fine-grained, crystalline structure, and sometimes in a pulverulent mass, known locally as "gray metal," which is easily powdered under the pressure of the fingers and in mining is carefully scraped and gadded out of the containing cavities and allowed to fall on canvas lying on the rock floor. Blasting is not practiced on this variety of ore, on account of its value and the danger of scattering and losing it. It contains from 200 to 500 ounces of silver per ton and 50 to 60 per cent of lead.

The Bunkhouse workings have yielded several distinct varieties of ore from different parts of the mine. The ore from the central portion of the body, taken out shortly after its discovery, varied in silver from 200 to 500 ounces per ton and was basic, containing a considerable proportion of cerargyrite. Subsequent workings developed large bodies of

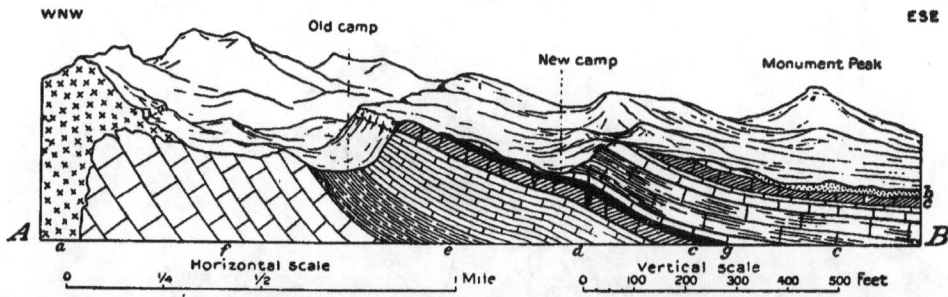

FIGURE 24.—Section across Lake Valley district (along line *A-B*, fig. 23). (After F. M. Endlich.) *a*, Rhyolite; *b*, andesite; *c*, Lake Valley limestone; *d, e*, Percha shale; *f*, Mimbres limestone; *g*, ore horizon.

chocolate-colored, manganiferous ore, with an average composition of 8 per cent silica, 12 per cent iron, and 24 per cent manganese, the remainder consisting of limestone and gypsum. The silver contents were 20 to 30 ounces per ton. Overlying this ore were considerable bodies of what became known as "fluxing ore," averaging 35 per cent excess of bases over silica, and carrying 5 per cent silica, 10 per cent iron, and 30 per cent manganese, the remainder consisting of limestone. The silver contents were always low, increasing and diminishing with the silica contents and averaging 5 ounces to the ton. The color of the fluxing ore was black.

Underlying the previously mentioned fluxing ore, at the base of the ore series, well into the Blue limestone, was a deposit of manganoferruginous ore, which was nearly neutral, with an average composition of 30 per cent silica, 12 per cent iron, and 18 per cent manganese. Its silver contents were low, seldom above 8 ounces per ton, and the ore was commercially useless. In appearance it is not unlike the Incline ore, having a blackish-brown color inclining to black.

Clark gives in his description the average contents of 4,000 tons of ore, a year's shipments by the company, during his administration. It is as follows: Silver, 47.7 ounces per ton; lead, trace; zinc, 4 per cent; silica, 28.5 per cent; iron, 13.9 per cent; manganese, 18.2 per cent.

An enumeration of the minerals found in the deposits was published by Benjamin Silliman, jr., in 1882, and is as follows:

> Native silver (in trifling quantity).
> Cerargyrite or horn silver (chlorite).
> Embolite (chlorobromide).
> Galena } (both rich in silver).
> Cerusite
> Vanadinite.
> Specular and red hematite.
> Limonite.
> Pyrolusite.

Manganite.
Psilomelane or wad.
Quartz, in the form of flint or chert; constitutes the gangue of the embolite and in places forms the cap rock of the ore bodies. Occurs also in drusy form with vanadinite.
Calcite, found coating the silver chlorides and also as crystallized calcite.
Ankerite (carbonate of lime, iron, and manganese).
Apatite (rare).

Geologic relations.—The ore deposits constitute pockets, chambers, or pipes in the lower beds of the Lake Valley limestone ("Blue" limestone of Clark). Some of the ore bodies appear to lie along the contact with an overlying bed of shale, but generally they are within the blue limestone or along the base at the contact with the "nodular beds" below. In shape and size

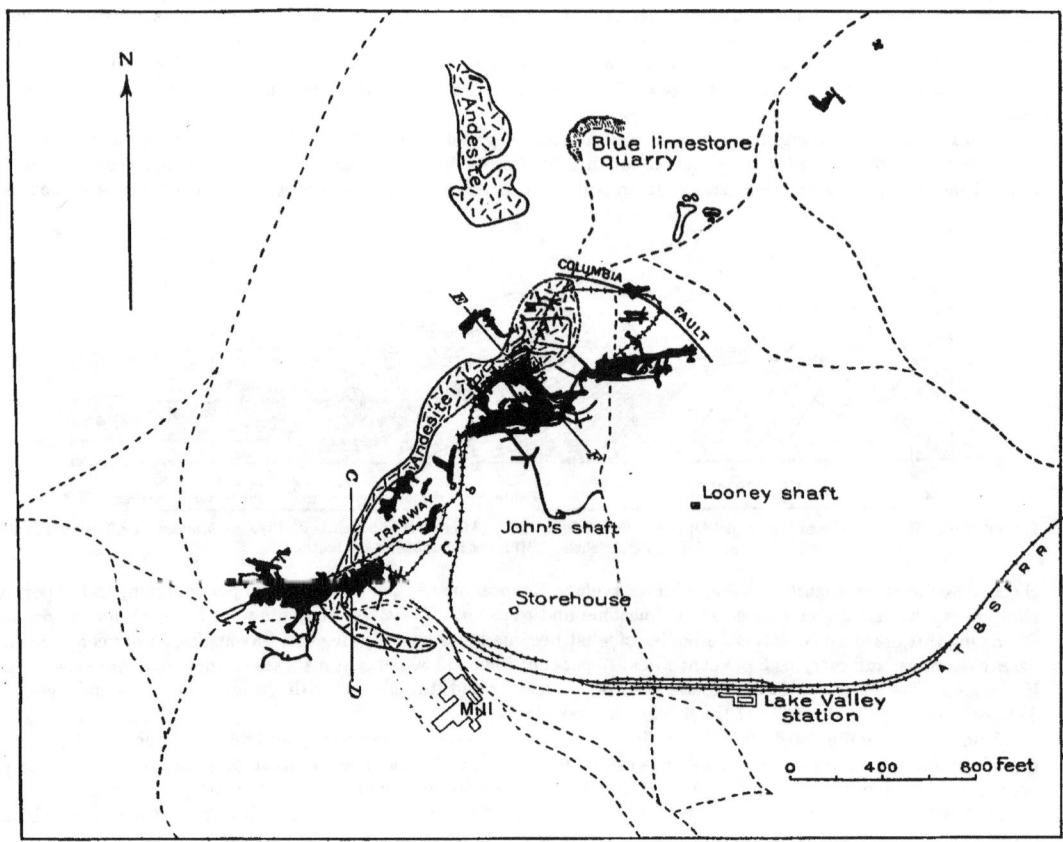

FIGURE 25.—Map of Lake Valley mines. (After Ellis Clark.) Black areas indicate ore bodies.

they vary greatly. The Bridal Chamber represents the largest single body of ore, the yield for this mine being over 2,500,000 ounces of silver. This deposit occurred at the top of the blue limestone just underneath the shale. In general the ore shoots appear to follow the planes of bedding.

Mine workings.—Three centers of mineralization are situated along the ridge to the north. The groups of mine workings representing these centers, from south to north, are the Grande, the Bella, and the Apache. (See figs. 25–26.) The yield follows the same order, being greatest at the south and decreasing northward, the Apache group having produced very little ore. Briefly stated, the workings are the following:

Grande group:
    Twenty-five Cut.
    Thirty Stope.

Grande group—Continued.
    Bridal Chamber.
    Carolina (this mine furnished little ore).
Bella group:
    Emporia Incline.
    Harrison.
    Bella Chute.
    Bunkhouse.
    Striby.
Apache: All northeast of the Bella group have yielded considerable iron flux and a comparatively small amount of commercial ore.

Operations were carried on chiefly from the surface by open cuts and inclines. In some places shafts have been sunk upon the ore. The deepest workings were those of the Emporia

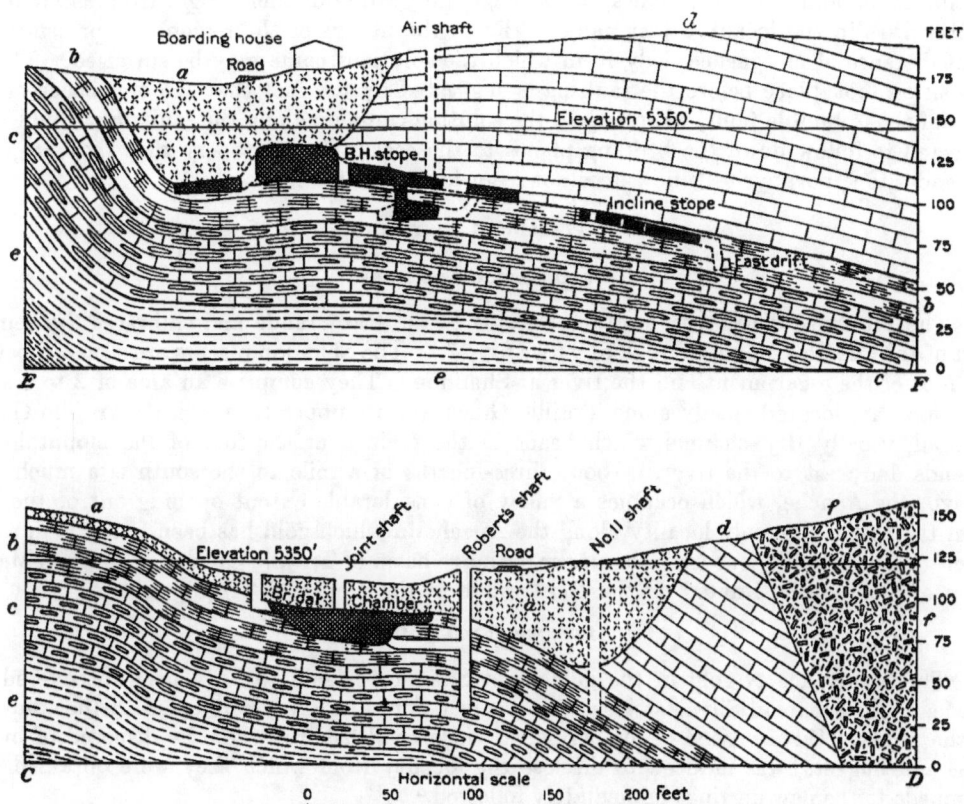

FIGURE 26.—Cross sections of deposits at Lake Valley (along lines C-D and E-F, fig. 25), showing position of ore bodies. (After Ellis Clark.) a, Andesite; b-d, Lake Valley limestone (b, blue limestone; c, nodular limestone; d, crinoidal limestone); e, green shale; f, rhyolite.

incline, which reached a depth of 150 feet below the surface. Below this level water was encountered, which, together with the low grade of the ore, has limited the workings to a depth of 100 to 150 feet.

Production.—No ore is mined at this locality at present. There has been practically no production from the Lake Valley mines from 1905 up to the present time (1909). Before 1905 operations were carried on in a small way, but the caving in of the water shaft caused a suspension until repairs could be completed. The best obtainable data are those given by Clark, who states that up to the time of the closing of the mines in 1893, the yield was approximately as follows.

Silver yield of mines in Lake Valley district to 1893.

	Ounces.
Bridal Chamber	2,500,000
Thirty Stope	1,000,000
Emporia Incline	200,000
Bunkhouse	300,000
Bella Chute	500,000
Twenty-five Cut	200,000
Apache and scattering	300,000
	5,000,000

Mode of formation of the deposits.—In their general character and relations these deposits correspond to those of the Hermosa, Kingston, and Tierra Blanca districts. The form and relations of the deposits, together with the siliceous replacement along the contact between the limestone and overlying shales, are strongly suggestive of their origin from ascending hot solutions, as in the localities just named. From the nature of the region, the presence at no great distance of an igneous body from which the solutions came may be surmised. That the deposition took place before the faulting is suggested by the fact that the ores occur on both sides of the Columbia fault, whereas had the solution come up along the fault they could not be expected to follow down the bedding planes to the east. Clark's theory that they are due to the andesite ("porphyrite") flow does not seem to be well founded.

PITTSBURG DISTRICT.

GEOGRAPHIC RELATIONS.

The Pittsburg district lies on the east side of the Rio Grande and embraces the southern part of the Caballos Mountain region. (See fig. 27.) The Shandon placers are situated between the foot of the escarpment and the river at Shandon. They comprise an area of 2 to 3 square miles and are located chiefly along Trujillo Gulch and its upper tributaries. Trujillo Gulch is a small, usually dry channel which heads in the plain near the foot of the mountains and extends due west to the river. About three-fourths of a mile to the south is a much larger arroyo, the Apache, which occupies a valley of considerable extent opening out of the range from the east. The only locality along the Apache in which gold has been found is in a small gulch (Union) leading out from the upper drainage basin of Trujillo Gulch. The small Mexican town of Shandon is situated on the Rio Grande at the mouth of Trujillo Gulch.

HISTORICAL STATEMENT.

The occurrence of gold in the gulches at Shandon is said to have been discovered by a Mexican, Encarnación Silva, at least two years before 1903, when it came to the knowledge of the public. Silva on one of his periodical visits to Hillsboro, where he disposed of his gold dust and nuggets, was induced to disclose the locality from which they were obtained and a stampede to the new diggings immediately followed.[a]

GEOLOGY.

The Caballos Mountains represent a block tilted to the east with a prominent fault scarp along its westward front. A high ridge bounds Apache Arroyo on the south side. This elevation is due to the tilting of a small block of Carboniferous limestone overlain by the Palomas gravel at a high angle (40°) to the north.

The valley plain north of Apache Arroyo is about 3½ miles wide and is bounded on the east by an escarpment 1,500 to 2,000 feet high and on the west by the bluffs of the present river channel. The present valley has been cut in the bottom of an older and wider valley, remnants of which appear as benches along the bluffs. In a view up or down the main valley the profile is that of a smooth plain with a decided slope from the foot of the escarpment on the east to the top of the river bluffs. On close inspection, however, this apparently plain surface is found

---

[a] Jones, F. A., op. cit., p. 87.

to be trenched deeply by steep-sided ravines and arroyos, which interpose serious obstacles to travel and whose presence may not be suspected a few hundred yards away. In the vicinity of Shandon the original plain surface has been partly base-leveled to the level of the benches marking the first cycle of erosion, leaving rounded hills along the border of this ancient stream valley. Of this nature are the two prominent hills called Apache and Esperanza, one on the south, the other on the north side of Trujillo Gulch.

The valley plain is underlain by the Palomas gravel, which along the foot of the escarpment rests upon a floor of granite and schist. The granite outcrops along the base of the escarpment and is overlain by the quartzites and limestones of the Cambrian system. A profile section from Shandon to the top of the range is shown on page 224.

From some point to the southeast flows of lava have come down the Apache Valley, covering the eroded surface of the Palomas gravel nearly to the river. The earliest of these flows was one of andesite, which is exposed in a few places by the dissection of the later flows of rhyolite.

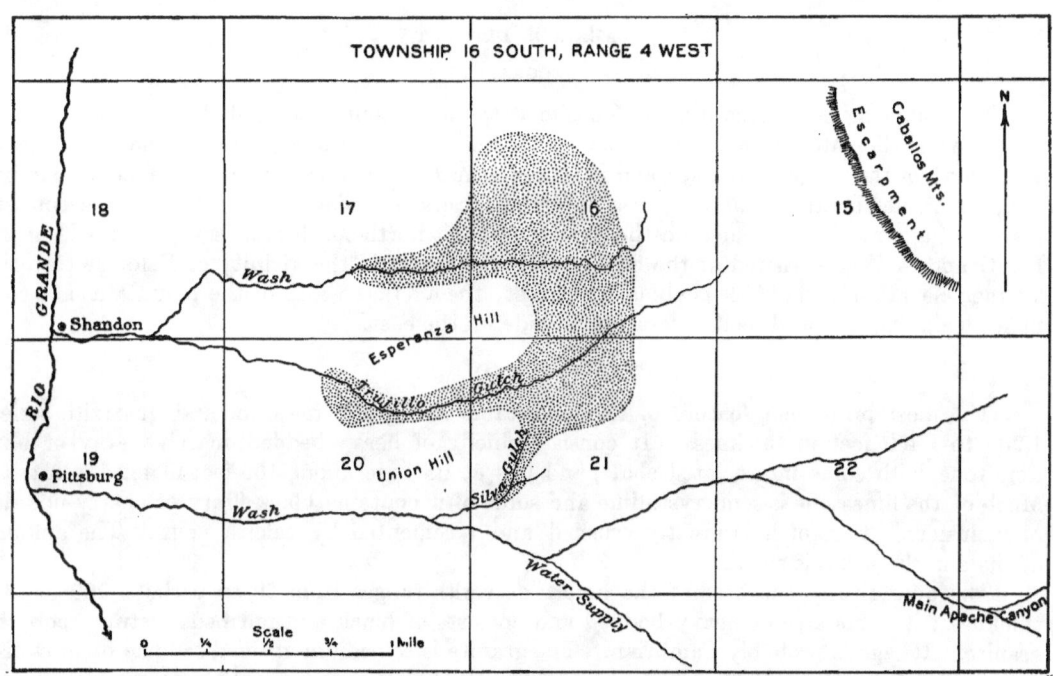

FIGURE 27.—Sketch map of Pittsburg mining district, showing location of Silva and Trujillo placers (dotted areas)

PLACERS.

As stated above, the gold-bearing gravels occur chiefly in Trujillo Gulch and the area drained by its upper tributaries. Above Esperanza Hill the surface is covered to a depth of several feet with a coarse sand derived from the decomposition of the granite that is exposed farther away along the foot of the mountain. For some distance above the hill this sand carries gold, but the colors gradually disappear on approaching the granite exposure. Around the east end of Esperanza Hill the sand rests on rhyolite, beneath which in places andesite is exposed. The lavas here evidently occupy a depression in the surface of the Palomas ("Mesa") gravel, for Esperanza and Apache hills are almost entirely made up of the gravel, and the canyon between them is entirely in this formation.

The Shandon Mining Company is now washing gold from the sand and gravel along the bed of Trujillo Gulch and the Union-Esperanza Mining Company is just beginning operations on Union Gulch. Water is obtained from wells sunk on the bank of the river 2 miles below and forced up to the placer field by powerful pumps.

Without attempting an authoritative statement as to the source of the gold, the writer believes it to be derived from quartz veins in the granite and schists just below the projecting headland east of Shandon. The facts on which this conclusion is based are (1) the occurrence and distribution of the gold in the granitic sand over a limited area west of the escarpment; (2) the coarse, unworn character of the gold, which, taken in connection with its distribution, indicates that it has not been transported far; (3) the occurrence of veins of quartz in the granite where exposed in that vicinity. Several well-marked veins were observed here running nearly north and south. It is not known that these carry gold. However, there may be others that do, or it is not improbable that the ore shoot may have been entirely carried away by erosion. The fact that no gold is found in the sand near the veins may indicate that it came from above the present surface.

Sierra County produced 1,111 fine ounces of placer gold in 1904 and 2,316 fine ounces in 1905. Most of this came from the Pittsburg camp. Since then the production has declined to nominal amounts, operations at the Shandon mines being suspended.

CABALLOS DISTRICT.[a]

GEOGRAPHY.

The following notes describe the deposits at the north end of the Caballos Range.

The Caballos Mountains trend about north and south on the east side of the Rio Grande. The length of the range is 30 miles, and it attains a maximum elevation of 10,000 feet at Timber Hill, which lies south of the copper and lead deposits described below. The range consists mainly of a monocline bounded on the west by a great north-south fault scarp overlooking the Rio Grande. Where visited at the copper and lead mines in the vicinity of Palomas Gap the average elevation of the crest is about 6,500 feet; the average width of the mountains is 4 to 6 miles, including a foothill belt several miles wide on the east.

GEOLOGY

The most prominent feature of the mountains is a great limestone and quartzite series, 1,200 to 1,400 feet in thickness. It consists chiefly of heavy-bedded, massive, gray or blue limestone, with some intercalated shale, and has at its base about 100 feet of hard quartzite. Much of the limestone is semicrystalline and some of it contains black flinty or cherty nodules or inclusions. Part of it is greatly crushed and recemented by calcite veins. The amount of shale in the series is small.

The quartzite at the base of the limestone series ranges from 50 to perhaps 200 feet in thickness; it is massive or heavy bedded and consists of black and red beds resting upon the granite. Its age is probably Cambrian. The granite is a medium-grained reddish or brownish rock, which from the uniformity of its contact with the overlying quartzite seems to be basal and older than the quartzite. In some places the two rocks appear to be perfectly welded, but this is probably due to subaerial disintegration preceding the deposition of the marine sediments.

Besides the reddish granite there is also present a medium or somewhat finer grained dark igneous rock which probably is an intrusive diorite.

South of the Marion mine, at the Lone Tree prospect, the granite and quartzite are cut by a pale brownish-gray porphyritic dike or intrusive rock.

ORE DEPOSITS.[b]

Occurrence.—The Caballos Range is crosscut by a system of steeply southward-dipping east-west fault fissures, some of which are filled with gangue and ore deposits. The pre-Cretaceous rocks are all more or less jointed, the granite by far the most, and the quartzite and limestone series is also gently folded or warped and in some places, particularly north of Palomas

---

[a] By F. C. Schrader.
[b] The ore deposits of this range have been briefly described by C. R. Keyes (Eng. and Min. Jour., vol. 80, 1905, pp. 149–151).

Gap, where the range is deeply cut by a tributary to the Rio Grande, it is on edge or overturned. The copper deposits are located mainly along the steep westward-facing slope of the range, 2 to 3 miles southeast of Palomas. Prospects are, however, also found north of Palomas Gap.

The copper ore seen at the Marion and Oohoo mines and the Lone Tree and other prospects occurs in steeply dipping east-west fissures as bornite, chalcopyrite, red copper oxide, malachite, azurite, and glance, all more or less mixed. The gangue is chiefly quartz, and the prevailing geologic horizon of the copper is in the quartzite just above the granite and quartzite contact; in places it reaches farther up into the quartzite or even into the limestone. The lead occurs as galena in a calcite and barite gangue in the overlying Carboniferous limestone, and is contained in veins similar in strike to those of the copper deposits. The ores are poor in gold and silver.

Copper deposits.—Several copper prospects have been developed in this district by the Victorio Chief Copper Mining and Smelting Company, the principal workings being at the Marion and the Oohoo mines. The Marion tunnel is situated about 3 miles south-southeast of Palomas at an elevation of about 5,000 feet, in the basal Cambrian quartzite of the range. It follows a fissure trending east and west and dipping 80° S. The hanging wall is fairly good, but the foot wall is poorly developed. The length of the tunnel in 1906 was about 600 feet. Bodies of copper ore, partly oxidized, partly also of chalcopyrite, have been found in the hanging and the foot walls of the fissure, associated with quartz, as well as with gritty and clayey gangue. The face of the tunnel is in quartzite.

The Lone Tree prospect is situated at approximately the same elevation.

The Oohoo incline is about a mile north of the Marion, at an elevation of 5,300 feet. It is contained in limestone which dips 25° E. The incline is now inaccessible, but is reported to be about 400 feet in length, following the dip of the limestone. The deposit seems to be connected with an east-west fissure like the Marion, but the walls are not regular. The limestone near the tunnel is hard and silicified. The Oohoo is reported to have produced a considerable amount of oxidized copper ore, some of which remains in sacks at the mine. The copper ore from this locality is said to contain $3 or $4 to the ton in gold and silver.

Lead deposits.—In the Carboniferous limestone of the Caballos Range lead prospects have been found at several places. At the Engle lead camp, 18 miles southwest of Engle station, on the Santa Fe Railway, the Southwestern Lead and Coal Company is engaged in development work on a series of veins in upper Carboniferous limestone. The veins, which are nearly vertical, carry rather small amounts of galena in a gangue of quartz and barite and extend about parallel with the dip of the beds, which is 15° N. 72° E. At the mouth of the tunnel the vein is about 7 feet wide. Exposures of the overlying Carboniferous "Red Beds" occur less than half a mile to the east. The development work consists of a tunnel, which in 1906 was 450 feet in length, and two shafts 80 and 120 feet in depth. Some picked ore has been shipped. About 5 miles north of the lead camp the same company is sinking a shaft on a small coal seam in the hope of finding a sufficient supply of this fuel to operate the power plant designed to generate electric power for use at the lead mine. The coal occurs in two seams 6 inches in thickness, separated by a 4-inch layer of shale and overlain by several feet of shale in which another seam of coal 2 or 3 inches thick appears. These beds occur in the Cretaceous sandstones, which dip to the northeast at a high angle (45°) and are cut at short intervals by basic dikes. A shaft sunk on the seam had reached a depth of 75 feet without showing any essential change in the thickness of the vein. Work was much impeded by water in the mine.

LUNA COUNTY.

By C. H. Gordon and Waldemar Lindgren.

GENERAL FEATURES.

Luna County, which was organized in 1901, contains 2,946 square miles. It is bounded on the west by Grant County, on the north by Grant and Sierra counties, and on the east by Dona Ana County. Mexico adjoins it on the south. (See fig. 28.)

In its extreme north end the county contains the southern termination of the great Mimbres Range, which is conspicuously indicated by the landmark of Cooks Peak, rising to an elevation of 8,330 feet. With this exception the county is a slightly undulating expanse of high plains, whose elevations range between 4,000 and 4,500 feet. Above these arid plains, dotted with yucca, rise many groups of hills or short ranges, but they are not nearly so well marked

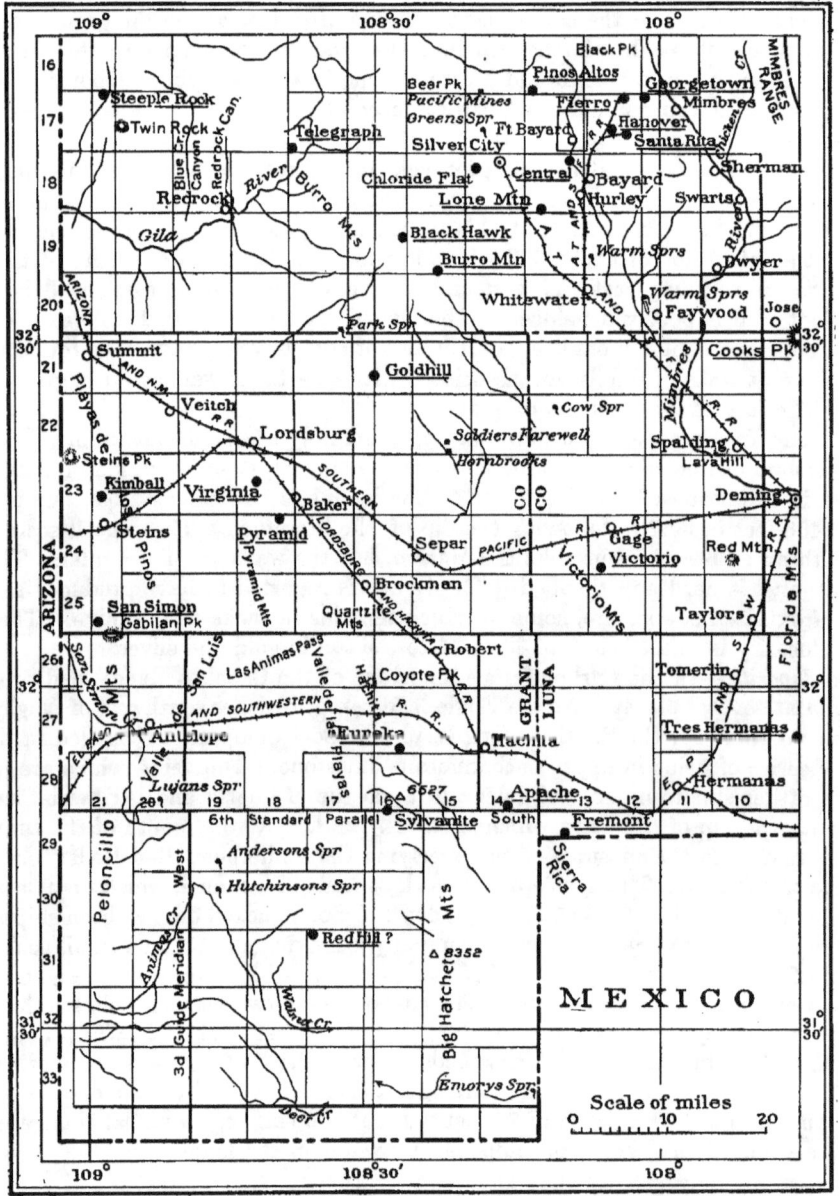

FIGURE 28.—Map of portions of Grant and Luna counties. (After land office map.) Names of mining districts underscored.

and regular in direction as in Grant County and in Arizona. The most prominent of these elevations is the Florida Range, whose jagged and picturesque outline is in full view to the southeast of Deming, the principal city of the county. The highest peak of the range attains 7,295 feet. The group of Tres Hermanas, 25 miles south of Deming, reaches 7,151 feet.

The drainage is irregular, in part into closed basins. Mimbres River, the principal watercourse, flows southward from the Mimbres Range, but soon sinks in the arid valleys and from

Deming southward its presence is indicated only by a sheet of fresh groundwater. This depression crosses the boundary line near Columbus and eventually merges into the closed basins of the Palomas Lakes in Mexico.

The rocks of the Mimbres Range, at Cooks Peak, are limestones of lower and upper Paleozoic age, intruded by porphyries. A similar series of Paleozoic rocks forms the Florida Range, but here the pre-Cambrian granitic basement is also exposed. Carboniferous limestones intruded by porphyries appear again in Tres Hermanas and at a few other places in the county, as in the Victorio Hills, west of Deming, but most of the purple, brown, or black hills and ridges rising above the high plains are basalts and andesites of Tertiary and even of Quaternary age.

Luna County has only four producing mining districts; the greater number of the volcanic hills appear to be barren of mineral deposits. At Cooks Peak lead ores have been mined since 1880 and work is still in progress. At Victorio Camp, 18 miles west of Deming, occur rich lead-silver-gold ores, but little production has been recorded of late years. In the Florida Mountains a little silver-lead ore was extracted many years ago. The Tres Hermanas yield lead-silver-gold ores as well as zinc ores; the deposits are in part of contact-metamorphic origin and in part fissure veins. All these districts contain lead or lead-zinc ores as replacements in Paleozoic limestones, and they are genetically connected with the older Tertiary porphyries rather than with the more recent andesites. Prospects of copper are reported from the Stonewall district, near the Mexican boundary line, and the railroad station of Hermanas.

COOKS PEAK DISTRICT.[a]

GEOGRAPHIC RELATIONS.

The Cooks Peak district is situated on the north side of Cooks Peak, about 18 miles southwest of Lake Valley and a like distance north of Deming. Cooks Range, of which Cooks Peak is the highest point, is a short range trending to the southeast and genetically a part of the Mimbres Range. Cooks Peak itself is a jagged peak rising to an elevation of 8,330 feet and constitutes a prominent landmark throughout the adjoining region. Cooks and Jose are two mining centers located about a mile apart and on opposite sides of the limestone ridge running northward from Cooks Peak.

DISCOVERY AND PRODUCTION.

The first discovery of ore in the Cooks Peak region appears to have been made in 1876, but no important locations were made until 1880, when Taylor and Wheeler located the principal producers of the district. During succeeding years there was much activity, and the rich oxidized lead ores were eagerly sought by the smelters. The present annual production ranges up to 1,500,000 pounds. The total production is estimated to be about $3,000,000, of which the Desdemona group is credited with $2,000,000, the Graphic group with $450,000, the Summit group with $350,000, and the remaining properties with about $200,000. Four-fifths of the production is lead and one-fifth silver.[b]

The production since 1903 is as follows, according to the tables in the "Mineral Resources of the United States:"

Production of Cooks Peak district, 1903-1908.

	Silver.	Lead.		Silver.	Lead.
	Fine ounces.	*Pounds.*		*Fine ounces.*	*Pounds.*
1903	5,748	1,343,361	1906	7,519	627,544
1904	4,401	576,795	1907	4,739	592,151
1905	5,198	463,956	1908	1,009	127,535

According to reports received in 1909 fluorite in commercial quantities is stated to occur in the foothills of the Cooks Range, 10 miles north of Deming.

---

[a] By C. H. Gordon.   [b] Jones, F. A., Mines and minerals of New Mexico, 1904, p. 181.

GEOLOGY.

Cooks Peak constitutes a massive intrusion of granodiorite porphyry. The ridge and elevations extending from it toward the northwest for several miles are carved out of the limestones and shales belonging to the Ordovician, Devonian, and Carboniferous systems; on the south, formations geologically higher appear at the surface. The limestones on the north are cut by dikes of similar porphyry which radiate from the main mass. Intrusions of granite porphyry and diabase occur also in the form of dikes and sills.

Under the microscope the granodiorite porphyry appears fresh, with prominent phenocrysts of andesine feldspar, many of which exhibit marked zonal structure; there are also a few phenocrysts of hornblende and biotite. The ferromagnesian minerals are partly altered to chlorite. Magnetite is relatively abundant. The groundmass is microcrystalline granular and composed of essentially the same minerals, together with quartz and orthoclase. The following analysis of this rock was made by George Steiger in the laboratory of the United States Geological Survey:

Analysis of granodiorite porphyry from Cooks Peak.

SiO_2	62.95	TiO_2	0.67
Al_2O_3	15.91	ZrO_2	None.
Fe_2O_3	3.30	CO_2	None.
FeO	1.37	P_2O_5	.18
MgO	2.18	MnO	.08
CaO	4.46	BaO	.03
Na_2O	4.05	SrO	.03
K_2O	2.95		
H_2O (−)	.72		100.07
H_2O (+)	1.19		

Between Cooks and Jose the rocks have been carved into bold relief, and the excellent exposures here furnish the following section, which corresponds well with that at Lake Valley:

Section between Cooks and Jose.

Character.	Thickness (feet).	Age.
5. Coarsely crystallized limestone, mostly in thick beds	75	Lake Valley limestone (Mississippian).
4. Shales with thin beds of limestone, fossiliferous	50	
3. Limestone, blue above, nodular cherty below	150	
2. Black fissile shales resting on uneven surface of limestone below. At bottom in places shales altered to black jaspery breccia.	200	Percha shale (Devonian).
1. Limestone, irregularly bedded, siliceous along surface at contact with shales above. Ores in upper part.	(?)	Mimbres limestone (Ordovician?).

The rocks have a general dip of 25° to 30° ESE., varied by some well-marked minor corrugations whose axes run approximately parallel with the range. The arrangement of the beds suggests two or more broad folds pitching slightly to the southwest.

ORE DEPOSITS.

Character of the ores.—The ores of the district consist chiefly of lead carbonate, associated with which are some zinc carbonate, limonite, galena, sphalerite, and pyrite.

Geologic relations.—The ore bodies occur in the upper part of the Mimbres limestone just below the Percha shale. As a rule they lie below the siliceous zone that follows the contact between the limestones and shale. Where denudation has removed the overlying beds, owing to the readiness with which the rock crumbles under the action of the weather, the siliceous formation may be largely removed, but patches of it here and there indicate its former wide extension. The silicification seems more generally to have affected the limestone; but one place where it affected the shales is in the gap just above Cooks post-office, where a remnant of black flint breccia occupies the position of the basal beds of the shale that once extended over this point. The ore deposits occur under broad arches in the limestone beds. One such mineralized belt occurs at Cooks and another at Jose. The former has been the chief producing locality.

The ore occurs in kidneys, pockets, and pipes, which are irregular in shape and size and in their distribution along the axis of the arching beds. The longer bodies or pipes vary much in their dimensions, in places becoming too small to admit the body of a man, but elsewhere expanding into large chambers, as in the big chamber of the El Paso mine, which is nearly 100 feet long and from 35 to 50 feet in cross section at the widest part. About $450,000 worth of ore was taken from this chamber.

Offshoots extend irregularly out into the limestone from the main bodies, and their relations give clear evidence of their origin through replacement of the limestone. Cross sections of these branching bodies may show a core of unaltered galena surrounded by a zone of lead carbonate and outside of this a shell of limonite. More or less quartz occurs with the carbonate ore. Individual ore bodies or kidneys are composed wholly of lead carbonate. Many of them, however, show at the center galena, calcite, and locally fluorite. The different deposits are connected by thin sheets of sulphide ores.

Some of the ore spaces are only partly filled with ore, leaving small caves, some above and some within the carbonate ore. These caves are usually lined with crystals of calcite or quartz. The ore spaces may be filled with clay alone, or partly with clay and partly with ore. In the large ore chamber referred to a cave of considerable size occurred over the ore. This cave was lined with sparkling crystals of calcite and stalactites of gypsum hung from the roof.

As stated above, dikes of porphyry and diabase are found in the vicinity. At one point a sill of diabase is intercalated in the Percha shale a short distance above the limestone. In general the sedimentary rocks along the dikes show little or no alteration.

Two parallel dikes of a doubtful basic rock occur about 6 feet apart just west of Mr. Poe's house. Mr. Poe states that these dikes have the appearance of cutting the ore bodies, the space between the dikes being in places entirely filled with ore. One body of zinc sulphide occurred wholly within the dike. Faulting has taken place to some extent.

Mine workings.—"The most important mines in the district are the Desdemona, Othello, and Monte Cristo, belonging to the American Smelting and Refining Company; the Graphic, belonging to the Graphic Mining Company; and the Poe mines, known as the Summit group."[a] On the opposite side of the mountain, at Jose, the most important properties are those of the Faywood Lead Company.

The oxidized ores lying near the surface have thus far been the only deposits to receive attention. These have been extracted mostly by open cuts, inclines, and tunnels into the face of the hill. At Cooks a few shafts have been sunk to reach the deeper-lying portions of the ore pipes. No active operations were in progress at either camp in 1905, though a little prospecting work was being done at Cooks. Since 1905, however, a small annual output has been recorded, as noted above under "Production." The ore is shipped without concentration.

Mode of formation of the ore.—In their occurrence and origin these deposits resemble many others in the Mimbres Range. It is thought that the granodiorite porphyry mass to the south supplied the heated solutions carrying the metals; and these solutions, rising along the arches of the limestone below the shale, were deposited through a process of interchange with the limestone along the fractures. The caves found in connection with the oxidized ores are clearly of later date and are connected with the oxidation of the ores.

FLORIDA MOUNTAIN DISTRICT.[b]

Deposits similar to those at Cooks Peak are reported to occur about the southeast end of the Florida Mountains, but as they were not visited no definite statement can be made concerning them. One of the most notable of these deposits was encountered in the Cave mine, which

---

[a] Jones, F. A., New Mexico mines and minerals, 1904, p. 181.   [b] By C. H. Gordon.

is said to have produced about $60,000 in 1905. No operations were then carried on, but prospecting was resumed in 1906.

On the northwest side of the mountain south of Capitol Dome, 10 miles south of Deming, an opening has been made on a vein carrying sulphides of copper in a shattered zone of andesite. Intrusions of granite porphyry were observed in the vicinity. The gangue consists of quartz and crushed andesite. The strike of the vein is about N. 62° E. and the dip 80° SE. Two drifts have been run—No. 1 for 280 feet; and No. 2, the lower, for 420 feet; they are 120 feet apart. A shaft starts 40 feet above the upper tunnel, extends down for 100 feet, and is connected by a horizontal tunnel with a raise of 60 feet from the lower tunnel. The deposits consist of small lens-shaped masses along the fracture and of impregnations of the altered andesite along the shattered zone. No important bodies have yet been opened.

To the northwest, below Capitol Dome, the west face of the mountain consists of several hundred feet of limestones overlying 135 feet of quartzites, which in turn rest on the irregular surface of red granite, in places much decomposed.

VICTORIO DISTRICT.[a]

GENERAL FEATURES AND GEOLOGY.

The high plains extend for a long distance west of Deming with an approximate elevation of 4,400 feet and are diversified only by a few small reddish or black volcanic hills. The Gage railroad station (elevation 4,478 feet) lies 18 miles west of Deming, and 4 miles south of it a somewhat conspicuous group of hills rises to a height of about 600 feet above the plains. Immediately south of the main mass is a low level-topped hill extending for about 1½ miles in a west-northwest direction, and on this is situated the Victorio Camp, named after the Apache chief, whose strongholds were in the Mimbres and Mogollon mountains.

The principal period of activity was from 1880 to 1886, when the big bodies of lead ore were being worked by Mr. Hearst, of San Francisco, under conditions which now would be considered very adverse. The Chance and the Jessie, the principal producers, are reported to have yielded $800,000 each in oxidized argentiferous lead ores. F. A. Jones[b] gives $1,150,000 as the total production up to January 1, 1904. In recent years renewed prospecting by Wyman & Corbett and Lesdos has shown that the camp may yet reenter the ranks of producers. Smaller shipments of partly oxidized lead ores have been made at intervals during the last few years. In view of the total production the developments in the camp must be considered slight.

The central hill of the group consists of a series of andesite flows, which almost entirely cover the underlying limestone. Specimens from the north bluff show a dark brownish-gray andesite, in places brecciated or tufaceous. It is considerably altered and the ferromagnesian silicates are largely replaced by secondary minerals, but outlines of black-bordered hornblende are still visible. The feldspars show in the thin section as short rectangular sections, most of which approach andesine in composition.

The smaller, southern hill extends southwest and northeast for about a mile and rises 250 feet above the plains. It is built up chiefly of fine-grained dark-gray dolomitic limestone or dolomite in heavy beds, dipping about 20° SW. In places the rock is cherty and the stratification generally obscure. The northern edge of the hill is marked by a dislocation. Several prospect holes sunk here disclose a peculiar coarse, loose quartzite with larger rounded fragments. This probably vertical fault is best shown in the 80-foot shaft on the Lesdos Rambler claim. Smaller limestone hills lie east and west of the main ridge just described.

MINERAL DEPOSITS.

General features.—The ore bodies are connected with a series of northeast-southwest trending, usually very tight seams, which in places open abruptly to form irregular bodies of galena, as much as 20 feet wide and locally contained in brecciated limestone. The outcrops

---

[a] By Waldemar Lindgren. [b] New Mexico mines and minerals, 1904, p. 184.

are as a rule inconspicuous. The ore is partly oxidized and accompanied by coarse calcite and siderite or by quartz as a gangue. A north-south vein of quartz, carrying tungsten minerals, differs markedly from the other deposits and is described below. No intrusive rock of any kind was found in the main limestone hill. The andesite in the largest hill of the group probably consists of flows later than the lead deposits and is thought to have no connection with their genesis. It is more likely that an intrusive body of porphyry or granite to which the ore owes its origin exists somewhere in the vicinity, buried under the detritus.

Detailed description.—A rough plat of the principal claims is shown in figure 29. The two claims which contained the principal ore body are the Jessie and the Chance, the former being owned by Wyman & Corbett, the latter by the Haggin interests. The developments consist of the Jessie tunnel, starting from the north foot of the hill as a crosscut which extends for 500 feet, and continuing along the line of the deposit to the south end line and through the Chance property. Near the south end of the Chance claim is a shaft 300 feet deep, the collar of which is 150 feet above the Jessie tunnel level. The total production had a value of over $1,000,000. Mr. Corbett states that $800,000 was taken out from the Chance and an equal amount from the Jessie.

The deposit shows on the surface at only one point, on the Jessie claim near the Chance line; otherwise the croppings are blind, and underground, outside of the ore bodies, the direc-

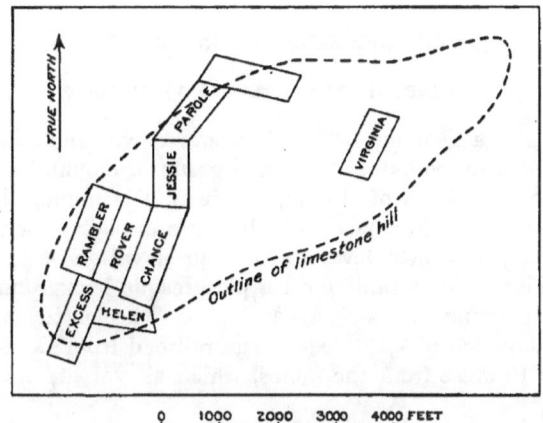

FIGURE 29.—Sketch map showing relative position of principal claims in Victorio mining district.

tion is indicated only by a tight, white seam. The strike of the fissure and the ore bodies is north-northeast; the dip is about 60° WNW. but is irregular and the ore bodies show offsets in dip in places. The seam opens rather abruptly in the southern part of the Jessie to a brecciated ore body, in places partly open and cemented by botryoidal calcite. This body continued for 1,000 feet into the Chance property. The largest part of the ore was found between the tunnel level and the surface, reaching up to a point 150 feet above the tunnel. No values were found in a 180-foot winze on the Chance in the ore body and it is said that the main shaft in the Chance did not open any ore bodies. Another ore body was found by Mike Burke, a lessee on the Chance some ten years ago, who in following a seam in a tunnel not far from the Chance shaft collar, striking N. 50° E. and dipping 70° NW., encountered shipping ore that yielded nearly $100,000. The shoot continued to a point within 30 feet of the Jessie ground and for 50 feet below the tunnel level.

The ore is a partly oxidized galena in calcite and locally in quartz gangue. The higher grade ores, with 15 to 22 per cent lead, contain about 15 ounces of silver and $6 in gold to the ton. Some of the poorer lead ores are richer in the precious metals, yielding about 50 ounces of silver and 2 ounces of gold to the ton. All the ores contain some arsenic.

The Helen claim, on the southwest point of the hill, contains several seams with ferruginous outcrops. One of them produced $30,000 between 1881 and 1886.

The Excess claim adjoins the Helen on the west and is said to be owned by the St. Louis Company. A shaft 200 feet deep, not accessible, is sunk at the foot of the hill, the dump showing fair quartzose lead ore with some silver. The deposit is stated to form an irregular body.

The Rambler, to the north of the Excess, seems to be located on the fracture or fault mentioned under "Geology." An 80-foot shaft discloses several seams with indications of lead and copper along the contact of limestone and quartzite, presumably the same dislocation that is indicated at the foot of the hill below the Jessie tunnel.

The Virginia, from which a little lead ore was shipped many years ago, lies about 3,000 feet east of the Jessie; it shows a northeastward trending fissure with quartz gangue in dolomite.

About 4,000 feet north of the Jessie, in low limestone hills across the flat, is an entirely different deposit, consisting of a north-south vein of white quartz, from 1 to 5 feet thick and with steep dip. Crystals of quartz show in places along the central seam. The quartz contains irregularly distributed black or brownish wolframite, probably near hübnerite, and a little pyrite and galena. A few foils of white mica were also noted in the quartz. Oxidized portions show wulfenite, probably also scheelite, derived by reaction between tungsten solutions and the limestone wall rock. The quartz is reported to contain a little gold. No alteration of the limestone was observed along the walls. The deposit is the property of George T. Brinkman, of Lordsburg, who has shipped a few tons of wolframite ore, hand picked by Mexicans. The only developments consist of surface pits.

TRES HERMANAS MINING DISTRICT.[a]

GENERAL FEATURES AND GEOLOGY.

About 25 miles south of Deming, on the Southern Pacific and Santa Fe railroad lines, and 10 miles north of the Mexican boundary the small group of mountains known as the Tres Hermanas rises from the general level of the high plains. The mining district is situated in the northern part of the group, the three peaks of which are in plain view from Deming. The road from Deming to the camp is almost level and has an elevation of 4,000 to 4,300 feet; on the east the crags of the Florida Mountains form a picturesque background to the plain, which is dotted by luxuriantly growing yuccas. According to Wheeler the highest peak of the Tres Hermanas attains an elevation of 7,151 feet. The railroad from Deming to Hermanas station is at the nearest point 10 miles from the mines, which as yet are most easily accessible from Deming.

The mines have been known for many years and have yielded a moderate production, possibly $200,000, principally from the Cincinnati vein and from the lead deposits in limestone near the present zinc mines. In 1904 zinc ores were discovered in the district, and shipments were made to smelters in the Mississippi Valley in 1905, but in 1906 and 1907 there was little activity. The geologic examination on which this paper is based was confined to the north end of the district. It appears, however, that the predominating rock in the mountain group is a granite porphyry, which forms the central peaks and is intruded in Carboniferous limestones of the foothills. Specimens collected near the zinc mines show a coarse light brownish-gray porphyry with phenocrysts of feldspar up to 15 millimeters in length, small foils of biotite, and small crystals of dark-green hornblende. The microscope shows that the phenocrysts consist of orthoclase with some oligoclase and that the crystals are embedded in a coarse micropegmatitic groundmass of quartz and orthoclase. Possibly the rock is more correctly to be classified as a quartz syenite porphyry.

The northern foothills at the zinc mines consist of bluish-gray fossiliferous limestones of Mississippian (lower Carboniferous) age, dipping to the northwest at low angles. At the junction with the granite porphyry strong contact metamorphism is noted.

---

[a] By Waldemar Lindgren.

MINERAL DEPOSITS.

The mineral deposits consist of normal fissure veins in granite porphyry and contact-metamorphic deposits of lead and zinc in limestone near the porphyry. Little work has lately been done on the veins, which are situated about 1 mile south of the zinc mines, but some information in regard to them was obtained from reliable sources.

The Cincinnati claim, owned by the Golden Cross and Eagle Company, is the largest producer and is said to have yielded $100,000. The deposit is stated to be a narrow vein in porphyry, very rich in lead and gold. The ore was found within 100 feet from the surface. Some ore has also been shipped from the Yellow Jacket claim which is reported to be an extension of the Cincinnati. The Hancock claim, owned by C. E. Burdick, of Deming, is said to contain a narrow vein in porphyry up to 8 inches wide. The shaft is 400 feet deep and some work was in progress in 1905. A thousand tons of rich lead ore, with some gold, is said to have been shipped.

The zinc mines are situated at the northern base of the first porphyry hills, close to the old surface workings where galena ores were formerly mined. Though the galena deposits appear to be exhausted, pieces containing this mineral are occasionally found. In 1904 attention was attracted to a heavy white massive or crystallized material occurring plentifully in the old workings, and it was found to consist of carbonates and silicates of zinc. The property, which consists of five claims forming a compact group, was worked by Thurman & Lindauer, of Deming, in 1905, and a considerable number of carloads of ore were shipped to Mississippi Valley smelters. Some difficulty was experienced in obtaining ore of the requisite high percentage, but the systematic and careful development which the property seemed to deserve was not undertaken and it was practically idle in 1906 and 1907. Another group of claims adjoining the property on the northwest is owned by Doctor Swope and associates, of Deming, and some ore containing zinc and lead was shipped in 1906. The railroad freight from Deming to the zinc-smelting works is $5 a ton, and the hauling to the station costs $2 a ton. Ore containing 29 per cent of zinc was paid for by smelters at about $11 a ton, and 37 per cent ore yielded approximately $16.50; obviously the lower grade left but little profit. The developments are slight, consisting of a few shallow inclines and surface cuts.

The traveler approaching Tres Hermanas from the north first encounters a low, broad limestone ridge, separated from the higher and more abrupt hills of porphyry on the south by a low gap. The principal zinc mines lies on the west side of this gap; the road leading up to it follows a broad gully, at first with a southeasterly direction, which close to the mines changes to due east. From the gap the plains are in full view on the east and west. The main contact between limestone and porphyry runs east and west about a quarter of a mile south of the gap. The office of the Thurman & Lindauer property is situated in the gulch just west of the gap. The summit of the flat-topped limestone hills, about 1,000 feet north-northwest of the office, shows a moderately thick bed of bluish-gray Mississippian limestone, highly fossiliferous in places. The strike is N. 80° E., the dip about 10° N. This series of limestones continues toward the north to the edge of the plains. In the little escarpment facing the gap the thin bed of fossiliferous rock is underlain by 5 feet of hornfels and below this lies a coarsely crystalline limestone. The gap is occupied by a similar coarse crystalline limestone with various dips and strikes; in one place, at some zinc prospects, a strike of N. 70° W. and dip of 30° NNE. were noted. Nearer the porphyry hills the dip increases to 60°. About 1,000 feet west-southwest of the office at the principal workings there is a small hill with limestone which lies flat on the top, but bends to dips of 20° on the north, west, and south. The actual contact with the porphyry south of the gap is not very well exposed, but near it the limestone contains several well-defined dikes, and in one place a projecting broad tongue of porphyry distinctly cuts across the strata.

Contact metamorphism is clearly manifest, the affected zone being in places over 1,000 feet wide. The principal complex of limestone northwest of the gap is, on the whole, not metamorphosed, but even here, more than half a mile north of the contact, a few thin beds of garnet rock and coarsely crystalline limestone are noted. In the little bluff facing southward from the highest point where the fossiliferous limestone outcrops the beds underneath this member are distinctly altered to calcareous hornfels and coarsely crystalline limestone. In the gap the crystalline limestone prevails, with intercalated beds of pure garnet rock and hornfels, the latter evidently representing a contact-metamorphosed lime shale. Some of the strata exposed here contain well-preserved specimens of *Spirifer* and *Fenestella* and look like limestone, but in fact the rock consists very largely of garnet.

The principal zinc workings, a quarter of a mile west of the gap, are in limestone in which the only sign of alteration consists of bunches of wollastonite.

The deposits contain oxidized zinc minerals and some galena; in part they lie parallel with the stratification, but in part also they follow short and ill-defined, nearly perpendicular fissures and veins cutting across the beds and varying in strike from northwest to southwest. The ores also in places form irregular bunches in the limestone. Several small deposits of the two last-named classes are found on the main limestone hill, in many of the prospect holes sunk there, but as yet they have not been proved to be of value. A short distance north of the principal gap oxidized zinc ores occur in kidneys and bunches between beds of coarsely crystalline limestone and garnet rock. A veinlike deposit striking N. 60° E. and standing nearly vertical also contains zinc ores at this place. Many of the prospect holes in the area of most intense contact metamorphism in the gap or immediately east of it have given assays high in zinc. The principal workings, about 2,000 feet west of the gap, consist of surface pits and small shafts with irregular workings. Here the oxidized zinc ores occur in the greatest abundance, in part parallel to the strata, in part along a vein with west-northwest strike. Little intelligent prospecting had been undertaken in 1905, the deepest hole having attained only 35 feet from the surface. The ore grades into limestone and great care must be exercised to avoid mixing with barren material.

About 1,000 feet north of this place, on the Contention claim, Doctor Swope and associate sunk an incline 150 feet long following the dip of the strata, which here was 30°. Some lead and zinc ores were shipped from this claim in 1906.

Systematic prospecting with drills would be necessary to develop these deposits. The principal ore bodies will no doubt be found intercalated between the flat-lying strata.

As noted above, the galena was discovered at an early date, but the zinc ores escaped detection until recently. A few green stains are the only indications of copper minerals found. Possibly exploration may develop more of this metal in beds at greater depth. The galena was seen at the principal workings accompanied by a little pyrite and intimately intergrown with wollastonite in a manner indicating simultaneous deposition. The zinc ores occur as dark-gray cellular masses and consist largely of the unusual mineral willemite, the pure anhydrous silicate. Smithsonite, the zinc carbonate, is also present in its usual mammillary form and is light gray or bluish in color. Hydrozincite was shown by Doctor Hillebrand to be present as an earthy incrustation. Tabular crystals of calamine, the hydrous silicate of zinc, were noted in crevices of the ore. Willemite appears to make up the bulk of the specimens collected. It forms small radial aggregates of slender hexagonal prisms, terminated by a flat rhombohedron, and is accompanied by a little of a dark material which looks like pyrolusite and gives a dark tinge to the ore. It also forms loose crystalline aggregates and crusts of needle-like crystals. This habit is unusual, the more common form recorded in text books of mineralogy being a short, stout hexagonal prism. The occurrence at Tres Hermanas seems to be paralled only by that of Moresnet, in Belgium, where the mineral is also accompanied by galena. The only other occurrences of willemite noted in this country are at Franklin Furnace, N. J., and as rarities at Clifton, Ariz.,[a] and at the Merritt mine in Socorro County, N. Mex. The zinc minerals of Tres Hermanas are beyond doubt derived from zinc blende by oxidation.

The principal zinc minerals of Tres Hermanas and their tenor are as follows:

Willemite (Zn_2SiO_4), zinc silicate, 58.6 per cent zinc.
Calamine ($H_2Zn_2SiO_5$), hydrous zinc silicate, 54.7 per cent zinc.
Smithsonite ($ZnCO_3$), zinc carbonate, 52.1 per cent zinc.
Hydrozincite ($ZnCO_3.2Zn(OH)_2$), basic zinc carbonate, 59.9 per cent zinc.

---

[a] Lindgren, Waldemar, The copper deposits of the Clifton-Morenci district, Arizona; Prof. Paper U. S. Geol. Survey No. 43, 1905, p. 11L. Genth, F. A., Contributions to mineralogy. Proc. Am. Philos. Soc., vol. 24, p. 37.

The zinc ores shipped from Tres Hermanas probably contained about 30 per cent zinc. The shipments of lead-zinc ores made by Doctor Swope are stated to have contained from 11 to 40 per cent of lead and 2 ounces of silver to the ton. They also contained up to 19 per cent of zinc, besides 4 per cent of lime, 4.2 per cent of iron, and 7.4 per cent of silica.

An analysis of picked material from a loose aggregate of yellowish-white minute crystals gave the result recorded below under 1. Analysis No. 2 represents a specimen that fairly illustrates the average ore but is probably of lower grade than the material shipped.

Analyses of zinc ores from Tres Hermanas.

[Analyst, George Steiger.]

	1.	2.		1.	2.
SiO_2	23.52	9.12	H_2O-	None.	None.
$(Fe,Al)_2O_3$	.26	.22	H_2O+	.57	.71
ZnO	65.18	24.00	MnO	None.	None.
CaO	5.78	37.44			
CO_2	4.59	29.05		99.90	100.54

The first analysis gives willemite (Zn_2SiO_4), 82.75 per cent; calamine ($H_2Zn_2SiO_5$), 6.93 per cent; calcite ($CaCO_3$), 10.32 per cent; smithsonite ($ZnCO_3$), 0.14 per cent; limonite, 0.31 per cent. The second analysis, the water being disregarded, gives rather closely 66.05 per cent of calcite and 32.77 per cent of willemite. As in the first analysis, the water above 110° C. indicates the presence of a small percentage of calamine. However, as in that analysis, the calamine must be calculated from the small quantity of water contained, which makes the exact amount uncertain.

GRANT COUNTY.

By L. C. Graton, Waldemar Lindgren, and J. M. Hill.

GENERAL FEATURES.

Grant County occupies the southwest corner of New Mexico, covering an area of 7,403 square miles. It is bounded on the north by Socorro County, on the east by Sierra and Luna counties.

Physiographically it falls into two distinct provinces. The northern part is drained by the headwaters of Gila River and on the east also by those of the Mimbres. This area is a mountain mass with few distinct ranges, the elevations ranging from 3,500 feet on Gila River to over 9,000 feet on the highest ridges; it forms in fact the southern edge of the great Plateau Province. The most pronounced range is the Mimbres, which lies on the boundary between Grant and Sierra counties and which may be said to form the margin of the Plateau Province on the east. The plateau character is in general not due to flat-lying sedimentary rocks, but rather to heavy Tertiary flows of rhyolite, andesite, and basalt which cover large areas. Into this irregular plateau Gila and Mimbres rivers have cut deep trenches. Near the southern edge of the plateau, underneath the lava flows near Silver City and Pinos Altos, erosion has exposed the underlying Cretaceous and Paleozoic sediments, which ordinarily dip at gentle angles but are broken by faults and intruded by granite porphyry or quartz monzonite porphyry.

The southern half of the county belongs to the Arizonan desert region and is characterized by a number of sharply marked narrow ranges, trending north and south and separated by wide, gently sloping arid valleys with elevations of 4,000 to 4,500 feet. Many of these ranges are low, but one of the highest points, Big Hatchet Peak, south of Hachita, reaches an elevation of 8,352 feet. Many of the valleys are closed basins with central playa deposits. From east to west the more prominent ranges are named Big Hatchet, Hachita, Pyramid, Animas, and Peloncillo.

The geologic structure of these ranges is complicated and, in the absence of detailed mapping, only the most general features can be indicated. The Big Hatchet, Hachita, and Animas ranges consist of tilted and faulted Paleozoic limestones and Cretaceous shales and sandstones, at many places intruded by granitic porphyries. The Paleozoic and later sediments appear at many places in the Peloncillo Range, but there are also in that range and the Pyramid Range large areas of rather basic igneous rocks, whose age is somewhat doubtful but which seem to represent lavas of an earlier period than those that cover so vast an area in northern Grant County. There is little evidence that the ranges were outlined by faults of the so-called Great Basin type. They seem to be erosional in origin.

The county contains a great number of mining districts and has generally occupied the leading position among the metal-producing counties of the Territory.

Among nonmetallic mineral resources the blue gem turquoise occupies the first place. The principal turquoise mines are situated within a small district in the Burro Mountains, although the mineral occurs also in the Hachita Range, about 10 miles west of the town of Hachita. The turquoise mines in the Burro Mountains have been described by E. R. Zalinski[a] and D. B. Sterrett.[b] Zalinski states that the Azure mine in the Burro Mountains has produced turquoise to the value of several million dollars since 1891. Here as elsewhere the turquoise veins stand in closest relationship to dikes of granite porphyry or quartz monzonite porphyry. A description of the geology of the district will be found on pages 321–322.

A large deposit of aluminous sulphates and also, it is said, of bauxite has been discovered in the canyon of Gila River about 25 miles north of Silver City.[c] It has not as yet been utilized. In the same vicinity and also at a place on Bear Creek, 12 miles northwest of Silver City, a material very similar to meerschaum has been found.[d]

The metallic resources of Grant County are more important. During the last decades of the nineteenth century the county yielded a considerable annual production, at first from gold placers, then from the easily worked and enriched upper parts of silver deposits and from gold-bearing, sulphidic quartz veins. To speak generally, the annual gold production has ranged from $100,000 to about $500,000, the higher figure being recorded in 1889. During the last few years the gold output has at times sunk below $100,000. In 1908 it was only $45,682. The silver production of early years is imperfectly known, but the recorded figures range up to a maximum of about 600,000 ounces in 1889. It dropped rapidly from those figures and during the ten years 1899 to 1908 has ranged from 48,000 to 231,000 ounces. In 1908 it was 95,477 ounces. The production of lead has fluctuated violently; since 1889 it has varied between 179,000 pounds in 1904 and 10,000,000 pounds in 1897. The average has perhaps been 1,500,000 pounds. In 1908 the production was 244,589 pounds. Little copper was recorded in the earlier years, but in 1897 the output attained 1,200,000 pounds and since then it has increased rapidly to 8,000,000 pounds in 1907, diminishing, however, to 5,240,000 pounds in 1908.

The most important silver and gold mining districts are located in the northeastern part of the county; in general, the deposits are fissure veins carrying gold and silver, in intrusive porphyries of early Tertiary age, or replacement deposits, carrying principally silver in Paleozoic limestones. There are also a few contact-metamorphic deposits. The copper is derived mainly from the old Santa Rita mines, known to the Mexicans before the American occupation, and from the lately developed mining district in the Burro Mountains. A considerable quantity of magnetic iron ore has been mined since 1899 at Fierro, near Hanover and Santa Rita.[e] From 1900 to 1908, inclusive, the production was somewhat more than 1,000,000 long tons.

The mining districts in the northern part of the county comprise the Pinos Altos, Chloride Flat, Santa Rita, Hanover, Central, Georgetown, Burro Mountains, Gold Hill, Lone Mountain, Blackhawk, and Telegraph and Steeplerock. The mining districts in the desert ranges in the southern part of the county are scattered and vary widely in their mineral characters, embracing almost every kind of deposit; silver, lead, and copper are the principal metals. The districts comprise the Kimball, San Simon, Pyramid, Shakespeare, Hachita, Apache No. 2, and Gillespie or Red Hill. The Gillespie district, which was not visited, lies in the extreme southwest corner of the county; galena and oxidized lead ores are mined here on a small scale.

---

[a] Econ. Geology, vol. 2, 1907, pp. 464–492.
[b] Mineral Resources U. S. for 1907, pt. 2, U. S. Geol. Survey, 1908, pp. 828–832.
[c] Hayes, C. W., The Gila River alum deposits: Bull. U. S. Geol. Survey No. 315, 1907, pp. 215–223.
[d] Sterrett, D. B., Meerschaum in New Mexico: Bull. U. S. Geol. Survey No. 340, 1908, pp. 466–473.
[e] Paige, Sidney, The Hanover iron-ore deposits, New Mexico: Bull. U. S. Geol. Survey No. 380, 1909, pp. 199–214.

GRANT COUNTY. 297

PINOS ALTOS DISTRICT.[a]

LOCATION AND HISTORY.

The Pinos Altos Mountains lie in northeastern Grant County, about 6 to 10 miles north-northeast of Silver City. They form a rather prominent ridge extending north and south for about 5 miles and reaching a maximum elevation of 8,036 feet. The line of the Continental Divide crosses the mountains at about their highest point, then swings northeastward between Bear Creek on the north and Whisky Gulch on the southeast. The town of Pinos Altos is situated on the eastern slope of the mountains, at an elevation of about 7,000 feet. The principal mines are located above the town on the east side of the mountains, but some development work has been done on the upper western slope. A narrow-gage railroad was begun in 1905 and finished in 1906, connecting the mines with the smelter at Silver City. In 1908 and 1909 mining received a setback from the failure of the Comanche Mining and Smelting Company and the closing of the concentrating and smelting plant at Silver City.

Placer gold is said to have been discovered in 1860[b] not far from the present location of the Mountain Key mine by a man named Birch, one of a party of California pioneers. The camp called Birchville, which immediately sprang up, grew rapidly. In the same year the first lode was discovered; it is still worked as the Pacific vein. Within two years of the date of discovery it was stated that 30 lode mines were being worked by 300 men. Some years later, when the locality was nearly abandoned by Americans, owing to the distractions of the civil war and to depredations by the Apaches, the Mexicans gave to the camp its present name, Pinos Altos.

After the war operations were resumed and in 1867 the first mill, of 15 stamps, was built. In 1883 a mill with a mechanical concentrator was erected, the first in the Territory. In 1896 the Hearst estate, of San Francisco, entered the camp and operated actively for some time, finally selling to the Comanche company in 1903 the mining property and the smelter at Silver City, which had been erected mainly for the purpose of treating the Pinos Altos ores. In 1903 a smelter was erected at the Silver Cell mine, but it was operated only a short time.

The total production of the camp, including both placer and lode mines, is stated by F. A. Jones[c] to have been about $4,700,000. Gold has been the principal product, but silver, copper, and lead have also contributed. Zinc deposits may in the future add somewhat to the total yield.

GEOLOGY.

The west side of the Pinos Altos Mountains appears to be made up mainly of quartzite, limestone, and sandstone. The crest and the eastern part of the mountains consists of igneous rocks, of which granodiorite, diorite, and diorite porphyry are the most prominent. These have penetrated part or all of the sedimentary rocks and have tilted them up so that the sediments in general have a strike ranging from north-northeast to north-northwest and dip at varying angles to the west. Local variations of both dip and strike are to be observed and are commonly related to smaller injected portions of the principal intrusive rocks. Strike faulting and much slide rock also obscure the structure.

Fine-grained dense banded quartzite furnishes abundant float near the summit of the ridge. It is in part conglomeratic, holding more or less angular fragments of dark quartzite. Its relation to the limestone was not determined, but it is regarded provisionally as equivalent to the quartzite of similar general appearance that lies above the Pennsylvanian limestones just west of Silver City. Limestone is the prevalent sedimentary rock. It is very much metamorphosed as a result of the intrusions, and though remains of corals and crinoid stems could be detected no fossils sufficiently well preserved to determine its horizon were observed. It was stated that a shell had been found, on the basis of which the rock had been called Ordovician. Until more detailed work can be done, however, the limestone can only be classed as Paleozoic, with the appearance and relations favoring Carboniferous rather than

---

[a] By L. C. Graton. [b] Jones, F. A., New Mexico mines and minerals, 1904, p. 47. [c] Op. cit., p. 52.

Silurian or Ordovician. On the lower western slope olive-colored sandstone is plentifully exposed and is provisionally regarded as Cretaceous, possibly Dakota. Whether or not intervening members of the geologic column are present was not ascertained.

The granodiorite is a reddish, granitic-looking rock, of average grain, locally called granite and generally regarded heretofore as the basement of the sediments. Fresh exposures are greenish gray in color. The rock consists of plagioclase (labradorite and andesine), orthoclase, quartz, green amphibole, and a few flakes of biotite. Magnetite, titanite, apatite, and zircon are present in grains of rather unusual size and abundance. The diorite is a dark granular rock closely related mineralogically to the granodiorite. It contains orthoclase and quartz only as accessory constituents, holds more hornblende and biotite, and contains some pyroxene; titanite is present and magnetite and apatite are plentiful. The diorite porphyry is dark and commonly fine grained, with its porphyritic character made evident through the development of stout prisms of pyroxene. Some plagioclase phenocrysts are also present. The groundmass consists chiefly of minute laths of plagioclase, a little orthoclase, and grains of augite, apatite, and magnetite. In places the groundmass is much bleached, leaving the pyroxene phenocrysts standing out prominently. A partial analysis of a specimen from the Cleveland mine was made by W. T. Schaller, as follows: SiO_2, 52.93; CaO, 8.50; K_2O, 4.71; Na_2O, 2.72.

These three types of igneous rocks are without doubt closely related as to origin and time of intrusion. Actual evidence as to their relative ages was not obtained, but is seems probable that the granodiorite, which makes up a large part of the lower eastern slope, is the oldest, and that the diorite porphyry, which appears to constitute the crest and such of the western slope as does not consist of sediments, was intruded a short time later. The diorite, which is plainly a basic facies of the granodiorite, was seen only in the southeastern part of the district, near the Silver Cell mine.

The age of these intrusive rocks is not known definitely. To the north of the Pinos Altos district they are covered by a thick flow of Tertiary rhyolite or dacite, and on the west slope of the mountains they cut the Paleozoic rocks. It seems justifiable, in the absence of other evidence, to class them with the late Cretaceous or early Tertiary intrusions that are represented in so many other parts of the Territory.

MINERAL DEPOSITS.

GENERAL FEATURES.

The mineral deposits of the Pinos Altos district fall into two classes—fissure veins in the intrusive rocks, and replacement lodes or pockets in limestone. Each possesses certain general characteristics, but the two classes show more points of resemblance than of difference. The principal distinction shown by existing developments seems to be that the veins in the intrusive rocks carry higher values in gold and silver and more lead.

The principal vein system of the district strikes N. 45°–60° E. and lies about halfway up the eastern slope of the mountains. Veins belonging to this system have been developed in the Deep Down-Atlantic, Mountain Key, Aztec, Pacific, and Tampico mines, named from northeast to southwest, and in several others. This vein zone crosses the contact of granodiorite and diorite porphyry, which strikes north or northwest near the Aztec mine, so that in general the mines to the northeast of the Aztec are in granodiorite and those to the southwest are in diorite porphyry.

PACIFIC MINE.

The Pacific mine, located on the eastern slope of the mountains southwest of the town, has been the largest producer in the district and was credited in 1905 with a total production of over $1,000,000. It is now owned by the Comanche Mining and Smelting Company, which operates the reduction works at Silver City. The mine is opened by two shafts, the Hearst and the Gillette, to a depth of 700 feet. A short crosscut tunnel intersects at a depth of 150 feet the northeastern or Hearst shaft, which is an incline sunk on the vein. Drifts connect with the Gillette shaft, which is now dismantled. A mill in the gulch near the Gillette shaft that was run during the Hearst control of the property is now dismantled also.

The vein cuts dark, fine-grained diorite porphyry, strikes about N. 60° E., and dips about 75° NW. Including developments in the Aztec mine on the northeast and in the Tampico on the southwest, the vein is said to be known for a length of over 4,000 feet. It averages about 2½ feet in width, but ranges from 1 foot to 10 or 12 feet. It consists of quartz, with some calcite, barite, and rhodochrosite and pyrite, chalcopyrite, galena, and sphalerite. The lead and zinc sulphides are only sparingly present in most places, but bunches rich in these minerals and associated with barite and calcite are found here and there. It is said that the "red spar," rhodochrosite, commonly accompanies the copper. The percentage of yellow sulphides, pyrite, and chalcopyrite, is as a rule rather large. The diorite porphyry is altered for an inch or two away from the vein to a softer olive-colored aggregate consisting chiefly of chlorite and sericite and containing crystals of pyrite that carry low gold values. Heavy rains during the winter of 1904–5 had filled the mine with water to the 300-foot level (300 feet below the tunnel, 450 feet below the collar of the incline) and when the water subsided the air was too bad to work below the 200-foot (350-foot) level. The shoot containing workable ore is said to be about 1,500 feet long and to pitch steeply to the southwest. Sulphides were first encountered at a depth of 60 feet. The oxidized portion above this level did not contain noticeable quantities of zinc carbonate or of copper. The upper levels have been pretty well stoped out, but it was stated in 1905 that a large tonnage was blocked out below the 200-foot (350-foot) level, awaiting the advent of the railroad for transportation to the Silver City reduction works. It was also stated that the bottom of the Hearst incline, at a depth of 700 feet, was in good ore. In general, however, the grade of ore declined somewhat with depth. The narrow parts of the vein were usually found to be richest; in the wider places zinc became more plentiful and the precious metals decreased. It is said that another vein to the west joins the main Pacific vein in the lower workings, the junction pitching to the southwest and carrying better than average values. The average value of the ore mined up to 1905 was reported to be at least $10 in gold and $5 in copper and silver to the ton. It is said that the galena is poor in silver and that the silver in the ore was difficult to save by concentration. Gold seems to be associated with the pyrite.

MOUNTAIN KEY MINE.

The Mountain Key mine is situated on the northeastern slope of the Pinos Altos Mountains, just west of the town. It has been worked more or less continuously since the earliest days of the district and is said to have produced over $500,000 to 1905. The developments comprise a 750-foot incline shaft on the vein and eight levels.

The country rock is granodiorite, though the contact with diorite porphyry is but a short distance to the west, and this rock is said to be encountered at the southwest end of the third (300-foot) level, which extends farthest in that direction. Five parallel veins are said to have been cut in the mine. The principal vein is the middle one; it strikes N. 53° E. and dips 58° NW. This is a quartz vein of 2 feet average width, carrying much pyrite, some chalcopyrite and calcite, small grains of light-colored zinc blende, and tiny cubes of galena. The granodiorite is altered for about an inch away from the vein, the products being sericite, quartz, kaolin, and pyrite. Ore was encountered at the surface and was followed down as a small shoot. At the 200-foot level a large shoot 150 to 250 feet long was encountered, and this continued to a point below the 500-foot level, the general pitch of the shoot being, as in the Pacific mine, steeply to the south. At the 500-foot level some extremely rich ore was found, $100,000 having been taken from one small stope. A little below this level the shaft went from good ore into broken ground containing much quartz but practically destitute of values. Here work was suspended on the supposition that the vein had either given out or been lost. Later work has shown that this crushed zone strikes about east and west, dips about 60° S., and has a width of only about 20 feet. Ore has been found on the underside of the crushed zone, and a shoot 400 feet long was said to be exposed in the lower levels in 1905. It was stated by the manager that the ore above the crushed zone contained much free gold, yielding two-thirds of its gold value on amalgamation, but that the ore in the lower levels carried the gold in the sulphides; consequently it was believed that the upper and lower ore bodies are on different veins, and that the vein of the upper ore body is east of that in the lower workings. In view of the fact that the crushed barren zone is said to cut but not to fault the vein developed in the lower levels, the grounds for assuming that the upper and lower ore bodies are parts of different veins seem not very well established. A fault south of the shaft in the lower workings is said to cut the vein off sharply, but the throw is only about 6 feet, and with that displacement the vein continues strong on the other side of the fault.

The ore at present exposed is said to be rather uniform in value, averaging $22 to $30 a ton, but reaching in places $80 or $90. Gold constitutes the principal value, but copper and silver are also present, the former averaging 3.5 to 4 per cent and ranging from 2 to 6 per cent, and the latter in some of the ore amounting to half as much as the gold. Copper is said to be increasing a little in the bottom. The finer-grained sulphide is said to carry the better gold values. As in the Pacific ore, the silver is said to have been difficult of recovery, having been largely carried off in a fine slime.

The easternmost vein is as large as the central or main vein. It has received a little development and is said to show some fair ore. Alongside the middle vein or as veinlets that come into it are streaks rich in zinc but low in gold. Where present as separate veinlets, these carry quartz, calcite, pyrite, chalcopyrite, and galena, in addition to the predominant dark-brown sphalerite. Plate III shows a 4-inch veinlet of this sort with the narrow band of altered granodiorite on each side. As is characteristic of many of the veins of this district, quartz lines both walls and extends as perfectly terminated prisms out into the central band of ore minerals.

CLEVELAND GROUP.

The Cleveland claims, where ores of the replacement type occur, lie on the upper western slope of the Pinos Altos Mountains. But little work has been done on the west side of the mountains, although it is claimed that soon after the discovery of the district a gold mine was worked on or close to what is now the Cleveland property. In addition to surface pits and trenches the development on this property consisted in 1905 of a 250-foot crosscut tunnel and a small amount of drifting from it. The output up to that time had been 200 to 300 tons of surface ores, hauled to the smelter at Silver City.

The workings are situated mainly in limestone, a short distance west of the quartzite contact. The bedding of the limestone is much obscured, as a result of alteration, but appears to strike north-northwest to northwest, and to dip not very steeply to the southwest. Diorite porphyry cuts this limestone as dikes and less regular bodies and in places exerts strong metamorphic action on the limestone, causing the development in it of garnet, specularite, quartz, and large crystals of calcite. At other places metamorphism is insignificant; at the portal of the tunnel, for example, a 3-foot dike of diorite porphyry cuts the blue limestone and produces almost no apparent metamorphic effect.

Fractures, along some of which faulting has evidently taken place, cut the limestone and have a strike of about N. 30° E. to N. 60° E. and a dip of 50° to 75° NW. A number of these fractures contain only clayey gouge material. Others, however, have served as passageways for ore-bearing solutions and now constitute lodes of quartz, calcite, and sulphides. Most of these lodes are replacements of the limestone walls, although the boundary between ore and country rock is sharp in many places. Dark zinc blende is the most abundant sulphide and is accompanied by more or less pyrite, with some chalcopyrite and a little galena. In places zinc is less plentiful and copper there tends to be higher. Four such lodes have been cut by the tunnel; they range from 2 to 10 feet in width. One has been followed for a few feet by a drift. Movements transverse or diagonal to the lodes appear to have disturbed some of them and may make the opening of the deposits somewhat uncertain.

Some of these masses of ore reach the surface and present strong outcrops of mixed limonite and silica, carrying malachite and a little native copper, also small drusy aggregates of calamine (hydrous zinc silicate) crystals. Assays of various lots of ore are said to show the following range of values: Zinc, 13 to 45 per cent; copper, 1 to 6.8 per cent; silver, 2 to 9 ounces to the ton; gold, a trace to 0.05 ounce to the ton. Most of the ore opened seems to average about as follows: Zinc, 20 per cent; copper, 2.5 to 3 per cent; silver, 5 ounces to the ton; gold, 0.03 ounce to the ton. Whether such ore will be found capable of profitable treatment and whether the encouraging appearance of the present meager workings will be maintained on adequate development will determine whether or not this property shall take its place with the producers of the district.

OTHER PROPERTIES.

The Deep Down-Atlantic mine, to the northeast of town, is said to have drifted on a vein in granodiorite out under the capping of rhyolite, and then to have raised; the vein is said to have been cut off when the rhyolite was reached. The probable deduction from such an occurrence is that the veins of the district are older than the rhyolite; this is substantiated by the other evidence available.

The Silver Cell mine is about 2 miles south of east from Pinos Altos. Rich native silver ore was discovered here in 1891[a] and considerable excitement resulted. The total production up to 1903 was placed at $100,000. Little has been produced since then. The vein lies in diorite, a phase of the granodiorite, and strikes about northerly, with a dip of probably 75° E. It is said to be narrow, ranging in most places from 2 to 8 inches in width. The inclined shaft is said to follow the vein and to be 400 feet deep. Native silver is reported to have been present in the deepest workings, also some chloride and the sulphide, argentite. No characteristic ore was to be seen at the mine at the time of visit. The gangue appeared to be mainly dolomite, with a little chloritic and siliceous alteration of the diorite wall. No work was in progress at that time.

Among other properties are the Gopher, Hirschberger, Ohio, Mina Grande, Kept Woman, Mogul, and Aztec; the last named, working on the extension of the Pacific vein, is said to have struck the contact of granodiorite and diorite porphyry. The Tampico was pumping out its shaft on the Pacific vein, to the south of the Gillette shaft. Prospects on the west side of the mountains, worked formerly by Houston & Thomas, furnished a little oxidized ore from small irregular deposits in limestone.

PLACERS.

If one may judge from the reports current in the region, the Pinos Altos placers have contributed about equally with the lode mines to the total output of the district. Bear Creek Gulch on the north and Rich Gulch, a tributary to it heading near the Mountain Key mine, Whisky Gulch on the east, and the gulch heading near the Gillette shaft were the principal producers, from an area about 1½ miles square. The unusually heavy rains of the winter of 1904–5 had worked over the stream gravels more thoroughly than was common, and in the early part of 1905 placer mining was carried on with more vigor than for many years previously.

CHLORIDE FLAT DISTRICT.[b]

LOCATION AND HISTORY.

The Chloride Flat district lies in a narrow shallow valley of that name, about 1½ miles west-northwest of Silver City. The mines are situated in the valley bottom at an elevation of about 6,300 feet. The narrow-gage line connecting Pinos Altos with Silver City passes close to the mines on the east side of the valley.

Silver ore was discovered in this district in 1871, but active production did not begin till some years later. All the mines were practically abandoned when the price of silver fell in 1893, and but little work has been done since. The total production up to that time is variously placed at from $3,250,000[c] to $5,000,000, making this one of the largest silver camps in the Territory. Among the principal producers were the Bremen, Seventy-six, and Baltic mines.

---

[a] Jones, F. A., op. cit., p. 51.   [b] By L. C. Graton.   [c] Jones, F. A., op. cit., p. 53.

GEOLOGY AND MINERAL DEPOSITS.

The geology of this region is comparatively simple, but is of interest in that there is here present one of the few sections of lower Paleozoic rocks known in New Mexico. The geologic section through Chloride Flat is shown in figure 30 and the fossils are tabulated on following pages.

The pre-Cambrian granite, which constitutes the core of the Bear Mountain ridge and outcrops to the west, is overlain successively and in apparent conformity by Cambrian quartzite, Ordovician and Silurian limestones, Devonian shale, Carboniferous limestone, and Cretaceous shale, all of which dip to the east. The Devonian rocks consist of a lower member of black shales about 200 feet thick, overlain by 260 feet of red shales. The outcrop of the black shales is practically coincident with the valley bottom at Chloride Flat. The red shales form the sloping east side of the valley, and just at the crest is a low cliff formed by the lowest strata of the Mississippian limestones. The west slope of the valley is made up of limestone, of which the uppermost portion, with an approximate thickness of 250 feet, is Silurian, and the lower portion is Upper Ordovician. At Chloride Flat the rocks strike about N. 35° W. and dip 25° NE. These rocks are cut by dikes and sills of porphyry, which is commonly gray, with prominent phenocrysts of nearly black hornblende. The principal mineral components are hornblende, plagioclase, pyroxene, quartz, orthoclase, and biotite, about in order of importance. The quartz is confined to interstitial spaces, whereas most of the other constituents assume somewhat more regular form. The rock may be classed as granodiorite porphyry or quartz monzonite porphyry. The intrusive rock is found most commonly cutting either Devonian or Cretaceous shale, probably for reasons depending on the structural characteristics of this fissile rock. It also, however, cuts the more massive sediments. On the whole, the porphyry produces little apparent metamorphic effect on the invaded rocks. A little pyroxene and epidote are developed in places in the shales at the immediate contact and the limestone is slightly silicified near the dikes.

The following is a detailed section of the rocks shown in figure 30:

Section west of Silver City.

Character.	Thickness (feet).	Age.
1. Main porphyry mass.		
2. Dark crumbly shale, with sandy layers	530	Cretaceous.
3. Porphyry intrusion.		
4. Shale, same as 2	350	
	880?	
5. Fault, with wedge of quartzite and chert conglomerate	(?)	(?)
6. Blue crystalline limestone (Pennsylvanian)	90	Carboniferous.
7. Fine-grained blue limestone (Mississippian)	370	
	460	
8. Red fissile shale	260	Devonian.
9. Black fissile shale	205	
	465	
10. Gray or pinkish limestone, iron stained	250?	Silurian.
11. Gray or pinkish limestone, iron stained	170?	Ordovician.
12. Very cherty limestone	185	
13. Gray cherty limestone	265	
	620	
14. Dark quartzite, impregnated with iron	155	Cambrian.
15. Gray cherty limestone	55	
16. Quartzite, same as 14	40	
17. Grayish shaly seam	4	
18. Quartzite, same as 14	95	
19. Limestone, same as 15	75	
20. Quartzite, same as 14	520	
21. Limestone, same as 15	35	
22. Quartzite, same as 14	75	
23. Shaly seam	3	
24. Quartzite, same as 14; conglomeratic at base	15	
	1,072	
25. Granite.		Pre-Cambrian.

The following determinations of fossils in the Silver City section have been made. The lot numbers correspond to those in figure 30.

Lot 16.—This lot contains a number of very imperfect specimens of the siphuncular sheaths of a *Cameroceras*. Siphuncles of apparently the same type of cephalopod are characteristic of the upper part of the Arbuckle limestone in Oklahoma and in the middle portion of the Pogonip limestone of Nevada. The bed in which they occur is regarded as of Beekmantown age.—E. O. Ulrich.

Lot 17.—
 Rafinesquina cf. alternata.
 Plectambonites sericeus (Richmond variety).
 Dalmanella tersa.
 Dinorthis subquadrata var.
 Platystrophia acutilirata.
 Rhynchotrema capax.
This is a typical Richmond fauna.—E. O. Ulrich.

Lot 18.—
 Favosites aspera.
 Stromatocerium pustulosum var.—E. O. Ulrich.

Lot. 18-A.—On ridge back of Silver City court-house. Close to section (fig. 30).
 Columnaria halli.
 Favosites aspera.
 Calapœcia canadensis.
 Stromatocerium pustulosum var.—E. O. Ulrich.

Lot. 19.—This lot contains separated valves of three apparently undescribed types of pentameroid brachiopods. The most abundant of these is a rather large plicated shell, the separated ventral valve of which recalls certain species of *Conchidium*. The dorsal valve being the more convex, however, making the shell one of the "reversed" type, it can not belong to that genus. The strong umbonal convexity of the dorsal valve suggests the peculiar Silurian genus *Capellinia*, but the more pronounced surface plication and the well-developed spondylium in the dorsal valve of our shell forbid its reference to that genus. Considering all its features the shell suggests closer relations to *Parastrophia* and *Anastrophia* than to the "nonreversed" true pentameroids. Of the latter only the "galeatiform" pentameroids (all Silurian and later types) present any striking resemblances to the Silver City shell, and among these, again, *Clorinda* (*Barrandella* H. & C.), having the sinus in the ventral valve, is nearer than *Sieberella*, in which the sinus is on the dorsal valve.

In my judgment the shell under consideration is a derivative from some late Ordovician *Parastrophia*, like *P. divergens*, the greater development of the umbonal region and the relative increase in length of shell being in line with the general trend of the evolution of the whole family. Hence, in my opinion, its stage of development is indicative of distinctly Silurian rather than late Ordovician (Richmond) age. Further, I think it highly probable that the bed from which Mr. Graton procured the specimens represents essentially the same Silurian horizon as that from which Mr. Gordon collected the corals referred to on page 227.

The second species is represented by two ventral valves of, apparently, a true *Pentamerus*, with subcircular and more evenly convex valves than in *P. oblongus*.

The third shell is represented by four very imperfect valves, with smooth exterior, well-developed area, and sessile spondylium. It may fall under *Stricklandinia*, but I suspect a better suite of specimens will show it to belong to an undescribed genus.

The geologist may refer to the fossils in this lot as three undescribed species of pentameroid brachiopods in developmental stages indicating early Silurian rather than latest Ordovician.—E. O. Ulrich.

Lot 22.—
 Menophyllum sp.
 Productus sp. indet.
 Productus sp. indet.

Lot 23.—
 Fusulina aff. F. cylindrica.
 Zaphrentis sp.
 Stenopora sp.
 Rhombopora sp.
 Derbya? sp.

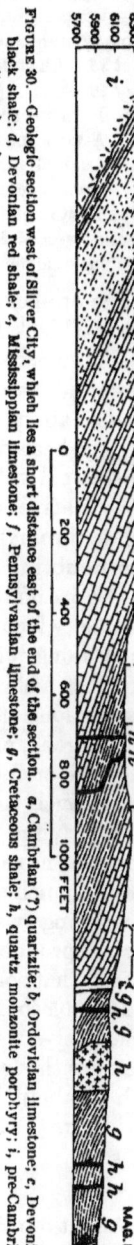

FIGURE 30.—Geologic section west of Silver City, which lies a short distance east of the end of the section. *a*, Cambrian (?) quartzite; *b*, Ordovician limestone; *c*, Devonian black shale; *d*, Devonian red shale; *e*, Mississippian limestone; *f*, Pennsylvanian limestone; *g*, Cretaceous shale; *h*, quartz monzonite porphyry; *i*, pre-Cambrian granite and gneiss.

Lot 23—Continued.
    Spirifer rockymontanus.
    Spirifer cameratus.
    Composita subtilita.
    Aviculipecten sp.
    Myalina swallowi.
    Conocardium sp.
Lots 22 and 23 represent the Mississippian.—G. H. Girty.
Lot 25.—Chætetes radians. This lot represents the Pennsylvanian.—G. H. Girty.
Lot 26.—
    Plicatula arenaria Meek.
    Astarte sp.
    Lucina? sp.
    Prionocyclus? sp. Imprint of a small ammonite.

The most abundant species is the *Astarte*, which is the same species that was collected by Mr. Lindgren in his lot 2, 1½ miles west of Morenci, Ariz., in 1902. The small ammonite, which is either a *Prionocyclus* or a *Prionotropis*, is sufficient evidence that the horizon is Cretaceous and within the Benton group.—T. W. Stanton.

The ore-bearing horizon is the uppermost part of the Silurian limestone, immediately underneath the black Devonian shale. Where ore occurs the limestone was much altered and such of the calcium carbonate as was not converted into crystals of calcite was replaced by minerals now represented by quartz, galena, argentite (?), and various oxidized compounds of lead and silver, as well as abundant hematite and pyrolusite and some magnetite and limonite. Other minerals may have been present in the large mines now inaccessible. Silver was the metal produced and the principal values were carried by silver chloride, from which the district derived its name; also by native silver and by a mineral which as described corresponds to argentite. Silver bromide (bromyrite) and silver iodite (iodyrite) are also said to have been present, but this seems rather doubtful. Lead was not plentiful enough to be of value.

The ore bodies are said to have been extremely irregular. Many of them were directly below the shale, which formed their roof, but by devious narrow shoots or streaks of ore that were formed along joints, bedding planes, or other fractures in the limestone the miner was led to pockets at a slightly deeper horizon. Although not occurring directly alongside the porphyry dikes for long stretches, the ore bodies are nevertheless undoubtedly related genetically to the intrusive rock, and seem to have been formed not very far away from the dikes. The localization of the ore bodies just below the shale may doubtless be attributed to the effect which that comparatively impervious rock had in collecting and confining the solutions in the upper part of the underlying limestone, thus causing precipitation at that horizon. Little exploration appears to have been done aside from the following of ore-bearing leads. In the aggregate, however, much underground work was done in the vicinity of the shale and limestone contact. The workings were not carried to great depth along the dip. The Seventy-six mine, one of the principal producers, was worked through an inclined shaft said to be 180 feet deep, although it is stated that some of the workings east of the shaft extended somewhat deeper.

In 1905 some excitement was created by a strike made on the Grand Central claim between the Seventy-six and the Baltic. A Mexican lessee named Manuel Taylor sunk a shaft through the shale near the locality where it wedges out and is succeeded on the surface by limestone. At a depth of about 35 feet he reached the limestone and just below the contact struck very rich chloride ore. As a result of this strike other prospects were being opened in the fall of 1905.

FLEMING CAMP.

About 5 miles northwest of Silver City, at a camp called Fleming, silver mining was carried on up to 1893. Some work was done on the slopes of Bear Mountain, the highest point in the ridge west of Silver City; also on Treasure Mountain, a ridge of northwest trend lying about 2½ miles southwest of Bear Mountain. One of the principal producers was the Old Man mine, at the south end of Treasure Mountain. It was discovered in 1882, and was worked as steadily as the Apaches would permit until about 1888. From then until 1893 it was worked at intervals. In 1905 one man was at work in the mine.

The ridge west of Silver City, which may be designated Bear Mountain Ridge, is composed mainly of sedimentary rocks, but an upthrust strip of pre-Cambrian granite and gneiss is exposed throughout much of the length of the ridge. This band of pre-Cambrian rocks does not exactly coincide with the axis of the ridge that resulted from its elevation. The ridge trends north-northwest, whereas the granite and gneiss outcrop strikes practically north. In consequence the pre-Cambrian rocks occupy the crest of the ridge for only a short distance near its middle; eastward-dipping sediments overlying granite form the crest in the southern part of the ridge and strata of westerly dip constitute the highest part of the ridge in its northern part. The camp of Fleming is situated on the west side of this northern part of the Bear Mountain Ridge, where westward-dipping sediments prevail.

The country rock in and about the Old Man mine on Treasure Mountain is quartzite, commonly of fine grain. In many places it is either much brecciated or else it is a conglomerate holding angular pebbles. It has a reddish-gray or reddish-brown color, and in some places shows closely spaced narrow red bands in the gray mass of the rock. It resembles in general appearance the brecciated or conglomeratic quartzite above the Pennsylvanian limestone of the Silver City section, a few miles to the southeast. This quartzite is likewise underlain by limestone, but the age of the latter was not determined. At the inclined shaft of the Old Man mine the dip, though confused, appears to be rather steep to the west or slightly south of west. Toward the north the strike changes from northerly to northwesterly, and even almost to west, with accompanying change in direction of dip from west to south.

The Old Man mine is developed by an inclined shaft said to be about 400 feet long, dipping about 45° to 50° W. The 75-foot level (about 50 feet vertically below the surface) runs northwestward and emerges on the slope of the hill. It is stated that no ore of importance was encountered below this level. From this level to the surface are many irregular pockety stopes which appear to lie mostly in a certain few beds of quartzite. A small chamber, about 20 by 20 feet close to the surface, is said to have furnished $40,000. This was extracted by open cutting. The ore is said to have been mostly silver chloride with some native silver and some argentite. None of the typical ore was seen. The quartzite near the old stopes is traversed by numerous drusy veinlets of quartz and carries finely disseminated pyrite in places, but the value of such material was not ascertained. The total production of the mine is reported as between $200,000 and $300,000.

Other developments in this locality were not visited, so it is not possible to state whether the occurrence at the Old Man mine is typical for this camp. The Pauline mine, higher up on the northwest side of Treasure Mountain, is said to have encountered good values in a quartz fissure cutting granite.

SANTA RITA AND HANOVER DISTRICTS.[a]

LOCATION AND HISTORY.

The Santa Rita and Hanover districts are situated in eastern Grant County. They are included in the territory known officially as the Central mining district, but as they are distinct geologically from the region about the town of Central, they are considered independently from it. The Santa Rita and the Hanover camps adjoin and show similar geologic features; they are therefore considered together. The division line between the districts is the low divide between Santa Rita and Hanover creeks, streams flowing respectively southwest and south, uniting just south of the districts and ultimately reaching Mimbres River. The average elevation is about 6,500 feet. The flat or rolling country which continues to the west is bounded at the north and south of the districts by rough escarpments of rhyolitic lavas, and on the east by the serrated edge of the limestone mesa that extends in this direction toward Mimbres River as far as the mining camp of Georgetown.

---

[a] By L. C. Graton.

The Santa Rita district is one of the oldest mining districts in the West, and ranks next to the Lake Superior district of Michigan as the earliest known important copper district within the territory now included in the United States. According to a pamphlet published in 1891 by J. Parker Whitney, then president of the company owning the Santa Rita mines, croppings of native copper were discovered by an Apache Indian in the latter part of the eighteenth century. The discovery was reported to Colonel Carrasco, an officer of the Spanish army, who was stationed near by. Through him was enlisted the assistance of a wealthy merchant and banker named Elguea, who lived in the city of Chihuahua. The mine proved to be profitable and Elguea soon bought the interest of Carrasco. A contract was made to supply copper for the Mexican coinage and all the copper produced was so used, amounting, it is said, to as much as 20,000 mule loads, or probably 4,000,000 pounds annually. The withdrawal of the Mexican troops early in the nineteenth century and the depredations of the Indians caused the mine to be worked only intermittently till almost the close of the century; but when worked good profits were generally realized.

In 1897 the Hearst estate took a lease on the property in order to furnish sulphide ores to its smelter at Silver City while a supply was being developed in its Pinos Altos mines. A considerable amount of rich ore was extracted. In 1899 the property was sold to the Santa Rita Mining Company, which is controlled by interests allied with the Amalgamated Copper Company. Of recent years, as in the early days, much of the mining has been done by Mexican lessees. Accurate records of production do not, of course, exist. It is stated by Mr. Whitney, in the pamphlet already referred to, that the most important mine was credited with a production of over $4,000,000 up to 1839. F. A. Jones,[a] who evidently gave considerable attention to the history of the district, estimated the total copper production from the time of discovery up to January 1, 1904, at 80,000,000 pounds. Since that time the annual production has averaged about 5,000,000 pounds, and it is known that the production for certain earlier years exceeded that amount.

The Hanover mine, which lies about 3½ miles north of the Santa Rita camp, was probably likewise known to the Indians at an early date, but record of its operation extends back only to about the middle of the nineteenth century. At that time it was said to be producing more than the Santa Rita mines. It was closed shortly after the beginning of the civil war, and for many years remained inactive. In 1902 the mine was purchased by Phelps, Dodge & Co., together with other property in the vicinity. (See Pl. VIII, B, p. 184.)

The iron deposits of Fierro, really part of the Hanover district, early attracted attention, and by 1891 had been reached by the railroad. The Colorado Fuel and Iron Company is still shipping several carloads daily from this place.

The other important operator in the region, the Hermosa Copper Company, began extensive prospecting and development operations about 1904. It is understood to be a subsidiary of the General Electric Company.

Smelting has been attempted in the district at the Santa Rita camp and at the Ivanhoe mine. There is now an old reverberatory furnace in the equipment of the Santa Rita Copper Company. On the whole, smelting in the district has been unsuccessful.

GENERAL GEOLOGY.

THE ROCKS.

The geology of the Santa Rita and Hanover districts is varied and complex. The rocks comprise representatives of sedimentary, metamorphic, intrusive, and effusive types. The sedimentary rocks consist chiefly of limestone and quartzite, though layers of shale occur interbedded with the limestone. The only limestones occurring in the immediate vicinity of the mines whose age could be established proved to be Carboniferous; both Pennsylvanian and Mississippian fossils were found. In the section at Georgetown, 4 miles northeast of Santa

---

[a] Op. cit., p. 38.

Rita, Devonian shales (and possibly some limestones) and Ordovician limestones occur. Fossils said to have been taken from limestone between Santa Rita and Georgetown were seen, and one of them was identified as Devonian by E. M. Kindle, of the United States Geological Survey. Some of the shaly limestone in the southwestern part of the district, for example at the Ivanhoe mine, suggests the Cretaceous in its general aspect, but fossils regarded by George H. Girty as probably Mississippian (lower Carboniferous) were found in it. The limestones that probably lie lowest in the column have almost everywhere suffered metamorphism which probably would have destroyed fossil remains. This metamorphism will be considered later. The limestone series has a thickness of at least 600 to 700 feet, to judge from the exposures on Hanover Creek. This thickness probably represents only a portion of the total thickness of the series.

The most conspicuous quartzite is found in the heart of the developed area in the Santa Rita camp and immediately to west and south. What is regarded as the same quartzite is present in places along the west side of Hanover Creek and in the Hanover mine and is said to extend on the surface for some distance farther northwest, though this statement was not verified. The rock is rather fine grained, gritty, decidedly cemented by silica, and commonly almost white, though in places it is almost black owing to the presence of black iron ore in small grains. The quartzite probably has a thickness of at least 200 feet; it occurs on the surface at the Romero mine and is present in a crosscut from the Santa Rita shaft 200 feet below; the intervening ground is inaccessible and it is not certain that the quartzite extends throughout that distance, as irregular intrusions of porphyry are numerous in the vicinity. The bottom of the quartzite is somewhere above the 300-foot level, for the rock is not seen at that depth. The quartzite along Hanover Creek underlies and apparently belongs at a lower horizon than the limestone; the quartzite at Romero Hill is believed to occupy a similar stratigraphic position[a] and in the absence of any other known quartzite below the Carboniferous is regarded probably as Cambrian and the equivalent of the thick quartzite series at the base of the Silver City section, 15 miles to the west. This view is strengthened by the similar character of these rocks in the two regions.[b]

Near the Ivanhoe mine, in the southwest corner of the area, quartzite is found close to the limestone already mentioned as probably Mississippian, and though the actual relations of the two were not ascertained, it is believed that the quartzite overlies the limestone. Much of this quartzite is gray, of finer grain and less cemented by silica than the quartzite previously described; it is conglomeratic and cherty in places like that occurring between the Carboniferous and the Cretaceous in the Silver City section. Quartzite believed to be related to this occurs on the Pedro claim on the hill east of the Santa Rita basin. A small thickness of quartzite is said to cap the high mesa east of and overlooking Hanover Creek, and quartzite has been reported recently from the hills west of Hanover Creek and below the town of Hanover. These four occurrences of quartzite are considered by the writer as presumably parts of one formation, which is to be regarded as of later age than the quartzite exposed in and around the principal Santa Rita mines and in the Hanover mine. Whether or not this younger quartzite formation belongs at the horizon of the similar material which is found at Pinos Altos, Silver City, Bear Mountain, and Telegraph, and which in the Silver City section occurs between the Pennsylvanian and the Cretaceous, can not be definitely stated, but such a position for it seems probable. This view is not in accord with those contained in private reports on the region made both before and after the present writer's visit to this district, and it must be admitted that the conclusions as to stratigraphic sequence here set forth are not regarded as thoroughly established. Further consideration will be given to this question on pages 309–311.

The intrusive rocks comprise two porphyries. The most important is granitoid in appearance and where fresh is of light-gray or greenish-gray color; locally it is called granite. The most noticeable constituents are dark biotite and some hornblende, white or pale-green feldspar, and quartz. In thin section the phenocrysts of feldspar are found to be mainly plagioclase, ranging from andesine to labradorite in composition, with some orthoclase; in addition to the larger

---

[a] See Mineral Resources U. S. for 1906, U. S. Geol. Survey, 1907, p. 402. [b] Compare description under Chloride Flat district, p. 302.

individuals of biotite and quartz there are in some specimens well-formed phenocrysts of green hornblende; titanite occurs in good-sized grains. The microgranular groundmass consists chiefly of quartz and orthoclase, with a little plagioclase, grains of magnetite, and small crystals of apatite, rutile, and zircon. Micropegmatitic structure is seen at many places in the groundmass. The common result of alteration is the kaolinization of the feldspar and the conversion of the hornblende and finally of the biotite into chlorite, epidote, and carbonate. Especially where weathered the resemblance to granite is pronounced.

The rock may be classed as a quartz monzonite porphyry. It is doubtless related to the granodiorite of Pinos Altos and to the granodiorite porphyry or quartz monzonite porphyry of the Silver City region. A partial analysis of a specimen from the 300-foot level of the Santa Rita mine, by W. T. Schaller, shows its most essential chemical features as follows: SiO_2, 65.15; CaO, 1.96; K_2O, 5.52; Na_2O, 2.81.

This rock has domed up the sediments and then been exposed in several places by their erosion. Its chief exposures are in the bottom of the Hanover Valley and in the so-called basin at Santa Rita. It also cuts the sediments as dikes and as apparently irregular intrusive projections from the main body. In places along its contact the rock is somewhat finer grained and apparently richer in silica. Near the limestone it is locally much epidotized. It has caused profound alteration of the limestones in its neighborhood and is to be regarded as the original cause of mineral deposition in the district.

The other intrusive rock is finer grained and darker and shows more plainly its porphyritic texture. In a rather dark gray groundmass it holds small rectangular feldspar phenocrysts, and in many specimens countless small needles of hornblende with axes lying in a common direction. The microscope shows that the feldspar phenocrysts are mostly plagioclase and that a few flakes of biotite and rounded grains of quartz are present, as well as apatite, zircon, and magnetite in usual development. The groundmass is cryptocrystalline; it is made up chiefly of feldspar, but whether plagioclase or orthoclase can not be determined optically; presumably the feldspar is plagioclase, in which case the rock would be a diorite porphyry. This rock was observed only as narrow dikes cutting the quartz monzonite porphyry and the sediments. Some of the dikes are much sheeted parallel to their strike and on weathering have a schistose appearance.

Both these porphyries undoubtedly belong in the post-Cretaceous intrusive group.

The effusive rock is represented by rhyolite of reddish or pinkish color. It forms a mass of considerable size immediately southeast of Santa Rita and is undoubtedly a remnant of the great flow that occurs to the north. The scarp bounding its upper edge is a prominent feature of the landscape; where it terminates on the north, a narrow erosion remnant standing upright a short distance from the cliff has been called the "Kneeling Nun," from a fancied resemblance to a woman kneeling before an altar. This landmark, visible for several miles, serves a useful purpose in the description of locations. The rhyolite is the youngest rock in the district, and is doubtless of Tertiary age.

The metamorphic rocks are of two kinds—sediments altered by the intrusion of porphyry into them, that is, contact-metamorphic rocks, and old schists that presumably represent the foundation of the sedimentary series and were probably formed by regional metamorphism. The limestones have, of course, been most affected by the contact metamorphism, though the included layers of shale have in some places been indurated and altered in mineralogical and probably chemical composition. The quartzite is little, if any, affected. The metamorphic zones in the limestone are similar to those now known in many other localities, but the action here seems to have been very intense and the accession of material from the magmatic agents must have been great. Beds of almost solid garnet, epidote, or pyroxene (diopside?) are found here and there; at other points mixtures of these minerals with calcite and some quartz occur. A fibrous or bladed pyroxene, which on weathering assumes a bronzy luster and is regarded as probably hedenbergite, is formed in places. Magnetite, pyrite, and zinc blende are locally developed in large quantities. The garnet and magnetite especially form great masses that outcrop strongly. Magnetite is the essential component of most of the iron ore that has been mined in the region. Bodies of the zinc sulphide have also been worked to some extent. The pyrite carries a little copper as chalcopyrite, and it is probable that the greater part of the copper that has been mined from this district existed initially in this disseminated state. It is evident, therefore, that this contact metamorphism of the limestone was of fundamental importance in the formation of mineral deposits. The limestone is in general most altered at and near the quartz monzonite porphyry contact, but metamorphism has extended through at least 200 or 300 feet of the limestone strata and laterally for much greater distances. As is commonly observed elsewhere, the composition and physical character of the individual beds have largely determined the amount of alteration, and in consequence the degree of metamorphism does not gradually diminish away from the intrusive rock but varies according to the position of the individual beds. In places near the intrusive rock and generally at some distance the principal change is simply a recrystallization of the limestone. It appears that where quartzite was present between the limestone and the porphyry much less alteration was produced in the limestone than where the molten rock came into actual contact with it.

The richness in iron of these metamorphic products is well shown by the deep rusty color which the weathered surface almost everywhere shows. The garnet, hedenbergite, and pyrite, as well as the magnetite to some extent, are converted into limonite. The pyroxene alters chiefly to serpentine, which perhaps is the material that has been found occasionally in the Jim Fair mine and elsewhere and called asbestos. Epidote appears to be the most stable of the metamorphic minerals, but it, too, finally yields and leaves a mass of limonite.

Schist was found at only two places—the Santa Rita mine in the center of the Santa Rita camp and the Modoc tunnel, near the north end of the Hanover district. The rock is very fine grained and either very dark brown or dark green, and where fresh shows the glistening surface of minute biotite flakes. Stringers and accretions of magnetite are common. At places in the Santa Rita mass are numberless tiny stringers of quartz. The microscope reveals the usual mosaic texture of the crystalline schists. Biotite is most abundant, followed by quartz and a feldspar that is not certainly determinable, but appears from its refractive index to be orthoclase. Magnetite, apatite, and zircon are present as accessories. Alteration produces serpentine abundantly, giving the rock a yellowish-green color, and as much of the rock exposed is somewhat altered it has received the local name "greenstone." The rock in the Santa Rita workings is similar to but a little coarser than that in the Modoc tunnel, and contains more magnetite. The supposition entertained at the time of its collection, that it might possibly be shale much metamorphosed by the near-by porphyry, was disposed of when microscopic study could be made. The rock is plainly the result of deep-seated regional metamorphism and is almost certainly pre-Cambrian. At both places the schist appears to occur as irregular bodies surrounded by the porphyry, as if torn from the parent mass and carried off in the porphyry melt. The mass in the Santa Rita mine is encountered on the 300-foot level and is at least several hundred feet in diameter. It is present under the quartzite found on the 200-foot level and is probably to be regarded as the pre-Cambrian basement on which the sediments were deposited. The mass in the Modoc tunnel is more clearly a detached piece; so far as exposed it was surrounded by quartz monzonite porphyry and its relation to the limestone, which outcrops on the hill above, was not ascertained.

STRUCTURE AND SEQUENCE.

Many of the facts regarding the structure and the attitude of the rocks have already been implied, and the question regarding sequence has been mentioned. The view that the quartzite found at the Santa Rita mines and at the Hanover mine lies below the limestones and is of Cambrian age is supported by the similarity of the rock to the known Cambrian quartzite west of Silver City and by the presence, probably immediately underlying it, of schist that must be regarded as pre-Cambrian. The conclusion that the limestone formerly overlay the ore-bearing rocks at Santa Rita had already been reached by S. F. Emmons. He says:[a] "The ore is found in a white quartz porphyry,[b] which was evidently once covered by horizontally bedded limestones, that still lie round the rim of the shallow basin where it occurs."

Since 1905, when the region was visited, a considerable amount of private geologic work has been done in the Santa Rita camp; the results of a part of these investigations have been published;[c] and recently the writer has been shown a private geologic report on the Santa Rita district made for the Santa Rita Mining Company in 1899. In both of these reports it is assumed that only one quartzite horizon is present in the region and that the quartzite in the heart of the Santa Rita camp is a down-faulted block. This view has been clearly presented in the older report, in which the opinion is expressed that the Carboniferous limestones underlie the whole Santa Rita property and that the quartzite lies above the limestones, separated from them by 150 to 280 feet of "iron formation." Concerning the relations of the limestone and quartzite in the heart of the Santa Rita camp the report says: "The change from the * * * limestone country on the north and east to the more diversified geology of Santa Rita [quartzite, 'iron formation,' porphyries, and rhyolite] takes place abruptly along a ragged line * * *. Along this line in various places both the iron formation and the quartzites are brought into lateral juxtaposition with the limestones under such circumstances that their present relations can be explained only by supposing that the rocks south of this line have dropped, relatively to the main limestone block, to the extent of from 400 to 600 feet." This reference to views not according with his own is made by the present writer because he feels that his own observations were not sufficiently detailed and extensive to establish his deductions conclusively, and it seems only fair to present all aspects of the question. The actual discovery and exact location of lines of faulting appear not to be mentioned in either of the private reports referred to, and, in both, the evidence of faulting seems to rest mainly on the assumption that the limestone is the lowest member of the column. Extending from the Kneeling Nun in a direction slightly west of north is a comparatively steep, serrated slope that forms the western boundary of the limestone mesa. This slope is to be regarded not as a fault scarp modified by erosion, but as the steep side of an anticlinal valley (Hanover Creek) formed by erosion of the elongated dome resulting from the intrusion of the quartz monzonite porphyry. The opposite or west wall of this valley is of gentler slope, but plainly exposes the same rocks forming the other limb of the anticline. This anticlinal structure is unmistakably evident through almost the entire length of the Hanover Valley. Santa Rita Creek has cut back into this steep east wall and formed a depression or pass, through which leads the road to Georgetown. The limestone recedes up Santa Rita Valley, exposing quartzite (and porphyry), the boundary line of the limestone tending to conform with the valley contours and the dip being away from the quartzite. It is evidently this situation that Emmons refers to when he says[d] that the limestones "lie around the rim of the shallow basin." This embayment in the limestone and the corresponding tongue of quartzite occupying the depression caused by Santa Rita Creek afford, perhaps, the best evidence in favor of the higher stratigraphic position of the limestone relative to the quartzite. The complicating feature of the "iron formation" in the private report referred to is now, of course, explained as the zone of contact-metamorphosed limestone, a type of deposit whose nature ten years ago was seldom recognized and poorly understood.

It seems probable, therefore, that the quartzite in the central part of the Santa Rita camp underlies the limestone. Nevertheless, the position in which it is found requires some explanation. If the quartzite is Cambrian, as seems probable, it lies of course at the base of the sedimentary series. But the full series is not present in the Santa Rita and Hanover districts, as is shown by sections at Georgetown, Lone Mountain and Silver City, nor is the quartzite present consistently between the limestone and the porphyry. The explanation of the presence of Cambrian and pre-Cambrian rocks in the Santa Rita camp and in the vicinity of the Hanover and Modoc mines in the Hanover district is believed to be that masses of these rocks have been

---

[a] Genesis of ore deposits, New York, 1902, pp. 449-450.
[b] The quartzite was apparently not separated from the altered porphyry that invades it promiscuously.
[c] Sully, John M., Report on the property of the Santa Rita Mining Company, Boston, 1909.
[d] Emmons, S. F., op. cit., p. 449.

torn off by the quartz monzonite porphyry magma and forced with it up through part of the overlying sediments to the abnormal stratigraphic position they now occupy. It may be that some of the lower limestones have likewise been brought up to a higher level and, being close to the magma, have been metamorphosed too much for determination. It may thus be that some of the shale exposed is from the underlying Devonian. If, therefore, as has been asserted by others, faults surround or bound this Cambrian quartzite mass, together with the pre-Cambrian schists and any older Paleozoic rocks that may have been forced up with it, the quartzite and adjacent rocks have probably been relatively raised rather than depressed along such faults.

Faults of minor importance are seen at several points in the district. The boundary between the calcareous shale or shaly limestone and porphyry in the Ivanhoe mine is evidently a fault plane, but the metamorphism which the sedimentary rock has undergone would indicate that the faulting took place at the time of the intrusion.

On the west side of Hanover Creek there is a double band of contact-metamorphosed limestone separated by a strip of porphyry. The question arises whether this may not be a repetition due to faulting. Careful study would doubtless settle the question, but definite evidence was not gained in the brief examination made. The contact between the upper or western porphyry band and the lower limestone mass was obscured, and it could not be determined whether that was a contact of intrusion or of faulting. From certain indications, however, it may be assumed that the upper porphyry band is intrusive—that is, a thick sill. In the first place, the lower band of altered limestone on the west side—usually referred to as the "first contact"— shows more intense metamorphism than the rock in corresponding relation to the porphyry on the east side of the creek. This might be explained if the limestone on the west side had been affected by an upper as well as a lower mass of the molten porphyry. Furthermore, on the east side there is also a "second contact" lying some distance above the lower or main zone of alteration, but here the band of porphyry is absent. The hypothetical sketch (fig. 31) may possibly explain the cause of these occurrences.

The quartz monzonite porphyry is much jointed, especially in two directions approximately vertical and at right angles. The northeast system is somewhat more pronounced than the northwest. Many of the diorite porphyry dikes follow one or the other of these joint systems, and have themselves been sheeted by later adjustment along the same plane.

ORE DEPOSITS.

GENERAL FEATURES.

The ore deposits of the Santa Rita and Hanover districts may be divided on the basis of origin into three groups—vein deposits in the porphyry; contact-metamorphic deposits occurring at or near the contact of limestone and porphyry; and concentrations as the result of oxidation and transportation—deposits of secondary enrichment. Of these the second group is most important as a source of iron and zinc and the third group is the principal producer of copper. Sharp lines of distinction can not be drawn in all cases.

VEINS.

The veins in quartz monzonite porphyry are fairly numerous, but have not been of much importance in the matter of production. The principal ones have been opened in the Hanover district by the Hermosa Copper Company. Nearly all the veins are but a few inches wide and represent both filling of narrow fissures and replacement of the porphyry through a zone comparable in thickness to the width of the open fracture. Sericite and chlorite result from the alteration of the wall rock. Quartz is the gangue, but the greater part of the vein is commonly made up of the metallic minerals. Of these pyrite and chalcopyrite are about equally important, with zinc blende generally less abundant and specularite in sparing amount. In some places, particularly in veinlets between two diorite porphyry dikes on the Honey Comb claim, molybdenite is present in clusters of small folia. In this Honey Comb material the quartz has insinuated itself through the porphyry so that the result suggests a rather coarse-grained pegmatite. It is said that these narrow streaks carry low values in gold and silver in addition to the copper. Exploration had been carried to a depth of 120 feet. At the Tourmaline claim, lying just north of the Honey Comb and south of the town of Fierro, on the east side of Hanover Creek, a shaft had been sunk 235 feet in porphyry. East-west stringers have been cut that carry heavy pyrite and chalcopyrite, with quartz crystals in druses. These stringers have been slightly faulted along a plane striking a little east of north. On the Mabel claim, about a mile south of the Tourmaline, a shaft was sunk 135 feet on a northeasterly sheeted zone in porphyry. This streak carried some pyrite and zinc blende, with low values in silver. The Wild Cat mine lies about a quarter of a mile south of Santa Rita Creek and about 1½ miles west of the Kneeling Nun. The mine was opened by a 410-foot vertical shaft, and was equipped with a concentrating mill which had been used on ore found in the upper levels. The later workings are from the bottom of the shaft, which is sunk throughout in porphyry. The mass of rock is altered, sericite being formed plentifully. Pyrite, with a small proportion of chalcopyrite, occurs in streaks and veinlets, also more or less disseminated through the rock. Specularite rosettes occur in little nests surrounded by much sericitized porphyry, shot through with small scales of the specular iron oxide. Values are said to average about 1 per cent copper and $1 to the ton in gold. In places the massive pyrite has been altered to glance. An interesting feature in this connection is that, in veinlets of pyrite that have been attacked by cupriferous solutions

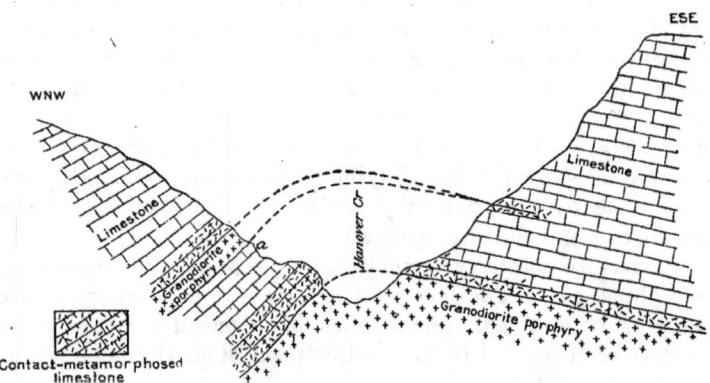

FIGURE 31.—Diagrammatic section across Hanover Creek, Grant County, showing relation of contact metamorphism to the porphyry. a, Relations not plain; no metamorphism actually observed at this point.

and partly converted into chalcocite, a fibrous or columnar structure, transverse to the walls of the veinlet, has been developed in the pyrite.

It is stated that some northeasterly veins in the porphyry, not far from the Ivanhoe mine, which have been prospected for copper, stop at the rhyolite capping.

Closely related to the veins in origin are the stockworks and irregular disseminations of pyrite, with specularite in places, in the porphyry and some of the adjoining rocks in the Santa Rita mines. Although the values found have not been sufficient to make these deposits important in their original state, later addition of copper from above has resulted in places in the formation of valuable ore bodies.

CONTACT-METAMORPHIC DEPOSITS.

The contact-metamorphic deposits have been broadly described in the section devoted to general geology. Those that have not been much modified by oxidation yield principally iron or zinc; copper is present as a rule only in sufficient quantity to constitute at the best but low-grade ore.

The iron deposits that have been chiefly worked lie on both sides of Hanover Creek not far from the porphyry contact.[a] (See fig. 31.) The principal operations have been at the Republic or Union workings of the Colorado Fuel and Iron Company on the west side of the

---

[a] A report on the Hanover iron deposits by Sidney Paige, based on more recent work, has recently been published by the Geological Survey in Bulletin 380, pp. 199-214.

creek about halfway between Hanover and Fierro. The iron ore outcrops in bold masses as magnetite mixed with a little limonite. It has been worked mostly in open cuts by quarrying. (See Pl. VIII, B, p. 190.) The ore is dropped down through chutes to underground workings and is hauled by electric motors through tunnels to the surface. A considerable amount of underground stoping has also been done. At the time of visit 350 tons daily were being shipped to Pueblo. Everything in the vicinity is so impregnated or stained with iron that the exact geologic relations are not evident. Porphyry lies on the east and a somewhat shaly limestone on the west. The dip of the iron-bearing zone must be pretty steep to the west. "The ore bodies are in the main irregular lenticular masses of magnetite; where exposed at the northern and southern pits of the mine they are decidedly long as compared with their width and depth. Besides these large bodies numerous outcrops may be observed, indicating the same mode of occurrence; that is, lentils of varying width and thickness, swelling and pinching in horizontal extension and, it may be safely presumed, presenting similar though uncertain irregularities in depth."[a] Large bodies have averaged from 60 to 70 per cent metallic iron, and are almost pure magnetite. In places the ore contains pyrite and chalcopyrite and these may carry up to 2 per cent copper. Such portions are left as horses, as the sulphur content is too high. In the workings near the surface the sulphides were oxidized to limonite and could be mined with the magnetite ore. Horses of more or less metamorphosed limestone are found here and there; at the edges of such limestone masses there is a gradual decrease of the amount of magnetite. Small bunches of garnet or of epidote are present in the ore, and a little serpentine-like material occurs all through the magnetite; probably it is an alteration product of one of the contact silicates. Much of the magnetite has a bronzy luster and a structure that simulates cleavage; in reality the magnetite appears to be pesudomorphic after coarsely crystalline calcite. Plates of specularite are seen rarely. According to analyses cited by Paige the average ore contains from 60 to 65 per cent of iron; the phosphorus is rarely above 0.07 per cent and often much less; the sulphur is about 0.02 per cent, though much higher in some parts of the ore body. There is no titanium and only small amounts of lime, magnesia, and manganese. A similar deposit has been worked at the Jim Fair mine on the east side of Hanover Creek just above Fierro, where a short, comparatively thick lens of magnetite is exposed at the contact. Another occurrence is in the vicinity of the Booth mine, near the terminus of the railroad in the Santa Rita camp, where the magnetite was evidently mixed with a considerable proportion of sulphides, which are now oxidized. Copper stains are common in this material, and it is possible that an appreciable amount of copper has been leached from the original sulphides and deposited elsewhere. It is believed that deposits of this nature, perhaps even a southern and eastern continuation of this very deposit, now eroded away, furnished most of the copper in the Santa Rita mines. At the Modoc mine, which is the property of Phelps, Dodge & Co., and lies well up on the western slope of Hanover Creek, about half a mile west of Fierro and an equal distance southwest of the Hanover mine, a zone of much metamorphosed limestone lies immediately above the quartz monzonite porphyry. The contact strikes about north and south and, to judge from the slope of an inclined shaft, dips to the west about 30°. The deposit is also reached by a tunnel, said to be about 800 feet long, that starts lower down the hill and continues in porphyry for most of its length. In this tunnel the schist described on page 309 was found. Massive yellow and reddish-brown garnet rock, with some epidote, carries magnetite, pyrite, and a little chalcopyrite; here and there a foil of molybdenite is seen. A considerable quantity of this material was shipped as flux by Phelps, Dodge & Co. to their smelter at Douglas, Ariz., but the copper values were too low to pay. Some of the material is rich in iron, closely resembling the ore of the Republic (Union) mine.

The Thundercloud (sometimes called Thunderbolt) zinc mine of the Empire Zinc Company is situated on the eastern slope of Hanover Creek just east of the town of Hanover. The deposits are in limestone close to the quartz monzonite porphyry contact. The beds here dip about 30° a little to the east of north. Bodies of almost solid sphalerite, containing a very little pyrite and galena, and associated with a bronzy mineral in radiating blades—probably

---

[a] Op. cit., p. 210.

the lime-iron pyroxene, hedenbergite—replace white, coarsely crystalline limestone. A considerable amount of ore is said to have been shipped from these bodies along the contact; part of the ore was oxidized to the carbonate, and less commonly to the silicate, calamine, but most of that is now exhausted.[a] It is stated that similar deposits of zinc ore occur near the limestone and porphyry contact on the west side of the creek a little southwest of the town of Hanover.

The Ivanhoe and Ninety mines are situated in the southern part of the Hanover district, just north of the Santa Rita branch of the railroad. The mines are thought to be on the same lode, which strikes about N. 50° E. and dips 45° to 50° SE. The lode is a combination of fissure filling and replacement along a contact between quartz monzonite porphyry and shaly limestone. The straightness of this contact across the sedimentary bedding indicates faulting, but, as already mentioned, the metamorphism of the limestone along the contact indicates that the faulting accompanied the intrusion. The porphyry lies on the southeastern or hanging-wall side and the limestone on the foot-wall side. The Ivanhoe mine was worked years ago and a smelter was erected for recovering lead and silver. A 200-foot vertical shaft gave access to the deposit, but the principal work was done in the upper workings, where oxidized material was found. It is said that one stope in the cerusite zone was 400 feet long. A new shaft, 360 feet deep in September, 1905, was being put down by the Hermosa Copper Company. It was farther from the outcrop than the old shaft, and passed through the lode at a depth of 275 feet. A crosscut to the lode from the 200-foot level of the new shaft shows a soft streak with little ore, but in the bottom workings from the old shaft, a short distance above, the lode is 15 feet wide. Pyrite probably predominates, but chalcopyrite is irregularly distributed throughout; narrow bands of fine-grained galena parallel to the plane of the lode are commonly associated with dark zinc blende. Some zincky streaks are as much as 6 inches wide. Quartz is not plentiful, and generally where it does occur it is as drusy, radiating crystals. The porphyry seems to be the most replaced, but bunches of limestone, unreplaced by ore but somewhat affected by contact metamorphism, are found in the ore, and in places the sulphides extend into the altered limestone of the foot wall. Ore was also encountered where the shaft crossed the lode. The crosscut from the second or 350-foot level had not reached the lode at the time of visit. Much of the ore is soft and oxidation is going on rapidly. The acid water coming down through the old workings reacts with the calcite of the limestone, producing gypsum, which forms in incrustations on the walls. This action is going on so rapidly that the air smells sour and good ventilation is required owing to the quantity of carbon dioxide set free. The ore is said to average 8 to 11 ounces of silver and about $1 in gold to the ton, with some copper and lead. It is reported that a concentrating mill has recently been built to treat this ore.

At the Ninety mine, a third of a mile to the northeast, a shaft being sunk by the Hermosa company had struck the lode at a depth of about 360 feet. The bottom of the shaft, at 370 feet, was just getting into material that resembled the crushed, altered limestone seen in the Ivanhoe mine. The ore consisted chiefly of chalcopyrite and quartz. Little lead or zinc was present, and pyrite was scarce. Good copper values appeared to extend over a width of several feet.

This Ivanhoe-Ninety deposit is probably to be regarded as most closely related to the veins in porphyry, but in certain respects it seems to be transitional in character between these veins and the contact-metamorphic deposits in limestone. The similarity of the ore minerals of the veins and the contact deposits is additional and better evidence that the two types of deposits are of similar origin. If those minerals whose formation was obviously due to the abundance of lime present are excluded from the contact deposits, the list then includes quartz, pyrite, chalcopyrite, sphalerite, galena, molybdenite, specularite, and magnetite. With the exception of magnetite, these minerals are the same as those present in the veins. It is probable that the source of materials forming both classes of deposits was the quartz

---

[a] For a description of the deposits worked some years ago, see Blake, W. P., Zinc-ore deposits of southwestern New Mexico: Trans. Am. Inst. Min. Eng., vol. 24, 1894, pp. 187-195.

A. SANTA RITA MINE, GRANT COUNTY.
Showing extent of dumps.

B. SYLVANITE CAMP, GRANT COUNTY.
Looking northeast across low gap in Hachita Mountains.

monzonite porphyry magma, and that whereas the deposits in limestone at the contacts were formed at the time of the porphyry intrusion, the veins were formed slightly later, after the rock had solidified and fractures had been developed.

DEPOSITS OF SECONDARY ENRICHMENT.

The copper deposits of the Santa Rita camp and the similar deposit of the Hanover mine, at the north end of Hanover Gulch, are related to either or both of the types of deposits already described, but their original character has been so modified by the results of secondary enrichment that in general it is that process and not the primary deposition that characterizes the ores. The principal deposits found at Santa Rita (see Pl. XXII, A) lie in an area about three-quarters of a mile square that is known as the Santa Rita basin and is really a local widening of the valley of Santa Rita Creek. This basin lies just west of the point where Santa Rita Creek breaks through the east wall of the anticlinal valley, as described on page 310. Quartzite and quartz monzonite porphyry constitute the floor of this basin; altered limestone and rhyolite, together with the porphyry, chiefly form the sides.

The greater part of the copper that has been mined undoubtedly existed as the native metal. It occurs mostly in the quartzite, but is also present in the porphyry, especially in portions that are much kaolinized. It is found in fractures in these rocks from the thinnest flakes up to slabs half an inch thick and several feet square. The richest ore was undoubtedly extracted in the early days and contained, according to report, much larger masses of the metal. The quartzite and altered porphyry are traversed in all directions by these cracks. Most of them are of small extent, but in many places, owing either to the distribution of the fractures or to the fact that copper is not deposited in all the fractures, there are zones or streaks of copper-bearing ground that can be followed for some distance. Rarely, specks of the native copper occur disseminated through the quartzite. In some of the softer porphyry native copper is found in irregular lumps and grains surrounded by kaolin. It is stated that near the surface large masses of native copper were found in the clayey alteration product of the porphyry. An interesting occurrence in the porphyry is as grains that have an outline corresponding to hornblende crystals and that have undoubtedly pseudomorphically replaced that mineral in the rock. These native copper pseudomorphs can often be found in the mill concentrates, and some of them have very perfect forms.

The surface or exterior portion of the pieces of native copper is coated with more or less of the red oxide, cuprite, which has been formed by direct oxidation. It is commonly crystalline, and these coatings consist in large part of numberless perfectly formed transparent cubes of beautiful color. In little cavities the capillary variety, chalcotrichite, is sometimes found. Much of the thinner plates of copper have now been completely changed into cuprite. Close to the surface some of the oxide has been converted into the green carbonate, malachite, but at moderate depths the carbonate is commonly present only as thin coatings, if at all. A little chrysocolla is found here and there in the quartzite, and azurite is present in little nests in the porphyry near the surface.

Next in importance to native copper and cuprite has been the cuprous sulphide, chalcocite, or copper glance. This mineral is confined almost entirely to the altered porphyry, but little of it having been found in the quartzite. The ore of this character that has been mined consists of stockworks and heavy impregnations of chalcocite in the kaolinized and silicified porphyry, similar to ore of that general nature found in so many districts. The presence in these ore bodies of cores and kernels of pyrite, and outside the ore bodies of pyrite in masses similar in distribution to the bodies of ore, as well as in a more disseminated state, and all more or less coated with chalcocite, makes it certain that at least a great part of the chalcocite was precipitated by and replaced pyrite.

In a few places were found crevices in the quartzite containing a central zone of glance and bordered by cuprite. Native copper could not with certainty be detected between the oxide and the sulphide, but in some of the lumps of glance occurring in porphyry films of native copper were found just within the coating of cuprite. Some of the chalcocite stringers have a fibrous structure across the veinlets similar to and possibly inherited from the structure in pyrite mentioned on page 312. A transverse fibrous structure, similar but more pronounced, is shown by many of the native copper veinlets. In some of them the fibers extend inward from each wall, meeting along an irregular line in the middle.[a] There is no doubt that much of the native copper has been derived by oxidation of the sulphur of chalcocite. It is not certain, however, that all of it was thus produced. The pseudomorphs of native copper after hornblende, for example, were probably formed by direct reduction of the copper-bearing solutions. Moreover, it is difficult to believe that oxidizing circulation could have been sufficiently complete to convert practically all the copper into native metal without having carried the oxidation to a much more advanced degree. No other explanation offers itself, however, than deposition originally as sulphide. It is certain, on the other hand, that much of the copper which now exists in the state of native metal or oxide was not precipitated on pyrite, but must have been deposited from solution in open spaces and in the porphyry as replacement along the fissures also. Whether or not a mingling of solutions containing hydrogen sulphide or some other soluble sulphide with those containing the copper caused this precipitation is unknown.

The principal bodies of native copper have been found in the uppermost 150 feet, and none has been found below the quartzite, which is between 200 and 300 feet thick where explored. Little workable chalcocite ore has been found in the porphyry below the 325-foot level. These facts, taken in connection with the character of the deposits, indicate that the copper came from above. It is interesting to note in this connection that in the private report written in 1899 for the Santa Rita Mining Company, already referred to, these deposits were recognized as secondary concentrations likely to give way at greater depth to leaner bodies of pyrite. This was before the papers of Emmons and Weed on the subject of secondary enrichment had appeared.

There is no zone of leached ground overlying the ore bodies of anything like sufficient thickness to have furnished the copper deposited below. On the other hand, it is believed that the metamorphosed limestones containing cupriferous sulphides that are present on several sides of the basin originally overlay the region of the present ore bodies, and that by means of the thorough oxidation which they must have suffered during degradation the copper that they held was dissolved and finally precipitated in the underlying rocks.[b] The present topography suggests that the commercial ore bodies may represent concentrations from an area of the altered limestone which was considerably greater than the existing one. Water rises to a level less than 100 feet from the surface in abandoned workings, and openings of less depth are comparatively wet.

Developments, both old and of recent years, are extensive. One chief source of native copper was the northeastern part of the basin just west of Santa Rita Creek. Here, at what are now the Romero workings, much mining by the Indians and Spaniards was done. The Romero shaft was sunk to a depth of 500 feet to explore these old workings, but encountered little ore below the 200-foot level. The Santa Rita shaft, near by, is 300 feet deep and gets below the quartzite into schist and porphyry that have supplied only scattered bodies of low-grade ore.

On the east side of Santa Rita Creek much of the early mining was done. Work in recent years has opened at lower levels very rich bodies of chalcocite ore in porphyry so altered that it was practically clay. The largest ore body was struck near the "air shaft" in a crosscut from the 300-foot level of the Hearst shaft. This was later connected by a long crosscut with the 300-foot level of the Santa Rita shaft. A stope about 200 feet long, 40 to 50 feet wide, and 100 feet high in maximum dimensions was taken out here, the whole mass averaging over 10 per cent and many carloads averaging 40 per cent. Large masses of solid glance were struck in places. The central portion of the body was shipping ore; the outer portions were of concentrating grade, gradually decreasing in value outward. This ore extended just below the 300-foot level. A little ore was found underneath, in the 400-foot level of the Hearst

---

[a] Compare the similar occurrence in the Morenci district, Arizona, described by W. Lindgren, Prof. Paper U. S. Geol. Survey No. 43, 1905, p. 101.
[b] Compare Emmons, S. F., op. cit., p. 450.

shaft. On the 300-foot level two other bodies of ore similar to that in the big stope were found en échelon with it. The southernmost of these bodies was about 150 feet south of the bottom of No. 7 shaft.

In the western part of the basin, as at the Chino, Wilson, and Head leases and the Lee mine, both chalcocite ores and native copper and cuprite ores have been worked, mostly above the 200-foot level. At the time of visit all the known bodies of rich chalcocite ore had been exhausted and production was coming chiefly from the operations of lessees.

Most of the ores that have been mined in the Santa Rita district have been of high grade. In the old days this was necessarily so, and in recent years, though concentration has been practiced, results have not equaled those attained in other camps, and consequently the minimum grade treated has been correspondingly high. It is claimed by some of the former operators that the combination of oxidized minerals, native metal, and slimy glance is very difficult of efficient concentration. A leaching process was tried at one time, but was not successful. The output of the lessee, who usually works only the native copper and cuprite ore, is a concentrate derived by picking and hand cobbing or by treatment in crude hand jigs. The company owns a mill which has been run intermittently, handling partly chalcocite ore and partly the cuprite and native copper ore. The greater part of the chalcocite ore has been shipped direct to the smelters. According to a recent report on the district made by a private engineer[a] as the result of extensive sampling there is still a very large tonnage of ore of milling grade, including old dumps and filling, as well as ground in place. The Chino Copper Company has been organized to work on a large scale these deposits of lower grade than have previously been mined, and has lately conducted extensive development operations, and had a detailed geologic survey made of its property.

The deposit at the Hanover mine was also plainly a result of secondary enrichment. The shaft is 200 feet deep, but no ore of importance was found below a depth of 150 feet. At the time of visit operations were confined to extracting some old filling and mining the peripheries of an old stope on the 100-foot level, and as the ground was very heavy, requiring the use of much timber, little information could be gained regarding the character of the deposit. The ore was apparently very soft, and most of that seen had "run," so that it looked simply like black mud. There was still to be seen in places earthy limonite impregnated with native copper and cuprite. This was said to be characteristic ore. In cavities chalcotrichite is sometimes found. On the northwest side of the ore body is quartzite; on the southeast side limestone. The ore body is said to have a length of about 400 feet in a northeasterly direction and to pinch at both ends and at the bottom like a canoe. The limestone wall, which is fairly regular and is probably a plane of stratification, dips steeply to the northwest. A dike of fine-grained porphyry cuts the quartzite and makes good ore near it. Where the limestone is cut by the porphyry it is metamorphosed, commonly to a white, sugary marble, but in places massive garnet is developed, carrying magnetite, pyrite, and chalcopyrite. A few narrow silicified streaks, carrying partly decomposed pyrite, cross through the ore.

Most of the ore probably represents enrichment of a contact-metamorphosed limestone deposit, the difference between this and the Santa Rita deposits being that here the enrichment has gone on chiefly in a portion of the original deposit itself, whereas in the Santa Rita mines the original deposits have been leached and the copper deposited in rocks at a lower horizon. This difference would explain the richness in limonite and the softness of the Hanover ore.

CENTRAL DISTRICT.

The town of Central lies about 7 miles east of Silver City and 1 mile south of the military post of Fort Bayard. The Central mining district is commonly made to include the region already described as the Santa Rita and Hanover districts, which lies to the northeast, as well as a number of camps and locations nearer the town of Central.

Discoveries were made at various points in the sixties of the nineteenth century, and at some of these points production resulted. In 1905 practically no work was being done. The

---

[a] Sully, J. M., op. cit.

San Jose mine, situated near the south end of the Hanover district, was before 1870 an important producer of copper, gold, and some silver from ore, said to be chiefly copper carbonates in porphyry. More recently concentration was attempted, but was said to be unsuccessful. Since 1905 this mine has produced some high-grade galena ore with a little chalcopyrite. About 2 miles southwest of Hanover, at Copper Flat, some excitement was aroused a few years ago over prospects of copper ore. Reports disagree as to the importance of the results of work done at that time, but operations were not long continued.

In Gold Gulch, which enters Whitewater Creek just west of Hanover Junction, are a number of claims on which considerable work was done in the early days. In 1903 rich gold ore was discovered, and a new period of activity began. A mill was erected and development vigorously carried on, but this excitement was of short duration; in October, 1905, the mill machinery had been taken away.

The rocks consist of the typical porphyry, related to that of Santa Rita and Silver City, and blocks of shale, probably of Cretaceous age. The veins cut both shale and porphyry. On the Lucky Bill claim an inclined shaft has been sunk on the vein, which strikes northeasterly and has an average dip of about 55° SE. The country rock is shale, which here is rather thickly bedded and practically horizontal; farther northeast this shale is cut off by porphyry. The ore contains pyrite, sphalerite, and galena with quartz and dolomite, also some gold. The Owl mine is located on a vein that dips about 70° SE., and cuts both shale and porphyry, but the workings are mostly, if not wholly, in the shale. The vein is marked by a strong iron outcrop, but the principal mineral in the unoxidized ore is zinc blende, with capillary veinlets of pyrite and chalcopyrite crossing it. In the porphyry, quartz is plentiful in the vein in well-formed prisms, but in the shale quartz seems to be present mainly as a product of partial silicification of the shale. Some of this ore was shipped for smelting and some was run through the mill.

At the Owen prospect, east of Gold Gulch, good gold values had at the time of visit recently been struck in an 80-foot shaft on a vein in shale striking northeast and dipping 75°–80° SE. The dump indicated that the bottom of the shaft had struck porphyry. Zinc blende was the most prominent mineral.

South of Gold Gulch, in Whitewater Creek, an attempt was being made to recover placer gold. After boilers, two powerful steam turbine pumps, and a Hendy hydraulic elevator had been installed, it was found that both water and gold were lacking.

Between the Owen prospect and the town of Central is a small exposure of sandstone, probably Cretaceous; aside from that the rock is porphyry. A number of pits have been sunk in this rock on narrow veinlets carrying quartz and, in some cases, considerable amounts of pyrite and zinc blende. The precious-metal values were evidently low.

The Texas mine, situated about 1 mile west of Central, is said to have produced a considerable amount of silver and a little gold. No one was to be found at the property when visited. The shaft, which has three compartments and is capped by a large head frame, is said to be 600 feet deep. The dump is small, so that either the underground workings can not be extensive or else much of the ground broken must have been ore. The vein strikes about east-northeast, and appears to have very steep dip. The country rock is porphyry, like that occurring all the way to Silver City. From specimens seen on the dump, the ore seems to be principally pyrite, with some galena and probably finely divided grains of black zinc blende in a dull horny quartz gangue. Druses are common, and in these the quartz is crystalline. The ore was treated in a mill, yielding a heavy pyrite concentrate.

GEORGETOWN DISTRICT.[a]

Lying about 4 miles northeast of Santa Rita and about 3 miles east of Fierro is the old silver camp of Georgetown, now virtually abandoned. The district lies on the steep slope of a tributary to Mimbres River, which is only a few miles to the east. Silver was discovered in 1866, but there was little development until about 1873. By 1875 the camp was booming,

---
[a] By L. C. Graton.

and work continued steadily till the slump in silver prices in 1892 caused the decay of the district. Very little has been done since that time. At the time of visit, in 1905, one Mexican lessee was gophering in old stopes. According to F. A. Jones[a] the production of the district has amounted to $3,500,000.

The ridge or mesa of Carboniferous limestone that borders the Santa Rita and Hanover districts on the east continues to Georgetown, where it ends in a sharp scarp. Underlying these limestones conformably is probably 200 feet of black crumbly Devonian shale. Under the shale is more limestone, the lowest rock exposed. Poorly preserved fossil remains found not far below the top of this lower limestone were identified by E. O. Ulrich as belonging to the Richmond horizon of the Ordovician. The dip of all these beds is 10° to 20° slightly west of south. Dikes of porphyry, probably granodiorite porphyry related to the intrusion at Santa Rita and Hanover, cut the two limestones and shale. Most of the dikes are approximately vertical and strike about N. 70° W. South of the district a sill of porphyry occurs near the bottom of the upper limestone. Some faulting with small displacement has taken place, and some of the dikes appear to occupy fault planes.

The horizon of the ore bodies was the upper part of the lower limestone—that is, just below the shale—and along or near the dikes. This is practically identical with their position at Chloride Flat near Silver City. The ore was mostly silver chloride, with some native silver and argentite. Silver bromide, silver iodide, and ruby silver are also reported to have been found. Certain pockets are said to have carried important amounts of lead carbonate with the silver. It is reported that on the Uncle Sam claim there was a deposit of galena carrying about $75 a ton in silver. Some vanadinite is said to have been found in the district.

The principal mines were the Naiad Queen, Commercial, MacGregor, McNulty, and Satisfaction. The first three were rich and large producers. All these were owned by the Mimbres Mining Company, a New York concern. The ore of too low grade to be shipped was treated in a concentrating mill containing five stamps and two Wilfley tables.

The Satisfaction mine was the only one open at the time of visit. The shaft is 160 feet deep, of which about 110 feet is in shale. The bottom portion of the black shale, a few feet thick, is heavier bedded and limy and forms a transition from the underlying limestone. The limestone is in general much silicified and seems to have shrunk in the process, leaving many narrow vuggs, though these may have been formed wholly by solution. Larger cavities are found, commonly lined with calcite crystals, and some large caves contain stalactites and stalagmites of calcite. Small quartz stringers are common. In other places a network of very narrow seams makes the rock look like a honeycomb. These seams contain a white mineral which under the lens is seen to be developed in groups of radiating fibers; examination by W. T. Schaller proved it to be a hydrous calcium silicate, possibly openite. The ore was very pockety; the stopes are of all shapes and run in any and every direction. The bunches of ore were in some places connected by narrow leaders of lower grade and in others not. The most regular stope seen in the mine was parallel to but 5 to 10 feet away from a porphyry dike about 15 feet thick. This stope was fully 60 feet long and 20 feet high and averaged 2½ to 3 feet wide. The limestone was perhaps a little more silicified near the dike, but no other alteration was visible. In this particular place the ground next the dike was of too low grade to work, but it is said that in some places the ore comes up against the dikes and that here and there good ore is found even in the dike itself.

The Naiad Queen ore body followed a dike. A shaft was sunk for a depth of 600 feet and is said to have followed good ore to the bottom.

The workings of the district are dry and the greater part of the ore has been oxidized. It is probable that the original ore was, in part at least, an argentiferous galena, but some of the silver may have been deposited as a rich silver mineral, like argentite, for lead was not noticeable in the ore seen. The silicification of the limestone and the deposition of the ores are believed to have been due to solutions genetically connected with and accompanying the porphyry dikes.

---

[a] Op. cit., p. 45.

LONE MOUNTAIN DISTRICT.[a]

An isolated group of three hills of low elevation, but rising conspicuously from the valley level, about 6 miles southeast of Silver City and nearly an equal distance southwest of Central, is known as Lone Mountain. According to F. A. Jones[b] the district was discovered in 1871 by Frank Bisbee, from whom the great Arizona copper camp took its name. Rich silver ore was found and a mill was erected. After two or three years of work, reported to have been successful, operations ceased, and little has since been done in the district.

The hills consist of Paleozoic strata that have been lifted up along a probable fault on the southwest and dip 25°–30° ENE. or NE. Quartz monzonite porphyry or granodiorite porphyry has intruded these sediments, and at one place the red pre-Cambrian granite is exposed. The rock lying probably at the bottom of the exposed sedimentary series is a ferruginous quartzite similar to that at the base of the Silver City section and probably Cambrian. A thickness of probably not over 60 feet is exposed, but the bottom is not seen. Near the top it alternates with limestone beds, and within a short distance gives way entirely to limestone. This limestone formation, which is several hundred feet thick, contains fossils concerning which E. O. Ulrich reports as follows:

No. 15. This consists of a slab of chert with poorly preserved casts of a small gastropod. It is probably a species of *Liospira*, which genus, excepting a depauperate species in the lower Clinton of Ohio, is confined to Ordovician rocks. A similar, perhaps identical, species occurs in the Richmond horizon of the Bighorn limestone in Wyoming. Except that the umbilicus seems to be a trifle larger it agrees, so far as casts of the interior admit of comparison, closely with the common Richmond variety of *L. americana*.

It is possible that there is a small thickness of Silurian rocks at the top of this limestone member, as at Silver City, but there is no evidence of it in the structure.

The northeastern slope of the hills practically coincides with the bedding of this limestone. Conformably overlying it and outcropping in the depression at the northeastern base of the hills is 200 to 250 feet of very fissile shale, most of which is black, but the upper portion is a little reddish. It resembles the shale of Chloride Flat, and is doubtless Devonian.

Next above, and also conformable, is a series of limestones, probably of Carboniferous age. With decreasing dip they extend away to the northeast and are covered. The thickness is unknown, but is several hundred feet at least.

At the base of this upper limestone, or else at the top of the shales, is a prominent sill of porphyry. Porphyry also cuts the rocks on the southeastern slope. A small exposure of granite was seen near the western base of the hills. The Ordovician limestones rest directly upon it, but were probably not so deposited. The creek just west of the mountain apparently follows a fault along which the rocks on the east have been raised; the main fault that has allowed the tilting of the strata can not lie far to the south. It seems probable that in the corner between these two intersecting fault planes some of the underlying granite has been thrust up into the overlying sediments.

West of the creek and probable fault are exposed limestones from which fossils, probably Pennsylvanian, were collected.

The principal mines are situated on the southeast side of the mountain. As at Chloride Flat and Georgetown, the main ore-bearing rock is the limestone below the Devonian shale. At Lone Mountain, however, the ores, instead of being found mostly in the beds immediately below the shale, occur mostly in fractures crossing the limestone strata. Some brecciation has taken place along these fractures and the limestone has been somewhat silicified. Limonite is plentiful and evidently resulted from oxidation of pyrite. Gypsum is also seen sparingly. Silver chloride was the most common ore mineral, but the richest ores carried curved bundles of native silver wires and, it is said, argentite also. A few specimens of ore found on a dump show minute specks of a soft black mineral that may be argentite. The veins are narrow and the values not persistent. It is said that most of the ore mined was very rich, though some concentrating ore was extracted.

---

[a] By L. C. Graton.   [b] Op. cit., p. 44.

The Monarch mine is said to have been the principal producer. The Monarch vein is fairly well defined and can be traced for several hundred feet. It strikes about north and averages about vertical in dip, though the inclination varies from one direction to the other in depth. The width ranges from 1 to 5 inches. The vein is said to be widest where a branch vein, striking N. 30° E., unites with it, and values are also said to have been good at this point. The deepest workings in the Monarch were at 100 feet. It is said that 40 to 70 ounce ore was present in the bottom when work was stopped. Near the Monarch are the Home Ticket, New York, and Eighty-Four mines. The Eighty-Four is stated to have penetrated porphyry in the bottom. These mines were on veins similar to that of the Monarch; none of them was worked at much depth. Near the northwest end of the mountain a vertical vein striking about northeast and possibly parallel to the supposed fault in the valley farther west has been opened at several points and has furnished some ore. For a distance of 2 miles northwest of Lone Mountain pits have been sunk at various points on brecciated and iron-stained zones in the limestone. Values appear not to have warranted further development. Southeast of the mountain ore of moderate grade is said to have been mined from an open cut and concentrated.

As at Chloride Flat and Georgetown, the ore deposition at Lone Mountain was probably related both in time and origin to the porphyry intrusion. Thorough oxidation has destroyed evidence of the character of the original ore minerals. The absence of much lead and the richness of the narrow veins, together with the reported presence of argentite, make it seem possible that that mineral may have been the principal silver carrier in the unoxidized ore. Pyrite is the only other sulphide known to have been present. Faint green stains seen occasionally are probably of copper.

BURRO MOUNTAIN DISTRICT.[a]

LOCATION AND HISTORY.

The Burro Mountains form a somewhat isolated group situated about 13 miles southwest of Silver City. Three rather prominent hills, of which the middle one has an elevation of 8,054 feet, constitute the heart of the mountains. Although there are prospects in several parts of these hills and their flanks, activity centers at the camp of Leopold, about 4½ miles northeast of the summits and about 12 miles by road from Silver City. The elevation is a little higher than at Silver City, a bench mark at the camp being 6,142 feet.

It is probable that Indians mined for turquoise in this district before it was known to white men. The first recorded discovery was in 1871. Both turquoise and copper were found, and considerable exploratory work done in the early eighties. Two smelters are said to have been erected in this period,[b] but they were unsuccessful. About 1902 interest in the region was revived by important discoveries of copper ore. In 1905 a considerable amount of exploration was going on and one company was producing regularly.

GEOLOGY AND MINERAL DEPOSITS.

GENERAL FEATURES.

The main mass of the Burro Mountains is a core of pre-Cambrian rocks consisting chiefly of granite, with associated pegmatite, and less abundant dioritic rock, in many places somewhat gneissic. Ancient sedimentary rocks may be present in places, for an undefined mass of rock apparently inclosed by granite near the Burro Chief mine appears to be quartzite and does not resemble any of the Paleozoic or later sediments known in this part of the Territory.

Monzonite porphyry of the general type so common in New Mexico breaks through the older rocks of the mountains. It is found on both the north and the south sides of the principal ridge and is exposed at intervals between Leopold and Silver City. The rock consists of phenocrysts of orthoclase and plagioclase feldspar in a microgranular groundmass of quartz and orthoclase. Augite phenocrysts were probably present, but are now altered to chlorite, epidote, etc. Small crystals of apatite are plentiful, and rutile is present in tiny grains. The

---

[a] By L. C. Graton. [b] Jones, F. A., op. cit., p. 56.

rock has undergone much alteration; some of it is silicified, some kaolinized, and some impregnated by pyrite. The weathered surface is commonly brown as a result of oxidation of the pyrite. Siliceous knobs and ledges are also seen.

The ore deposits occur mainly in the monzonite porphyry and are doubtless related to that rock in origin.

COPPER DEPOSITS.

The most important deposits in this district are impregnations and replacements along irregular fractures and zones of crushing. Silica and cupriferous pyrite have been introduced and constitute the primary lodes. Practically all the workable ores are the result of enrichment of this lean pyritic material. Rich oxidized minerals—native copper, cuprite, melaconite (?), chrysocolla, melanochalcite (?), malachite, and azurite—are found at or near the surface, but at greater depth chalcocite and chalcocite-coated pyrite are the ore minerals, and at a still lower level these give way to unaltered pyrite.

The principal mine in 1905 was that of the Burro Mountain Copper Company at Leopold. The development consisted of an old inclined shaft, said to be about 500 feet deep, and a new vertical shaft about 220 feet deep. The inclined shaft was filled with water within 105 feet of the collar, which was the third or 160-foot level of the vertical shaft. Drifts and cross-cuts at four levels from the vertical shaft constitute several thousand feet of workings. A 125-ton mill equipped with crushers, rolls, Huntington mills, Wilfley tables, and Frue vanners was being enlarged to a daily capacity of 250 tons.

The country rock is of monzonite porphyry, and the deposit is marked at places on the surface by jagged, silicified, and iron-stained outcrops.[a] The ore zone has a general east-west extension, but is irregular and appears to be related to east-west fractures and to northeast fractures, possibly also to a northwest fracture system. The dip is likewise irregular, but inclines rather flatly to the south. The fractures which determine the location of the ore bodies are only a few inches wide at most, but they occur in series and the intervening rock is in many places considerably crushed. A heavy gouge of kaolin is found in some of these fractures and much of the adjoining rock is bleached, softened, and kaolinized, but in places it is silicified. The primary ore mineral is pyrite, carrying a little copper. Some apparently unenriched pyrite yielded 0.36 per cent of copper on assay. This mineral occurs as fillings in the main fractures and in the smaller crevices of the shattered porphyry, and also is disseminated in grains through the rock by replacement. All the workable ore has been formed by enrichment of this pyritic material, and the enrichment consists chiefly in partial or complete replacement of the pyrite by the cuprous sulphide, chalcocite, or copper glance. In places there is a suggestion that the chalcocite has been precipitated not by pyrite but by kaolin, but this can not be definitely established. The richest ore commonly occurs in the more definite fissures or fractures, associated with kaolin. In such places masses of practically solid glance are sometimes found, and shipments of ore running as high as 50 per cent of copper have been made. The most important ore bodies, however, are much larger and commonly extend from one fracture to another of the same series. In several places the workable ore is bounded by these kaolin-bearing fissures or "slips," thus causing the ore body to be regarded as a "vein;" but the values are not actually confined to these limits, and here and there, especially on the underlying side of the foot-wall slips, the copper content is high enough to make extraction profitable. In such places the limits of the ore are indefinite, the percentage of copper gradually decreasing, and the extent of workable ore can be determined only by assay. The ore bodies are large, reaching 50 feet or more in width and several times that in length. Lang[b] shows that the largest ore bodies are formed at the intersection of important members of different fracture systems.

The zone of impoverishment or leaching is in most places less than 50 feet deep. The ore on the first, or 100-foot level, contains more iron and less copper than that on the 130-foot and 160-foot levels; in fact, it appears less enriched. It was reported in 1905 that the lowest, or

---

[a] S. S. Lang, formerly geologist and engineer of the company, has given a good description of the mine and of other properties in the district, in Eng. and Min. Jour., September 1, 1906, pp. 395–396.
[b] Op. cit., p. 395.

210-foot, level then being driven had encountered low-grade pyritic ore under the enriched ore of the upper levels, but since then important discoveries have been announced.

The average grade of ore mined in 1905 was about 4 per cent, ranging from 2 per cent up. Gold is practically absent and but very little silver is present. The richest ore was shipped direct and averaged over 20 per cent copper. The other ore was concentrated about 8 or 10 into 1, with a reported recovery of 75 to 80 per cent. At the present time the mine is one of the most important producers of the Territory. It has recently been acquired by Phelps, Dodge & Co.

The Alessandro Mining Company was in 1905 continuing development of a property on which a 240-foot shaft had formerly been sunk and foundation for a mill erected. In September of that year the shaft was 255 feet deep and had encountered stringers of chalcocite in altered porphyry. Copper oxide and carbonate were also found in seams and partly impregnating the rock, forming a low-grade ore. No production has since been made.

Development work was being carried on at three claims by the Comanche Mining and Smelting Company, of Silver City. These were the Klondyke, Boone, and Maquon claims.

Among other properties then idle was the Virginia, east of the mine of the Burro Mountain Company. A northeast lode dipping 50° NW. had been explored by a vertical shaft said to be about 200 feet deep. The ore is partly chalcocite and partly chrysocolla, malachite, and cuprite. Some good ore had been taken out previous to 1905 and the mine has made some production since that time. On the Copper Gulf claim, east of the Virginia, several shallow shafts have been sunk, disclosing carbonate and oxide minerals in much-jointed porphyry.

An east-west lode dipping 70° to 75° N. has been explored on the Colwell claim, near the Virginia. On the Boston claim, west of the Virginia, a north-south lode dipping east shows a chrysocolla outcrop, and a lode similar in attitude and in character is exposed on the Niagara claim, still farther west. Development on none of these three claims had reached below the zone of oxidized minerals. At the Burro Chief property a lode striking N. 13° E. and dipping about 60° E. contains good-looking ore carrying malachite, azurite, and cuprite. Most of the surrounding rock is granite, but on the foot-wall side is found a fine-grained siliceous rock that seems probably to be quartzite. It carries malachite and cuprite in tiny fractures. Not far to the northeast a vein of fluorite, carrying also some crystalline calcite, cuts through the granite. This deposit, which is from 2 to 15 feet wide, was worked for flux when smelting was attempted in the district.

According to recent reports (in 1909) prospecting operations had been considerably extended. The Burro Mountain Copper Company remained the only producer. The depth attained is 350 feet; the developments aggregated 30,000 feet and the ore reserves were estimated at 3,000,000 tons. The Chemung Copper Company has about 26,000 feet of development work and has attained a depth of 600 feet, on which level the workings are still in ore according to reports. The shaft is down to the 800-foot level, but on this level the ore has not yet been cut. The tonnage developed is said to be about 3,000,000 tons. The Savanna Company is developing, with about 20,000 feet of openings; the depth attained is 450 feet. The ore reserves are believed to be about 2,000,000 tons. The National Company is also developing, with about 8,000 feet of openings, a depth of 300 feet, and a considerable tonnage of developed ore.

On the south side of the Burro Mountains, in what is known as the White Signal district, some prospecting has been done for gold and copper, as well as turquoise. A number of quartz veins occur in granite, but quartz monzonite porphyry is present in places, probably as dikes. The strike of most of the veins is west-northwest, and the dip steep toward the north. The veins are pretty much oxidized, but contain remains of pyrite, some chalcopyrite and galena, and locally a little sphalerite or zinc blende. Fair values in gold are said to have been found in places, and a little ore has been shipped. The Neglected mine lies on the southern flank of the Burro Mountains. A quartz vein striking N. 75° W. and dipping about 75° S. has been opened by shallow workings. It varies in width from 5 to 35 feet, according to report. Granite is present on both sides of the vein, but a decomposed gouge or kaolinized material several feet wide occurring on the hanging-wall side may possibly be an altered porphyry dike. The principal ore mineral is chalcopyrite, which occurs in plentiful small grains disseminated fairly evenly through the rock. Assays for gold indicate an average value of $5 to $10 a ton. The ore is said to concentrate well at about 15 to 1, and the concentrates carry most of the gold and good values in copper. No work was being done at the time of visit. Work was in progress at the Bisbee mine, farther east, and it was reported that good gold and copper values had been encountered, but the property was not visited.

TURQUOISE DEPOSITS.

Turquoise was mined in the Burro Mountain district in prehistoric times; old excavations and dumps and stone implements mark the site of these operations and have led to the rediscovery of some of the deposits. The principal deposit is the Azure mine, which lies about 1½ miles north of Leopold, overlooking the Mangas Valley. The mine lies outside the main mass of quartz monzonite porphyry in which the principal copper deposits are situated, but dikes of this rock cut the granite in which the Azure deposit lies. The turquoise lies in a distinct fracture zone which strikes about northeast and dips about 50° SE. The surrounding granite is traversed by many narrow parallel and cross fractures. The walls are cut by a perfect network of small quartz stringers, but close to the principal fractures the granite is considerably kaolinized. The turquoise lies in or close to these fractures and is commonly associated with the kaolin. The mine has been worked by tunnels at several levels, but in recent years the upper portion has been worked by an open cut. The production is reported as several million dollars.

A good description of the Burro Mountain turquoise occurrences and of the Azure mine in particular has been given by E. R. Zalinski.[a] Turquoise has been found at a number of places near the Azure mine; near the Burro Chief copper mine,[b] where the adjoining granite is much silicified and pyritized; a few miles northeast of Leopold, along the road leading to Silver City; and at several places in the White Signal district, on the south side of the Burro Mountains. At one place in the White Signal district the turquoise lies along the side of a 2-foot porphyry dike that cuts the granite. Only the Azure mine and one prospect in the White Signal district were being worked in 1905.

The formation of the turquoise is undoubtedly related to the intrusion of the quartz monzonite porphyry, as also the accompanying kaolinization and pyritization. It is probable that the phosphoric acid and copper of the turquoise were supplied by gases or solutions coming from the intrusive mass, but the alumina may have been derived from the rock in which the turquoise was deposited.

BLACK HAWK DISTRICT.[c]

The deserted little camp of Black Hawk, or Alhambra, lies near Bullards Peak, a few miles northwest of the Burro Mountains and just outside of the western boundary of the Silver City quadrangle as mapped by the Geological Survey. According to F. A. Jones,[d] rich silver float was discovered in 1881 and shortly afterward the Alhambra mine was located. Other mines were soon started. The district was booming in 1885, 1886, and 1887. The Hobson and Alhambra mines were worked up to 1893. No work of consequence has been done in the camp since that time. The total production of the camp amounts to perhaps $1,000,000.

The general country rock is a coarse-grained dark gneiss carrying quartz. It is locally called syenite. A light-gray porphyry which cuts it in places as dikes is similar to the porphyry of the Burro Mountain camps, except that it contains but little quartz, and therefore is a monzonite porphyry. Other dikes, apparently more numerous, are rather fine grained and of a dark-greenish color. Although considerably decomposed, making determination uncertain, the rock appears to be a diorite porphyry. Both dike rocks are probably related in origin and are of similar age and source to the quartz monzonite porphyry of the Burro Mountains.

---

[a] Econ. Geology, vol. 2, 1907, pp. 464–492; Eng. and Min. Jour., vol. 86, 1908, pp. 843–846.
[b] See Snow, C. H., Am. Jour. Sci., 3d ser., vol. 41, 1891, pp. 511–512.
[c] By L. C. Graton.
[d] Op. cit., p. 54.

Silver was the only metal produced in important quantity by the camp. The deposits are veins cutting the gneiss, some of them parallel to or otherwise associated with dikes of the diorite porphyry. The Black Hawk mine was the largest producer of the district, being credited with a production of nearly $600,000.[a] The vein, which strikes about N. 67° E. and dips about 65° NW., is developed by a 600-foot incline, with a winze 140 feet deeper from the lowest level. A second inclined shaft was started and had attained a depth of 100 feet when work was suspended. It is stated that much water was encountered in the deeper shaft. The ore is said to have been "spar" (probably dolomite and barite), with much native silver, a considerable quantity of argentite, and a little chloride. The vein was irregular, ranging from an inch to a foot or more in width. The values were erratically distributed through the vein, rich pockets being scattered through almost barren portions of the vein. The ore that was mined was of high grade. Large pieces are said to have been taken out that were almost solid native silver. Much trouble was experienced with ore stealing and the writer was told that the company probably received not over half the silver taken from the mine. Little could be learned at the time of visit as to the horizontal extent of the stopes. It was stated, however, that when the mine was closed there was exposed in the bottom of the mine a shoot several inches wide and about 60 feet long containing good values as native silver. Surface exposures and material found on the dumps did not throw much light on the character of vein. The principal alteration of the rock is carbonatization. Barite in flat tables was plentiful in places, intergrown with some material, probably pyrite, which is now represented by limonite. Slight stains of copper are present. In other pieces dolomite constitutes the bulk of veins 2 to 4 inches wide. It forms druses and contains narrow irregular bands of siderite. Through the dolomite are scattered minute specks of a black metallic mineral, possibly argentite. Whether these materials are characteristic of the pay ores could not be ascertained definitely.

At the Albambra mine a 400-foot incline is sunk on the vein, which strikes about east and west and dips steeply to the south. Breccia containing fragments of schist and monzonite porphyry is found on the dump, also dolomitic vein stuff like that at the Black Hawk. Much good ore is said to have been taken out. The Rose mine is said to have had two veins, known as the "white lead" and the "red lead," which crossed at a low angle. A 200-foot shaft was sunk, but little ore was found in depth, most of it being obtained near the surface, at the intersection of the two veins. It is said that $100,000 was taken out here. The Hobson mine, located on an east-west vein dipping north, produced some good ore, mostly argentite with a little native silver. The shaft is stated to be about 200 feet deep. Rich pockets of native silver ore were found in the Good Hope workings, which have a maximum depth of about 120 feet.

TELEGRAPH DISTRICT.[b]

The Telegraph district lies northwest of Black Hawk and Bullards Peak. It is situated among steep rugged hills, about a mile west of Gila River, midway between Cliff and Redrock. According to F. A. Jones[c] silver ore was discovered in the district in 1881, and in 1885 the Telegraph Mining Company erected a 15-stamp mill at the river. "The enterprise was first attended with marked success; the ore body, however, being small, was soon exhausted and the concern collapsed." A 5-ton leaching plant was erected near the old mill site in 1903. This was not in operation and the camp was deserted when visited in 1905.

The prevailing rock in the region surrounding the Telegraph district is pre-Cambrian granite, with some irregular areas of gneiss. In places remnants of sedimentary rocks are found. Quartzite and unevenly bedded carbonaceous shale appear to be most common, but limestone is present near the mine workings in the Telegraph district. The sedimentary masses seem to be comparatively narrow blocks faulted down into the granite and thus protected from erosion. Three or four miles southeast of the Gila, on Wild Horse Creek, a block of shale dipping 10° to 15° W. is present on the east side of the creek, but ends abruptly against a hill of granite a short distance east of the stream. Granite forms the west wall of the canyon, but high above the stream it is capped with quartzite. At the Telegraph camp the sedimentary rocks occur

---

[a] Jones, F. A., op. cit., p. 55. [b] By L. C. Graton. [c] Op. cit., p. 71.

for the most part in a depression surrounded by granite. Two blocks of the sediments resting on granite dip rather sharply southward toward probable fault planes, which limit them. The hill to the south is capped by flat-lying quartzite.

Time was not available for a systematic search for fossils, and no organic remains were found. The shales resemble, in lithologic character, those just west of Silver City, in which Cretaceous fossils were found, and the quartzite, which probably everywhere underlies the shale, contains undefined masses of characteristic chert conglomerate, similar to that in the sandstone that occurs between Cretaceous shale and Carboniferous limestone in the Silver City section. Where limestone was found, it is intercalated with the quartzite and shale. It is possible, therefore, that these sediments of the Telegraph district are of Cretaceous age. Both the sedimentary rocks and the granite are cut by dikes of gray porphyry similar to that present in the Burro Mountain and Black Hawk districts.

In the granite between the inclined blocks of sediments on the north and the flat quartzite capping on the south, a tunnel has been driven along a vein that strikes about N. 35° E. and dips about 45° SE. To judge from the timbers in the stope, the vein was 2 to $2\frac{1}{2}$ feet wide near the mouth of the tunnel, but appears to have pinched rapidly as the tunnel advanced. Such material as is now visible is simply fractured, much silicified granite, with drusy iron-stained quartz and in places minute black specks, which probably carried the silver. Its value is not known. The vein could not be detected in the shale, which reaches nearly to the mouth of the tunnel, and perhaps does not enter the sedimentary rocks, though this seems unlikely. In the granite on the north side of the gulch, a little beyond the northernmost sedimentary block, there are small workings on a fairly well-defined north-south vein about 3 or 4 feet wide. A stope of considerable size has been taken out from the surface. The ore is apparently similar to that seen in the workings farther south, except that it is intimately associated with fluorite. The presence of fluorite veinlets in the limestone indicates that the veins are later than the sediments and are probably related in time and origin to the porphyry of the dikes. The fluorite is present mainly as octahedrons. It is mostly purple and, as is characteristic of the mineral, the distribution of the color is in some places zonal and in others blotchy and irregular.

Circumstances made it impossible to visit the Anderson camp, which lies on the east side of the Gila, about 6 miles southeast of Telegraph. According to F. A. Jones[a] there are a number of copper deposits in the district and a large fluorite vein.

Up Gila River from the Telegraph district the granite walls become lower and the valley widens. Scattered masses of porphyry are seen. In the vicinity of the mouth of Mangas River are partly consolidated grits, cross-bedded, but practically horizontal. In places a dense rhyolite capping is present. Rhyolite, carved by erosion into fantastic shapes, occurs locally for several miles up Mangas River. It probably is a now detached portion of the great lava series represented at Santa Rita, north of Pinos Altos, and at intervals between Gila River and the Mogollon Mountains. About 3 miles up the Mangas from the point where it enters the Gila is the Cora Miller mine, which is said to have encountered some good silver ore several years ago. On account of high water it could not be reached at the time the vicinity was visited. The mine is apparently in granite.

GOLD HILL DISTRICT.[b]

The Gold Hill district is situated 12 miles northeast of Lordsburg and 15 miles south of Leopold. The camp is nearly surrounded by a circular rim of hills, called the Little Burro Mountains.

Gold is said to have been discovered in the district in 1884, and shortly afterward two stamp mills were erected and were operated till the oxidized ores were exhausted. More recently some of the old dumps have been cyanided, with reported success.

The Little Burro hills, like the Burro Mountains to the north, are composed of pre-Cambrian rocks. Garnetiferous gneiss and amphibolite schist are cut by a rather fine-grained

---

[a] Op. cit., p. 70. [b] By L. C. Graton.

granite and by dikes of pegmatite and diabase. In places the amphibolite is much epidotized. Some of the pegmatite dikes appear to pass into quartz veins.

The veins are of massive quartz, with some calcite and in places much pyrite or its oxidation product, limonite. Galena is sparingly present. Druses are not uncommon. The wall rock is silicified and pyritized. The ores carry gold and some silver. Silver is said to have been found more plentifully in the northern part of the camp. The veins are believed to be of pre-Cambrian age.

Among the principal producers were the Standard, California, Reservation, and Snyder. A typical vein is that of the Reservation, which is 2 to 4 feet wide, strikes N. 55° E., and dips 80° SE. The average ores are said to have run $15 to $40 a ton, but $125 ore was found in some of the mines. The total production of the district is not known.

STEEPLE ROCK DISTRICT.[a]

The Steeple Rock district lies in northwestern Grant County, close to the Arizona line. It is most conveniently reached from the town of Duncan, Ariz., on the Arizona and New Mexico Railway, from which it is distant about 17 miles in a northeasterly direction. The district lies at an elevation of about 4,500 to 6,000 feet, and the surface of the country is rough. The drainage flows southward into Gila River.

It is stated that the first prospecting was done in the region about 1880. A little later the advent of English capital stimulated activity in the district. The Carlisle mine, the most important of the district, was vigorously worked, and is reported to have produced $3,000,000 up to 1897, when it closed. In October, 1905, no work whatever was being done in the district.

The region about Steeple Rock exposes chiefly effusive rocks, of which the most important is a reddish, medium-grained variety that resembles in appearance the soda rhyolite of the Mogollon Mountains. Quartz occurs in medium-sized grains, however, and under the microscope it appears that orthoclase is present in unimportant amount, whereas plagioclase (oligoclase and andesine) and greenish-yellow biotite, in addition to the corroded grains of quartz, are the principal determinable minerals. The dull reddish-brown groundmass, which owes its color to limonite, is cryptocrystalline and microspherulitic. If the rock may be classified by the character of its phenocrysts, it is a dacite, with orthoclase a little more abundant than in the normal rock of that name.

In some of the canyons and mine workings other rocks are found. At the Jim Crow mine, southwest of the Carlisle mine, is found a dark fine-grained rock of probably surficial character. It is considerably decomposed but appears to be intermediate in composition between andesite and basalt. The dump of the Carlisle mine shows a dark-gray, very fine grained and dense porphyry that corresponds to diorite porphyry. The sharply formed phenocrysts of feldspar are mostly labradorite. Pyroxene is now represented by aggregates of chlorite and epidote. The finely trachytic groundmass is composed almost entirely of feldspar laths, most if not all of which are plagioclase. A purplish porphyry of coarser grain, with plentiful and distinct white feldspar phenocrysts, exposed in the canyon just below the Carlisle mine, is also a diorite porphyry. It has the same feldspar phenocrysts as the specimen just described, the pyroxene is a little altered, and the groundmass is similar but somewhat more coarsely crystalline. The two specimens are probably phases of the same rock.

The natural supposition would be that the diorite porphyry is pre-Tertiary and older than the probably Tertiary flows, and this may actually be the case. But the general aspect of the two classes of rocks suggests the possibility that the diorite porphyry may be the somewhat deeper seated and intrusive representative of the effusive rock series.

The ore deposits of the Carlisle district belong in the class of Tertiary deposits and are similar to those at Mogollon, Cochiti, and elsewhere. Quartz is the predominant gangue mineral and shows the radial arrangement with the intervening druses characteristic of veins of this type. Mingled with the quartz is more or less calcite, some of which is developed in small barrel-shaped crystals. Pyrite, zinc blende, chalcopyrite, and galena, named about in the

---

[a] By L. C. Graton.

order of abundance, are present, mostly in fine grains sprinkled through the gangue. The zinc especially tends toward segregation in streaks, however. In the richer ore are dark stringers and patches made up of very finely divided sulphides. All those above mentioned appear to be present, and in addition there is probably some silver mineral, possibly argentite. Much of the quartz is dull and horny and has doubtless replaced wall rock either at the sides of the vein or included in it as fragments. As in similar ore in other districts, deposition of the sulphides seems to have been especially favored in those places where the wall rock has been silicified.

The veins are marked in many places by rather strong siliceous outcrops, which are not so prominent, however, as those in the Mogollon or the Kimball district. The vein system is rather complex; most of the veins that have been worked strike northwestward and many of them curve and branch. The location of the properties that have produced does not seem to indicate that the intersection of veins are ore shoots. In most of the ore mined the gold value a little more than equaled that of the silver.

The Carlisle mine of the Steeple Rock Development Company is controlled by the Exploration Company (Limited), of London, but the greater part of the production was made while it was the property of the Carlisle Gold Mining Company. The vein strikes west-northwest and dips about 65° SSW. The mine was developed by a vertical shaft about 600 feet deep. The reduction equipment consisted of a 60-stamp mill with plates and concentrating tables. Roasting was tried for a short time, but did not prove successful. The mill is now dismantled. The best ore was found in the upper 300 feet; below that the values were lower and recovery more difficult. The principal ore shoot was found west of the shaft, and here a stope was taken out 100 feet long and 40 feet wide. The roof of this stope caved in and the stope is now filled with water. North of the main vein and parallel with it was a stringer, a few inches wide, of ore which was the only exception to the general type of ore found in the district. It consisted of granular quartz with fine particles of free gold scattered through it and was very rich. It is said that the old tailings from the mill could now be cyanided with profit.

The Imperial mine, lying southwest of the Carlisle, is on a north-south vein that dips about 75° W. The gold exceeded the silver in value, and it is asserted that good ore is now present in the bottom of the shaft, which is said to be about 200 feet deep. The Jim Crow and New Year mines lie to the northwest, on a northwestern branch from the Imperial vein; both the vein and the branch outcrop plainly. Other producers of similar ore were the Laura, East Group, Homestake, and Clear Lake.

A few miles northwest of the Carlisle mine, on the Arizona line, similar ore is found, but it contains more copper.

KIMBALL DISTRICT.[a]

The Kimball or Steins Pass district lies north of Steins Pass station on the Southern Pacific Railroad, and close to the Arizona line, in the northern extension of the Peloncillo Mountains. Prospecting is said to have begun about 1875, but it was not until about 1883 that real mining began. Silver has been the chief product of the district but some gold ores have recently been found. The total production of the district is not known but is probably under $500,000.

In the vicinity of the Kimball district the Peloncillo Range consists of rather low hills of smooth contour and moderate slopes. On both sides these hills gradually merge into the plateau level at an elevation of about 4,500 feet. The climate is arid and adequate water supply is obtained only from deep wells.

The principal rocks of the district are a dark-gray, moderately fine grained diorite porphyry, rather similar to that at Steeple Rock, to the north, and a flow rock, probably rhyolite. In the vicinity of the Beck mine prominent dikes of monzonite porphyry cut through what is probably the diorite porphyry, much altered. In places, as at the Wyman mine, the diorite porphyry is bleached and colored pinkish. Some faulting has taken place. The character of the ores can best be shown by a description of typical occurrences.

[a] By L. C. Graton.

The Volcano mine, in the north end of the district, was the largest producer, having turned out, it is said, several hundred thousand dollars' worth of silver. The mine was worked through an inclined shaft 200 feet deep, and the ore was treated in a 10-stamp mill with pans. No work has been done for a number of years. The vein is marked by a brecciated and silicified zone 50 feet wide that forms a prominent outcrop and is called locally a dike. It is said that this vein can be traced in a nearly north-south course for about a mile, but at the mine the strike is N. 25° W., with a dip of 75° E., that becomes steeper at the bottom of the shaft. The vein is a fault fissure with diorite porphyry on the east or hanging-wall side, and rhyolite (?) on the west. Intense brecciation and slickensides in the gouge are additional evidence of faulting. Pay ore was found generally in a band of quartz on the hanging-wall side of the brecciated zone. The quartz was bunchy and not continuous and in places pay values extended into the silicified zone itself. Good ore was also taken from a surface trench along the foot wall of the vein. The stopes were a series of irregular chambers ranging from 2 to 10 feet wide. Most of the ore was taken out above the 100-foot level, and it was mainly oxidized, consisting of films of greenish silver chloride in the many seams. Some of this ore was very rich. Below a depth of 100 feet less ore was found and the values were lower; sulphides were also encountered. The unoxidized ore is very similar in appearance to the silver ore of the Mogollon region. The quartz is commonly hackly through the solution or replacement of thin plates of calcite. The sulphides are in fine particles grouped in patches through the quartz and appear most abundantly near fragments of wall rock included in the vein. Pyrite is plentiful but is exceeded by a black mineral which probably carries the silver.

The Federal group of the National Gold and Silver Mining Company, situated south of the Volcano mine, was but little developed in 1905. The country rock, which is much altered, is probably silicified rhyolite but may possibly be diorite porphyry. Zones of brecciation, much silicified, project prominently above the surface, and prospecting along these revealed the presence of gold and silver. The principal "vein" strikes about N. 5° W. and dips 60° to 80° E. It is 5 to 20 feet wide at the outcrop and can be traced for more than 3,000 feet. As distinguished from the Volcano vein, little massive quartz is present; the brecciated country rock has been converted into a flinty hornstone and impregnated with sulphides similar to those at the Volcano mine. Gold constitutes the principal value of the ore, and where the material is oxidized fine specks of free gold can be seen. Some high assays are reported. Minute hexagonal plates of honey-yellow color are found in narrow cracks in the oxidized ore. The mineral may be vanadinite. Silicification evidently spreads out at the surface, for the outcrops are generally somewhat wider than the vein is found to be on development.

The Beck mine, also owned by the National Company, lies about 3 miles southwest of the Federal group. The mine is developed by a 300-foot inclined shaft, and the ore is treated in a small concentrating mill. The country rock is chloritized porphyry and is cut by dikes of monzonite porphyry having a strike to the northeast. The lode at the shaft strikes about N. 70° E., and a little farther northeast turns still more eastward; it dips steeply northwest. The ore consists of pyrite and zinc blende, with some galena, chalcopyrite, bornite, and chalcocite accompanied by calcite and quartz. Some of the ore, especially that in which zinc is plentiful, is said to carry high values in silver. The ore is said to have been rather richer in the upper workings. In some places the walls are sharp and in others the values simply decrease gradually. The chloritized wall rock carries much pyrite in rosettes of intergrown cubes; most of this is barren, but in some gold is found.

The ores of this district are believed to be of Tertiary age, associated with volcanic rocks.

SAN SIMON DISTRICT.[a]

GENERAL FEATURES AND GEOLOGY.

The San Simon district is situated in the Peloncillo Range only a few miles from the Arizona line and 12 miles north of Antelope station, on the El Paso and Southwestern Railroad. Animas Valley, in New Mexico, on the east side of the range, has a considerably higher elevation than

---

[a] By Waldemar Lindgren and L. C. Graton.

the San Simon Valley on the west, in Arizona, and several low passes in the Peloncillo Range connect the valleys. The railroad crosses at Antelope; 10 miles farther north is Cowboy Pass and 2 miles still farther north is Granite Gap. The elevation at Antelope is 4,415 feet and Granite Gap is but slightly higher. The mining camp lies on the western slope about 150 feet lower than the pass and overlooks the broad débris fan leading down to the San Simon Valley. Across the San Simon rise the rugged Chiricahua Mountains, and the rhyolite bluffs near the mining camp of Paradise in that range are in full view. Up to Cowboy Gap the Peloncillo Range is fairly high and appears to be capped by Tertiary lavas. Between Cowboy and Granite gaps it is a low ridge rising but a few hundred feet above Animas Valley, and falling off toward the San Simon as a steep bluff. The road crosses Granite Gap but little above the level of the valley and then descends to the camp. The ridge is built up of heavy benches of compact blue cherty limestone dipping from 10° to 30° W. No fossils were found and the horizon is perhaps below the Carboniferous. The gap is located on an east-west fault line bringing up a block of coarse-grained reddish granite on the north, in juxtaposition with the limestone on the south. It can not be stated definitely whether this granite is a post-Carboniferous intrusive or a part of the pre-Cambrian basement. At first glance distinctly intrusive rocks seem to be missing in the mineralized area, but closer search reveals several very irregular dikes of granite porphyry of no great width in the limestone bluff south of the gap.

GRANITE GAP MINE.

The Granite Gap lead mine was, in 1905, the only property of importance in the district, though a few prospects are scattered in the limestone as far south as Antelope or Railroad Gap. The mine at present consists of several consolidated properties, and the extensive workings, probably aggregating several miles, are located chiefly on the west side of the limestone bluff south of the gap. Work has been carried on for twenty-five years, but on a larger scale only since 1897, when Corbett & Wyman obtained control of the property. A consolidation of the Louise, San Simon, and Granite Gap claims was effected in 1903, and for two years the mines were operated under the direction of S. C. Pratt. Difficulties and lawsuits of various kinds followed, interfering with the production, and at present the control rests with the United States and Mexico Development Company. The total production is large, its value being estimated to be at least $600,000. The ore is shipped to El Paso and Deming and is much sought after by the smelters. In 1905 thirty Mexican miners were employed. Shipments were continued in 1906. There is no water at the mine; it is hauled to the camp from a well in the San Simon Valley. The workings are extensive and complex; a complete lack of maps rendered difficult the working as well as the study of the deposits.

In 1905 the principal production was derived from the Louise claim, at the south end of the property. This deposit is opened by a tunnel about 50 feet higher than the camp. A dike of fine-grained, light-colored granite porphyry, containing small phenocrysts of quartz and orthoclase, is shown at the tunnel, where it is 5 feet wide and dips 60° N. No contact metamorphism appears in the adjoining blue and compact limestone. Somewhat above the tunnel the dike widens, but within a short distance it pinches entirely. In the workings the dike is fairly regular. The ore occurs exclusively in the limestone, but always near the dike and generally on its hanging side. The workings extend above tunnel level and the ore has been followed for 100 feet below it. The workings are characterized by extreme irregularity and the development has proceeded simply on the basis of following the ore. Some of the stopes are 10 feet wide; but most of the ore bodies form smaller bunches, pockets, or stringers. The ore is almost wholly oxidized to cerusite, but some of it contains kernels of solid, probably recrystallized galena. A little copper, zinc, and arsenic is present. The only gangue noted is calcite in coarse aggregates; the ore contains also much limonite and manganese oxide. Very little gold is present, the shipped ore contains from 10 to 15 ounces of silver to the ton and up to 20 per cent of lead, besides a little copper. It is probable that the original ore consisted of galena and tetrahedrite. A few hundred feet north of the Louise are the middle workings, which extend through the entire hill from east to west. Practically no work was done here in 1905. The ore seems to follow a westward-trending fissure, and much lead carbonate has been extracted. Here, too, the deposit is opened by a tunnel, but at a somewhat higher level than in the Louise workings, and the ore has been followed down for a depth of 200 feet. Dikes are present, and it is stated that none of the best ore occurs far away from the porphyry. At the north end of the hill some oxidized ore containing a little copper and lead has been extracted. It occurred along a small dike of granite porphyry.

The intimate relation of the irregular bodies of lead ores in the limestone to the porphyry dikes and the absence of contact-metamorphic phenomena are the most interesting genetic facts in relation to these deposits. A careful and scientific study would probably show that the possibilities of the mine are not exhausted.

A deposit of fire clay having the appearance of a dark-gray shale (Cretaceous ?) is mined near Antelope and utilized at Bisbee for converter lining.

DEPOSITS BETWEEN GRANITE GAP AND STEINS PASS.

The Peloncillo Range continues northward from Granite Gap for about 20 miles. Twelve miles north of Granite Gap the Southern Pacific Railroad crosses it at Steins Pass. Between the two localities mentioned the range contains a few deposits which were visited by Mr. Graton in 1905. His description follows:

The Mineral Mountain mine lies about 4 miles southwest of Steins Pass. It is developed by an inclined shaft 200 feet deep, sunk on a vein that strikes N. 80° E. and dips 65° S. The country rock is much decomposed, but is probably close to diorite porphyry in composition. The vein consists of quartz carrying a considerable percentage of galena, most of it fine grained. In places the ore carries up to 20 ounces of silver to the ton. In October, 1905, the mine had been closed for over a year and was half full of water.

The Lizzie Paul prospect, 1½ miles south of the Mineral Mountain mine, shows a decomposed basic porphyry with a system of fractures running N. 55° E. and dipping 85° SE., in which occur decomposed pyrite and some chrysocolla and chalcocite. Some of this material is said to give fair assays in gold.

The Johnny Bull mine, situated about 8 miles southwest of Steins Pass, near the Arizona-New Mexico boundary line, is said to be controlled by Phelps, Dodge & Co. There are two inclined shafts; the deeper is said to have a depth of 150 feet, and is cut at a depth of about 40 feet by a short crosscut tunnel. When visited the mine had been closed for some time and only this tunnel was accessible. The prevailing rock is limestone. Fossils are present but are too poorly preserved to be of use in determining the age of the rock; it is probably Carboniferous. Structure has been obscured by metamorphism and alteration, but the strike of the bedding seems to be about north and south, with dip to the east. The limestone is cut by dikes of a rock that is probably monzonite porphyry. A large dike a little east of the mine is said to be traceable southward for several miles. The ore body is a typical contact-metamorphic deposit in limestone. The limestone is somewhat marmorized and contains numerous cavities lined with calcite crystals. Garnet and quartz are formed abundantly but in irregular bodies and patches. Cavities in the garnet contain crystals of quartz coated with opal. Epidote is not plentiful, but there is much of a nearly white mineral in radiating blades that is probably wollastonite. A dull-greenish material of dense texture is also probably a pyroxene. Bunches of massive pyrite are present in places and chalcopyrite, with an occasional speck of bornite, is sprinkled through the garnet. Limonite or a similar hydrous oxide of iron is formed near the surface, also coatings of malachite and azurite, and on the Copper Queen claim, just to the north, lumps of chrysocolla are found associated with quartz, calcite, and opal.

At the 40-foot level a stope about 40 feet long, 15 feet wide, and 15 feet high has been made. The garnet zone which this followed strikes about north and south, but appears to be steeper than the bedding of the limestones, having a dip of 75° to 80° E. Some stoping is said to have been done below. It is stated that the foot wall of this zone is porphyry, but this could not be determined from surface outcrops or the meager exposures seen underground. The copper content of the ore was evidently low.

It is said that some copper-lead ore was taken from shallow workings in limestone close to the large dike east of the Johnny Bull mine.

LORDSBURG DISTRICT.[a]

GENERAL FEATURES.

The Lordsburg mining district lies 3 to 8 miles south-southwest of the town of that name in central Grant County. Embraced in this district are two camps—Pyramid, at the south end, and Shakespeare, or Virginia, at the north. Lordsburg, the junction point of the Southern Pacific and the Arizona and New Mexico railroads, is situated on the high plateau that forms the floor from which rise the isolated desert ranges. One of these short ranges, known as the Pyramid Mountains, ends just south of Lordsburg; the mining camps are situated in the low hills at its north end at elevations of 4,500 to 5,500 feet.

F. A. Jones[b] states that prospecting began about 1870 and that in following years, especially just before and after the building of the transcontinental railroad, the district was active. As in so many of the early camps in this part of the Territory, silver was the metal chiefly sought. Later copper and some lead were produced, and gold was recovered incidentally. Much of the ore was treated in pan-amalgamation mills. The largest of these, the Leidendorf, in the Pyramid camp, was built in 1883 and was operated intermittently till 1893. Smelting was attempted in the early days, as is witnessed by remains of an old adobe furnace near the camp of Shakespeare. A few years ago a smelting furnace and a concentrating mill were erected at Lordsburg, but evidently not in any serious attempt to treat the ores of the region. In 1905 development was in progress at several places, but little production was being made. It has been stated[c] that up to 1907 the total production amounted to about $750,000 in silver, copper, gold, and lead, but from what could be learned this figure seems somewhat high.

The main part of the Pyramid Mountains south of the district is said to be composed of volcanic rocks, chiefly andesite. Exposures of fresh rock are rare within the district. The hills are mostly covered with residual soil and the intervening flats consist of characteristic desert wash. The hard rocks that are exposed are considerably decomposed. The type most in evidence is a grayish or greenish-gray rock whose porphyritic texture is brought out in some places by small white feldspar crystals. In other places, however, phenocrysts are not seen, the rock consisting of a cryptocrystalline aggregate. If the few specimens found sufficiently fresh to permit determination are representative of the mass, this most prevalent rock is a diorite porphyry. Pyroxene and a rather acidic plagioclase feldspar appear to have been the principal constituents. Chlorite and epidote are the chief products of alteration and doubtless account for the color of the rock; sericite is formed abundantly near some of the veins.

A noticeable feature of the geology and topography of the western part of the district consists of bold, wall-like ledges, locally called dikes. Most of these have about the same trend—east-northeast. Lees Peak is at an intersection of such ledges at a low angle. They are in reality brecciated and much silicified zones in the porphyry, and owe their prominence to their superior resistance to erosion.

As distinguished from the western part of the district, where the diorite porphyry is common, the eastern part shows more numerous exposures of effusive rock, probably andesite. The freshest of this rock that was seen consists of laths of plagioclase with a little pyroxene and numerous patches of limonite that by their outline suggest that they result from olivine. Aggregates of chlorite, carbonate, and epidote constitute the more altered varieties. Owing either to the small exposures or to the advanced alteration of the rock the structure generally can not be discerned. East of the Aberdeen mine, however, near the south end of the Shakespeare camp, the purplish and brownish rock shows distinct bedding. Though the beds are somewhat

---

[a] By L. C. Graton. [b] Op. cit., p. 58. [c] Jones, F. A., Eng. and Min. Jour., vol. 84, 1907, pp. 444–446.

contorted, the most prevalent attitude corresponds to a strike of N. 75° E. and a dip of 25° to 30° N. The rock here looks tufaceous.

Most of the information regarding the ore deposits was gained either from the statements of miners or from an examination of the dumps, as few of the workings of the district that had encountered ore were accessible at the time of visit. The ore deposits are fissure veins, and these are believed to cut both the andesite and the diorite porphyry, though in no single case could it be concluded with certainty that the wall rock was andesite instead of much altered diorite porphyry. A subdivision of the veins on the ground of mineral character seems warranted, although examination of all the veins might show transitional types. The kind of vein most exploited has quartz gangue, with calcite commonly in subordinate amount, but here and there constituting a large part of the gangue. Tourmaline is sometimes found in dark patches made up of rosettes of tiny needles. Chalcopyrite generally predominates over pyrite, and the richer sulphide, bornite, is commonly present, though not by any means so abundantly as thought by the miners, who mistake the bluish or purple tarnish of the chalcopyrite for bornite. Galena is occasionally seen in small grains, and here and there is rather plentiful. The ordinary zinc sulphide, sphalerite, is present in one or two veins. Molybdenite is sparingly present, but in such small foils that as a rule it is detected only by means of the hand lens. Some mineral in dark patches made up of extremely fine particles is present in much of the ore. It may be finely divided pyrite, but is more probably a dark sulphide.

The other type of vein has been worked in the Pyramid camp, at the south end of the district. The gangue consists of quartz, with plentiful barite and carbonates. Some of the carbonate is the pinkish rhodochrosite, called manganese by the miners, and some is probably siderite; calcite is also present. Cerargyrite (silver chloride) was most important, though some argentite and native silver are said to have been found in the richest ore. Chalcopyrite occurs sparingly and pyrite is rare. Molybdenite is present as in the veins of the other type, and to a less extent the finely divided dark sulphide, especially in association with rhodochrosite. No galena or other lead mineral was seen.

The veins are generally rather narrow, ranging in width from a few inches to a few feet. They are notably drusy, and in the first or copper type of veins there is a tendency toward radial arrangement of the quartz around centers of deposition which may represent replaced fragments of the wall rock. Between such radial clusters the druses are found. Minute crystals of secondary quartz have been deposited in some of these druses. The sulphides are present mostly in rather small grains, especially in the copper veins near the centers of the radial quartz masses, though narrow streaks of solid sulphide are found in some places.

Water is encountered in most of the mines at depths usually not much greater than 100 feet. In many places oxidation has not extended to water level, but in others it has been practically complete considerably below the present water level. Limonite is the most plentiful product of oxidation of the veins. Malachite is present in the copper veins, with chrysocolla here and there. Galena alters to cerusite and less abundantly to wulfenite. The silver chloride found in the silver veins is undoubtedly a product of oxidation. The rhodochrosite becomes covered with a sooty film of manganese oxide.

Most of the veins strike northeasterly, though some trend nearly or quite east and west. Probably the most common dip is toward the southeast. Some of the croppings are very prominent, being, in fact, the dikelike siliceous ledges already mentioned. As a rule, however, these silicified zones have not been found to carry good values, and most of the best producers have been marked by inconspicuous outcrops.

Little can be stated regarding the average value of the ore mined in the district. The range has been from the lowest value permitting extraction up to 20 to 30 per cent of copper and 100 or more ounces of silver to the ton in rich spots. The best silver ore is said to have averaged 48 ounces to the ton by the carload, with the average for shipping ore considerably under that and for the milling ore lower still. Lead and gold have been of some importance in certain mines.

Conclusions as to the genesis of these veins are rendered uncertain because of the unsatisfactory nature of the observations concerning both the general geology and the ores. It is believed, however, that the veins of the Lordsburg region belong in the class with those at Red River, Cochiti, Mogollon, and numerous other camps and that they are probably related to the effusive rock magma, which in this district is represented by andesite. The absence of certain features indicating formation at shallow depth commonly found in veins of that class may perhaps be accounted for by the circumstance that the greater part of the andesite that probably once covered the district appears to have been removed by erosion, and the portions of the veins that are now exposed were therefore at some distance from the surface at the time of their formation.

MINE DESCRIPTIONS.

The Leidendorf mine, also known as the Viola or Venus, was probably the largest producer of the district and is reported to have turned out between $150,000 and $200,000, mostly in silver. It is situated in the Pyramid section, at the south end of the district. There are said to have been several parallel veins; the main one, opened at points along the surface, strikes about N. 40° E. The shaft is about 350 feet deep, but in 1905 held much water. A 20-stamp pan-amalgamation mill was operated for over ten years before the great fall in the price of silver, since which time the property has been idle. It is claimed that the mill recovery was low, and reworking of the tailings has been considered.

The Last Chance mine adjoins the Leidendorf on the south. A shaft 180 feet deep was sunk to explore a vein striking a little more to the east than that at the Leidendorf and dipping about 60° SE. The mine was worked from 1882 to 1890, producing about $100,000 in silver. It was then idle for fifteen years, but was unwatered in 1905. The vein was said to be $3\frac{1}{2}$ to 4 feet wide. The ore had been taken from a shoot pitching to the east. Most of the ore had been mined out down to the 100-foot level, where the shoot was said to have a stoping length of 300 feet of shipping grade and still more of concentrating grade. The work in progress at the time of visit was aimed to extract ore from the lower levels that could not be profitably worked in the earlier days.

The Nellie Bly and Robert E. Lee mines, lying north of the Leidendorf and Last Chance, are owned by the Pyramid Peak Consolidated Mining Company. They are located on a vein that strikes northeastward and dips to the southeast, the Nellie Bly being southwest of the Lee. In a chloritized zone of rock that is probably the diorite porphyry are found veins of quartz and calcite carrying chalcopyrite and bornite. The Nellie Bly mine is developed by a vertical shaft 130 feet deep. At water level, 70 feet, one ore shoot was found, 50 feet long and nearly 5 feet wide; it carried chiefly chalcopyrite. On the same level a smaller mass of rich bornite ore was said to have been found. Chalcopyrite ore with $2 in gold to the ton was also found in two shoots on the lower or 118-foot level. It was stated that similar ore had been found in the Windmill shaft, 350 feet farther southwest. The production of the Nellie Bly up to 1905 was said to be $20,000 in copper, silver, and gold. The ratio of silver to copper was about 1 ounce of the former to each 20 pounds, or 1 per cent of the latter.

The Lee mine was said to be over 200 feet deep, but water stood up to the 70-foot level. A production of $110,000 is said to have been derived from a low-grade copper-silver ore, mostly through the operation of a Huntington mill, during the years 1894 to 1898.

The Aberdeen or Morningstar mine lies about 3 miles north of the Nellie Bly, in the southern part of the Shakespeare or Virginia camp. A quartz vein striking approximately east and west carries chalcopyrite with some galena and zinc blende. The ore had been concentrated in a small mill and a considerable production of copper, gold, silver, and lead is said to have been made. No work was being done at the time of visit. A newer inclined shaft, called the Atlantic, had been sunk on another vein striking N. 85° W. and dipping 80° N. This shaft, 150 feet deep, was caved and inaccessible below a depth of 90 feet. The vein varies from a few inches to 2 feet in width and in places carries a large proportion of chalcopyrite (locally called bornite).

The Bonnie and the Lena or Miser's Chest mines lie about a mile northwest of the Aberdeen and are said to open several parallel veins of northeasterly trend. The Lena, 300 feet deep, produced gold, silver, and copper from a quartz-chalcopyrite vein, but had been abandoned when visited. A shaft on the Bonnie had been sunk 200 feet, but as the ground was soft it had caved and a new shaft was being started. Chrysocolla and malachite were found in the upper part of the vein; below, chalcopyrite disseminated through the quartz vein carried, in addition to copper, gold and in places a fair amount of silver. Some high-grade ore had been shipped.

The North American Mining Company was sinking a new shaft at its Cobre Negro mine, about 1 mile northwest of the Bonnie. According to the manager, the old shaft was sunk to a depth of 270 feet on the incline, following a vein that dips 35° SE. Down to water level, 170 feet on the incline, the ore was mostly oxidized copper carbonate with a little sulphide and carried copper, silver, and gold. For 40 feet below, the vein was practically barren. Then came a narrow seam of sulphides which increased in width to 4½ feet, but again pinched to 6 or 8 inches in the bottom of the shaft. In this sulphide ore the gold values especially were good and there were fair values also in silver and copper. The development of too much water for the pump to handle caused this shaft to be abandoned. The new shaft, having an inclination of about 70° SE., was 150 feet deep at the time of visit. It followed a streak showing sparing amounts of malachite and partly decomposed pyrite. No good assays had been obtained up to that time. Some pockets of very rich gold ore were said to have been found in another shaft on the Cobre Negro property.

The Superior mine is situated about a mile east-northeast of the Cobre Negro. The vein, which strikes northeast and dips southeast, is developed by a shaft 405 feet deep. The ore is rich in quartz, with chalcopyrite, some bornite, and a little galena and molybdenite. It is stated that in the working of the mine only the ore carrying pay values in the precious metals was saved. In 1905 copper ore from the dump was being shipped, and it was said that a zone of low-grade ore 30 feet wide was present in the bottom of the shaft. The total production was reported to be $35,000.

The Dundee and Eighty-Five mines, between the Superior mine and the little town of Shakespeare, are possibly on the same vein. The strike, shown by exposures on the surface, is east-northeast. The ore is similar to that of the Superior, and some good copper ore has been mined. The Atwood mine, east of the Eighty-Five, is credited with a considerable production of gold, silver, and copper, extracted by concentration. The ore was taken from one of the strongly outcropping silicified zones. West of the Eighty-Five mine, on one of these silicified zones, the Century mine was worked some years ago. Lead was the chief product, and most of the ore was oxidized. Specimens seen on the dump are mostly of porous cerusite, or carbonate of lead, but some wulfenite, or lead molybdate, is present in thin, yellow plates, arranged like a honeycomb, with porous limonite filled in between.

A number of other properties in the district have been worked with more or less satisfactory results.

EUREKA (HACHITA) DISTRICT.[a]

GENERAL FEATURES.

The old Eureka camp is situated 8 miles nearly due west of Hachita railroad station, in the eastern foothills of the Hachita Range, at an elevation of about 4,800 feet. The locations extend north and south for a distance of several miles and the district includes the Copper Dick mine, 6 miles to the south of the central part. A few miles to the northwest, on the opposite side of the range, near Pothook station, are other deposits said to be on the contact of limestone and porphyry and carrying lead and some gold. These were not visited; a small production is reported. The Eureka district was discovered about 1880, the principal mines being the King, the American, and the Hornet. At the Hornet a water-jacket lead smelter was erected in 1883, but evidently did not run for a long time. The value of the metallic output of the district probably does not exceed $500,000; exact records are not obtainable. For many years after the boom in the early days the camp remained quiescent, but the advent of the El Paso and Southwestern Railroad in 1902 stimulated mining operations, and during the last few years

---

[a] By Waldemar Lindgren.

the value of the ores shipped annually from the district has sometimes reached $30,000. During 1907 new discoveries of gold deposits were made on the west side of the Hachita Range. Although the production is small, it can not be said that the possibilities of the camp are exhausted.

GEOLOGY.

The northern and lower part of the Hachita Range extends from the railroad 10 miles southward to a low pass whose elevation is about 5,200 feet. South of this pass rises Hachita Peak, to an elevation of 6,527 feet. The eastern foothills of this northern part are marked by the prominent bluff of a frontal ridge of limestone (probably Carboniferous), the beds of which dip 30° W. The deposits are situated just west of this ridge and are chiefly in Paleozoic limestone of similar dip. Abundant dikes and sheets of a quartzose porphyry, probably quartz monzonite porphyry, cut the limestone within the mineralized area. In contrast to conditions in the Hachita Valley, where the water stands several hundred feet below the surface, the water level in these foothills, in places at least, is within 60 feet from the surface.

MINES AT OLD HACHITA.

The most northerly and most productive deposit is opened in the King mine. According to local authority the ore extracted had a value of $300,000, but authentic figures are not available. The mine is said to be owned by the Owl Mining Company, of Boston. Intermittent work has been done on it, but at the time of visit it was idle. The shaft is said to be 400 feet deep. The water stood 60 feet below the surface. The deposit forms a vein in quartzose porphyry striking N. 52° E. and dipping steeply to the northwest. The width of this vein is at most a few feet. The intrusive rock seems to be a dike confined by limestone on both sides. On the north side this limestone is 40 feet distant and no distinct alteration could be seen at the contact. The ore contains copper and silver, principally as copper glance and silver chloride, in quartzose gangue.

Half a mile farther south is the American mine, which was likewise idle in 1905, but, like the King, it is worked at intervals. The total value of the production is reported to be about $100,000. The shaft has an elevation of approximately 4,800 feet and is 100 feet deep. The deposit is a regular vein striking N. 42° E. and dipping 70° NW. and is traceable for about 400 feet. Its width is at most a few feet. The country rock is a fine-grained limestone. Porphyry appears, however, in the lower part of the workings. Adjoining the vein the limestone is altered, but the decomposed croppings do not exhibit the character well. The dump shows abundant brownish and greenish limestone containing garnet, pyrite, and zinc blende. Thin sections show a normal contact-metamorphic rock in which, however, the zinc blende appears to be somewhat later than the garnet. Veinlets of calcite and pyrite cut across this altered limestone. The ore consists of galena, rich in silver as well as of pyrite, much zinc blende, and a little stibnite. It apparently occurs both in garnet and in a gangue of coarse calcite. The facilities for examination were very poor, but the deposit seems to be an interesting link between a vein and a contact-metamorphic deposit.

A prospect owned by A. C. Young was next visited. It is contained in hard and dense bluish-gray limestone which shows no trace of alteration, and no porphyry was observed in the vicinity. Some tons of very rich lead-silver ore containing wolframite are stated to have been shipped.

The Hornet mine, owned by the Elliott Mining Company, is situated three-fourths of a mile south of the American mine. The old water-jacket smelter marked Union Iron Works, San Francisco, 1883, still stands, but no information could be obtained as to the production. The mine was idle in 1905, but some production was reported in 1906. The country rock is unaltered limestone with probably Carboniferous fossils, and the beds dip 30° W. The deposit forms irregular, pockety masses, and the workings are said to extend westward for several hundred feet, following the dip of the sedimentary rock. The ore consisted of lead carbonate, stained black by manganese, running about 25 ounces of silver to the ton, and of heavy galena, said to be much richer in silver. The only gangue seen consisted of some coarse calcite. Near the water level zinc blende appeared with pyrite. Below the mineralized stratum a sill of porphyry 20 feet in thickness is reported to occur, which in turn is underlain by limestone.

A short distance south of the Hornet is the Silver Bell mine, in which development work was carried on in 1905 by A. C. Young, one of the pioneers of the district. Two carloads of lead ore were shipped in 1903. The deposit is reported to consist of a vein in limestone opened by shaft to a depth of 150 feet.

COPPER DICK MINE.

Ten miles southwest of Hachita, on the northern slope of Hachita Peak, is situated the copper mine known as the Copper Dick. It lies just south of the low pass in the range, at an approximate elevation of 5,200 feet, 6 miles south of the group of mines described in the preceding paragraph. It is stated that almost 20 carloads of copper ore were shipped a few years previous to 1905. Shipments of a lower grade of ore, averaging about 10 per cent of copper, were reported in 1905 and 1906. The total output is said to be about 1,500 tons. No work was being done at the mine at the time of visit in October, 1905. The developments consist chiefly of an incline following the croppings of an ore-bearing stratum at an angle of 10° for a distance of 100 feet; the ore body is blocked out by a few drifts of no great aggregate length.

Behind a frontal ridge of westward-dipping Carboniferous (?) limestone, cut by a few dikes of porphyry, lies a series of hard sandstones and shales which dip southward at gentle angles and which appear to be separated from the limestone by a fault. This series makes up the northern slope of Hachita Peak. No fossils were found, but the rocks probably should be assigned to the Cretaceous, the presence of which in this part of the country has not been hitherto known. The total thickness of the beds exceeds 600 feet. A remarkable feature is a fairly general contact-metamorphic alteration which is most pronounced at the mine. No large masses of intrusive rocks, to whose influence this alteration could be ascribed, were seen, but no thorough examination of the vicinity was made. The shales are black and compact, the quartzitic rocks very hard. The ore-bearing stratum is almost entirely composed of a yellowish-brown garnet with a little coarse calcite, indicating derivation from a limestone. Below it lies a white, hard, quartzitic rock in which the microscope reveals much fine-grained colorless pyroxene and garnet. Shales and garnet rocks overlie the deposit. The only intrusive rocks seen consist of three dark-colored lamprophyric dikes of slight thickness, which are hardly adequate to account for the general metamorphism. The microscope shows typical panidiomorphic structure with acicular brown hornblende, orthoclase, and andesine, and some secondary pyrite; the rock appears to be a vogesite. Cropping at the surface with but slight oxidation, the ore-bearing garnet bed is about 3 feet thick and has been followed for about 100 feet on each side of the incline. A fairly large tonnage of ore containing about 10 per cent of copper was blocked out, and the possibilities of the deposit were perhaps not exhausted. The ore is said to contain 2 ounces of silver and half a dollar in gold per ton. The copper is fairly evenly distributed as chalcopyrite intergrown in large grains with the garnet in such a manner as to indicate beyond doubt the contemporaneous origin of the two minerals.

SUMMARY.

The ore deposits of the Eureka district occur as veins or bedded or irregular deposits in Paleozoic limestone or Cretaceous (?) strata; more rarely as veins in a quartz-bearing monzonite porphyry. The district contains abundant minor intrusive bodies of the latter rock, which in places have contact-metamorphosed the sediments. The ores occur with calcite gangue, but locally also in closest genetic association with garnet. Galena (argentiferous), zinc blende, pyrite, chalcopyrite, chalcocite, and stibnite are the principal ore minerals. Wolframite is also reported from one locality. The whole occurrence points to the formation of the deposits by eruptive after-effects, and some of them appear to form connecting links between contact-metamorphic and vein deposits.

338

SYLVANITE DISTRICT.[a]

LOCATION.

The Sylvanite district covers an area about 6 miles long and 3 miles wide in the central portion of the Little Hachita Mountains. The town of Sylvanite, on the west side of the mountains near the west center of the district (see fig. 32), is about 12 miles due southwest of Hachita,

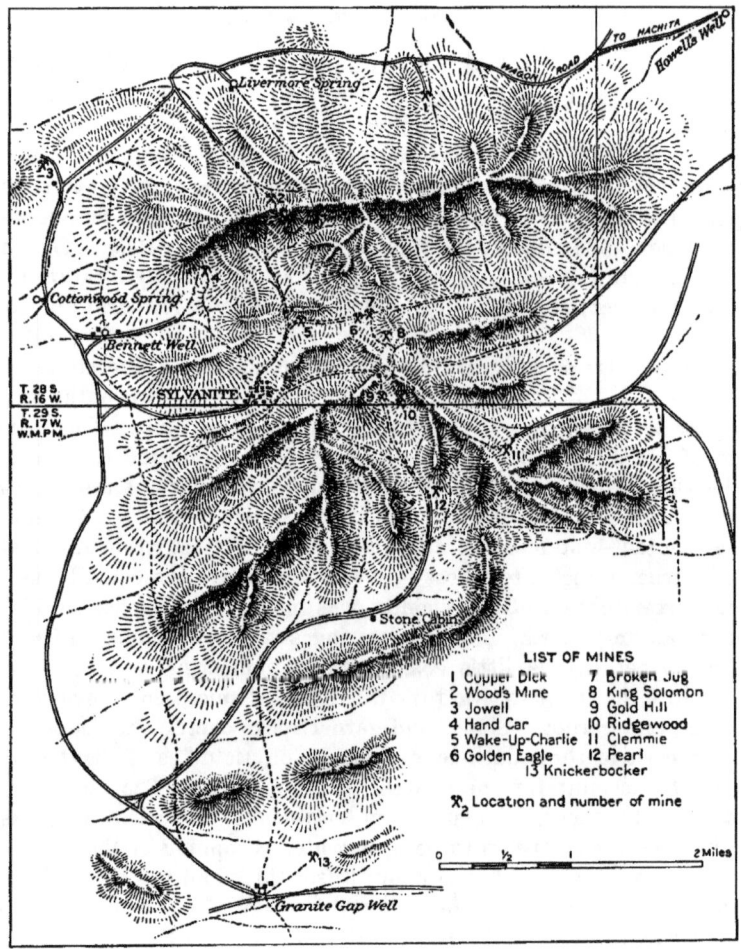

FIGURE 32.—Sketch map of Sylvanite district.

a junction point on the El Paso and Southwestern Railroad. The 18-mile wagon road between these two towns passes through a low gap 2½ miles north of Sylvanite.

HISTORY.

In the early eighties prospecting for copper in the region now known as Sylvanite resulted in several locations. While at work on one of these claims in February, 1908, "Doc" Clark discovered placer gold in a small gulch west of Livermore Springs. This started the Sylvanite boom. Gold in small quantities was obtained from many of the gulches on the west side of the mountains, though most of the washing was confined to the region north of Cottonwood and west of Livermore Springs. Dry washers and a few rockers were used, as water was scarce,

---

[a] By J. M. Hill.

and it is said that from $3 to $15 a day was recovered by a few men. The total placer production is variously estimated from $2,000 to $3,500. In the early part of March, 1908, placer work had largely been abandoned and prospecting for gold-bearing ledges began. Almost all of the region about the Sylvanite town site was located within a few days. From the first of March, 1908, to the end of the year Sylvanite had an average population of 500 persons. In June, 1909, only 70 remained in the tent village, with possibly as many more at the various prospects in the mountains.

GEOGRAPHY.

The Little Hachita Range rises about 2,500 to 3,000 feet above the desert valleys on either side—to the west the Playas and to the east the Moscow Wash. The slopes are steep on the east and northeast, with more gradual descents to the southwest. The valleys are usually short and of high gradient, with the exception of Stone Cabin Gulch, which drains practically all the southern and western parts of the region south of Sylvanite. There are four low east-west gaps through the range, two crossed by roads and two by trails.

Water for Sylvanite is obtained from several wells situated about a mile west of town, in a wide wash, and is extremely hard. There are two other places on the west side of the range where water is found near the surface. Livermore Springs, at the north, furnishes hard water at a depth of a few feet, and at Granite Gap a supply of fairly soft water is obtained from a 10-foot well. Howell's well, at the extreme northeast corner of the district, is about halfway between Sylvanite and Hachita. At the Hand Car mine (No. 4[a]) water is developed 40 feet underground, but none of the other workings in the higher mountains show the slightest indication of moisture. In the Playas Wash, about 3 miles west of Sylvanite, two large, shallow pools collect in wet seasons, and it is possible that a water supply could be obtained by pumping from wells near these lakes.

GEOLOGY.

In the Sylvanite district a thick series of thin-bedded quartzites, dark shales, limestone, and dolomite of Paleozoic or younger age is found. These rocks are considerably faulted, have been intruded by several different types of porphyry, and show some contact-metamorphic alteration. The sedimentary beds cover at least two-thirds of the area shown in figure 32. They are part of a monocline dipping to the southwest. No fossils were found in the area by which the age of the beds could be determined. Structural relations, however, apparently place them in the upper Paleozoic or later. Limestones north of the Sylvanite-Hachita road, tentatively referred to the Carboniferous (p. 336), are overlain by a series of thin-bedded shales and limestone that is similar to the series southeast of Sylvanite and is apparently connected with it on the east side of the range. They may prove to be of Cretaceous age.

In the region north of a line extending east from Bennett's well the western portion of the mountains is composed of a large mass of slightly quartzose monzonite, which appears to be laccolithic, as remnants of sedimentary beds are found on a few of the higher ridges. This body is cut by numerous small dikes of kersantite or closely related rock, along which considerable epidote has been developed. A few quartz-filled fissures in this mass, with prevailing east-northeast strike, carry a little malachite and iron oxides. East of the monzonite thin-bedded sedimentary rocks are exposed, forming the great mass of the mountains, lying almost flat near the monzonite, but dipping steeply to the west-southwest on the east. It would appear that the tilting of the beds to the southwest had been followed by flattening or raising of the northwest section by the intrusion of the monzonite. The laccolithic mass extends southward under the flat on which Sylvanite stands, but the hills both east and west of the town are capped on their south sides by sedimentary rocks. (See Pl. XXII, B.)

One-half mile north of Sylvanite there is a very prominent red porphyry dike that strikes east and west and is traceable for at least 1½ miles east of the point where it emerges from the wash near Bennett's well. Fresh surfaces of this dike show a pink to buff groundmass in which

---

[a] Numbers after names of mines refer to the location numbers on figure 32.

are fairly large crystals of feldspar and biotite. The groundmass is composed of minute crystals of orthoclase with a few grains of quartz. The rock called "trachyte" by the miners is a porphyritic phase of syenite, dikes of which are found in other parts of the camp. Along this dike the monzonite has been considerably altered with the development of sericite, biotite, and chlorite. As a rule this alteration results in a dark-green to black rock, but at the Golden Eagle mine (No. 6), which lies just south of the dike, the resulting rock is chocolate colored, owing to the development of a brown mica and another mineral whose exact character has not yet been determined.

The Hachita Mountains to the south and east of Sylvanite are made up mainly of sedimentary rocks. The general dip is to the southwest, the angles being 50° at the north and 15° to 20° at the south end of the belt. There are at least two intrusive porphyries in these beds. They occur as dikes and sheets, the latter possibly rather numerous, as those seen were thin and could not be distinguished from the dark shales and limestones without close inspection. Two dikes in particular were noted. One, a porphyritic diorite containing a little quartz, is exposed along the crest of the ridge between the Gold Hill and Clemmie properties (Nos. 9 and 11). The other, opened at the Gold Hill mine, is a very fine-grained buff-colored syenite dike that under the microscope is seen to be composed of orthoclase crystals, a few grains of quartz, with sericite and chlorite as alteration products. Along these intrusions a small amount of sericite is developed in the limestones, which are also somewhat silicified for short distances from the contacts. No garnet was seen at any of the contacts noted, but pieces of float and the sand in a few of the gulches indicate its presence in some localities.

The area has suffered two postmineralization movements of comparatively slight throw. The first, an east-west displacement, manifests itself in clay zones along some of the veins and in the center of the big Gold Hill dike. The second, a north-south movement with throw of about 2 feet, is shown by the displacement of veins at the Gold Hill and Pearl.

Just north of the Granite Gap well is an east-west fault of considerable vertical displacement, which brings a coarse pink granite into contact with thin-bedded shales and quartzites. This contact is traceable for at least 4 miles across the range and may be considered the southern limit of the Sylvanite district. Two sets of dikes are seen in the granite south of this fault. The older are very fine-grained buff syenites, very similar to the Gold Hill dike, and the younger are dark basic dikes which resemble the kersantite dikes of the monzonite area.

ORE DEPOSITS.

The ore bodies of the Sylvanite camp have not been opened to a depth sufficient to warrant any assertion as to their ultimate character or extent. Most of the mines are located on quartzose veins that have a prevailing easterly strike or are associated with intrusive rocks having that strike. The surface ores that have so far been exploited have high gold values, found free with tetradymite in a sericitic limonite-stained quartz. At depths of 30 feet the veins are oxidized but contain some sulphides with the carbonates. Very few of the shafts or tunnels have reached more than 60 feet below the surface. From the rapid increase of chalcopyrite and pyrite and the decrease of gold values in this depth, it seems probable that with further development bodies of comparatively low grade auriferous sulphides will be found in place of the free gold and tetradymite combination of the upper portions of the veins.

Four varieties of quartz veins were noted—in monzonite, in altered monzonite near intrusive syenite, entirely in syenite dikes, and in sedimentary beds without apparent igneous connection. Besides veins, two other forms of ore occurrence were noted—a bedded deposit between limestone and intrusive diorite and a quartzite impregnated with sulphides.

The few veins in monzonite contain white banded quartz, somewhat iron-stained directly at the surface, that strike east-northeast and dip at very high angles to the southeast. In the upper few feet these veins are somewhat stained with malachite and azurite, but at depths of 10 to 12 feet pyrite and chalcopyrite appear to be the only ore minerals present. These sulphides are reported to carry small quantities of gold. Some rich ore taken from the surface of one of these veins was largely sericitic, iron-stained quartz with free gold.

About 250 feet south of the big east-west syenite dike north of Sylvanite there are a series of quartz stringers in the altered monzonite. These, so far as seen, strike in two directions, east and northwest; the plane of the intrusion is east and west. The vein filling is massive quartz containing disseminated free gold, tetradymite, and a little pyrite. The veins, so far as development shows, stand almost vertical, vary from mere seams to veins 4 inches in width, and are much richer at the surface, where they are rusty, than in depth. The quartz is invariably separated from the walls by a thin film of chlorite gouge and is easily sorted. At the Golden Eagle mine a small amount of orthoclase feldspar is intergrown with the quartz.

In the syenite dike at the Gold Hill mine numerous somewhat rusty quartz stringers cut the dike in all directions. An east-west shear plane in the center of the dike is filled with about half an inch of limonite, malachite, and azurite, which in a few places opens out into small bodies of pyrite and chalcopyrite. A second series of small fractures in north-south directions contain, besides a film of gouge, a little limonitic quartz. On the surface the limestone adjacent to the dike is somewhat siliceous, and in a few places bodies of sericite are intimately mixed with quartz, unaltered limestone, syenite, tetradymite, and free gold.

Small fissure veins cutting sedimentary beds are opened here and there, but so far as seen only one had any particular value. The Pearl vein strikes east and west and dips to the south at steep angles. It averages 18 inches in width and is filled with quartz, calcite (some of which is brown), and very finely disseminated free gold. The gold was found largely within 10 feet of the surface and was said to be accompanied by tetradymite. A little free gold was visible at a depth of about 30 feet in one place, but the greater part of the vein is barren below 15 feet.

Only one example of a deposit between sedimentary and intrusive rock was noted. The igneous rock is a quartz-bearing diorite sheet, varying from 18 inches to 12 feet in thickness, lying under a dark, unaltered limestone. The diorite contains a little disseminated pyrite and chalcopyrite, particularly near places where it is cut by quartz-calcite stringers that show the same minerals. Immediately above the intrusive rock there is about 6 inches to 1 foot of limonitic quartz and calcite, much of which is stained by malachite and azurite. In some specimens a little dark mica, probably biotite, is seen intergrown with the gangue minerals.

At one place a 15-foot bed of quartzite lying between dark limestones is very much jointed in east-west and north-south directions. These joints are filled with one-sixteenth to one-half inch of quartz, which is said to carry good gold values. In the same bed about 30 feet underground the interstices between the quartz grains are filled with pyrrhotite and a little chalcopyrite, forming a low-grade auriferous copper ore.

DETAILED DESCRIPTIONS.

The Wood or Buckhorn property (No. 2), situated just north of the crest of the highest ridge west from Little Hachita Peak, is the oldest location in the district; it was first worked in 1881. The property has produced 2,200 pounds of ore averaging $129 a ton, but the general tenor of that in the mine at present is said to be about $15. The deposit is an open fissure which strikes N. 65° W. and dips steeply to the northeast. The hanging wall is composed of thin-bedded silicified sedimentary rocks. The foot wall is a very much sheared, altered basic igneous rock that was probably originally monzonite. The vein is partly filled with white vuggy quartz, the openings being coated with limonite, hematite, and pyrolusite. On panning this material some magnetite, a little pyrite, and a few colors of gold were visible. There is a 300-foot tunnel and a 75-foot shaft on the vein which varies from 2 to 15 feet in width, with an average of 3 feet.

The Jowell property (No. 3) is situated in some low limestone hills about 1 mile north of Cottonwood Springs, on a quartz vein that strikes N. 60° E. and is traceable by discontinuous outcrops for about 2,200 feet. The two shafts on the property are said to be 65 and 90 feet deep, but in June water stood in both of them within 20 feet of the collar. It is said that the vein is 4½ feet wide between good walls. The material on the dumps is a soft sheared chlorite and quartz, impregnated with pyrite and a little chalcopyrite. Some native silver is said to occur in this gangue, but none of it was visible.

The Hand Car claim (No. 4), the property of "Buck" Bennett, is located about three-fourths of a mile east of his well near the top of the mountains. The development work consists of a 200-foot crosscut tunnel, some open cuts, and a 20-foot shaft exposing a quartz vein that strikes northeast and southwest and dips 85° SE. The country rock is sheared and altered quartz-bearing monzonite. The surface ore is said to have contained some sericite, free gold, and tetradymite and to have given very high gold assays. The ore in the tunnel, about 180 feet under the crest of the hill, is largely pyrite, with very small quantities of chalcopyrite that is reported to run 0.5 per cent of copper and $16 in gold to the ton.

The Golden Eagle (No. 6) is worked by a 30-foot shaft in a brownish altered monzonite that contains disseminated pyrites. Two small quartz stringers are exposed, one striking east and west and the other N. 65° W., both of which carry tetradymite and free gold in a massive white quartz gangue. Some specimens of this ore show crystals of white orthoclase intergrown with the quartz. The property has produced 3 tons of $25 gold ore, and the present lessees expect to make a small shipment in a few months.

The Gold Hill or Martin & Norton mine (No. 9), at the head of Stone Cabin Gulch, is a little less than a mile due east from Sylvanite. The mine was located by J. M. Wilcox and I. E. Predmore, who shipped 120 tons of $19 gold ore and 880 pounds of sericitic surface ore that is reported to have averaged 7 ounces in gold to the ton. The mine is now the property of L. G. Carlton, of Cripple Creek, who has a small force of men at work. The ore occurs in stringers of quartz, some of which carry a little chalcopyrite, malachite, azurite, and limonite, and which cut a very light buff syenite dike in the usual east-west directions. The dike strikes practically east and west and dips at very high angles to the south. Three postintrusive movements are indicated. The first was an east-west strain producing cracks that were later filled with quartz; this was followed by another east-west movement along the center and sides of the dike-producing gouge; and last there was a north-south movement that has displaced the veins about 2 feet. The gouge formed during the second movement carries $40 to $60 a ton in gold in sheared limonite, malachite, and azurite, with chalcopyrite and a little pyrite in places. The whole dike, however, is reported to average about $4 a ton in gold even where metallic minerals are not visible. North of the syenite are a few outcrops of porphyritic diorite apparently connected with the main dike of this rock seen on the top of the ridge to the east. The general country rock is a series of thin-bedded silicified limestones, dark shales, and quartzites that dip 40° to 50° SW. The contact with the syenite is in many places frozen, showing no alteration other than the silicification except in one place on the east hill, where a large body of quartz and sericite is developed that carries tetradymite and free gold.

The development work consists of a long open cut and three tunnels, all on the faulted or sheared zone in the syenite dike. The two lower tunnels start from the same elevation in a gulch and run east for 160 feet and west for 250 feet. Seventy-five feet above, on the west side of the gulch, there is a 140-foot tunnel which starts just at the lower end of the open cut. The cut is 150 feet long and about 15 feet deep. From it was taken most of the 120 tons that went into the largest shipment.

The Ridgewood property (No. 10) joins the Gold Hill on the southeast and is owned by the Eureka Sylvanite Mining Company. Some rich gold ore was taken from surface cuts and a shallow shaft on the hill southeast of the tunnel mouth, but no ore was encountered in the tunnel, which is 40 feet long. The deposit is bedded, dipping southwestward between overlying limestone and a sheet of porphyritic diorite. Between these rocks there is from 6 inches to 1 foot of quartz and limonite that has a little copper carbonate stain in the richer portions and contains some dark mica (biotite?) near the intrusive. Some of this ore from the surface ran high in free gold.

The Clemmie property (No. 11), on the east slope of the mountains, is 2 miles by trail east of Sylvanite. It is owned by W. T. Holcomb, who states that 1,800 pounds of siliceous gold ore that assayed $40 a ton has been shipped from surface cuts. The values are found in rusty joint planes and quartz stringers that cut a fairly heavy bed of quartzite in approximately north-south and east-west directions. The quartzite strikes N. 45° W., dips 50° SW., and is traceable up along the trail toward Sylvanite for about 900 feet. There is a 50-foot tunnel on the foot-wall side with 20 feet of drifting to the south on a fault striking S. 65° W. This fault is 30 feet from the mouth. Beyond this plane the quartzite becomes coarser and has fewer quartz veins cutting it. The interstices of the coarse quartzite are filled with pyrrhotite and chalcopyrite, forming an ore that is said to assay 6 per cent of copper and to carry a little gold.

The Pearl or Monte Cristo mine (No. 12) is in Stone Cabin Gulch, 1½ miles southeast of Sylvanite and about three-fourths of a mile south of the Gold Hill. The four claims are controlled by Texas people. They were located in March, 1908, but no discovery of mineral was made until December of the same year. Nine tons of gold ore has been shipped and is said to have yielded returns of $1,000.

The deposit is a quartz-calcite vein, varying from 1 inch to 8 feet in width, with an average of about 18 inches, that strikes N. 85° E. and dips about 80° S. This vein cuts thin-bedded black shales and blue limestone that strike N. 55° W. and dip 40° to 45° SW. No croppings of either dike or sheet were noted. The quartz is locally frozen to the walls, where it is in comb form, but there has been postmineral movement along the vein, as a clay gouge is present on the walls in most places. A north-south postmineral movement is shown in the east tunnel by a fault displacing the vein 2 feet.

The development consists of several short tunnels and an open cut, together with a short tunnel on the east side of the ridge on which the mine is located. It is reported that in the open cut good values were found to a depth of 10 feet, below which the vein was almost barren, carrying $2 a ton at the face of the lower tunnel, which is about 35 feet underground. In the east tunnel the vein is about 18 inches wide and contains considerable brownish calcite with the quartz. Some of this material carries beautiful specimens of free gold, but its scarcity and irregular distribution make the value of the ore uncertain.

The Knickerbocker or Quartzite mine (No. 13) is at the extreme south end of the district, about half a mile northeast of the Granite Gap well. It is reported that some very rich silver ore, running as high as 1,120 ounces to the ton, was shipped from this mine in the early eighties. The deposit is an iron-stained quartz vein that strikes almost due east and west and apparently dips to the north at high angles. This vein cuts thin quartzite and shale beds, and some of the quartz is found following the bedding, which dips about 20° SSW. The sedimentary series is cut off by a fault about 300 feet south of this mine, with granite on the south side of the fault plane. The Knickerbocker vein has been traced for nearly 1½ miles to the east, maintaining an almost constant distance to the north from the fault. It is opened by shallow pits and apparently most of the ore was taken from the seams parallel to the bedding rather than from the vertical vein.

APACHE No. 2 DISTRICT.[a]

TOPOGRAPHY.

A small group of bare rounded hills rise from the plains 6 miles south-southeast of the town of Hachita, on the El Paso and Southwestern Railroad. The elevation at their base is but slightly higher than that of Hachita, which is 4,504 feet above sea level, and their summit probably scarcely attains 5,500 feet. Broad débris fans stretch out from the hills in every direction. The Apache No. 2 district lies at their southwestern base, and its principal mine, the Apache, has an elevation of 4,655 feet (aneroid). The locations extend for some distance in an easterly direction and the mineralized belt is said to be continued in the Sierra Rica group of hills in Mexico, just southwest of the corner in the boundary line.

GEOLOGY.

The mass of the hills appears to consist of a quartz-bearing porphyry. The rock near the mines is greenish gray and contains phenocrysts of quartz up to 5 millimeters in diameter; also abundant small white prismatic sections of plagioclase. The ferromagnesian silicate probably was hornblende, but this is now replaced by epidote. The groundmass is light gray,

---
[a] By Waldemar Lindgren.

microcrystalline, consisting of quartz and orthoclase and not very plentiful. No ore minerals were noted in the porphyry. Intrusive rock of this type is common in the mining districts of New Mexico, and an analysis would probably show its affiliation with the quartz monzonite porphyries.

The first rocks seen in the extreme southwestern foothills on the road from Hachita are bluish-gray, unaltered, evidently Paleozoic limestones with gentle dip. Swinging around this southwestern spur the road enters a gulch leading up to the main peak of the hills, and in the lower part of this gulch the Apache mine is situated. The irregular contact of porphyry and limestone continues eastward along the southern foothills of the group. Within a few hundred feet of the contact the limestone becomes contact metamorphosed, but the change is not uniform in all places. At the main shaft the limestone is altered to an extremely coarse-grained calcite; nearly the same limestone is converted into greenish or brownish garnet mixed with calcite.

MINERAL DEPOSITS.

The Apache mine is the only producing property in the district and is worked on a deposit of contact-metamorphic type. It lies on the west side of the little gulch already referred to and on the east side of a tongue of limestone projecting northward for a few hundred feet into the porphyry. The deposit was discovered by Robert Anderson and has been worked by him or by lessees for many years. In the early days some rich horn-silver ore was shipped from the property, but for the last few years the main product has been a low-grade oxidized copper ore rich in calcite and hence in good demand at the smelters. The ore is hauled to the railroad at Hachita and several thousand tons have been shipped during the last few years.

The strata of limestone at the mine strike in a general northerly direction and dip from 40° to 70° E. They are locally metamorphosed to an unusually coarse aggregate of calcite, many of the cleavage pieces being 6 inches in diameter, and the ore shipped consists almost wholly of such coarse calcite with associated copper minerals. The outcrops are indicated by heavily iron-stained rocks from which the calcite has been partly leached; they lie directly at the contact of limestone and porphyry.

Near the surface much ore, forming irregular bodies, has been stoped and the shoot pitches somewhat to the south. A vertical shaft 150 feet deep has been sunk a short distance south of the main croppings, and a winze sunk 100 feet below the deepest workings shows good oxidized ore in the bottom. There is no water in the mine. Some of the stopes are as much as 30 feet in width. The developments have been confined to the eastern contact, along which the porphyry adjoins the limestone without faulting. No cross-cutting has been done to the western contact, which probably is not more than 300 feet distant. The ore follows a certain stratum which has been changed by contact metamorphism to coarse calcite instead of to garnet. There is no ore in the porphyry. The calcite contains but little iron and almost no magnesia; apparently the ore is oxidized throughout and the predominant mineral is malachite, mostly in tufted crusts deposited with secondary calcite crystals in cavities of dissolution. There is also a little chrysocolla. A little chalcocite was noted in the deepest part of the mine. The iron appears partly as limonite, but this has been largely dehydrated to red or steel-gray hematite. Pseudomorphs of this mineral after pyrite are common.

Examination of the ore in the mine revealed some pyrite and chalcopyrite in small grains disseminated through the coarse calcite. In places remaining masses of such probably primary and very low grade sulphide ore are surrounded by outer crusts of limonite and hematite and a narrower inner crust of malachite, indicating a progressive process of concentration exactly analogous to the "kernel roasting" in metallurgy. This important occurrence, which throws light on the processes of oxidation, is described in more detail on page 55. As shipped the ore contains 3 to 4 per cent of copper and about $1 to $1.50 in gold and 6 ounces in silver to the ton.

About 800 feet east of the main shaft are rather extensive shallower workings from which the silver ore was extracted. This ore, also oxidized, was contained in a bed of limestone which has been more or less completely converted into garnet.

FREMONT DISTRICT (LUNA AND GRANT COUNTIES).[a]

GENERAL FEATURES.

The following notes on the Fremont district are the result of a day's visit in the latter part of June, 1909. The district lies 15 miles southeast of Hachita and about 5 miles south of Victorio siding (both on the El Paso and Southwestern Railroad); and is in the northwestern foothills of the Sierra Rica, along the United States and Mexico line at the point where the boundary turns to the south (fig. 33). Most of the mineralization is in Mexico, only the outliers being in the United States. There is no settlement in the region except the small camp at the International mine. At present water is hauled from Victorio for domestic use, but a well is being sunk about 3 miles west of the International and 2 miles north of the Eagle mine. Fuel is obtained from mesquite, scrub oak, and other desert shrubs. There are three good roads into this camp—one from Victorio and two from Hachita. Of the latter the better

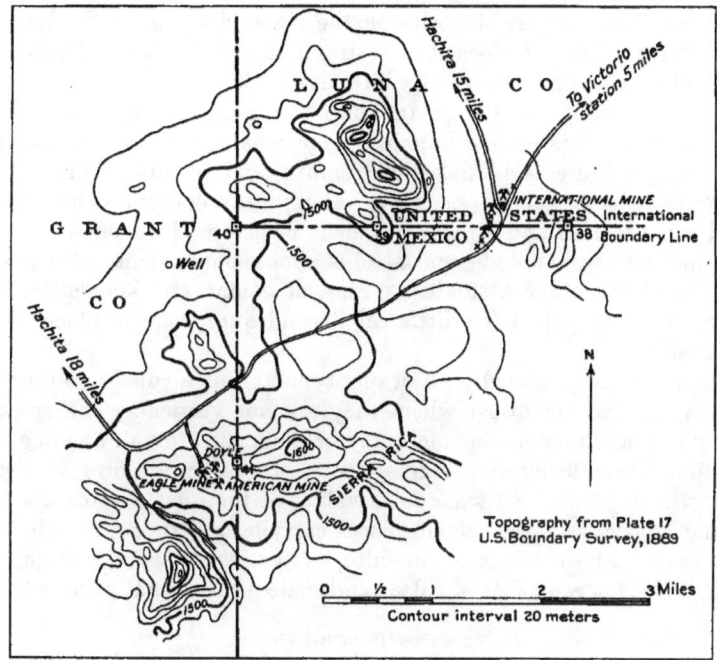

FIGURE 33.—Map of Fremont district.

and shorter is the road to the International mine, which skirts the north side of the Apache Mountains. The other goes around the west end of the same hills, within half a mile of the Anderson-Apache mine.

The principal properties of the district are the International, with a reported production of $750 in silver-lead ores; the Eagle, whose total production of argentiferous galena is stated to be about 200 tons; and the Daisy, with a reported production of $8,000 worth of copper carbonate ores carrying silver. Other properties are the Silver Fox, American, Doyle Zinc, and Barnett.

GEOLOGY.

The country rock of the region is largely limestone, ranging from thick blue strata to thin gray and buff beds, including in one place a heavy bed of red limestone. Along the east-west boundary the bedding is usually thin with a prevalent dip of 30° to 45° WNW. A vertical dislocation that cuts across the corner of Mexico with a strike of N. 30° E. forms the vein at

---

[a] By J. M. Hill.

the International mine. Parallel to this main fault and north of it are several smaller zones of movement. Along the north-south boundary the formations are heavier-bedded blue to gray limestones, probably Paleozoic, that dip to the west at low angles (15° to 20°). These beds are also cut in several places by small northeast-southwest faults.

There are evidences of the intrusion of at least two types of porphyry, though in the area seen outcrops of these rocks were very indistinct and relations uncertain. It is said by operators on the Mexican side that in the central part of the Sierra Rica the intrusive rocks are much more important, occurring as stocks, dikes, and sheets. Of these rocks one is a granite porphyry with a very fine grained light-buff groundmass, containing phenocrysts of quartz, usually about one-eighth inch or less in diameter; the other is a dark, finely crystalline lamprophyric rock of the kersantite family containing a few long needles of hornblende. No evidences of contact metamorphism were seen along these dikes.

ORE DEPOSITS.

There are two classes of ore deposits on the American side of the line—replacements in limestone and quartz veins in dislocations cutting limestone. Both of these forms of deposits seem to have a close connection with the intrusive rocks.

In the replacement deposits the greater part of the ores so far exploited are oxidized, and of these there are two types—(a) a limonitic, somewhat quartzose ore containing malachite, tenorite(?), copper pitch ore, some flakes of bismuth, and possibly a little brittle silver; (b) a quartz-limonite-manganese gangue occurring along watercourses, with cerusite, very little malachite, and some masses and streaks of galena coated with carbonates. Two replacement deposits of sulphide ores were noted—one a bed of almost pure galena, with a very minor amount of chalcopyrite, and the other a small exposure of galena and sphalerite. Neither of these deposits has much gangue, though a little calcite and quartz are in places present, intimately mixed with the sulphides.

The quartz veins are practically all of one type and as a rule have a northeast-southwest strike, with a dip to the northwest where they are not vertical. The croppings are stained almost black by iron and manganese and many of them show rough banding due to deposition. Sulphides are found practically at the grass roots, but carbonate ores are associated with the sulphides to depths of at least 60 feet. The values of the quartz veins are found in argentiferous galena and chalcopyrite, at certain places carrying a little gold. The surface ore occurs with a vuggy quartz, and consists of limonite, pyrolusite, cerusite, malachite, a little azurite and horn silver, and a few remnants of galena and chalcopyrite usually incrusted with carbonates.

MINE DESCRIPTIONS.

International mine.—The International mine is situated just north of the international boundary between the United States and Mexico, half a mile west of boundary post No. 38. The first claim was located in 1880 by Volney Rector, the present owner. It is reported that the mine has produced about 50 tons of lead-silver ore, valued at $25 a ton, all of which was shipped previous to 1889. The best shipment, a small one of 10 tons, ran 40 per cent of lead and $62 to the ton in silver, when the latter metal had a value of 95 cents. At present there are about 80 tons of galena ore of medium grade on the dump and as much more in one of the stopes.

The deposit is a banded quartz vein from 2 to 10 feet wide, with a usual average of 3 feet, occupying a fault that strikes N. 30° E. and has a high dip, varying from 85° to 90° NW. The flat in which the mine is located is underlain by probably Paleozoic limestones that strike N. 50° E. and dip 30° to 40° NW. The north or hanging side of the vein is composed of thin gray to buff beds separated from the vein quartz by a coarse red grit, the grains showing enlargement somewhat similar to that in a quartzite; this grit is very constant throughout the mine. The foot wall is a massive red limestone.

The vein is traceable for 5,000 feet, 2,000 feet of which are in the United States. In this distance there are numerous pits and shafts. Just north of the line some stoping has been done from shallow shafts and open cuts. The ore is in a limonitic quartz gangue; it consists of malachite, azurite, some cerusite, and a few kernels of chalcopyrite and galena, and is said to run 10 per cent of copper and 60 ounces of silver to the ton. About 500 feet farther north a 150-foot shaft was sunk, starting on the vein. There was good galena ore in the upper levels of this shaft, but the vein where cut at 150 feet is filled with barren quartz. At 250 feet north of this opening is a 50-foot shaft with 60 feet of drift southwest on the vein. In this stope there are two grades of ore. The hanging-wall ore is limonitic quartz, with a little galena, but the reported large silver values apparently come from chloride and cerusite. On the foot wall there are 2 to 3 feet of massive galena ore, said to run about $35 a ton in lead and silver.

Eagle mine.—The Eagle mine, half a mile west of boundary post No. 41, is about 3½ miles southwest of the International mine and 18 miles south-southeast of Hachita. The mine was worked in the late eighties and possibly to some extent before that time. The owner is Volney Rector. About $5,000 has been expended for development and a total of 200 tons of argentiferous galena has been sold that averaged 40 per cent of lead and 20 ounces of silver to the ton. Shipments of 80 tons each were made in 1906 and 1907.

The country rock is blue (Paleozoic?) limestone and dolomite, rather thick bedded, with a north-south strike and flat westerly dip. There does not appear to be any intrusive rock present and the limestones show no metamorphism except that represented by the ore bodies. The deposit is clearly a selective replacement in limestone along what appear to be watercourses.

The work on the claim consists of an old irregular shaft 30 feet deep from which some ore has plainly been taken. This was found in open watercourses largely occupied by quartz, calcite, and limonite, but containing cerusite and a few pockets and stringers of galena. The new shaft, 100 feet south of the old workings, is sunk 40 feet through massive dark-blue limestone, showing no metamorphism. At the bottom of this shaft a 2-foot bed of galena with a very little quartz and calcite was cut, below which there is apparently unaltered limestone. A 35-foot drift to the east shows the ore body to be widening in that direction and at the face of the drift there is 6 feet of almost pure galena. The roof is hard, firm limestone, very even, and showing no metallic minerals. The floor, on the other hand, is irregular and contains masses and impregnations of ore. This ore body, its position and dip being taken into consideration, should be found below the old workings at no very great depth.

Daisy mine.—The Daisy mine, in the extreme northern part of the Fremont district, lies on the eastern slope of the Apache hills about 7 miles southeast of Hachita. There are evidences of very old surface workings on this property and it is rumored that the mine produced some ore in the late sixties. The claim is owned by the estate of Dave Whaley. D. B. Davidson, who is working a small force of men, estimates the total production of the mine as about $8,000.

The country rock is blue, fairly massive limestone without fossils, striking east and west and dipping to the south at low angles. About half a mile north of the mine the sedimentary rocks are capped by basalt, a few remnants of which are to be seen near the workings. At this mine the limestone is cut by a fault which strikes N. 40° E. and dips 60° SE. Along this fault a limestone breccia of considerable thickness is found in which the ore occurs as "bowlders" and small stringers. The ore is, as far as developed, soft limonite, malachite, tenorite (?), and copper pitch, with a little quartz, chalcopyrite, native bismuth, and possibly some brittle silver that is reported to average, when sorted for shipment, 18 per cent of copper, 18 ounces in silver, and $1 to $5 in gold to the ton.

The workings consist of a number of open cuts and shallow, irregular shafts grouped on a flat within an area of 200 by 75 feet. On a low hill to the southwest there are some shallow pits showing a continuation of the fault in that direction. None of the workings are more than 40 feet deep, and most of them are less than 25 feet. They follow small, irregular stringers of iron and copper carbonates which, in a few places, open out into lenses of ore 2 or 3 feet in largest dimension. The development gives a very inadequate idea of the kind or extent of the deposit.

Other prospects.—There are various small prospects throughout the region in the limestone. In the basalt, at the north end of the district, a few shallow pits on east-west joint fissures show a small amount of copper carbonates that are said to run fairly well in silver.

Within a mile from the Eagle mine there are a number of prospects in limestone. Most of these are iron-quartz veins cutting the sedimentary beds, usually in northeast-southeast directions, and standing almost vertical.

The Silver Fox mine, one-half mile southwest of the Eagle, is developed by a 90-foot inclined shaft on a 2-foot iron-quartz vein, that strikes northeast and southwest, dips 30° NW., cuts limestones, and is near a mass of dark porphyritic kersantite. It is reported to have produced some ore in the early days of the camp, and at present there is some galena ore on the dump that is said to average 20 per cent of lead and 15 ounces of silver to the ton.

The American joins the Eagle claims on the southeast. On this location there are two quartz veins, much iron stained, that are exposed by shallow pits at several points. One strikes east and west, the other north and south. Both are vertical and show a little copper carbonate ore that is reported to run 4 per cent of copper and 10 ounces of silver to the ton.

The Barnett property, adjoining the Eagle on the east, is developed by a 90-foot shaft on a quartz vein cutting limestone that carries copper carbonates of a reported value of 20 per cent of copper and 15 ounces of silver and one-half ounce of gold to the ton. The shaft is not accessible, but there is about 5 tons of siliceous limonitic malachite ore on the dump.

The Doyle property, one-fourth mile east of the Eagle, is in a replacement deposit in flat, westward-dipping limestones that contains a large percentage of zinc with slightly argentiferous galena. Two shallow shafts are used for development and there is about 20 tons of sphalerite-galena ore on the dump, said to run 20 per cent of zinc, 10 per cent of lead, and a very small amount of silver.

The Fitch property, at the extreme south end of the district, 20 miles south-southeast of Hachita, is reported to be on a 20-foot quartz ledge in limestone and is developed by a 50-foot shaft and 25 feet of drifts. The ore, largely lead carbonate with some silver values. is said to be of low grade and requires careful sorting.

OTHER COUNTIES.

Many counties of the eastern part of New Mexico have not been mentioned in this account of the mining districts. Some of them contain prospects or indications of metal deposits.

In Union County, occupying the northeast corner of the Territory, deposits of alum are mentioned as occurring on Ute Creek, and indications of gold, silver, copper, and lead are said to exist near Folsom.[a] Deposits of copper, probably in sandstone, have been prospected by the Fort Pitt Company near Clayton.

Quay, Guadalupe, and Roosevelt counties are principally stock-raising regions of the Great Plains. Indications of copper ores in Mesozoic sandstones are reported to occur in the Pintada Canyon, Guadalupe County.

The large county of Chaves lies in the southeastern part of the Territory, adjoining Lincoln County on the west and the State of Texas on the east. It is traversed by Pecos River, along which, at Roswell, important irrigation districts have lately been developed.

Eddy County, in the southeast corner of the Territory, contains in the southwest a part of the Guadalupe Range, mainly built up of eastward-dipping Carboniferous beds. Prospects of copper ores in limestone are said to occur in this vicinity near the Texas boundary line and southwest of the town of Carlsbad.

---

[a] Rept. Governor of New Mexico, 1903, p. 246.

INDEX.

A.

	Page
Abe Lincoln mine, discovery of	19
Aberdeen mine, description of	334
Abiquiu, copper deposits near	75, 77, 124, 149
placers near	124
rocks near	32
Abo Pass, deposits at	77
Abo sandstone, character and distribution of	31, 77, 134
deposits in	77
Accuracy, degree of	15
Admiral group, description of	131–132
Ajax mine (Colfax Co.), description of	101–102
Ajax tunnel (Taos Co.), description of	87
Alabama mine, description of	100–101
Alamogordo, deposits near	184
Albemarle group, ores of	158
rocks near	151
Albemarle mine, description of	158–159
Albuquerque, rocks near	44
Alessandro Mining Company mine, description of	323
Alhambra mine, description of	324, 325
Allerton claim, location of	161
Alum, occurrence of	348
Alum Gulch, mining on	86, 87
Ambrosia mine, ores of	250, 252
American claim (Colfax Co.), location of	99
American Flag mine, production of	267
American mine (Grant Co.), description of	336, 348
American mine (Lincoln Co.), description of	177–178
Amizett, deposits near	84
rocks near	83
Anaconda group (Taos Co.), description of	86
Anaconda mine (Colfax Co.), description of	108
Ancho, gypsum at	175
Anchor district, deposits in	60, 82, 88–89
location of	82, 88
rocks of	88
Anderson camp, deposits of	326
Andesite, analyses of	43
character and distribution of	43–44, 238
deposits in	47–48, 68–69
Andy Johnson claim, location of	269
Animas Range, rocks of	295
Anniseta Gulch, description of	100
Antimony, occurrence of	47
Antisell, Thomas, on New Mexico mines	16, 206–207, 210
Apache Arroyo, deposits on	282
rocks near	282–283
rocks of	28
Apache district (Sierra Co.), copper of	218, 263
deposits of	18, 68–71, 218, 262–266
description of	260–261
map of	261
rocks of	55, 262
silver of	218, 263
Apache group (Sierra Co.), description of	276, 281
Apache Hills, rocks of	39
Apache mine (Grant Co.), description of	344
Apache No. 2 district (Grant Co.), copper of	344
deposits of	51–53, 55–56, 344
description of	343
ore of, section of, figure showing	78
rocks of	343–344
Archean rocks, occurrence of	27
Arrastre mine, rocks of	126

	Page
Arroyo Cobre, deposits in	149
Arroyo de la Parida, view at	242
Arroyo Hondo district, deposits in	49, 50–51, 82
location of	82
rocks in	28, 51
Arroyo Salada, rocks of	229
Arsenic, occurrence of	47
Atwood mine, description of	335
Aztec mine (Colfax Co.), description of	96–97
location of	92, 96
ores of	89, 95, 96, 97
production of	97
Aztec mine (Grant Co.), rocks of	301
Aztec Ridge, rocks of	97
Azure mine (Grant Co.), production of	296, 324
Azure shaft (San Miguel Co.), ores in	116

B.

	Page
Baldy district. See Ute Creek district.	
Baldy Peak, deposits at	92, 97–105
rocks of	94, 97, 248
section of, figure showing	93
Baldy tunnel, description of	103
Bancroft, H. H., on New Mexico mines	18, 20
Barilla mine, deposits of	206–207
Barnett mine, description of	348
Basalt, age of	45
analyses of	45–46
character and distribution of	44–46, 238
effusion of	26, 33, 42
Bauxite, occurrence of	296
Baxter Mountain, mines near	179
Bear Creek, meerschaum on	296
Bear Circle Gulch, placers on	301
Bear Mountain Ridge, rocks of	305
Beaubien and Miranda grant, prospecting on	93
Beck mine, description of	328, 329
Bella group, description of	276, 281
Bent, deposits at	187
Benton claim, description of	169
Bernal Creek, deposits on	117–118
Bernalillo County, description and deposits of	162–163
production of	21–23
Bibliography of New Mexico	15–17, 214–217
Big Hatchet Range, rocks of	295
Birchville. See Pinos Altos.	
Bisbee, Ariz., rocks at	30, 32, 226
Bitter Creek, mines on	88
Blackbird mine, description of	199
Black Colt, claim, location of	269
Black Copper mine, description of	86
Black Hawk district, rocks of	38
Bismuth, occurrence of	47
Bitter Creek, mining on	88
Black Girl claim, description of	161
Blackhawk district, deposits of	57–58, 61, 325
description of	324
production of	324
rocks of	324–325
silver of	325
Blackhawk mine, description of	325
Black Horse mine, description of	97–98
Black Mountains, mining on	86, 87
rocks near	209

349

INDEX.

	Page.
Black Range. *See* Mimbres Mountains.	
Black Range district, copper in	218, 263
deposits in	68, 69-70, 71, 218, 262-266
description of	220, 260-261, 268
maps of	214, 261
rocks in	35, 37, 43, 44, 227, 229, 232, 234, 236-238, 246, 262, 268-269
silver in	218, 261, 263
view in	258
Blake, W. P., on New Mexico	16
Blake mine, description of	117-118
ores of	118-120
section through, figure showing	118
Bland, location of	150-151
mines near	159-161
Bland district. *See* Cochiti district.	
Bliss sandstone, occurrence of	35, 226
Bluebird claim, description of	145
Blue Canyon, rocks of	240
views in	240
Blue Rock claim, description of	87
Bolson plains, description of	221
Bonanza mine (San Miguel Co.), description of	120-121
Bonanza mine (Sierra Co.), description of	275
rocks of	77
analyses of	43
Bonito, deposits near	179
Bonito district, location of	176
Bonnie mine, description of	334-335
Booth mine, ores of	313
Boston claim (Grant Co.), ores of	323
Boston claim (Rio Arriba Co.), ores of	133
Bradley mine, ores of	112
Brady, F. W., on guano in New Mexico	214
Bridal Chamber. *See* Lake Valley.	
Brinsmade, R. B., on San Pedro Mountains	170-175, 214
on Socorro County	242
Brögger, ——, on intrusion	41
Bromide district, copper of	124, 126, 131-133
deposits of	49, 50-51, 124, 126, 131-133
description of	124-125
mines of	126, 131-133
rocks of	27, 125-126, 130-131
silver of	126, 132-133
Bromide mine, description of	132-133
Bromide No. 1 district. *See* Tierra Blanca district.	
Brown, Ross, on metal production	20
Brush Heap claim, location of	269
Buckeye group, ores of	258
Buckhorn mine (Grant Co.). *See* Wood mine.	
Buckhorn mine (Rio Arriba Co.), location of	130
Bueno claim. *See* Highland Chief claim.	
Bull of the Woods claim, location of	98
Bunker Hill mine, description of	87
Burro Chief mine, ores of	323
Burro Hill, deposits in	120, 121
Burro Mountain Copper Company mine, description of	322-323
Burro Mountain district, copper in	19, 61, 62, 322-324
deposits of	17, 19, 61, 62, 296, 322
description of	321
rocks of	27, 28, 38, 321-322
turquoise in	17, 296, 321, 324
Butler, B. S., work of	15, 37

C.

	Page.
Caballos district, copper of	218, 284
deposits in	74, 218, 284-285
description of	284
lead of	218, 284, 285
rocks of	27, 28, 29-30, 229, 232, 238, 283-285
Calamity Jane claim, location of	269
Caledonia claim, location of	269
Cambrian rocks, character and distribution of	29-30, 34-35, 225, 226
Canyon cito district, copper of	218
deposits of	74, 241
lead of	218
location of	240
rocks of	218, 240
Capitan, coal at	175, 176
Capitol dome, rocks near	290
Capulin prospect, location of	148
Carbonate Creek, rocks on	269
Carboniferous rocks, character and distribution of	245-246
See also Mississippian; Pennsylvanian.	
Carlisle, deposits at	69
Carlisle mine, description of	327, 328
Carlsbad, deposits near	184
Carpenter district, deposits of	64, 65, 272
description of	272
lead of	218, 272
map of	268
rocks in	37, 272
zinc of	218, 272
Cash Entry mine, description of	167
rocks of	130, 165-166
analysis of	39, 165
Cashier group, description of	89
Casina claim, rocks at	152
Cave mine, production of	290
Cavern mine, location of	251
ores of	254-255
Central City, rocks near	38
Central district, copper of	318
deposits of	57, 61, 318
description of	317-318
Central mine, description of	167
Century mine, ores of	335
Cerrillos del Coyote. *See* Coyote Buttes.	
Cerrillos Hills district, coal of	163
deposits of	57-58, 60, 163
lead of	163, 164, 166
map of	164
rocks of	35, 36, 54, 164-166
analysis of	39, 165
silver of	163, 164, 166
turquoise of	163, 164, 167
zinc of	164, 166, 167
Chalcocite prospect, description of	149
Chama River, placers on	75, 124
Champion claim (Taos Co.), description of	90
Champion mine (Socorro Co.), description of	192
Chance mine, description of	290, 291
Chemung Copper Company mine, description of	323
Chester claim, ore of	103
Chicago claim, ores of	263
Chihuahua Gulch, rocks in	246, 247
Chloride, deposits at	214
rocks at	262
Chloride Creek, deposits on	264-265
view on	258
Chloride Flat, deposits of	62, 63-64, 65, 304-305
description of	301
fossils of	303
production of	63, 301
rocks of	38, 65, 302-304
section of	302
figure showing	303
silver at	18, 301
See also Fleming Camp.	
Chupadero Mesa, ores at	205
Cieneguilla, placers near	32
Cimarron Canyon, deposits in	101
rocks in	94
Cimarroncito district, deposits of	51, 53, 75, 106-108
location of	92, 105
mines of	107-108
rocks of	36, 105-106
thin section of, figure showing	107
Cimarron Range, deposits of	89, 92
rocks of	35, 88-89, 91-94
section of, figure showing	93
Cimarron River, rocks on	105
Cincinnati claim, description of	292, 293
Clark, Ellis, on Lake Valley ores	66, 214, 277-280
Clarke, F. W., analysis by	165
Clayton, deposits near	348
Clemmie mine, description of	342-343

INDEX. 351

	Page.
Cleveland group, description of	300-301
ores of	64, 65, 300-301
rocks of	42, 300
analysis of	39
Clifton, Ariz., rocks at	30, 32
Clinoplains, definition of	222
Coal, occurrence of	31
occurrence of. See also Colfax County; San Miguel County; McKinley County; Rio Arriba County; Valencia County; Santa Fe County; Lincoln County; Socorro County; Sierra County.	
production of	19
Cobalt, nonoccurrence of	47
Cobre Negro mine, description of	335
Cochiti district, deposits of	68, 69-70, 71, 141, 153-158
description of	150-162
mines of	158-162
ores of	19, 153-158
production of	150, 157
rocks of	35, 36, 151-152
silver of	153, 157
zinc in	155
Colfax County, coal of	92
copper of. See Elizabethtown district.	
description of	91-92
iron of. See Elizabethtown district.	
mining districts of	92-107
placers of	92
production of	20-24
rocks of	91-92
silver of. See Elizabethtown district.	
See also Elizabethtown district; Cimarroncito district.	
Colla Canyon, deposits in	158
Colorado mine, ore of	202-203
Colorado rocks, occurrence of	32, 34
Colossal claim, production of	265
Columnar section in Sierra and Socorro counties, figure showing	223
Colwell claim, location of	323
Comanche Canyon, rocks at	93
Commercial mine, production of	319
Compromise lode, description of	138
Comstock claim, location of	269
Conejos Mountains, deposits in	149
Confidence mine (Colfax Co.), ores of	101
Confidence mine (Socorro Co.), description of	199
Contact-metamorphic deposits. See Metamorphic deposits.	
Contention mine (Colfax Co.), description of	108
Contention claim (Luna Co.), description of	294
Continental claim, ores of	133
Cooks, deposits at	289
fossils from	231
rocks at	228
Cooks Peak district, deposits of	62, 63-64, 288-289
description of	287
lead of	287
production from	63, 287
rocks of	37, 65, 239, 287, 288
analyses of	37, 39, 288
section in, figure showing	288
silver of	287
Cooney, deposits near	18
location of	191
production of	19
See also Mogollon district.	
Cooney, J. C., ore discovered by	191
Cooney Canyon, rocks of	194
section of, figure showing	194
views of	190, 204
Cooney mine, description of	200-201
Cope, E. D., on Santa Fe marl	237
Copper, discovery of	17
distribution of	47
production of	19-24
sources of	57-58, 69-70, 74
See also Sandstone; Red Beds; particular counties.	
Copper Belle claim, rocks at	173
Copper Boy, ore of	203
Copper City, deposits at and near	144, 149

	Page.
Copper Dick mine, description of	335, 337
Copper Flat, deposits at	318
Copper Glance mine, description of	144-145
Copper Hill district (Taos Co.). See Picuris district.	
Copper Hill district (Valencia Co.), deposits in	139-140
location of	135
Copper Hill Mining Company's mine, description of	89, 90
Copper King mine, description of	86
Copper Mountain, veins on	91
Copper ores, enrichment of, by surface waters	61-62
gold and silver in	24
occurrence of	43-51, 53
oxidation of	55-56, 344
views of	78
Copper Park, deposits of	95, 104
Copper Queen claim (Grant Co.), ores of	331
Copper Queen mine (Socorro Co.), ores of	202, 203
Copper Queen mine (Taos Co.), description of	86
Copper Queen mine (Valencia Co.), description of	139
Copperton district, deposits of	77, 135, 138-140
location of	135
Copperton, rocks of	28, 135-170
Copperton Mountain, rocks of	137
Copper veins, exceptional, occurrence and character of	74
Cora Miller mine, ores of	326
Correlation table, figure showing	226
Coyote Buttes, location of	234
view of	238
Coyote district, copper in	109
Cretaceous rocks, character and distribution of	32, 34-35, 41, 236, 246
deposits in	48
Cristobal Range, rocks of	238
Criterion claim, location of	89
Croesus mine, description of	129, 130
Crown Point mine, description of	161
Cuchillo Negro district, gold in	18
Cuchillo Range, rocks of	239, 262
Culebra Range, rocks of	83
Cunningham Gulch, deposits in	131-132, 168

D

	Page.
Daisy mine, description of	345, 347
Dakota (?) sandstone, character and distribution of	34
Dalton prospect, location of	113
Danbury mine, rocks of	126, 133
Deadwood claim, description of	87
Deep Down-Atlantic mine, description of	301
Deep Down mine, description of	199-200
Deformation, occurrence and character of	25-26, 32-33
Deming, smelter at	19
Denver mine, description of	100
Desdemona mine, location of	289
Devonian rocks, occurrence of	30-31, 34, 228
Dewey mine, description of	241
Dikes, occurrence and character of	40
Dillon tunnel, description of	133
Dixie Queen mine, description of	128-129
rocks in	126, 128
Dolores district. See Old Placers.	
Doming, cause of	40-41
Dona Ana County, copper of	205
deposits of	205
description of	205
lead of. See Organ Mountains.	
production of	20-24
silver of. See Organ Mountains.	
See also Organ Mountains; Hembrillo district.	
Dona Dora mine, ores of	210
Douglas, Ariz., smelter at	19
Doyle mine, description of	348
Dry Gulch, deposits in	176, 177
Duck mine, ores of	130
Dundee mine, description of	335
Dutton, C. E., on New Mexico rocks	16, 31, 32, 34, 137, 214, 219, 234, 248-249

E

	Page.
Eagle Creek, deposits on	176
Eagle mine, description of	345, 347

352 INDEX.

	Page.
Eddy County, deposits in	348
production of	23
Edison claim, ores of	89
Effusive rocks, analyses of	275
See also Igneous rocks.	
Eighty-five mine, description of	335
Elephant Butte, rocks at	44, 236
Elizabethtown district, copper in	95, 96, 104
deposits of	18, 51–54, 57–58, 60, 95–105
iron of	95, 96, 102–103
location of	92
mines of	96–105
placers of	75, 93, 104–105
production from	93
rocks of	32, 36, 41, 59, 93–95
section at, figure showing	93
silver of	95
water supply in	92–93
Elliott & Kennedy claim, ores of	133
El Paso, Tex., rocks near	44, 227
smelter at	18, 19
El Paso limestone, occurrence of	35, 227
El Paso mine, description of	65
Embudo, deposits near	75
Emerald prospect, description of	130
Emmens, N. W., on Jones Camp ores	204–205
Emmons, S. F., on copper in sandstones	76, 78, 149
on ore deposits	310
Emmons, W. H., on ore deposits	78
Emperor group, ores of	139
Empire mine, location of	275
Endlich, F. M., on New Mexico rocks	214
Enterprise mine, location of	201
Epigenetic deposits, definition of	48
Erosion, work of	26, 33, 34
Eruptive rocks. *See* Igneous rocks.	
Esperanza Hill, rocks of	283
Estalina Spring, rocks at	260
Estancia, salt lakes at	163
Estey district, copper of	201–202
deposits of	77, 78, 201
description of	201
rocks of	201–202
Eureka Creek, placers on	100
Eureka district. *See* Hachita district.	
Eureka Gulch, rocks in	126, 128–129
Eureka mine (Lincoln Co.), ore of	183
Eureka mine (Sandoval Co.), description of	145–146
Eureka mine (Socorro Co.), description of	266
Excelsior prospect, description of	212
rocks near	209
Excess claim, description of	292
F.	
Fahlbands, definition of	50
Fairbanks, H. W., on New Mexico rocks	215
Fairview, rocks near	239
Faulting, occurrence and character of	25, 33
Federal group, description of	329
Fierro district, iron deposits of	47, 55, 296, 306, 312–313
rocks in	55
Fitch mine, description of	348
Flagstaff mine, production of	267
Fleming Camp, deposits of	304–305
Florida Mountains, deposits of	27, 28, 30, 62, 225
rocks of	225
Florida Mountains district, deposits of	287, 289–290
rocks of	287
Floride claim, description of	201
Folsom, deposits near	348
Fort Stanton, gold at	18
Franklin Mountains, rocks of	29, 30, 35, 208
section in	35, 208
Frazer Mountain Copper Company, mine of	84
Fremont district, deposits of	58, 60, 346–348
description of	345
map of	345
rocks of	345
Freeport claim, rocks in	126

	Page.
French Henry claim, description of	99
French Henry Mountain, deposits on	99
rocks on	99
Fusselman limestone, occurrence of	35, 227
G.	
Galena claim, ores of	101
Galisteo Creek, deposits on	75, 104
rocks on	32
Galisteo sandstone, character and distribution of	38
Gallina district, copper deposits of	148
description of	147–148
Gallinas district, description of	176
Galloway mine, rocks at	207
Gallup, coal near	123
rocks near	32
Garnet, analysis of	185
origin of	41, 56
Garnet mine, description of	187
Garst mine, description of	108
Geologic history, relation of, to topography	25–26
Geologic nomenclature, discussion of	234–236
Geology, columnar section showing	223
correlation of, plate showing	226
description of	25–81
See also particular counties.	
Georgetown district, deposits of	62, 63–64, 319
description of	318
production	63, 319
rocks of	30, 38, 65, 306–307, 319
silver at	18, 318–319
Gila River, bauxite on	296
rocks on	27
Gilbert, G. K., on New Mexico	16
Gilmore mine, coal of, analysis of	111
Girty, G. H., fossils determined by	228, 230, 266, 269, 304, 307
on Manzano group	234
Glacial epoch, imprints of	26
Glenwoody district, deposits of	82, 91
rocks of	27, 82, 91
view in	82
Glorieta district, deposits of	77
iron of	107, 112
rocks of	111–112
Glorieta Mesa, rocks of	32, 110–112
Gold, discovery of	17
distribution of	47
occurrence of	49, 51
production of	19–25
sources of	24, 57–58, 63–64, 69–70
Gold Bug group, description of	263
Gold Camp, deposits of	210
Gold Dollar claim, ores of	101
Golden, deposits at	75
Golden Eagle mine, description of	341, 342
Golden Treasure mine, description of	87
Gold Gulch, deposits in	318
Gold Hill district, deposits in	49, 50–51, 326–327
description of	326–327
mining in	86, 340
rocks of	327, 340
Gold Hill mine, description of	342
rocks at	340, 341
Gold Stain mine, ore of	183
Good Hope claim, location of	161
Good Hope mine (Grant Co.), ores of	325
Good Hope mine (San Miguel Co.), description of	115
Good Luck mine, ore of	183
Gordon, C. H., acknowledgments by	214
on New Mexico rocks	16, 29, 31, 45, 215
on Sierra and central Socorro counties	213–285
Gordon, C. H., and Lindgren, W., on Luna County	285–295
Gorman, M. A., aid of	141
Grafton, deposits near	263–264
Grand Canyon series, correlation of	29
Grand Central claim (Grant Co.), ores of	304
Grand Central group (Sierra Co.), ores of	272
Grande group, description of	276, 280
Grand View group, ores of	272

INDEX.

	Page.
Granite, deposits in	47
Granite Gap, deposits at	62, 63–64
production from	63
rocks at and near	330, 340
Granite Gap mine, description of	330–331
ores of	64, 330–331
Granite porphyry, thin section of, figure showing	107
Grant County, bauxite of	296
copper of	296
See also Santa Rita; Burro Mountains; Hanover; Central; Burro Mountains.	
copper of. See Lordsburg; Sylvanite; Apache No. 2.	
deposits in	49
description of	295
iron of. See Hanover.	
lead of. See Gillespie; Kimball; Lordsburg.	
meerschaum in	296
mining in	19, 296
production of	20–24, 296
rocks of	295–296
silver of. See Pinos Altos; Chloride Flat; Georgetown; Lone Mountain; Telegraph; Lordsburg; Hachita; Sylvanite.	
turquoise of	296, 321, 325
zinc of. See Pinos Altos.	
See also Apache No. 2; Blackhawk; Burro Mountains; Chloride Flat; Central; Georgetown; Gillespie; Gold Hill; Hachita; Hanover; Kimball; Lone Mountain; Pinos Altos; Pyramid; San Simon; Santa Rita; Steeplerock; Shakespeare; Telegraph.	
Graphic-Kelly limestone. See Kelly limestone.	
Graphic mine (Luna Co.), location of	289
Graphic mine (Sierra Co.), description of	255
history of	241–242
location of	241
ores of	250–251
view of	252
ock of and near	54, 247, 248
analyses of	39
view of	254
Graton, L. C., on Colfax County	91–107
on Lincoln County	175–183
on New Mexican rocks	30, 32, 78, 215
on Otero County	184–190
on Rio Arriba County	124–133
on Socorro County (eastern and western)	190–205
work of	15
Graton, L. C., and Lindgren, W., on Mora and San Miguel counties	108–123
on Taos County	82–91
Graton, L. C., and Schrader, F. C., on Sandoval County	140–162
Graton, L. C., Lindgren, W., and Hill, J. M., on Grant County	295–348
Gray Eagle claim, location of	265
Great Basin structure, character and occurrence of	33
Great Plains, character of	26
ore deposits of	46
Great Republic mine, description of	263, 266
Great Western mine, description of	201
Grey Hawk mine, production of	200
Grouse Gulch, placers in	104–105
rocks in	95, 100, 101
Guadalupe County, deposits in	123, 348
Guadalupe Mountains, deposits of	184
Guano, occurrence of	191
Gypsum, occurrence of	175, 184

H.

	Page.
Hachita district, deposits of	56, 57–58, 59, 60, 335, 336–337
description of	335–336
production of	335–336
rocks of	55, 336
silver of	336–337
smelter in	18, 335
Hachita Range, deposits of	51, 53, 335–336
rocks of	32, 38–39, 40, 295, 336, 340
silver in	17, 336
turquoise in	17, 296
Hadley, mining near	116
Hagens Peak, deposits near	265
Hamilton mine, description of	113
ores of	49, 50, 109, 113–114
rocks near	110–111, 113–114
section of	111
Hancock claim, description of	293
Hand Car mine, description of	339, 342
Hanover Creek, rocks on	307, 311
Hanover district, copper of	18, 306, 315–317
deposits of	18, 51–53, 55, 57–58, 61, 306, 311–317
description of	305
iron of	306, 312–313
rocks in	3–5, 38, 41, 43, 55, 306–311
section of, figure showing	312
zinc of	314
Hanover mine, description of	306, 307, 309, 317
Hansonberg, deposits at	77, 203
Hardscrabble mine, production of	251
rocks at	246–248, 250–252
Harry Bluff claim, location of	99
Harry Lyons claim, location of	99
Hawkeye mine, ore of	183
Headstone district. See Bromide district; Hopewell district.	
Helen claim, description of	291
Helen Rae mine, description of	177–178
Hembrillo district, deposits of	205
Hermosa district, deposits of	18, 62–67, 77, 267–268
description of	266
production from	63
rocks in	55, 65, 232–233, 266–267
analysis of	67
silver in	218, 267
Herrick, C. L., on New Mexico	16, 31, 74, 215, 222, 233–236, 245, 249
on ores	84
Hidden Treasure mine (Colfax Co.), ores of	101
Hidden Treasure mine (Rio Arriba Co.), description of	130
Highland Chief claim, deposits on	87
Hill, J. M., work of	15, 20, 176
Hill, J. M., Graton, L. C., and Lindgren, W., on Grant County	295–348
Hillebrand, W. F., analysis by	73
Hillsboro district, copper of	218, 275–276
deposits of	18, 62, 64, 65, 68, 69–70, 214, 218, 275–276
description of	272–273
fossils of	227, 228
iron of	218, 275
lead of	218, 275
manganese of	218, 275
map of	273
placers at	75
production of	273
rocks of	30, 31, 37, 228, 238, 239, 273–275, 277
analysis of	39, 43, 275
silver of	218, 275–276
History of New Mexico	17
Hobson mine	324
Homestake claim, ores of	126
Honey Bee mine, ore of	183
Honey Comb claim, ores of	311
Hop Canyon district, deposits of	259
description of	258
rocks of	255, 258
Hopeful mine, description of	178
Hopewell claim, location of	161
Hopewell district, deposits of	49, 50–51, 124, 126–130
description of	124–125
iron of	128
placers in	75, 124, 130
rocks of	27–28, 125–126
Hopewell Mountains, rocks of	124
Hornet mine, description of	335, 336–337
Howell, Edwin, on Zuni Mountains	135
Hueco limestone, occurrence of	31, 35
Humbug Gulch, mining in	93
placers in	104

I.

Entry	Page
Ibex claim, rocks of	177
Igneous rocks, character and distribution of	47, 247-248
effusion of	26, 33, 233, 237, 238, 248, 262, 274-275
metamorphism by	213
intrusion of	25, 27, 32, 238-239, 247-248, 274
Illinois claim, location of	269
I mine, description of	186
view of	184
Imperial mine (Grant Co.), description of	328
Imperial mine (Socorro Co.), location of	251
Independence mine, description of	88
ore of	86, 88
Indians, culture of	17
hostility of	18, 191-192, 259, 262, 277
revolt of	17
International mine, description of	345, 346-347
Intrusive rocks, deposits associated with	47
See also Igneous rocks; Tertiary rocks.	
Investigations, history of	16-17
Ione tunnel, description of	86
Iowa and New Mexico mine, description of	178
Iron, deposits of	51, 53
deposits of, age of	49
See also Colfax County; Santa Fe Range; Rio Arriba County; Valencia County; Santa Fe County; Lincoln County; Socorro County; Sierra County; Grant County.	
Iron Hill, rocks of	125
Iron King mine, description of	160
Iron Mountain, deposits of	102-103
rocks of	96, 102
section of, figure showing	103
Iron Mountain mine, description of	128
rocks of	126, 128
Ivanhoe mine (Grant Co.), description of	314-315
rocks near	307
Ivanhoe mine (Sierra Co.), description of	263-264
rocks near	125, 126, 133

J.

Jarilla, smelting at	19
Jarilla district, copper in	184
deposits in	56, 60, 75, 185-187
description of	184-185
mines of	185-187
rocks of	35, 185
analysis of	185
Jarossa prospect, description of	148
Jaw Bone mine, description of	128
rocks in	125, 126, 128
Jay Hawk mine, description of	87
Jemes, deposits at	150
hot springs at	157
sulphur at	141
Jenks, William, on copper at Jemes	150
Jenks Draw, works in	151, 152
Jessie mine, description of	290, 291
Jewett, J. J., on New Mexico	215
Jicarilla district, deposits of	51-53, 57-58, 59, 60, 75, 183-184
description of	176, 183
iron in	176
rocks of	39, 183
Jim Crow mine, rocks at	327, 328
Jim Fair mine, ores of	313
Jim Fisk mine, location of	211, 212
Joe and Jenny mine, description of	115
Johnny Bull mine, description of	331-332
Johnson, D. W., on New Mexico rocks	16, 32, 33, 164-166, 168, 215, 249
Jones, F. A., on ancient mines	17, 166
on New Mexico rocks, mines, etc	16, passim
Jones Camp, iron deposits of	51-53, 203-204
rocks at	203-204
Jones prospect, location of	113
Jornada del Muerto, location and character of	218, 221-222
section through, figure showing	226
Jose, deposits at	289
rocks at	288, 289
Jowell mine, description of	341
Juanita Extension mine, description of	257
Juanita mine, description of	256-257
ores of	250, 252-255
Jurassic (?) rocks, character and distribution of	34
Just Before claim, ores of	202

K.

Entry	Page
Kangaroo claim, location of	269
Kelly, mining at	19
rocks at and near	229, 232, 244, 246, 247, 248
section at, figure showing	244
view of	254
Kelly limestone, character and distribution of	231, 245-246
Kelly mine, description of	255-256
ores of	55, 250-252
plan of, figure showing	256
production of	242
rocks near	245
Kendall break, location and nature of	266-267
Keyes, C. R., on New Mexico	16, 27, 205, 207, 216, 235-236, 245, 249
Key mine, description of	255, 257-258
location of	241
ores of	250, 251-252
rocks in and near	245, 246, 249
Keystone district. See Anchor district.	
Keystone mine (Sierra Co.), description of	263, 266
Keystone tunnel (Taos Co.), description of	89
Kiawa Mountain, rocks of	125, 133
Kimball district, deposits of	68, 69-70, 329
description of	328
production of	328
rocks of	39, 328-329
Kindle, E. M., fossils determined by	307, 328
King mine, description of	335, 336
King Richard claim, ores of	133
Kingston, deposits near	272
fossils from	231
rocks near	31, 34, 234, 239, 268-269
section at	34
figure showing	224, 269
Kingston district, copper of	218
deposits of	18, 62, 64, 218, 269-270
description of	268
fossils of	228
lead of	218
map of	268
production from	63, 270
rocks of	31, 34, 37, 65, 225, 226, 228, 232, 245, 268-269
section in	269
silver of	218, 269
Kneeling Creek, location of	308
Knickerbocker mine, description of	343
Kunklin prospects, copper at	112

L.

La Belle district, deposits of	89
location of	82, 89
rocks of	89
La Cueva Creek, copper on	112
rocks on	32
Ladrone Gulch, rocks of	269
Ladrones Range, rocks of	31, 229, 245
Lady Franklin claim, location of	269
Lakes, formation of	26, 33
Lake Valley district, deposits of	62, 63-64, 66, 218, 277-282
description of	276-277
lead of	218
manganese of	218, 279
maps of	278, 280
production from	63, 281, 282
rocks of	30, 31, 43, 227, 228, 277
section of	229
figures showing	279, 281
silver at	18, 218, 277, 279-280, 282
Lake Valley limestone, character and distribution of	34, 228-231
fossils from	230-231
Lamy, gypsum near	163
Lanoria quartzite, character and distribution of	29

INDEX.

355

	Page.
Las Animas district. *See* Hillsboro district.	
Last Chance mine (Grant Co.), description of	33
Last Chance mine (Socorro Co.), description of	192, 199
view of	190
Las Tusas Mesa, rocks of	137, 140
section of, figure showing	135
view of	148
Las Vegas Range, rocks of	109
Latite, analysis of	193
character and distribution of	44, 238
deposits in	68–69
Laura S. claim, description of	161
Lavas, association of, with mineral deposits	46
deposits associated with	48
effusion of	26, 33, 42
succession of	42
veins connected with. *See* Veins.	
See also Igneous rocks; *particular lava rocks.*	
Lazarus Gulch, deposits of	172, 174
rocks of	171
Lazarus mine, deposits of	51, 54
Lead, discovery of	18
distribution of	47
production of	20–24
sources of	57–58, 62, 63–64, 74
See also Taos County; Santa Fe County; Socorro County; Sierra County; Luna County; Grant County.	
Lead and zinc ores, oxidation of	55
Lead ores, gold and silver in	24
Lead veins, exceptional, occurrence and character of	74
Lee, W. T., on New Mexico rocks	16, 31, 32, 45, 216, 221, 234, 235–238
Lee mine, ores of	317
Lees Peak, rocks of	332
Legal Tender mine, description of	100
Leidendorf mine, description of	334
Lemitar Mountains, rocks of	233
Lena mine, description of	334–335
Leopold, deposits at and near	322–323, 324
Lillian tunnel, description of	89
Limestone, metamorphism of	41–42
oxidation in	55
replacement deposits in. *See* Replacement deposits.	
veins in. *See* Veins.	
Lincoln County, coal of	175
description of	175–176
gypsum of	175
mining districts of	176–184
iron in. *See* White Oaks district; Jicarilla district.	
production of	21–24
rocks of	175
See also Nogal district; White Oaks district; Jicarilla district.	
Lincoln Lucky mine, ores of	64–65, 172
Lincoln mine, rocks near	170–171
Lindgren, Waldemar, on Bernalillo and Torrance counties	162–163
on Dona Ana County	205–213
on New Mexican rocks	31, 154–155
on San Juan and McKinley counties	123
on Santa Fe County	163–175
work of	15, 30, 134
Lindgren, W., and Gordon, C. H., on Luna County	285–295
Lindgren, W., and Graton, S. C., on Mora and San Miguel counties	108–123
on Taos County	82–91
Lindgren, W., Hill, J. M., and Graton, L. C., on Grant County	295–348
Literature of New Mexico	15–16
of the ore deposits	16–17
Little Buck claim, ore of	212
Little Casina claim, description of	161
veins of	153–154, 156
Little Charlie claim, location of	201
Little Fanny mine, description of	192, 200
Little Gem claim, ores of	88, 89
Little Mack mine, ores of	180
Little Mollie claim, location of	161
Live Oak claim, ore of	169
Livermore Springs, placers near	338–339
Lizzie Paul prospect, ores of	331
Log Cabin mine, ores of	271

	Page.
Lone Mountain district, deposits of	62, 63–64, 65, 320
description of	320
fossils of	320
rocks at	30, 38, 65, 320
silver of	320
Lone Star mine, description of	159–160
rocks at	152, 160
Lone Tree prospect, ores of	285
rocks at	284
Long Canyon, deposits in	87
Lookout claim (Hop Canyon), location of	259
Lookout mine (Tierra Blanca), ores of	271
Lordsburg district, copper of	335
deposits of	57–58, 60–61, 333–335
description of	332
lead of	335
production of	332
rocks of	39, 332–333
silver of	333
Loring mine, description of	115
Los Cerrillos, smelter at	18
Los Cerrillos Hills, silver in	17
turquoise in	17
Los Organos, view of	204
Louise claim, description of	330
Lucero Creek, placers on	82
Lucky Bill claim, rocks of	318
Lucky mine (Otero Co.), description of	186
Lucky mine (Socorro Co.), ores of	202
Lucy claim, location of	259
Luna County, deposits of	287–295
description of	285–287
lead of. *See* Cooks Peak district; Tres Hermanas district.	
map of	286
mining districts of	287
production of	23–24
rocks of	286–287
silver of. *See* Cooks Peak district; Victorio district.	
zinc of. *See* Tres Hermanas district.	
See also Cooks Peak; Tres Hermanas; Victoria; Florida Mountain.	

M.

	Page.
Mabel claim, ores of	312
McCaffery, P. S. *See* Yung and McCaffery.	
MacGregor mine, production of	319
Macho creek, coal near	111
deposits on	113
rocks of	110
McKinley County, coal in	123
copper in	123
description of	123
Madera limestone, character and distribution of	31, 233, 246
Madrid, coal near	163
rocks at	168
Madrid formation, character and distribution of	34
Magdalena, location of	241, 243
rocks near	243
Magdalena district, copper of	242, 353
deposits of	18, 19, 47, 51–53, 55, 250–258
description of	220, 241–243
history of	241–242
lead of	18, 214, 218, 242, 250
map of	252
mining in	53, 241–242
production of	242
rocks of	243–250, 253–254
section of, figure showing	240
silver of	242
structure of	248–250
zinc of	214, 218, 242, 250–255
See also Magdalena Mountains.	
Magdalena group, character and distribution of	31, 34, 231, 232–233, 246
subdivisions of	31
Magdalena Mountains, description of	242
rocks of	27, 28, 31, 35–37, 54, 229, 231–234, 238, 239, 243, 248–250, 258
analyses of	39
view of	242
Magma, emanations from	56

356   INDEX.

	Page.
Mahoning group, description of	264
Mailleuchet prospect, description of	113
Manganese, occurrence of	218, 275
Mangas River, rocks on	326
Manzano group, character and distribution of	31, 34, 231, 233–234, 246
subdivisions of	31
Manzano Mountains, deposits of	134, 163
rocks of	163
Map of New Mexico	46 Pocket
See also particular counties, etc.	
Marble, origin of	41–42, 56
Marcou, Jules, on New Mexico	16
Marguerite claim, location of	259
Marion mine, ores of	285
Martin & Morton mine. See Gold Hill mine.	
Matthews-Whiteside mine, description of	139
Maud S. mine, description of	199–200
Medio Dia Canyon, copper in	162
Meerschaum, occurrence of	296
Memphis mine (Dona Ana Co.), description of	212
rocks at	209
Memphis mine (Taos Co.), description of	88
Mercury, nonoccurrence of	47
Merrill, G. P., on Stephenson-Bennett mine	211
Merrimac mine, description of	212
rocks near	208, 212
analysis of	39
Merritt mine, location of	239, 240
Mesa Blanca Capulin, deposits at	148
view of	148
Mesozoic rocks, occurrence of	32
Metal deposits, age of	48
distribution of	46–47
map showing	46
geologic distribution of	47–48
types of	47–48
Metal production, details of	19–25
Metals, character and distribution of	47
Metamorphic deposits, character and distribution of	51
development of	53
genesis of	56
location of	57
map showing	52
minerals in	51–53
ores of	51–53
oxidation of	55–56
structure of	53–55
Metamorphism, relation of, to deposits	48
Mica, occurrence of	124, 163
Midnight district. See Anchor district.	
Midnight mine, description of	89, 265
Mikado prospect, description of	113
Mill Canyon, deposits of	68, 69–70, 259
description of	258
rocks of	258
Miller, S. A., on Lake Valley fossils	216
Mimbres limestone, character and distribution of	34, 226–227, 232
Mimbres Mountains, deposits of	11, 68–69, 272
description of	26, 220
rocks of	26, 30, 35, 66, 226, 229, 287
Mineral Creek, deposits on	200
Mineral deposits, relation of, to intrusives	40
relation of, to lavas	46
Mineral Mountain mine, description of	331
Mineral Point mine, description of	129
rocks of	126, 129
Minerals, character and distribution of	80–81
Mines, number of	24
Mining, history of	17–19
Mining districts, location of	46
location of, map showing	46
Minnehaha mine, description of	263
Misers Chest. See Lena mine.	
Mississippian rocks, character and distribution of	30–31, 34, 228–231, 246
Modoc mine (Dona Ana Co.), description of	213
ores of	51, 52, 209, 213
rocks at	207, 213
view of	210

	Page.
Modoc mine (Grant Co.), ores of	313
Mogollon, location of	191
Mogollon district, copper of	192, 194, 198–199, 201
deposits of	18, 68, 69–71, 194–199
description of	191–192
mines of	192, 194, 199–201
ores of, description of	195–197
genesis of	198
mineralogy of	197–198
value of	198–199
production from	69, 192
rocks of	42, 43, 44, 192–194
analyses of	192, 193
silver from	47, 192, 194, 198–199
veins of	195–198
figure showing	196
views in	199, 204
Monarch mine, description of	321
Monero, coal near	124
Montana group, character and distribution of	34
Monte Cristo mine (Grant Co.). See Pearl mine.	
Montecristo mine (Luna Co.), location of	289
Montezuma claim, description of	98
Montezuma district, deposits of	139
location of	135
Montezuma Point, rocks of	171, 173
Montoya limestone, occurrence of	35, 227
Mora County, copper in. See Coyote district, Santa Fe Range.	
description of	108–109
mining districts in	109, 114–116
production of	21, 22, 24
rocks of	109
See also Santa Fe Range; Pecos River.	
Mora Mountains, description of	83
rocks of	31, 32, 33, 34, 83
Mora Range, rocks of	117, 122
Moreno Centennial mine, description of	100
Moreno River Valley, placers in	75, 93, 104–105
rocks in	94
Mormon mine, ores of	210
rocks at	207
Morningstar mine. See Aberdeen mine.	
Morrison (?) formation, character and distribution of	34
Mountain Key mine, description of	299–300
Mountains, magnitude of, relation of, to minerals	46–47
Mount McKenzie, rocks of	165
rocks of, analysis of	165
Mount Taylor, rocks of and near	134

N.

	Page.
Nacimiento district, copper of	77, 78, 143–147
description of	143–144, 147
map of	144
ore of, figure showing	78
Naiad Queen mine, ores of	319
Nambe, mica near	163
Nana mine, ores of	264–265
Nannie Baird mine, description of	185–186
view of	184
National mine, description of	323
Neglected mine, ores of	323–324
Nelly Bly mine, description of	334
Newberry, J. S., on New Mexico	16, 143
New Era mine, description of	265–266
New Placers district, copper of	171
deposits of	171–175
description of	170
discovery of	17, 75
lead of	163, 172
origin of	76
production of	75, 174–175
map of	164
rocks of	170
New Year mine, location of	328
Nickel, nonoccurrence of	47
Ninety mine, description of	314–315

	Page.
Nogal district, deposits of	57–58, 75, 177–179
description of	175, 176
mines of	177–179
production of	176
rocks of	39, 176–177
subdistricts in	175, 176
North Homestake mine, ores of	181
rock of, analysis of	39

O.

	Page.
Ocate Creek, rocks near	44, 92
Ocean Wave claim, work at	267
Ogilvie, I. H., analyses by	40
Ojo Caliente, deposits at	17, 72
hot springs at	72–74
water of, analysis of	73
Old Abe mine, description of	181–182
ores of	59, 180–181
Old Baldy Mountain, description of	258
Old Man mine, description of	304, 305
Old mine, rocks in	173
Old Placers district, description of	167–168
discovery of	17, 75
lode deposits of	169–170
origin of	76
production of	75, 168–169
rocks of	168
Old Reliable mine, deposits of	51, 54, 169–170
Only Chance claim, view of	101
Oohoo mine, ores of	285
Opportunity mine, ores of	273, 275
Ora claim, ores of	133
Ordovician rocks, occurrence of	30, 34, 35
Ore deposits, literature on	16–17
Organ Mountains, deposits of	51–53, 57–58, 60, 62, 63–64, 209–213
description of	205–206
lead in	18, 209–210
map of	206
production from	63, 210
rocks of	29, 35, 37–38, 41, 54, 206–209
analyses of	38, 39, 207
silver of	209–210, 211
views in	204, 210
Ortiz mine, description of	169
discovery of	17
rocks at	168
Ortiz Mountains, deposits of	17, 51, 53, 57–58, 60
map of	164
rocks of	35, 36, 41, 54, 59
analyses of	40
See also Old Placers.	
Oscura Range, deposits in	77–78
rocks of	27, 35, 45
Oshkosh claim, location of	259
Otero County, copper in	184
description of	184
gypsum of	184
placers of	187
production of	23–24
Othello mine, location of	289
Owen prospect, deposits on	318
Owl mine, ores of	318
Oxide claim, description of	90

P.

	Page.
Pacific mine, description of	297, 298–299
Paleozoic sediments, character and distribution of	29–32
deposits in	47
Palomas, deposits at and near	267–268
rocks at	232, 266–267
Palomas Chief mine, description of	267
Palomas district. See Hermosa district.	
Palomas gravel, character and distribution of	33, 237
Palomas-Pelican mine, ores of	65–66
Paragon claim, production of	98
Parsons district, location of	176
Pat Collins claim. See Old Reliable.	
Pauline mine, location of	305

	Page.
Pay Roll mine, description of	131
Pearl mine, description of	343
Pecos district, deposits in	49, 50–51
Pecos River, copper deposits on	110–114
rocks on	28, 31–32
Pelican group, location of	267
Peloncillo Range, deposits in	328
rocks of	27, 39, 41, 42, 295–296, 328–331
Penacho Peak, deposits near	112
Penalta Canyon, deposits in	162
Pennsylvanian rocks, character and distribution of	29, 31, 34, 35, 231–236
correlation of	236
relations of	29, 31
subdivisions of	31, 231
Percha shale, character and distribution of	30, 34, 228
Petaca, mica at	124
Phillipsburg, deposits near	263
Picuris district, copper in	90–91
deposits in	49, 50–51, 90–91
location of	82, 89–90
rocks in	27, 28, 50–51, 90–91
silver in	90–91
Pinal schist, character and distribution of	29
Pino Canyon, deposits in	160–161
rocks in	152
Pinos Altos, mills at	18
Pinos Altos district, deposits of	18, 57–58, 59, 61, 62, 64, 298–301
description of	297
placers of	75, 297
production from	61, 297
rocks of	35, 38, 43, 59, 295, 297–298
analysis of	39
silver of	301
vein of, section of, figure showing	58
zinc of	297, 300
Pintada Canyon, deposits of	123, 348
Pioneer Creek, deposits on	87
Pittsburg district, deposits of	19, 218
description of	282
map of	283
placers of	283–284
rocks of	282–283
Placer Creek, placers on	88
Placer Mountain, gold at	18
Placers, discovery of	17
occurrence and character of	74–76
production of	24, 75
Placitos, copper at	141
Polar Star mine, location of	266
Polvadera Mountains, rocks of	238
Poñil Creek, location of	92
placers in	75, 92, 97
Porphyry, deposits in	47
Porphyry Hill, rocks of	277
Posey claim, location of	161
veins of	154
Pothook, deposits near	335
Poverty Creek, rocks on	263
Pre-Cambrian deposits, character and distribution of	48–49
developments of	49
types of	47–48, 49
genesis of	51
metasomatism of	50–51
minerals of	50–51
ores of	49–50
Pre-Cambrian rocks, character and distribution of	26–29, 225–226, 243–245
deposits in. See Pre-Cambrian deposits.	
intrusions in. See Tertiary rocks.	
copper in	138–139, 149–150
petrography of	27–29
Preston, L. S., work of	99
Prince, L. B., on gold production	75
Prince Albert mine, ore of	183
Puerco River, coal on	134
Puzzle claim, location of	161–162
Pye, Harry, lode discovered by	260–261, 262
Pyramid district, deposits of	57–58, 59, 60–61
Pyramid Mountain, rocks of	41, 42, 240

	Page.
Pyramid Range, deposits of	332
rocks of	296, 332

Q.

Quartzite mine, description of	343
Quaternary deposits, character and distribution of	237
Quaternary system, lavas of. *See* Lavas.	
Quaternary time, events of	33

R.

Railroads, construction of	18
Rambler claim, description of	292
Ransome, F. L., on Bisbee rocks	30, 32, 226
Raton, rocks at	33
Raymond, W. R., on New Mexico	15, 20
Ready Cash, location of	267
Ready Pay mine, location of	275
rock of, analysis of	39
Rebel Chief group, description of	98–99
ores of	95, 98–99
section through, figure showing	99
Recent time, deposits of	238
events of	33–34
Reconnaissance, objects of	15
Red Bandana mine, description of	100
Red Beds, character and distribution of	31–32
copper deposits in	48, 77–79, 137, 143–149, 163, 202
origin of	79, 149
minerals in	149
See also Manzano group; Sandstones.	
Red Fissure claim, rocks of	126
Red Jacket mine, description of	129
rocks near	125, 126, 129
Red River district, copper in	86
deposits of	68, 69–70, 71, 82, 85–86
lead in	86, 87
location of	82, 84
mines of	86–88
placers of	84, 88
rocks of	35, 36, 42, 43, 85
silver in	86, 87
Replacement deposits, character and distribution of	62–63
genesis of	67
metasomatism in	66–67
ores of	63–64
production from	63
relation of, to intrusive rocks	62, 63
structure of	64–66
Republic mine (Grant Co.), description of	312–313
view of	184
Republic mine (Sierra Co.), description of	263
Reservation mine, ores of	327
Rhyolite, analysis of	192
character and distribution of	42–43, 238
deposits in	47–48, 68–69
Richardson, G. B., on rocks of Abo Canyon	28
on Franklin Mountains	29, 30, 31, 35, 208
Rich Gulch, placers on	301
Richmond mine, location of	275
Ridgewood mine, description of	342
Rincon, rocks near	31, 44
Rio Arriba County, coal in	124
copper of. *See* Bromide district.	
deposits of	124
iron of. *See* Hopewell district.	
description of	124
mining districts of	49, 124–133
placers in	124
production of	23–24
rocks of	124
silver of. *See* Bromide district.	
See also Hopewell district; Bromide district.	
Rio Grande valley, bolsons of	221
formation of	25
rocks of	33, 34, 44, 45
Rio Hondo district, copper in	84
deposits in	84
placers in	75
rocks of	83–84

	Page.
Rio Palomas, rocks on	45
Rio Percha, deposits on	275
rocks on	238, 274
Rio Ruidosa, deposits on	176
Rising Sun shaft, ores in	116
Robert E. Lee mine, description of	334
Rociada, prospects near	115–116
Rociada district, deposits in	49, 50–51, 109, 115–116
description of	114
rocks of	27, 114–115
Rock Cliff Creek, deposits in	264
Romeo Hill, rocks of	307
Rosedale district, deposits of	68, 70, 218, 260
description of	259
rocks of	259–260
Rosedale mine, ores of	259, 260
Rose mine, ores of	325
Roseta claim, description of	89
Rubicon prospect, rocks near	125
Ruby claim, ores of	210

S.

Sacramento Mountains, rocks of	184
Sacramento Valley, formation of	25
Salt, occurrence of	191
Sam Adams claim, description of	115
Sammock prospect, ores in	116
Sampson claim, ores in	132
Sampson tunnel, description of	87
San Acacia, rocks at	219, 238
San Andreas limestone, character and distribution of	31, 234
San Andreas Range, deposits of	74, 205
rocks of	35, 234
San Augustin Peak, deposits on	210
rocks on	209, 212–213
San Augustin Plain, description of	221
San Cristobal Range, rocks of	27, 28
Sandia formation, character and distribution of	31, 232–233, 234
Sandia Range, deposits at	77, 141
rocks of	27, 162–163
Sandoval County, coal of	141
copper of. *See* Sierra Nacimiento; Abiquiu.	
deposits of	141–162
description of	140–141
production of	23
silver of. *See* Cochiti district.	
zinc of. *See* Cochiti district.	
See also Sierra Nacimiento; Cochiti district.	
Sandstone, copper deposits in	48, 76, 79
copper deposits in, character and distribution of	76–77
distribution of, map showing	76
genesis of	78–79
ores of	77–78
view of	78
Sangre de Cristo Range, rocks of	26–27
San Jose mine, ores of	318
San Juan County, description of	123
San Lorenzo district, description of	241
San Marcial, rocks at	44, 219, 238
San Mateo Mountains, description of	220
map of	214
rocks of	43, 238, 259–260
view of	258
San Miguel County, coal in	109
copper in. *See* Rociada district; Santa Fe Range; Pecos River; Tecolote district.	
deposits in	109, 112–114
description of	109
mining districts of	114–123
production of	23–24
rocks of	109
See Santa Fe Range; Pecos River; Rociada district; Tecolote district.	
San Miguel mine, deposits at	144, 147
description of	146–147
rocks near	142, 146
view from	148
San Pedro district. *See* New Placers.	

	Page.
San Pedro mine, description of	172-174
development at	53, 173
rock of	54, 174
analysis of	39
San Pedro Mountains, deposits of	51-53, 55, 57-58, 60, 62, 64, 171-174
rocks of	35, 36
analyses of	39
smelting in	19
San Simeon district, deposits of	51-53, 330-332
description of	329-330
lead of	330-331
rocks of	330
Santa Fe County, coal in	163
copper in	163
deposits in	49, 163
description of	163
iron in	112
lead in. *See* Cerrillos Hills; New Placers.	
mica in	163
mining districts in	110
production of	20-24
rocks in	28, 110
silver in. *See* Cerrillos Hills.	
turquoise in	163, 164, 167
zinc in	164, 166, 167
See also Cerrillos district; Old Placers; New Placers.	
Santa Fe Creek, rocks on	110
Santa Fe marl, age of	45
deposition of	33
occurrence of	83, 237
Santa Fe Range, copper of	110-114, 163
gold of	112-114
iron of	112
mica in	163
rocks of	109
Santa Rita Creek, deposits on	315, 316-317
Santa Rita district, copper of	17, 18, 61-62, 306, 315-317
deposits of	306, 311-317
description of	305
rocks of	306-311
Santa Rita mine, description of	315-317
history of	306
rocks of and near	309
analysis of	39
view of	314
Sardine claim, ores of	133
Satisfaction mine, description of	319
Savanna mine, description of	323
Schaller, W. T., analyses by	39, 44, 179, 192, 193, 298, 308, 319
Scheelite, occurrence of	47
Schelerville, deposits of	179
Schelerville district, location of	176
Schrader, F. C., on copper of Sierra Nacimiento	141-150
on New Mexico rocks	28, 32
on Valencia County	134-140
work of	15, 141
Schrader, F. C., and Graton, L. C., on Sandoval County	140-162
Sefrina prospect, description of	148
Senorito Creek, deposits on	145
Shakespeare camp, deposits of	332, 334
rocks of	332-333
Shafer, M. K., on New Mexico rocks	34
work of	141
Shandon, placers at	75, 282-284
rocks at	225
section near	225
Shandon quartzite, character and distribution of	34, 225-226
Shumard, G. G., on New Mexico rocks	217
Sierra Blanca, rocks of	176
Sierra Blanca region. *See* White Mountain district.	
Sierra Colorada, location and character of	137
Sierra County, bibliography of	214-217
bolson plains in	221-222
climate and vegetation	217
columnar section in, figure showing	223
culture	218
coal in	213
copper in	213
deposits in	213-214, 239, 260-285

	Page.
Sierra County, description of	213-214
drainage of	218-219
iron in. *See* Hillsboro.	
lead in. *See* Hillsboro; Kingston; Carpenter; Lake Valley; Caballos.	
manganese in. *See* Hillsboro; Lake Valley.	
mining districts in	214, 218
mountains of	220-221
production of	21-24, 284
rocks in	222-239
See also particular formations.	
silver in. *See* Black Range; Apache; Hermosa; Hillsboro; Kingston; Tierra Blanca; Lake Valley.	
zinc in. *See* Carpenter district.	
See also Black Range; Apache; Hermosa; Kingston; Hillsboro; Carpenter; Tierra Blanca; Lake Valley; Pittsburg; Caballos.	
Sierra Nacimiento, coal near	141
copper in	19, 77, 124, 141-150
description of	141-142
rocks of	28, 32, 141, 142
See also Nacimiento district; Gallina district; Abiquiu.	
Sierra Oscura, copper in	19, 203
iron in	203-205
rocks of	28-29, 201-203
Sierra San Andreas, rocks of	28
Silliman, Benj., jr., on New Mexico rocks	217, 265, 279-280
Silurian rocks, occurrence of	30, 34, 35
Silver, discovery of	17
distribution of	47
occurrence of	49-50
See also Taos County; Colfax County; Rio Arriba County; Santa Fe County; Socorro County; Sierra County; Luna County; Grant County.	
production of	20-25
sources of	24, 57, 58, 62, 63-64, 69-70
Silver Bell mine, ores of	337
Silver Cell mine, ores of	61, 301
rocks near	298, 301
Silver City, rocks near	30, 31, 32, 35, 38, 41, 295, 297
silver at	18
smelter at	19, 297
Silver Creek, deposits on	199-200
Silver Fox mine, description of	348
Silver Monument mine, description of	265, 266
Silver Mountain district, deposits of	258
description of	258
See also Water Canyon.	
Silver Pipe limestone, occurrence and character of	54, 231, 246
Sixteen to One mine, description of	131
Skeed, W. J., aid of	134
Smelter Gulch, deposits of	139-140
Smelting, history of	18-19
Smith, L. R., on Cochiti ores	128
Snake Gulch, deposits near	273
Snake mine, work on	275
Snow Flake claim, rocks near	125, 126
Snyder tunnel, description of	89
Socorro, rocks at	44
smelter at	18, 19
Socorro County, bibliography of	214-217
bolson plains in	222
climate and vegetation	217
coal of	191
columnar section in, figure showing	223
copper of. *See* Mogollon district; Estey district; Canyoncito district.	
culture of	218
deposits of	191, 194-205, 214, 239-260
description of	190-191
drainage of	218-219
guano of	191
iron of. *See* Jones Camp; Sierra Oscura.	
lead of. *See* Magdalena district; Canyoncito district.	
mining districts of	191, 194-205, 214, 218
mountains of	220
production of	21-24
rocks of	222-239
See also particular formations.	

	Page.
Socorro County, salt of	191
silver of. *See* Mogollon district; Magdalena district; Socorro district.	
zinc of. *See* Magdalena district.	
See also Mogollon; Magdalena; Mill Canyon; Water Canyon; Socorro Peak; Rosedale; Estey; Jones; Canyoncito.	
Socorro Mountains, rocks of	36, 44, 233, 238
views in	238, 240
Socorro Peak, view of	238
Socorro Peak district, deposits of	68, 69–71, 218
description of	239–240
production of	239
rocks of	239–240
silver of	218, 240
Soldado Canyon, deposits of	206
South Fork, deposits on	84
rocks on	83–84
South Homestake mine, ores of	180, 182
view in	178
South Juanita mine, location of	250
work on	255
South Mountain, rocks of	35
Springer, rocks near	92
Springer, F. M., on New Mexico rocks	217
Springs, hot, occurrence of	71
Stanton, T. W., fossils determined by	304
Steeple Rock district, deposits of	68, 69–71, 327–328
description of	327
rocks of	327
Steiger, George, analyses by	39, 43, 44, 67, 165, 185, 240, 295
Steins Pass district. *See* Kimball district.	
Stephenson-Bennett mine, description of	210–211
ores of	64, 206, 209
rocks near	208
view of	210
Stevenson, J. J., on New Mexico rocks	16, 26–27, 31, 32, 34, 110, 217
work of	83
Stonewall district, deposits of	287
Storrs, L. S., on New Mexico rocks	217
Strawberry mine, description of	133
Sullivan, E. C., analyses by	39, 165, 171, 173–174
Sulphur, occurrence of	141
Summit group, location of	289
Superior mine, description of	335
Sylvanite camp, view of	314
Sylvanite district, copper of	338
deposits in	58, 60, 75, 338–339, 340–343
description of	338–339
map of	338
placers of	338–339
rocks of	339–340
silver of	343
Syngenetic deposits, definition of	48

T.

Tampa mine, description of	132
Tampico mine, workings of	301
Taos, gold near	17
Taos County, copper in. *See* Rio Hondo district; Red River district; Picuris district.	
description of	82
lead in. *See* Red River district.	
mining districts of	83–91
placers in	82
production of	20–23
rocks of	82
silver in. *See* Red River district; Picuris district.	
See also Rio Hondo; Red River; Anchor; La Belle; Copper Hill; Glenwoody.	
Taos Range, rocks of	83
view of	82
Taylor, Mount, rocks of	45
Tecolote district, copper in	77, 109, 117
description of	116–117
mines of	117–123
rocks of	117
section of, figure showing	118

	Page.
Tecolote Mountain, rocks of	117
Telegraph district, deposits of	57–58, 61, 326
description of	325–326
rocks of	38, 325–326
silver of	325
Tellurides, occurrence of	47
Tertiary rocks, character and distribution of	237
deposits in	48, 237
doming of	40–41
intrusion in	237
metamorphism by	41–42
mode of	40–41
intrusives in	35–46
analyses of	39–40
composition of	36–40
distribution of	35–36
relation of, to mineral deposits	40
veins connected with. *See* Veins.	
lavas of. *See* Lavas.	
Tertiary time, events in	32–33
Texas claim (Dona Ana Co.), ores of	210
Texas mine (Grant Co.), description of	318
Thompson, Gilbert, map prepared by	17
Three Bears mine, description of	186–187
Thundercloud mine, description of	313–314
Thunder mine, description of	107–108
Tierra Amarilla grant, placers on	124
Tierra Blanca district, deposits in	64, 218, 271
description of	271
map of	268
rocks in	37, 66, 271
silver of	218, 271
Tijeras Canyon, copper in	163
Tilted blocks, formation of	25
Tip Top mine, description of	161
Tomboy tunnel, description of	87
Tom Paine mine, description of	167
Topography, relation of, to geology	25–26
Torpedo mine, description of	211–212
ores of	54, 209, 211–212
rocks near	208
Torrance County, description and deposits of	162–163
Torrance mine, description of	239, 240
Tourmaline claim, ores of	312
Trachytes, so-called, identification of	44
Treasure Mountain, deposits on	304–305
Tres Hermanas district, deposits of	51–53, 55, 57–58, 287, 293–295
description of	292
lead of	293, 294
production of	292
rocks of	38, 41, 54–55, 287, 292
zinc of	292–295
Triassic rocks, character and distribution of	32, 34
Tres Piedras, rocks at	125
Triassic rocks, character and distribution of	236
Trujillo Creek, deposits on	271
Trujillo Gulch, deposits on	282, 283
Tuerto, placers at	75
Tuerto Creek, placers on	174–175
Tularosa Canyon, rocks in	187
Tularosa district, copper of	184, 188–190
copper of, origin of	189–190
deposits of	77, 78, 188
description of	187
rocks of	187–189
Turkey Creek, deposits on	178
Turner, W. H., fossils identified by	201
on Estey district	202–203
Turkey Creek, deposits on	263
Turquoise, early use of	17
matrix of	165–166
occurrence of	163, 167, 296, 321, 325
Tusas Peak, rocks near	125, 132–133
Twining, deposits at	84
rocks near	83–84
Twining district. *See* Rio Hondo district.	

INDEX.

U.

	Page.
Ulrich, E. O., fossils determined by	227, 277, 303, 319, 320
Uncle Sam claim, ores of	319
Union claim, location of	161
Union County, deposits in	348
Union Gulch, placers on	283
Union mine. *See* Republic mine.	
United States claim, location of	269
Uplift, occurrence and character of	25–26, 33
U. S. Treasury mine, location of	265
ores of	266
Ute Creek, alum on	348
deposits on	101
placers in	97, 105
rocks on	105

V.

Valencia County, coal of	134
copper of	49, 134, 138–140
deposits of	134–140
description of	134–136
iron of	140
production of	23
rocks of	45, 136–137
See also Zuni Mountains.	
Valles Mountains, rocks of	42–43, 124, 141, 151
Vanadium, occurrence of	274, 276
Veins, lead and copper. *See* Lead veins; Copper veins.	
Veins connected with intrusive rocks, age of	59
character and distribution of	56–57
metasomatism in	59
minerals in	57–58
ores of	57–58
oxidation in	59
relations of	58
structure of	58–59
Veins connected with volcanic rocks, character and distribution of	67–69
distribution of, map showing	68
genesis of	71
metasomatism in	71
ores of	69–70
oxidation of	70
structure of	69
Veins in limestone, character and distribution of	62–63
genesis of	67
metasomatism in	66–67
ores of	63–64
production from	63
relation of, to intrusive rocks	62, 63
structure of	64–66
Venus mine. *See* Leidendorf mine.	
Vera Cruz district, location of	176
Vera Cruz mine, ore of	178
Victor claim, deposits on	87
Victoria district, deposits of	18, 62, 63–64, 287, 290–292
description of	290
map of	291
production from	63, 290
rocks of	43, 287, 290
silver of	291
Viola mine. *See* Leidendorf mine.	
Virginia camp. *See* Shakespeare camp.	
Virginia claim (Luna Co.), description of	292
Virginia mine (Grant Co.), description of	323
Virginia mine (Otero Co.), description of	188, 190
rocks near	187
Volcanic rocks. *See* Lavas.	
Volcano mine, description of	329

W.

	Page.
War Eagle claim, rocks of	125, 126
Washington mine, description of	150
Water Canyon, mines near	242
rocks near	36–37, 245, 246, 247
analysis of	39
See also Silver Mountain district.	
Whale mine, description of	131
Wheel of Fortune mine, ores of	259
Whip-Poor-Will claim, deposits on	87
Whisky Gulch, placers on	301
White, C. A., on Lake Valley	217
White Eagle claim, location of	265
White Mountains, gold in	18, 62, 78
rocks in	35, 39
White Oaks, coal at	175
White Oaks district, deposits in	57–58, 66
description of	19, 175–176, 179
iron in	176
mines in	179–182
placers in	179
production from	60, 176, 179
rocks of	39, 179–181
analysis of	39, 180
view in	178
White Signal district, deposits of	323–324
Whitewater Creek, mining on	318
Whitney, J. P., on Santa Rita mine	306
Wicks Gulch, deposits near	273
Wicks mine, description of	276
Wild Cat mine, description of	312
Wild Horse Creek, rocks on	325
Willemite, occurrence of	294
Willow Creek, mines on	100, 101, 113
placers on	92, 105
rocks on	95
Wilson mine, description of	90–91
Wislizenius, A., on New Mexico	16
Wolframite, occurrence of	47, 336–337
Wood mine, description of	341
Wyman mine, rocks at	328

Y.

Yellow Jacket claim, description of	293
Yeso formation, character and distribution of	31, 234
Young America mine, location of	251
Young prospect, ores of	336
Yung, M. B., and McCaffery, R. S., on ore deposits	28, 109, 171, 172

Z.

Zalinski, E. R., on turquoise mines	296, 324
Zinc, distribution of	47
occurrence of	49, 51
ores of, analyses of	295
See also Sandoval County; Santa Fe County; Socorro County; Sierra County; Luna County; Grant County.	
production of	20, 23–24
See also Lead and zinc ores.	
Zuni Mountains, copper of	134–140
deposits in	49, 50–51, 77, 134–140
description of	25–26, 134–135
iron in	140
ore of, section of, figure showing	78
rocks of	27, 28, 31, 32, 123, 134, 136–137
section of, figure showing	135

O

Other Publications by Miningbooks.com

- Placer Gold Deposits of Nevada
- Placer Gold Deposits of Utah
- Placer Gold Deposits of Arizona
- Gold Placers of California
- Browns Assaying
- Arizona Gold Placers and Placering
- Arizona Lode Gold Mines and Gold Mining
- Dredging for Gold in California
- Metallurgy
- Gold Deposits of Georgia
- Placer Examination: Principles and Practice
- Geology and Ore Deposits of the Creede District, Colorado
- Gold in Washington
- Placer Mining in Nevada
- Gold Placers and their Geologic Environment in Northwestern Park County, CO
- Placer Mining for Gold in California
- Geology and Ore Deposits of Shoshone County, Idaho
- Gold Districts of California
- Gold and Silver in Oregon
- The Porcupine Gold Placer District Alaska
- Gold Placer Deposits of the Pioneer District Montana
- Economic Geology of the Silverton Quadrangle, Colorado
- The Ore Deposits of New Mexico

- Roasting of Gold and Silver Ores, and the Extraction of their Respective Metals without Quicksilver
- Geology and Ore Deposits of the Summitville District San Juan Mountains Colorado

www.ingramcontent.com/pod-product-compliance
Lightning Source LLC
Chambersburg PA
CBHW082104230426
43671CB00015B/2605